Sustaining the Global Agriculture Supply Chain

Sneha Kumari
Symbiosis School of Economics, Symbiosis International University, Pune, India

V. G. Venkatesh
E.M. Normandie Business School, Le Havre, France

K. K. Tripathy
Ministry of Health and Family Welfare, Government of India, India

Hemachandra Padhan
Indian Institute of Management, Sambalpur, India

M.P. Sukumaran Nair
Center for Green Technology and Management, India

IGI Global
Scientific Publishing
Publishing Tomorrow's Research Today

Vice President of Editorial Melissa Wagner
Managing Editor of Acquisitions Mikaela Felty
Managing Editor of Book Development Jocelynn Hessler
Production Manager Mike Brehm
Cover Design Phillip Shickler

Published in the United States of America by
IGI Global Scientific Publishing
701 East Chocolate Avenue
Hershey, PA, 17033, USA
Tel: 717-533-8845
Fax: 717-533-8661
Website: https://www.igi-global.com E-mail: cust@igi-global.com

Library of Congress Cataloging-in-Publication Data

Names: Kumari, Sneha, 1988- editor | Padhan, Hemachandra, 1993- editor | Tripathy, Kamal K., 1973- editor | Sukumaran Nair, M. P, 1950- editor | Venkatesh, V. G., 1976- editor
Title: Sustaining the global agriculture supply chain / edited by Sneha Kumari, V.G. Venkatesh, K.K. Tripathy, Hemachandra Padhan, M.P. Sukumaran Nair.
Description: Hershey, PA : IGI Global Scientific Publishing, [2025] | Includes bibliographical references and index. | Summary: "The book will try to provide justice with a stakeholder perspective at every functional level, from quality inputs to post-harvest management, storage, and processing"-- Provided by publisher.
Identifiers: LCCN 2024053595 (print) | LCCN 2024053596 (ebook) | ISBN 9798369343302 hardcover | ISBN 9798369352120 paperback | ISBN 9798369343319 ebook
Subjects: LCSH: Agricultural industries | Food industry and trade | Supply chain management
Classification: LCC HD9000.5 .S84 2025 (print) | LCC HD9000.5 (ebook) | DDC 338.1/9--dc23/eng/20250207
LC record available at https://lccn.loc.gov/2024053595
LC ebook record available at https://lccn.loc.gov/2024053596

British Cataloguing in Publication Data
A Cataloguing in Publication record for this book is available from the British Library.

Editorial Advisory Board

Table of Contents

Detailed Table of Contents

The diverse agricultural landscapes across the country, influenced by factors like climate and support systems, demand a more tailored approach. Unfortunately, the emphasis remains on measuring productivity rather than farm incomes, leading to a mismatch between national and regional needs. To address these issues, there is a need for a shift towards demand-driven policies, where programs are designed to meet the sector's needs effectively. However, the existing setup lacks flexibility and often leads to states merely implementing central directives without considering local conditions. Therefore, the present chapter aims to design structural changes for innovative approaches to the global agriculture value chain. A paradigm shift is required where countries collaborate based on the evolving agricultural landscape. The chapter has tried to discuss different cases of bringing innovative changes in the global agriculture value chain.

This chapter talks about how agro-processing and value chains are significant for making agriculture a successful and competitive sector.We will learn how these linkages help advancing the quality of agricultural produce and make them more tempting to buyers around the world. We will discuss how transforming agriculture is essential, and value chains play a big role in this by taking everyone work collectively and for better. We will also talk about the importance of supporting local supply chain. These are the networks which are important for producers as well as consumers because they ensure there's enough food and create new economic opportunities in agricultural sector. With the highest population in the world and the ability to export goods, India plays a major role in the global agricultural sector. By understanding and supporting agro-processing and value chains, we can make sure farming continues to flourish, promoting both local communities and the global market.

Since agriculture as an enterprise can sustain, only if it can yield net positive returns in favour of the farmer-producers, robustness of agricultural value chains assumes critical importance and this is feasible when efficient scales of operations support the system. In the given context of predominance of small & marginal farmers, the solution lies in promoting collectives of farmers and enabling aggregation of produce. It is in this context that the Government has consciously adopted the approach of farmers' mobilisation and aggregation of agri-produce, which will bring in efficiency of higher scale and facilitate the nation's commitment to establishing robust supply chains

There is a great deal of potential for improving quality assurance procedures when blockchain technology is integrated into agricultural supply chains. This review investigates how blockchain technology can enhance quality control, traceability, and transparency in agri-supply chains. Blockchain technology provides a transparent and traceable record of each transaction and movement inside the supply chain, offering a decentralized and immutable ledger system that can address these issues. This work explores a number of uses cases for blockchain technology in agriculture, with a particular emphasis on how it makes it possible to track and monitor products in real time from farm to table. Furthermore, blockchain improves transparency by giving customers access to comprehensive data regarding the provenance, care, and caliber of agricultural goods Agri-supply chain case studies and real-world blockchain applications are analyzed to show how blockchain affects productivity, sustainability, and the development of trust.

Supply chain management (SCM) is a crucial aspect of the global economy. It also plays a significant role in the agriculture sector. Key factors for success include liberalization of a 1.4 billion consumer market. India's industrial growth is largely driven by the manufacturing sector. However, the growth rate of the manufacturing sector could not be adequate for the requirement of an employable population. The service sector, which accounts for over half of India's GDP, has seen a high growth rate after liberalisation. Agribusiness supply chain management (SCM) involves managing relationships between businesses responsible for efficient production and supply of products to meet consumers' requirements. Government agencies are implementing legal reforms to encourage private investment in agricultural marketing infrastructure and promote coordinated supply chains and traceability. Industry 4.0, powered by big data, cloud computing, blockchain, IoT, AI, and machine learning, requires companies to rethink supply chain design, leading to faster, more efficient supply chains

This research explores how cooperatives are essential to increasing agricultural exports in Bangladesh, India, and France, three very different economies. In light of the COVID-19 epidemic and geopolitical unrest, this study investigates the ways in which export cooperatives promote sustainability, diversity, and resilience in global supply chains. The study sheds light on cooperatives' contributions to value addition, market access, and quality control by analyzing their governance structures, operational dynamics, and market strategies. It reveals best practices and obstacles through a comparative study, providing stakeholders and policymakers with insightful information. The results highlight the cooperatives' revolutionary potential in promoting fair economic growth and strong international trade networks. This study offers a thorough grasp of the ways in which cooperative businesses and worldwide agricultural exports interact, encouraging readers to delve deeper into the complex relationship between the two.

This study investigates the transformative impact of digitalisation on the agri-food supply chain (AFSC), with a focus on operational sustainability. Through a comprehensive bibliometric analysis of 438 documents from the Web of Science (WoS) and SCOPUS databases, key research streams within AFSC literature are identified. The study highlights how digitalisation enhances efficiency, traceability, and sustainability, while also presenting challenges such as technological barriers and cybersecurity risks. Complementing the literature review, 24 interviews were made with AFSC members and stakeholders (including raw material producers, processors, manufacturers, retailers, logistic service providers and trade associations) to provide practical insights into the adoption and impact of digital solutions. The findings underscore the necessity for tailored digital strategies across different AFSC sub-sectors, emphasizing the role of digitalisation in improving operational resilience and sustainability.

20% ethanol blending is targeted for petroleum fuel (EBP-20) by countries. However, traditional feedstocks like sugarcane, rice, and maize are facing constraints due to climate change and food security concerns. To bridge this gap, this research explores that Sweet Sorghum as an alternative potential feedstock for bio-ethanol production. While having lower conversion efficiency and ethanol price compared to traditional crops, Sweet Sorghum offers advantages such as multiple cropping cycles per year, high temperature tolerance, and low water requirements. This makes it a sustainable choice for bio-ethanol production and augmenting the ethanol supply.

Agriculture gradually depletes the soil of the essential nutrients for the growth of plants resulting in low crop productivity and yields. Soil is to be continuously fortified by the application of fertilizers to maintain food production. Therefore, an unbreakable supply chain is essential for continuous production of food from the fields. The supply chain gets disrupted following global economic downturn, pandemic or war situations. Due to this disruption supplies are affected, farmers are overburdened by rise in fertilizer prices beyond their reach and it could possibly lead to global food insecurity. India is a major foodgrain producing country and nearly 55% of its 1.44 billion population depends on agriculture for survival. This chapter discuss importance of al fertilizers in Indian agriculture and how the consumption is managed through domestic production and imports. The disruption in global supplies due to Covid 19 and the Ukraine war and problems encountered in making adequate supplies and government efforts to streamline the fertilizer supply chain are discussed.

food security. But the technology adoption has its own challenges. It is necessary to address issues including data protection, farmers' digital literacy, and the large upfront costs associated with implementing new technologies.

The authors in this chapter have made a detailed study upon the usage of Artificial Intelligence in Supply Chain Mechanism System. Firstly, the authors have gone through the existing literature and practices to know about the current situation. Furthermore, the review of existing literature also helped the authors in identifying the loopholes in the present system. Apart from that the authors have made a comparative study of United States, Africa and India to know how various countries have used this technology to revolutionize agriculture in their respective countries. Lastly the authors have provided solution to the existing problem.

Agri-food sustainability has grown to be the globe's most important issue throughout the years. Technologies that promote environmentally friendly agricultural farming have proven effective in reducing poverty brought on by increased food demand. Agri-food system uses have been significantly altering the global landscape recently as consequence of internal and external variables. The forthcoming contribution of study is that we can utilize it for enhancing economic efficiency, better control mechanism and improved decision making. This study will prove beneficial to the users and other stakeholders like technology providers, experts and society at large. Furthermore, this study provided guidelines for further research on the proposed concept. The conclusions drawn in the study will be helpful to researchers, experts, smart technology suppliers and policy makers who ultimately are responsible for supporting the evolution to extra sustainable and cultured agri-business practices.

The economy is significantly influenced by agriculture. Globally, the primary issue and burgeoning topic is agriculture mechanization. The population is growing at an exponential rate, which is simultaneously driving up demand for jobs and food. For the farmers to meet these needs, the conventional techniques they employed were insufficient. New automated techniques were consequently introduced. The utilization of Artificial intelligence technique has protected agricultural productivity from various threats, such as rising populations, changing climates, employment difficulties, and problems with food safety. Architectures based on machine learning have been confirmed to be among the best regarding precision and resilience. By using systems based on AI, crop yields can be increased while using less water, fertilizer, and pesticides. Besides improving worker safety, intelligent expertise can support in reducing its consequence on expected ecosystems, which will help keep food costs low and ensure more crops are grown to feed the world's expanding population.

Foreword

In an era where food security is increasingly under threat from climate change, market volatility, and systemic disruptions, the need for robust and sustainable agricultural supply chains has never been more critical. This book provides a comprehensive exploration of Agricultural Supply Chain Management (ASCM) and its pivotal role in promoting food security while fostering sustainable practices across the globe.

As we delve into the intricacies of ASCM, we encounter not only the challenges it faces but also the innovative solutions emerging from the intersection of technology, data analytics, and regenerative agricultural practices. The shift towards climate-smart and waste-reducing supply chains is essential for addressing the social and environmental pressures that threaten the integrity of our food systems. This book thoroughly examines the multifaceted nature of supply chain management, emphasizing the need for coordinated efforts that encompass planning, monitoring, and adaptation.

Through an insightful selection of key chapters, the contributors shed light on different aspects of global supply chain that can enhance resilience and efficiency within agricultural supply chains. The integration of machine learning, artificial intelligence and data analytics not only informs stakeholders about past events but also equips them with the foresight to navigate future uncertainties effectively.

Importantly, this book advocates for a stakeholder-centric approach, ensuring that voices from all levels of the supply chain—from farmers and agricultural cooperatives to policymakers and academics—are included in the discourse. Each chapter serves as a valuable resource, offering practical insights and best practices that are applicable in both developed and developing contexts.

As we explore the global landscape of agricultural supply chains, this work stands as a testament to the power of collaboration and knowledge sharing. By learning from sustainable models and initiatives worldwide, we can collectively pave the way toward a more resilient and equitable food system.

In a world where food security is intricately linked to environmental stewardship, the insights gathered in this book are not just timely but necessary. It is my hope that readers will find inspiration and guidance within these pages, as we strive together for a future where agriculture can thrive sustainably and effectively meet the needs of all communities.

Otaviano Canuto
Policy Center for the New South, USA

Otaviano Canuto is former Vice-President and Executive Director, World Bank; former Executive Director at the IMF; and former Vice-President at the Inter-American Development Bank.

Preface

INTRODUCTION

Agricultural supply chains play a pivotal role in ensuring food security and promoting the sustainable development of agriculture. Agricultural Supply Chain Management (ASCM) encompasses the planning, design, coordination, organization, and monitoring of the flow of agricultural products from farm to fork. This process is essential for the efficient functioning of global food systems. However, these supply chains are frequently challenged by a range of factors, including disruptions in supply, market volatility, and the impacts of climate change, all of which pose significant risks to their stability and sustainability (Routroy & Behera, 2017). The transition towards a climate-smart, regenerative agricultural supply chain holds the potential to mitigate the negative effects of social and environmental pressures. To reduce waste, optimize resource utilization, and enhance the sustainability of livelihoods, the agricultural sector has redefined its supply chain strategies. In addition, Agricultural Supply Chain Management (ASCM) now increasingly involves the management of big data, which is characterized by its volume, veracity, variety, value, and velocity. This data-driven approach is critical for improving decision-making and fostering resilience within agricultural supply chains (Panetto et al., 2020).

Volume – The agricultural supply chain generates massive data, encompassing a large scale of information across various stages of the process.

Velocity – The rate at which data is produced within the agricultural supply chain is rapid, with continuous updates occurring in real-time concerning waste, processing, and movement of agricultural goods.

Variety – The data generated by the agricultural supply chain is diverse, covering a wide spectrum of variables including production, procurement, waste management, inventory control, and processing. This diversity of data requires robust systems for collection, analysis, and integration to optimize decision-making across the supply chain (Sharma et al., 2020).

Veracity – While data is generated and stored at each stage of the agricultural supply chain, concerns regarding its accuracy and reliability persist. Ensuring data integrity is critical for effective decision-making and management. This challenge can be addressed through coordinated monitoring of the supply chain coupled with advanced data analytics, which can enhance the dependability and precision of the data, thereby facilitating informed decision-making and process improvements (Kamble et al., 2020).

Descriptive analytics can effectively manage the supply and demand of agricultural goods by providing insights into past events, detailing what happened and how it unfolded with respect to agricultural commodities. Predictive analytics, on the other hand, leverages historical data to forecast the likely trajectory of future events within the supply chain. Moving beyond predictive analytics, prescriptive analytics enhances decision-making by evaluating potential outcomes and recommending optimal actions. It assists in addressing critical questions such as what the best possible outcome may be and how it can be achieved within the supply chain. The integration of Artificial Intelligence (AI) into agricultural supply chain management has significantly contributed to overcoming these challenges, offering advanced solutions for optimizing efficiency and mitigating risks (Nayal et al., 2021). The transition to sustainable supply chain management is a complex process that requires a critical examination of each function within the supply chain. Every component must be carefully analyzed to ensure its alignment with sustainability objectives and to identify opportunities for improvement and optimization (Kumari et al., 2023b). Studies must explore elements that support agricultural supply chains (ASC) globally, including institutional support, organizational restructuring, network and symbiosis, training, agri-input facilities, knowledge dissemination, quality storage, capacity building, promotional activities, data analytics, and monitoring (Kumari et al., 2023a). Initiatives focused on supply chain networks and industrial symbiosis within the global agricultural supply chain are essential for sustaining the supply chain and fostering its regeneration. It is equally important to examine how by-products and their potential uses can be integrated into the regeneration of ASC processes. The effectiveness of global agricultural supply chain management largely hinges on how well stakeholders are informed about emerging developments and how these innovations can be applied to enhance supply chain efficiency. Planning for adequate storage and warehousing infrastructure is also vital for prolonging the shelf life of agricultural commodities and ensuring a more robust supply chain. This book aims to offer a unique contribution by proposing a forward-looking roadmap to strengthen agricultural supply chains, particularly in developing and underdeveloped countries.

Target Audience

The book will help the students, practitioners, and policymakers to understand the initiatives and strategic interventions required in every function of the Agriculture Supply Chain around the globe. This will also help the agricultural supply chain strengthen its functional levels. The book will benefit stakeholders in risk assessment and risk management in the agricultural supply chain. In addition to risk management, the book will also provide measures and learnings from sustainable models of the agriculture supply chain across different parts of the world. The book will be a unique contribution towards proposing a future road map for developing and underdeveloped countries to strengthen their agriculture supply chain. The book will help the students, practitioners, agriculture consultants, agriculture cooperatives, farmers, farmer clusters, procurement centers, and policymakers understand the initiatives taken in every function of the Agriculture Supply Chain worldwide. This will also help the agricultural supply chain strengthen its functional levels.

The book has 18 chapters that contribute to sustaining the global agriculture supply chain. The chapters focus on the approach to the global agriculture supply chain in different countries. The chapter has been carefully selected based on the theme of the book. The book highlights the structural changes in the agriculture supply chain, agro-processing for sustainable growth, collectivization approach for supply chain, transformation of agri supply chain through blockchain, agricultural supply chain management strategies in industry 4.0, enhancing agricultural export dynamics, evolving agri-food supply chain through digitalization, agricultural warehousing, supply chain through machine learning and artificial intelligence, agricultural performance by leveraging digital technologies in agriculture 4.0, and applications, advantages, and challenges of artificial intelligence in agriculture. The book also highlights the cases on the supply chain of commodities like agri inputs, namely fertilizer, onion, and sweet sorghum, and cases of the agriculture supply chains from Namibia, Africa, Global South, and India. The chapter discusses the cases on the value chain for selected agriculture commodities like French bean value chain from Kenya to the European Union, tomato value chain in the European Union, onion value chain in New York, potato value chain in the Netherlands, grapes value chain in Tanzania, pomegranate value chain in Afghanistan and South Africa, mango value chain in Ethiopia, banana value chain in Ecuador, pulses value chain, apple value chain in Ethiopia, peach value chain, coffee value chain at Toraja, orange value chain in South Africa, avocado value chain in India, and aloe vera value chain in Greece. The book will help in understanding the strategic interventions required for the global agriculture supply chain.

Chapter 1: Structural Changes for Innovation in Farmer-Centric Global Agricultural Value Chain

The diverse agricultural landscapes across the country, influenced by factors like climate and support systems, demand a more tailored approach. Unfortunately, the emphasis remains on measuring productivity rather than farm incomes, leading to a mismatch between national and regional needs. To address these issues, this chapter aims to design structural changes for innovative approaches to the global agriculture value chain. A paradigm shift is required where countries collaborate based on the evolving agricultural landscape. The chapter has tried to discuss different cases of bringing innovative changes in the global agriculture value chain.

Chapter 2: Enhancing Agricultural Competitiveness - Harnessing Value Chains and Agro-Processing for Sustainable Growth

This chapter explores the pivotal role of agro-processing and value chains in enhancing the competitiveness and sustainability of the agricultural sector. It examines how these linkages contribute to improving the quality of agricultural products, thereby increasing their appeal to global markets. The chapter further discusses the necessity of transforming agriculture, with value chains acting as a key driver of this transformation by fostering collaborative efforts among stakeholders. Additionally, the chapter highlights strengthening local supply chains, as these networks are crucial for both producers and consumers, ensuring food security while creating new economic opportunities within the agricultural sector. Given the increasing population and its significant capacity for agricultural exports, the global agricultural landscape occupies a central position. By supporting and understanding agro-processing and value chains, it is possible to promote sustainable farming practices that benefit both local communities and the global economy.

Chapter 3: A Collectivization Approach to Agricultural Supply and Value Chains

The collectivization aims at establishing in the country a multimodal and seamless infrastructure.

Supply chains will also need to be supported by offering the stakeholders operational, technical, financial, and support services. The cooperatives, for example, provide agricultural inputs and loans to farmers, as well as support the production, processing, transportation, storage, and marketing of agricultural produce.

This chapter links the activities of co-operatives and other forms of collectives with the agricultural supply chain models and offers managerial implications to cooperative societies/other collectives, policymakers, and agricultural supply chain actors.

Chapter 4: Transforming Agri Supply Chains- A Blockchain Approach for Quality Assurance

The chapter presents the integration of blockchain technology into agricultural supply chains with significant potential for enhancing quality assurance procedures. The chapter examines the ways in which blockchain can improve quality control, traceability, and transparency across agri-supply chains. By providing a decentralized, immutable ledger system, blockchain enables a transparent and traceable record of each transaction and movement within the supply chain, effectively addressing challenges related to information accuracy and reliability. The chapter explores several use cases of blockchain technology in agriculture, with particular emphasis on its capacity to facilitate real-time tracking and monitoring of agricultural products from farm to table. Moreover, blockchain enhances transparency by granting consumers access to detailed information regarding the provenance, handling, and quality of agricultural goods. Case studies and real-world applications of blockchain in agri-supply chains are analyzed to demonstrate its impact on productivity, sustainability, and the building of trust among stakeholders.

Chapter 5: Feasibility of Agricultural Supply Chain Management Strategies in Industry 4.0

The chapter discusses the key factors driving success in SCM and focuses on managing the relationships between businesses to ensure the efficient production and distribution of agricultural products that meet consumer demand. To support this, government agencies are implementing legal reforms designed to foster private investment in agricultural marketing infrastructure while also promoting coordinated supply chains and traceability mechanisms. The advent of Industry 4.0 technologies—such as big data, cloud computing, blockchain, the Internet of Things (IoT), artificial intelligence (AI), and machine learning—has necessitated a rethinking of supply chain design. These technologies enable the creation of faster, more efficient supply chains that are better equipped to address the evolving challenges of the agricultural sector.

Chapter 6: Enhancing Agricultural Export Dynamics: A Comparative Analysis of Cooperatives Across Economies

This chapter examines the critical role of cooperatives in enhancing agricultural exports across three distinct economies Bangladesh, India, and France. In the context of the COVID-19 pandemic and ongoing geopolitical instability, the study investigates how export cooperatives contribute to the sustainability, diversification, and resilience of global supply chains. By analyzing the governance structures, operational dynamics, and market strategies of these cooperatives, the study highlights their contributions to value addition, market access, and quality control. Through a comparative analysis, it identifies both best practices and challenges, offering valuable insights for stakeholders and policymakers. The findings underscore the transformative potential of cooperatives in fostering equitable economic growth and strengthening international trade networks. This study provides a comprehensive understanding of the interaction between cooperative enterprises and global agricultural exports, encouraging further exploration of the complex relationship between these two key elements of the agricultural sector.

Chapter 7: Evolving Agri-Food Supply Chain Through Digitalisation: Challenges and Responses for Operational Sustainability

This study investigates the transformative impact of digitalization on the agri-food supply chain (AFSC), with a focus on operational sustainability. Through a comprehensive bibliometric analysis of 438 documents from the Web of Science (WoS) and SCOPUS databases, key research streams within AFSC literature are identified. The study highlights how digitalization enhances efficiency, traceability, and sustainability, while also presenting challenges such as technological barriers and cybersecurity risks. Complementing the literature review, 24 interviews were made with AFSC members and stakeholders (including raw material producers, processors, manufacturers, retailers, logistic service providers, and trade associations) to provide practical insights into the adoption and impact of digital solutions. The findings underscore the necessity for tailored digital strategies across different AFSC sub-sectors, emphasizing the role of digitalization in improving operational resilience and sustainability.

Chapter 8: Agricultural Warehousing: Functional insights, Challenges, and Best Practices

The chapter reviews the current state of agriculture warehousing worldwide, this is particularly important in reducing postharvest losses and increasing food safety. It points to the economic and social consequences of faulty storage solutions in developing countries. It stresses the necessity for upgrading agricultural storage, sharing technologies, incorporating online storage, and applying modern storage facilities, especially among smallholder farmers. The issue of agri-warehousing, however, runs into numerous challenges in infrastructure, technologies, and regulatory frameworks in both developed economies as well as in emerging economies. Given the emphasis on theory, policy, and best practices, it is hoped that this chapter will provide a necessary foundation for a better understanding of the agricultural warehouse system from a more practical perspective.

Chapter 9: Overcoming Supply Chain Barriers in Onion Through Technology

The onion economy is characterized by major price volatility due to seasonal production, while there is daily consumption by almost every household. The crop is highly perishable and depends upon stored produce for about 6-7 months as there is no harvest between May to November. The crop, besides suffering from considerable storage losses, is also susceptible to unfavorable monsoons, which can lead to huge crop loss and high prices. Price rise is addressed through government agencies releasing buffer stocks, fixation of Minimum Export Price and even regressive measures such as export ban. This chapter examines the price fluctuations and the status of demand, supply, and exports of onion crops. The chapter highlights that the supply shocks must be addressed through technology, which can increase yields and also be leveraged for creating advanced storage structures in place of traditional ones, often used by farmers. This serves as the real solution rather than export bans.

Chapter 10: Sweet Sorghum Supply Chain for Bio-Ethanol Production

Many countries have set a target of 20% ethanol blending (EBP-20) for petroleum fuels; however, traditional feedstocks such as sugarcane, rice, and maize are increasingly constrained by the impacts of climate change and concerns over food security. To address these challenges, this chapter explores the potential of Sweet Sorghum as an alternative feedstock for bioethanol production. Although Sweet Sorghum exhibits lower conversion efficiency and ethanol yield compared to traditional crops, it offers significant advantages, including multiple cropping cycles per year, high-temperature tolerance, and minimal water requirements. These characteristics make Sweet Sorghum a promising and sustainable option for bioethanol production, potentially augmenting the global ethanol supply while mitigating the pressures on food crops.

Chapter 11: Indian Agriculture-Overcoming Fertilizer Supply Chain Disruptions

Agriculture gradually depletes the soil of the essential nutrients for the growth of plants resulting in low crop productivity and yields. Soil is to be continuously fortified by the application of fertilizers to maintain food production. Therefore, an unbreakable supply chain is essential for the continuous production of food from the fields. The supply chain gets disrupted following global economic downturns, pandemics, or war situations. Due to this disruption, supplies are affected, farmers are overburdened by a rise in fertilizer prices beyond their reach, and it could possibly lead to global food insecurity. This chapter discusses the importance of all fertilizers in Indian agriculture and how consumption is managed through domestic production and imports. The chapter highlights disruption in global supplies due to Covid 19, the Ukraine war, problems encountered in making adequate supplies, and government efforts to streamline the fertilizer supply chain are discussed.

Chapter 12: Sustaining Livestock Supply Chain in Namibia: Global Demand and Supply Scenario, Prospect of Lucrative Livestock Farming in Namibia

Namibia has tremendous potential in the livestock sector, which has come a long way in developing its competitive advantage with a good animal identification and traceability system. This potential is likely to be harnessed with Namibia's progressive policies, plans, and actions to integrate the existing three livestock marketing systems -small communal farmers in the north, emerging small-scale farmers, and existing commercial farming in the south, to gain access to high-end markets, particularly by supporting small farmers in the north part of the country on integrated crop-livestock systems, range land management plans, and strengthen knowledge and capacity of the farmers. Integration of the three systems is expected to provide ample opportunities for grower models, central processing facilities like Agro Processing Hubs, and Agri Transformation Centers nearer to the smallholder farmers as one-stop shops to do business and bring prosperity.

Chapter 13: Microfinance Serving Agriculture Supply Chain in Dapaong, Togo in Western Africa

This chapter examines the crucial importance of microfinance in the agricultural supply chains of developing countries. The study aims to determine the impact of microfinance on agricultural activities in Dapaong City in northern Togo. The chapter uses a case study approach by conducting semi-structured interviews with a representative group of 15 local farmers. Microfinance plays a key role in funding new projects and improving farmers' living and working conditions. However, agricultural financing is still limited due to the risks and specific challenges of rural areas. Additionally, microfinance hasn't yet helped farmers shift to more eco-friendly practices, as many still use chemical inputs. Consequently, microfinance does not fully contribute to the sustainability of global supply chains. Therefore, additional funds are necessary to enable this transition.

Chapter 14: A Case of Climate Financing in Agriculture in the Global South-- Imperative to Develop Innovative Financing Instruments

Climate change is a threat to the agricultural sector; rising temperatures, altered precipitation patterns, and increased frequency of extreme events lead to reduced crop yields, and food security. Further, the agriculture sector contributes to greenhouse gas emissions that lead to climate change. The sector requires substantial investments in climate-smart practices and technologies to address these challenges. However, it has traditionally been underserved by the financial system, so farmers and other stakeholders don't have access to the financing to follow climate-smart agriculture practices. This chapter identifies financing challenges and explores critical role innovative financial mechanisms can play in alleviating these challenges. This includes the development of tailored financial products, the use of risk-sharing instruments, and the promotion of public-private partnerships. It highlights the importance of the capacity building of FIs and farmers to design and implement climate-smart interventions and develop robust metrics to measure the impacts of these investments.

Chapter 15: Application of Digital Technologies in Agri-Supply Chain -The Story of India and Comparative Narrative

Agricultural supply chains are undergoing a huge change due to the advancements in digital technology. These technologies have triggered the unknown concepts of sustainability, efficiency, and transparency. The integration of digital technologies like big data analytics, blockchain, artificial intelligence (AI), and the Internet of

Things (IoT) in agricultural supply chains is discussed in the chapter. IoT devices provide for precision farming and less resource waste. Blockchain technology improves transparency and traceability guaranteeing food safety. It also enables to confirm the calibre of agricultural products. Big data analytics is helpful in the better decision-making. All these technologies address the major issues facing the agriculture industry. The issues include resource management, climate change, and food security. But the technology adoption has its own challenges. It is necessary to address issues including data protection, farmers' digital literacy, and the large upfront costs associated with implementing new technologies.

Chapter 16: Supply Chain through Machine Learning and Artificial Intelligence

This chapter presents an in-depth analysis of the application of Artificial Intelligence (AI) in supply chain management systems. Initially, the chapter presents a comprehensive review of existing literature and practices to assess the current state of AI integration within supply chains. This literature review not only provided insights into the prevailing trends but also helped identify gaps and limitations within the current systems. Additionally, a comparative study of the United States, Africa, and India was undertaken to explore how different countries have leveraged AI technologies to transform agricultural supply chains in their respective contexts. Finally, the chapter proposes solutions to address the identified challenges, offering recommendations for enhancing the efficiency and effectiveness of AI-driven supply chain mechanisms.

Chapter 17: Driving Agricultural Performance by Leveraging Digital Technologies in Agriculture 4.0

Agri-food sustainability has grown to be the globe's most important issue throughout the years. Technologies that promote environmentally friendly agricultural farming have proven effective in reducing poverty brought on by increased food demand. Agri-food system uses have been significantly altering the global landscape recently as a consequence of internal and external variables. The forthcoming contribution of the study is that we can utilize it to enhance economic efficiency, better control mechanisms, and improve decision-making. This study will prove beneficial to the users and other stakeholders like technology providers, experts and society at large. Furthermore, this study provided guidelines for further research on the proposed concept. The conclusions drawn in the study will be helpful to researchers, experts, smart technology suppliers, and policymakers who ultimately are responsible for supporting the evolution to extra sustainable and cultured agri-business practices.

Chapter 18: Applications, Advantages, and Challenges of Artificial Intelligence in Agriculture

The economy is significantly influenced by agriculture. Globally, the primary issue and burgeoning topic is agriculture mechanization. The population is growing at an exponential rate, which is simultaneously driving up demand for jobs and food. For the farmers to meet these needs, the conventional techniques they employed were insufficient. New automated techniques were consequently introduced. The utilization of Artificial intelligence techniques has protected agricultural productivity from various threats, such as rising populations, changing climates, employment difficulties, and problems with food safety. Architectures based on machine learning have been confirmed to be among the best regarding precision and resilience. By using systems based on AI, crop yields can be increased while using less water, fertilizer, and pesticides. Besides improving worker safety, intelligent expertise can support reducing its consequences on expected ecosystems, which will help keep food costs low and ensure more crops are grown to feed the world's expanding population.

The chapter has undergone multiple rounds of review and editing, ensuring its comprehensive and refined content. It is poised to contribute significantly to driving structural changes within the global agricultural supply chain. This book will serve as a valuable resource for all stakeholders, providing a deeper understanding of agricultural supply chains at the global level and offering insights into their complexities and potential for improvement in bringing sustainable global agriculture supply chains.

Sneha Kumari
Symbiosis School of Economics, Symbiosis International University, Pune, India

V. G. Venkatesh
E.M. Normandie Business School, Le Havre, France

K. K. Tripathy
Ministry of Health and Family Welfare, Government of India, India

Hemachandra Padhan
Indian Institute of Management, Sambalpur, India

M.P. Sukumaran Nair
Center for Green Technology and Management, India

REFERENCES

Kamble, S. S., Gunasekaran, A., & Gawankar, S. A. (2020). Achieving sustainable performance in a data-driven agriculture supply chain: A review for research and applications. *International Journal of Production Economics*, 219, 179–194. DOI: 10.1016/j.ijpe.2019.05.022

Kumari, S., Jeble, S., Venkatesh, V. G., Nagarajan, C., & Shi, Y. (2023a). Antecedents of agriculture supply chain performance during COVID-19: An emerging economy perspective. *Operations Management Research : Advancing Practice Through Research*, 16(1), 489–510. DOI: 10.1007/s12063-022-00295-3

Kumari, S., Venkatesh, V. G., & Shi, Y. (2023b). Assessment of Risks and Risk Management for Agriculture Supply Chain. In *Supply Chain Risk and Disruption Management: Latest Tools, Techniques and Management Approaches* (pp. 155–172). Springer Nature Singapore. DOI: 10.1007/978-981-99-2629-9_8

Nayal, K., Raut, R. D., Queiroz, M. M., Yadav, V. S., & Narkhede, B. E. (2021). Are artificial intelligence and machine learning suitable to tackle the COVID-19 impacts? An agriculture supply chain perspective. *International Journal of Logistics Management*, 34(2), 304–335. DOI: 10.1108/IJLM-01-2021-0002

Panetto, H., Lezoche, M., Hormazabal, J. E. H., Diaz, M. D. M. E. A., & Kacprzyk, J. (2020). Special issue on Agri-Food 4.0 and digitalization in agriculture supply chains-New directions, challenges and applications. *Computers in Industry*, 116, 1-5. DOI: 10.1016/j.compind.2020.103188

Routroy, S., & Behera, A. (2017). Agriculture supply chain: A systematic review of literature and implications for future research. *Journal of Agribusiness in Developing and Emerging Economies*, 7(3), 275–302. DOI: 10.1108/JADEE-06-2016-0039

Sharma, R., Kamble, S. S., Gunasekaran, A., Kumar, V., & Kumar, A. (2020). A systematic literature review on machine learning applications for sustainable agriculture supply chain performance. *Computers & Operations Research*, 119, 1–42. DOI: 10.1016/j.cor.2020.104926

Acknowledgement

The publication of "Sustaining the Global Agriculture Supply Chain" marks the culmination of collaborative efforts among researchers, scholars, academicians, and organizations. As editors, we extend our deepest gratitude to all those who have contributed, both directly and indirectly, to the realization of this book.

First and foremost, we express our sincere appreciation to our esteemed authors. Their profound expertise, invaluable insights, and unwavering commitment have been essential in shaping this volume into a comprehensive examination of the latest advancements in food security and agricultural supply chain management at the global level.

We are equally indebted to our reviewers and editorial advisory board members. Their meticulous attention to detail, constructive feedback, and scholarly input have significantly enhanced the quality, rigor, and academic integrity of each chapter. Their contributions have been indispensable in ensuring the relevance and depth of this work.

We would also like to extend our heartfelt thanks to IGI Global publishing team for their consistent support throughout the publication process. Their dedication to excellence and professionalism has been evident at every stage, from concept development to the final product. A special note of gratitude goes to Ms. Bina Sinha, Mr. Raj Kishor, Dr. Rameshwar Dubey, Dr. R. Jayalakshmi, and Dr. Yogesh Patil for their invaluable encouragement, guidance, and continuous support during this journey.

Finally, we acknowledge the broader academic and industry communities across the globe, whose ongoing research and innovations continually push the boundaries of knowledge in this field. We hope this book will contribute meaningfully to the discourse on food security and agricultural supply chains and inspire further scholarly inquiry and exploration in this dynamic and critical area.

Sneha Kumari

Symbiosis School of Economics, Symbiosis International University, Pune, India

V. G. Venkatesh
E.M. Normandie Business School, Le Havre, France

K. K. Tripathy
Ministry of Health and Family Welfare, Government of India, India

Hemachandra Padhan
Indian Institute of Management, Sambalpur, India

M.P. Sukumaran Nair
Center for Green Technology and Management, India

Chapter 1

Structural Changes for Innovation in Farmer-Centric Global Agricultural Value Chain

Sneha Kumari
https://orcid.org/0000-0002-7228-1986
Symbiosis School of Economics, Symbiosis International University, Pune, India

V. G. Venkatesh
https://orcid.org/0000-0002-5340-4637
E.M. Normandie Business School, Le Havre, France

K. K. Tripathy
https://orcid.org/0000-0001-7456-0287
Ministry of Health and Family Welfare, India & Government of India, India

Binod Anand
World Cooperation Economic Forum, India & High Power Committee, India

Sonam Kumari
Symbiosis School of Economics, Symbiosis International University, Pune, India

ABSTRACT

The diverse agricultural landscapes across the country, influenced by factors like climate and support systems, demand a more tailored approach. Unfortunately, the emphasis remains on measuring productivity rather than farm incomes, leading to a mismatch between national and regional needs. To address these issues, there is a need for a shift towards demand-driven policies, where programs are designed to

DOI: 10.4018/979-8-3693-4330-2.ch001

meet the sector's needs effectively. However, the existing setup lacks flexibility and often leads to states merely implementing central directives without considering local conditions. Therefore, the present chapter aims to design structural changes for innovative approaches to the global agriculture value chain. A paradigm shift is required where countries collaborate based on the evolving agricultural landscape. The chapter has tried to discuss different cases of bringing innovative changes in the global agriculture value chain.

INTRODUCTION

Food security is now coupled with nutrition security globally, especially after COVID-19. It is connected to prosperity and good health and significantly impacts other economic sectors. With the Farmers as producers, a new Cooperative Economic Framework is on the rise for the farmers. With continuous innovation and efforts towards the Cooperative Sector, countries will create a Democratic and Farmer-centric Value chain globally, provided a slew of structural changes promptly be executed. Increasing production and productivity may not ensure a healthy global agriculture supply chain. Surprisingly, many farmer suicides occurred after periods of surplus in crops like cotton, tobacco, chilies, potato, and onion despite record yields.

The population worldwide is expected to grow to 2.3 billion by 2050. According to the estimates, the global population is expected to reach 9 billion by 2050 and 10.1 billion by the end of the twenty-first century (Bastein et al., 2013). The increase in population will lead to increased demands globally and will impose tremendous pressure on the Earth to meet those particular demands. (Franklin-Johnson et al., 2016). The consumption of natural resources continues to rise, and by 2030, it will require resources equivalent to two to three planets by 2050. (U.N. Sustainable Development Goals, 2015). Therefore, it is essential to focus on the productivity and supply of agricultural produce. The supply chain of agricultural produce will not be sufficient to meet the demand. Thus, there has been a shift from the agriculture supply chain to the value chain.

The agriculture value chain is dependent upon global value export. Value-added export consists of five components: direct domestic value-added, indirect domestic value-added, foreign value-added, the share of foreign value added from a direct importing country, and the share of indirect value-added that returns home. Global Value Chain (GVC) participation index constitutes domestic value added from export, foreign value added (FVA), and pure double counting (arises when intermediate goods cross borders back and forth multiple times). The gross value addition (GVA) index indicates a country's participation in GVC (Balié et al., 2019). The expansion of GVC has led to a productive agriculture and food sector. GVC has transformed from traditional commodities to high-value processed commodities (Montalbano & Nenci, 2020). Inter-country input-output tables and a matrix of bilateral trade flow determine the trade-in value added. Import tariffs act as a barrier to GVC participation. Quality control, logistics, storage facilities, packaging, and service quality determine efficient GVC. Tariffs, nontariff measures, barriers to service trade, and domestic agriculture policies influence the international trade of agro-food products. The relation between GVC and economic growth depends on structural factors like geography, market size, level of development, and industrial structure.

The study is based on a literature review of research manuscripts, reports, agriculture statistics, and websites, focusing on successful cases of agricultural value chains in different countries. The cases have led to building structural changes in the agriculture value chain and propose a framework for an innovative global agriculture value chain. Recent shifts in the global trade landscape and corresponding adjustments in foreign trade policies have significant implications for the country's agricultural sector, especially regarding exports. The Global Agriculture Value Chain has faced challenges in maintaining its competitive edge in farm exports. While the focus of global agriculture has traditionally been on achieving self-sufficiency, there is now a growing need to transition towards export-oriented value production. Low productivity, high logistics costs, limited value addition, export promotion hurdles, non-tariff barriers, and quality issues constrain the Global Agriculture Value chain.

The chapter discusses a more dynamic and responsive approach needed in the agricultural supply chain, and people need to think about the global agriculture value chain. This involves recognizing the complexities of the agricultural sector, empowering countries to address national challenges, and fostering better coordination for agriculture processing. The chapter discusses the country's approaches towards global agriculture value chains.

WORLD AGRICULTURE SUPPLY CHAIN

Economic diversification for agricultural commodities can be seen in the context of the global energy transition. The agriculture supply chain needs to be climate-friendly; therefore, the focus is more on the energy transition. Agriculture 2.0 has seen a paradigm shift that requires a traceable food system. We have seen a significant shift with the gifts of technological marvels like blockchain and artificial intelligence increasingly becoming prominent globally. This tracing technology will help a better and more robust food system in India. Deriving optimum and better-quality output from our resources has clear, broader objectives. Digital and climate-friendly technology can be implemented to make the supply chain user-friendly. The agricultural commodities supplied need some primary processing to be procured longer. In this regard, Agriculture-dependent countries may undergo primary processing for commodities like turmeric and refined chocolate.

Foreign direct investments in agriculture are helping to build food security. The top 3 agri-food projects in 2022 for developing countries have been the Sudan Agri Project of 1600 million dollars, the FGV Chuping Agro Valley Integrated Dairy Farming Project of 1074 million dollars in Malaysia, and the Morocco Dakhla Agri Project of 213 million dollars (UNCTAD, n.d). Russia -Ukraine war has led to food security concerns in developing economies. International investments in agriculture and value chain projects have been slow since adopting sustainable development goals (SDGs). There has been an increasing trend of investment in agriculture in developing countries. In 2022, Mexico, the United Arab Emirates, and China were the top destinations for investment in food processing (UNCTAD, n.d). Growing demand for fertilizer and agrochemicals has led to high interest in international projects. Brazil has been at the top for research and development (R&D) investment. Agrifood will require an additional 300 billion dollars annually to eliminate poverty and hunger, meeting SDG 1 and SDG 2.

Global Production and Supply Chain for Fruits over the last 20 years

The figure shows the average export, import, production, and supply chain waste generated over the past 20 years. The global production of fruits has had a high trend over the years (Figure 1). The world's fruit production has grown at a rate of 59 percent over the last 20 years. The highly produced fruits are bananas, apples, watermelons, oranges, and grapes. India supports 17.84 Percent of the World's Population and 15 percent of Livestock, with merely 2.4 percent of the World's share. India is the world's leading producer of cereals, milk, sugar, fruits, vegetables, spices, eggs, and seafood.

Figure 1. Global supply chain for fruits

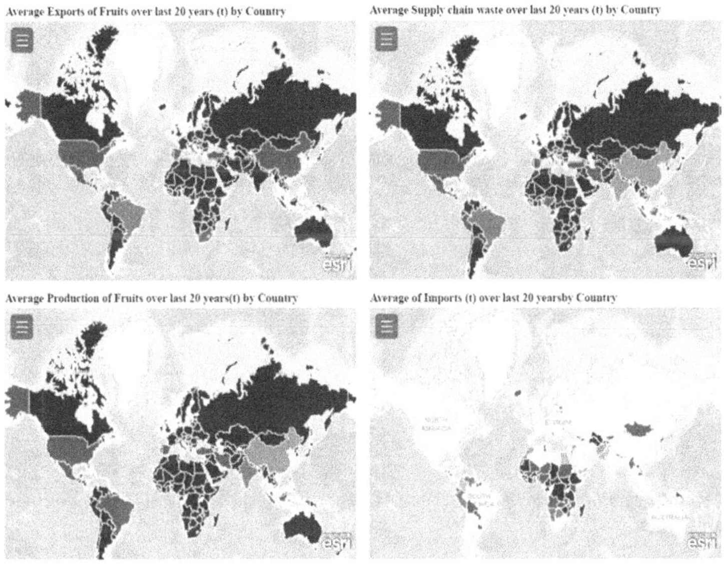

Source: Created by Authors From UN Food and Agriculture Organization (FAO) https://ourworldindata.org/agricultural-production

Figure 2. Average production and supply chain waste

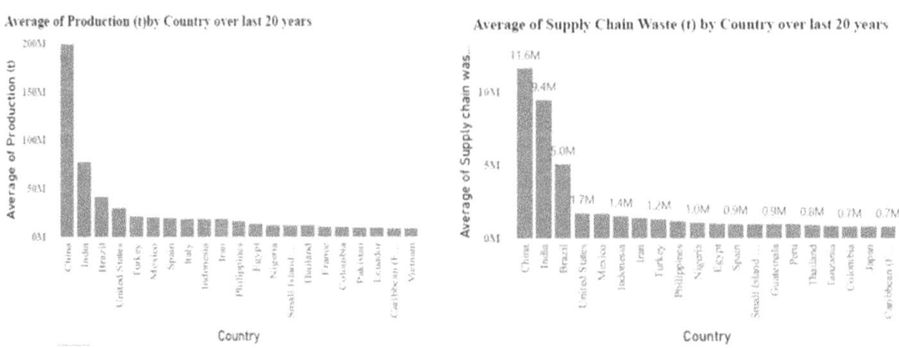

Source: Created by Authors From UN Food and Agriculture Organization (FAO) https://ourworldindata.org/agricultural-production

The data on the avcragc production of fruits over the last 20 years shows that China, India, and Brazil are the top fruit-producing countries in the world. Being high in production, the countries also contribute to high supply chain waste (Figure 2). This is because fruits are perishable in nature and demand value chain strategies.

Figure 3. Average imports and exports

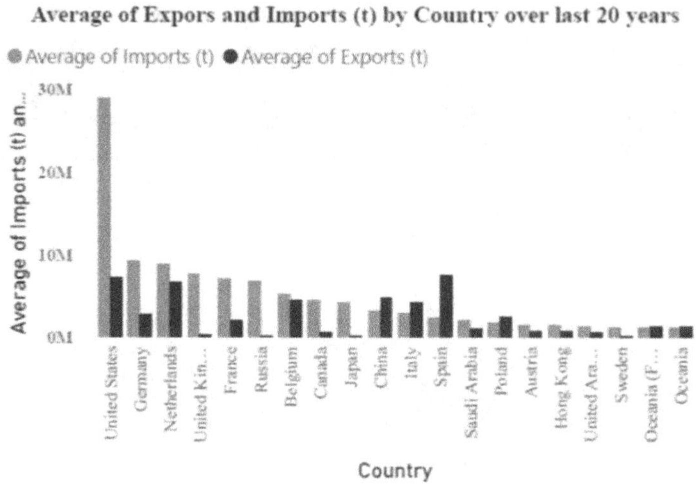

Source: Created by Authors From UN Food and Agriculture Organization (FAO) https://ourworldindata.org/agricultural-production

The data over the past 20 years shows that the United States, Germany, Netherlands, and France are the countries with the highest importation of fruits globally.

Global Production and Supply Chain for Vegetables over the last 20 years

The figure shows the average export, import, production, and supply chain waste generated over the past 20 years. The global production of vegetables has had a high trend over the years (Figure 4).

Figure 4. Global supply chain for vegetables

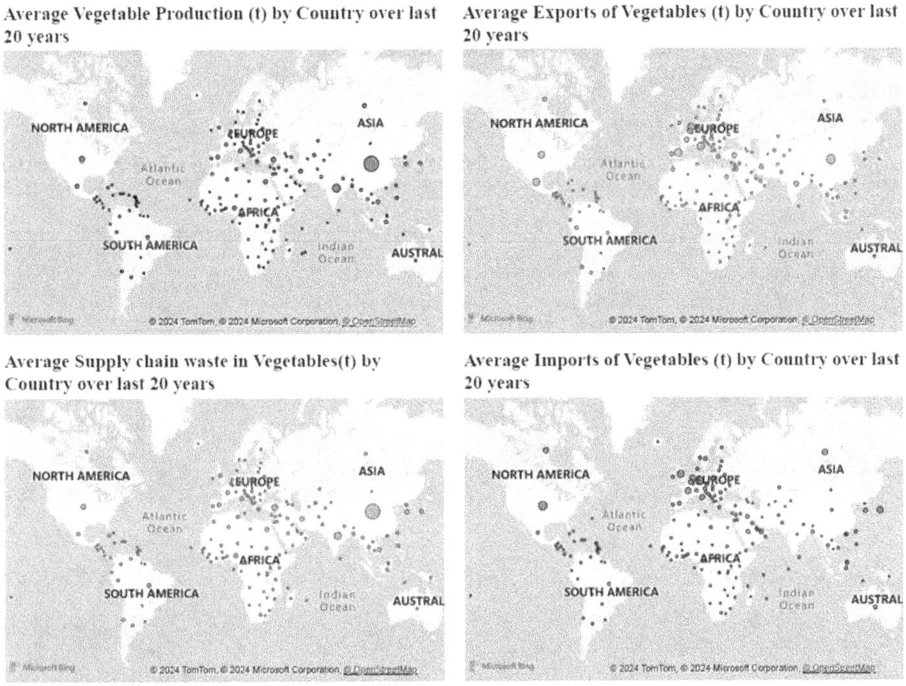

Source: Created by Authors From UN Food and Agriculture Organization (FAO) https://ourworldindata.org/agricultural-production

Figure 5. Country-wise Food Metrics in Selected 10 Countries in 2021

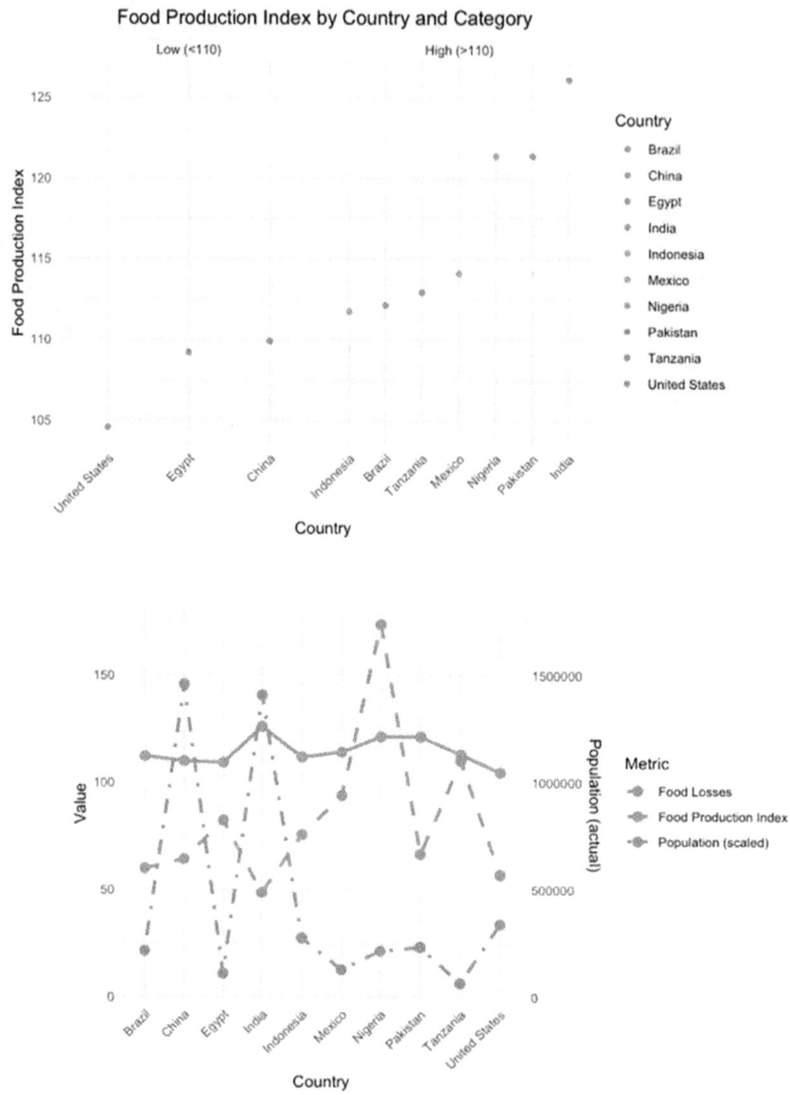

Created by authors from Source: Food and Agriculture Organization of United Nations

Application of Different Theory in Supply Chains

Porter's Value Chain refers to the forces for the attractiveness and competitiveness of the industry (Porter, 1979). The bargaining power of suppliers refers to the suppliers driving prices. The bargaining power of buyers refers to the driving power of buyers. Inter-competition refers to the number of competitors. The danger of substitution refers to the substitute products and New Entries.

The value chain theory is relevant, applied to fruits and vegetables, and classified into four categories. They are mainly Intercompany Relationship theory, Institutional Economics theory, Social Network theory, and Supply Chain management theory (Table 1).

Table 1. Application of theory in agriculture supply chain

Sr.	Construct	Variables	Author (Year)
1	Tangible Resource (Resource Based Theory)	Technology Infrastructure Financial capital Organizational Resources (Warehouse)	Barney, 1991 Shih and Pierson, 2017 Chomilier, 2002
2	Intangible Resources (Resource Based Theory)	Knowledge Skills Competency	Barney, 1991 Chomilier, 2002 Grant, 1991
3	Social Capital (Social Capital Theory)	Trust Network Coordination Commitment Social Participation	Schniederjans et al., 2016 Pateman, 2013, Samii et al., 2002, Kassar and Singh, 2017, Bozionelos and Singh, 2017
4	Contingent Factors (Contingent Based Theory)	Government Economy Technology Geography	Havisto and Kovacs, 2014 Overstern et al., 2017 Kunz and Gold, 2017, Singh et al., 2018
5	Cost Benefit (Transaction Cost Theory)	Effectiveness Efficiency Equity	Gupta et al., 2017; Young-Ybarra and Wiersema (1999); Williamson (2008), Salvado et al., 2018
6	Social Network (Social Network Theory)	Social Media Information Technology Analytics tool	Gupta et al., 2017, Jones et al. (1997), Rowley (1997), Wu et al. (2012), Griffith et al., 2017, Jeble et al., 2018, Jeble et al., 2016

continued on following page

Table 1. Continued

Sr.	Construct	Variables	Author (Year)
7	Stakeholder (Stakeholder Theory)	Supply Chain Partners Farmers Cooperatives Government Supporters	Gupta et al., 2017, Freeman (2001), Lantos (2001)
8	Agility and Trust (Complexity Theory)	Information Sharing Behavior uncertainty Coordination Flexibility	Gupta et al., 2017, Crozier and Thoenig (1976), Guide and Van Wassenhove (2009), Gunasekaran et al. (2014), Dubey et al. 2017
9	Mimetic pressure (Institutional Theory)	Uncertainty Norms Values	Gupta et al., 2017, DiMaggio and Powell (1983), Liang et al. (2007)
10	Supply Chain Management	Strategic Planning Monitoring Controlling Dynamic	Kabra and Ramesh (2015) Haavisto and Kovacs, 2014 Beamon and Balcik, 2008 Schulz and Heigh, 2009
11	Sustainable Environment Practices	Resource Conservation Pollution Reduction	Gupta et al., 2017, Harhara et al., 2017; Kumari and Patil, 2018
12	Intercompany Relationship theory	The relationship theory refers to the collaborative efforts among the stakeholders and their network for the value chain. This theory considers the trustworthiness and coordination among the stakeholders	Kulwijila, 2018
13	Institutional Economics theory	The Institutional Economics theory discusses the governance and organization of transactions. The theory is classified into transaction cost economics and agency theory.	Kherallah & Kristen, 2002
14	Transaction cost economics theory	Transaction cost economics theory discusses the costs in the value chain.	Zaridis et al., 2021
15	Agency theory	Agency theory refers to the engagement of principals and agents in the value chain. This theory refers to the governance of the stakeholders in the value chain.	Zaridis et al., 2021
16	Social Network Theory	The social network theory focuses on the interactions of social and economic factors required to link stakeholders in the value chain.	Patil et al., 2023
17	Supply Chain Management Theory	The theory refers to managing operations and activities to improve product quality.	Wang et al., 2023

continued on following page

Table 1. Continued

Sr.	Construct	Variables	Author (Year)
18	Collective Action Theory	In ASC, a collective effort is required to increase the industry's income. The requirement of quality standards, technology, financial support, and management of the agriculture commodities drives multiple interests and collectivization of stakeholders' efforts.	Reinecke & Donaghey, 2017
19	Agency Theory	Agency theory refers to the effective monitoring of the supply chain performance. The theory mentions the variation in the incentives of human resources to monitor and protect the shareholders' interest. The agriculture supply chain involves different stakeholders who need to be aligned together for the interest of the supply chain. The top management is responsible for better allocating and managing the work across the functions of the supply chains. The gains incurred are distributed according to the interests and work in each function.	Hillman & Dalziel, 2003
20	Resource Dependence Theory	The resource dependence theory is applicable in the supply chain as monitoring resources is very much required for the supply chain performance. The ability of the top management to bring resources results in an efficient supply chain performance.	Hillman & Dalziel, 2003
21	Contingency Theory	Contingency theory suggests that no ideal structure fits all organizations but that the optimal structure is contingent upon several contextual factors and results from an adaptation process. According to the political scenario, the government may focus on supply chain management.	Kunz and Gold, 2017
22	Stakeholder Theory	Stakeholder theory refers to the involvement of different tiers of the supply chain with the cooperative and collaborative efforts of the stakeholders —the collaborative strategies in the supply chain result in productive output. Firms must consider all stakeholders' opinions and expectations for a successful ASC.	Kayikci et al., 2022
23	Theory of Planned Behavior	The behavior of interest for any function in the agriculture supply chain is determined by the intention to perform and the perceived behavior control. The function approach towards ASC, like processing, quality maintenance, and production, depends on stakeholders' intentions and behavior. The intention to perform is determined by attitude, subjective norms, trust, and perceived behavioral control. The intention of the stakeholders is essential for the ASC's performance.	Giampietri et al., 2018
24	Resource Based Theory	Tangible resources include land, buildings, natural resources, and financial Capital; intangible resources include knowledge, skills, business processes, etc. The intangible resources like knowledge, skills, and competency may also impact a sustainable supply chain.	Barney, 1991 Grant, 1999 Becker, 1990

continued on following page

Table 1. Continued

Sr.	Construct	Variables	Author (Year)
25	Innovation Diffusion Theory	Innovation decisions result from knowledge, persuasion, decision, implementation, and confirmation. ASC functions require the implementation of innovative ideas. The innovations in production, processing, grading, sorting, or packaging are driven by the knowledge level of the stakeholders to implement, their perception to adopt, and confirmation for the same. The diffusion of innovative ideas in the ASC helps it function properly.	Qadar et al., 2023
26	Commitment Trust Theory of Relationship	The agriculture supply chain involves different partnerships for the flow of agricultural commodities. Commitment and Trust help preserve stakeholder relationships. Trust among the stakeholders leads to relationship commitment to meet the demand in the agriculture supply chain.	Morgan & Hunt, 1994
27	Dynamic Capability Theory	The agricultural market has been unpredictable, with a shift in the demand for perishable commodities like fruits and vegetables. Firms need to reconfigure innovative practices and technology management in supplying agricultural commodities. For example, few agricultural commodities require a cold chain, while a few require hi-tech processing. The firms can undergo a dynamic environment based on the market's requirements.	Wilson & Daniel, 2007
28	Social Capital Theory	Social Capital is defined as a function of trust, social norms, participation, and network. These functions affect social and economic activities. The influence of Social Capital can be seen more prominently in groups or small communities. Social Capital comprises social theory and economic development, families and youth behavior, education, community life, work and organization, governance, and collective action problems.	Nakagawa and Shaw, 2004 Krishna, 2002 Woolcock, 1998
29	Complexity theory	Complexity is a characteristic of a system. The complexity designs how a system behaves over time and adopts new methods. The theory is appropriate for the study as it will help to understand the complex systems in the ASC and how the functions from Production to distribution have been adopting changes. The theory will also help to understand the interactions of stakeholders and how the organization adjusts to the processes.	Vachon and Klassen, 2006 Pacheco and Paul, 2023
30	Organization design theory	Organization theory has been mainly designed to achieve strategic goals. The theory pertains to the information processing, resource-based, dynamic capability, and contingency theories. It refers to an organization's limited capacity of human resources. Organization design theory is appropriate for the study as ASC needs to integrate tasks at every function. The proper allocation of work at each function of ASC from production to distribution requires a flow of proper information and integration of work.	Piekkari & Westney, 2017 Kenis & Raab, 2003

Source: Created by Authors

Existing Cases on the Value Chain for Selected Agriculture Commodities

The agricultural trade has seen transformations in production, be it wheat in Australia and Ukraine, processing into flour in Indonesia and Turkey, and exporting to China to make noodles and to Africa and the Middle East to make bread. Value chain connects producers, processors, and consumers across the world. A few cases on the value chain of selected commodities are discussed below.

French Bean Value Chain from Kenya to the European Union

Kenya, is located in East Africa along the Indian Ocean coastline. Thc country is renowned for its rich wildlife. Kenya's agriculture includes tea, coffee, cotton, fruits, vegetables, maize, and wheat.

Kenya's geography, climate, and population exhibit significant diversity. The variation in climate and terrain contributes to Kenya's ecological and demographic complexity. Kenya is a major global producer of tea and coffee and ranks third in the export of fresh produce, including cabbages, onions, and mangoes. Smallholder farms are the primary producers of staple crops such as maize, as well as potatoes, bananas, beans, peas, and chillies.

French beans are named after the popularity of the commodity in France during the 19th century. French beans are primarily exported from Kenya. European retail outlets drive the French bean value chain. Smallholder farmers lack the technical or financial capability to comply with the standards demanded by the European Union. French beans are a highly perishable commodity (Usaid-Kaves, 2015). It suffers from high postharvest losses. The beans are handpicked and collected in crates. The collected beans are protected from the sun by a cold storage facility. The pre-cooling and packaging of the beans are done in compliance with the quality standards. The beans that comply with good agriculture practices and residue limits are fit for export, while others are rejected. The rejected beans are sold to traders or at local markets.

Tomato Value Chain in the European Union

The European Union (EU) comprises 27-member states, primarily located in Europe. In the European Union, common winter crops include wheat, rapeseed, rye, and triticale, whereas maize, sunflowers, rice, soybeans, potatoes, and sugar beet are typically grown as summer crops. Barley is cultivated in both winter and spring varieties. Tomatoes are recognized as one of the most carbon-intensive food products. According to research conducted in European Union (EU) countries, 63 percent of tomato waste occurred at the processing and consumption stage, and 51 percent of greenhouse gas emissions were from production. The European Union explored the potential of mitigation strategies for greenhouse gas emissions, resulting in tomato food waste reduction and process optimization (Xue et al., 2020). The European Union is expected to become the net exporter of processed tomatoes by 2030. In Southern European countries, most fresh tomatoes are grown in an open field, while in other countries, they are grown in soil or soilless greenhouse productive systems. The consumption of fresh tomatoes is expected to decline while the demand for processed tomatoes will increase (European Commission, 2017). Processed tomatoes have higher levels of lycopene than fresh tomatoes. The high level of lycopene protects against some forms of cancer (Dominguez et al., 2018). Tomato skin, seeds, pomace, and culled tomatoes are the byproducts of tomato. The tomato and tomato products are transported by trucks within the EU, and they are shipped outside. The tomatoes are prevented from wastage at the retailing, distribution, processing, postharvest, and production stages. The waste is also managed through composting, anaerobic digestion, energy recovery, and landfilling. The EU countries focus on the supply chain of tomatoes and synthesize alternate products from byproducts like peel, seed, and pomace.

Onion Value Chain in New York

New York is located in the North eastern United States and is classified as one of the Mid-Atlantic states. It shares borders with the Atlantic Ocean, New England, Canada, and the Great Lakes. New York is a significant agricultural state, ranking among the top ten in the production of 30 different commodities. It holds the leading position nationally in the production of cream cheese and cottage cheese. It ranks second in the production of apples, maple syrup, cabbage, and yogurt, and maintains high rankings in the cultivation of tart cherries, green peas, grapes, squash, and other crops. New York has been leading in the United States onion supply chains for onions, popularly known as black dirt onions. New York soil is rich in sulphur, resulting in the state's production of yellow onions. The Black dirt onion in New York has a unique taste due to the high sulphur in the soil (Jeanneaux, 2023). The variety commands a specific market segment, resulting in a competitive advantage and resilience in the value chain. France has Roscoff pink onion, which is famous for its fresh eating properties and beautiful color (Jeanneaux, 2023). The black dirt onion faces competition with the Roscoff pink onion, but the distinguishing feature of New York onion makes it unique among consumers. Onion processing includes dehydration, drying, pickling, and frozen onions. The value-added produce from onions is sold in flakes, powder, chutney, paste, animal feed, natural dyes, onion oil, onion pulp, and pharmaceutical ingredients.

Potato Value Chain in the Netherlands

The Netherlands, located in northwestern Europe, is renowned for its flat landscape, which features extensive canal systems, tulip fields, windmills, and a network of cycling routes. The Dutch agricultural sector is primarily focused on the production of cereals, particularly wheat, as well as feed crops such as fodder maize and potatoes. The horticultural sector is centered on the cultivation of vegetables and flower bulbs. Greenhouses in the Netherlands are predominantly used for the production of vegetables, including sweet peppers, and flowers, such as roses. The Netherlands is the leading global market for potato seeds. Around 8 lakh tonnes of potato seeds are exported to countries. Since potatoes are vegetatively propagated, there is a high probability of disease and infection transmission through soil (Louwaars, 2023). Packaging, labeling, and quality assurance innovation create an efficient potato value chain. New products like sweet potato chips and orange-fleshed sweet potato flour increase the market potential in Central Uganda (Horton et al., 2023). The Participatory Market Chain Approach (PMCA) has resulted in innovative drivers, an increased intensity of innovations, and improved outcomes. The innovative drivers include agroecology and livelihood, policies and market conditions, history, culture, and institute support potential for value addition, commodity, and its market, fidelity, leadership, innovation, and follow-up. Smallholder produces native potatoes for home consumption and sells a few potatoes to local markets in Bolivia Altiplano. A local source organization, CAD, facilitated the PMCA application, leading to product development and market testing. A new product named Miskipapa was developed, where the organization selected, bagged, and washed fresh potatoes for sale in supermarkets, hotels, fairs, and company stores. The project to improve the vegetable market in West and Central Java was implemented by the Indonesian Vegetable Research Institutes and led to 13 distinct innovations that involved processed products such as potato chips.

Grapes Value Chain in Tanzania

Tanzania is situated in East Africa. Agriculture crops grown in Tanzania include maize, sorghum, millet, rice, wheat, beans, cassava, potatoes, bananas, cashew nuts, coffee, cotton, tea, and tobacco.

Grapes are rich in nutrients. Grapes are helpful for the treatment of dyspepsia, high cholesterol, uric acid, stones in the urinary tract, bile duct, liver problems, and digestion issues. According to the Food Agriculture Organization (FAO, 2010), seventy-one percent of the world's grape production is used as wine, two percent is processed into raisins, and twenty-seven percent is used for fresh fruit consumption. The peel of the grapes is used for pectin and essential oils. The peel of the grapes is also used as cattle feed and candies. The juice produced from raisins is used in cosmetics for bleaching and skin-softening. The grape value chain requires adequate market accessibility, investment in post-har infrastructure, like storage and packaging facilities, and comprehensive training for handling and improving grape quality. Grape Value Chain requires mapping the linkages between actors, activities, and product flow (Kulwijila, 2018). Factors influencing grape losses at a farm level include the age of the grapes at harvest, quantity harvested, credit flow, unreliable market, and experience of the harvesters (Kulwijila, 2021).

Pomegranate Value Chain in Afghanistan and South Africa

Afghanistan is situated at the intersection of Central Asia and South Asia. Before the widespread cultivation of poppies, the majority of arable land was devoted to cereal crops, with wheat being the primary staple. Other commonly grown food grains include maize (corn), rice, and barley. Cotton also plays a significant role in textile production. Additionally, fruits and nuts are important commodities in the country's export sector. Afghanistan is known as the country of the pomegranate fruit due to its high quality and scientific production methods. Afghanistan pomegranate production expanded in 2009 when pomegranate processing was also given prior importance. A million-pound juice factory was constructed in Kabul for processing (Daqiq, 2021). High production and processing led to an increase in the exports of pomegranate. The value chain led to providing credit facilities to the farmers for pomegranate production. Kandahar Chamber of Commerce reported in 2015 that 7681 metric tonnes of pomegranates were exported to Pakistan, Tajikistan, India, and Europe. Afghanistan Central Statistics Organization 2019 reported 99871 metric tonnes of pomegranate production on 9721 hectares of land in Afghanistan. The pomegranate value chain in Afghanistan focuses on a successful pomegranate value chain emphasizing access to modern technology for production, processing, and sorting. The South African pomegranate industry is export-oriented. The Pomegranate Association of South Africa (POMASA) is the pomegranate industry body established in 2019 to address industrial progress for pomegranate processing (Pienaar & Barends-Jones, 2021).

Mango Value Chain in Ethiopia

Ethiopia located in the Horn of Africa, is a landlocked country characterized by rugged terrain and divided by the Great Rift Valley. The majority of agricultural production is subsistence-based, with a significant portion of commodity exports derived from the small-scale agricultural cash-crop sector. Key crops include coffee, pulses (such as beans), oilseeds, cereals, potatoes, sugarcane, and vegetables.

Agricultural commodities dominate Ethiopia's exports, with coffee serving as the largest foreign exchange earner. Additionally, Ethiopia is the second-largest maize producer in Africa. The country's livestock population is believed to be the largest. Mangoes are harvested at the ripening stage, identified by their color, texture, and smell. Further, they are washed to remove pesticides, bacteria, and dirt particles. For processing, the mangoes are peeled with machines, followed by deseeding. The flesh of mangoes is pulped. The pulp is further concentrated to reduce the water content. The mango puree is sterilized and frozen for long-term storage. Ethiopia's mango market suffers from a lack of market to absorb the production, low price, poor product handling and packaging, lack of transparency, lack of knowledge and skills, limited mango variety, and prevalence of pests and diseases (Honja, 2014). The cluster approach in the mango value chain has led to overcoming constraints. Consumer perception of the mango value chain is driven by the mangoes' food safety, price, marketing, quality, and shelf life (Ayyaz et al., 2023). Therefore, managing mangoes' quality and shelf life is essential to improve the market. Mango is processed into juice, jam, nectar, pulp, chutney, and jellies. The mango byproducts have diverse uses. Mango peel is a rich source of fiber and antioxidants used to make mango powder, mango leather, and chutney. The mango seeds are a good source of protein and fiber. It is used in the form of oil, powder, and soap.

Banana Value Chain in Ecuador

Ecuador is located on the equator along the western coast of South America. Its varied geography includes the Amazon rainforest, the Andean highlands, and the ecologically diverse Galápagos Islands. The cultivation of subsistence crops such as maize (corn), potatoes, beans, and cassava (manioc) plays a crucial role in the rural economy. India is a leading producer of bananas, and Ecuador is a leading exporter of bananas. The European Union, the United States, the Russian Federation, China, and Japan are the leading importers of bananas. Banana traders have interlinkages across different regions (Gebre et al., 2020). Bananas are a delicate commodity, estimating losses at different value chain stages. Bananas are reaped, washed in treated water, and packed for loading within 5 hours. Temperature management system washing, air drying, and proper coordination can reduce losses in the value chain (Mvumi et al., 2016). The Ecuadorian banana value chain has been highly dynamic and known as the world's leading banana exporter. Eight thousand eight hundred fifty-one banana producers cultivating bananas in less than 30 hectares of land sell bananas to 200 exporting countries. The Banana value chain is driven by the legal framework, voluntary sustainability standards, financial crisis, markets, innovations, natural hazards, and value chain organization (Coral and Mithofer, 2023).

Pulses Value Chain

Pigeonpea are harvested once they turn yellow or brown. The peas are separated from the pods by undergoing threshing. The pigeon peas are dried, shelled, and cleaned. The cleaned peas are rinsed to remove any dirt and then dried. The hulled peas undergo polishing to remove any residue and improve their appearance. Further, the peas are distributed and may undergo processing to form canned pigeon peas, frozen pigeon peas, or even pigeon pea flour. The byproducts of pigeon peas include soaps, curries, and flour for baking and cooking. Husks of pigeon peas are used as animal feed or for composting—stems of the pigeon pea plant are used for biomass energy or biochar.

Apple Value Chain in Ethiopia

Ethiopia located in the Horn of Africa, is a landlocked country characterized by rugged terrain and divided by the Great Rift Valley. The majority of agricultural production is subsistence-based, with a significant portion of commodity exports derived from the small-scale agricultural cash-crop sector. Key crops include coffee, pulses (such as beans), oilseeds, cereals, potatoes, sugarcane, and vegetables. Agricultural commodities dominate Ethiopia's exports, with coffee serving as the largest foreign exchange earner. Additionally, Ethiopia is the second-largest maize producer in Africa. The country's livestock population is believed to be the largest. Ethiopia's highlands, with low temperatures, are suitable for apple production (Getahun et al., 2018). Quality management for the right market is one of GVC's concerns. Unlike other countries, Ethiopia faces risks in terms of apples like contamination, disease, etc. Apples are harvested once they are ripe. They are processed into applesauce and apple juice. The byproducts of apples include apple pomace and apple core, used as livestock feed and fertilizer; apple seed used as a source of oil; and apple pectin, used in jams and jellies.

Peach Value Chain

The quality of peach is determined by the physicochemical parameters for value addition (Diezma et al., 2021). Peach is a highly perishable commodity and hence demands cold storage. The fruit's pulp is juicy, has nutritional benefits, and is rich in vitamin A, potassium, and minerals—peach syrup and puree-rich ingredients for infant food, jams, jellies, and yogurt. Peach byproducts contain oil in its seed, anthocyanins in pomace, pectin in the peel, and protein in the seed. Peach kernel oil is used as a flavoring agent. It is also used in liqueur production. Peach skin is processed into fruit-based filling for the bakery. Waste generated in sorting, peeling, and processing is used for livestock feed. Peach vinegar and fermented produce add flavor to pickles, chutney, etc.

Coffee Value Chain at Toraja

Toraja is located in the mountainous region of South Sulawesi, Indonesia. The Toraja region is renowned for its high-quality coffee, particularly Toraja Robusta and Toraja Arabica varieties. Organizations such as the Toraja Coffee Farmers Association are actively involved in promoting sustainable coffee production and processing practices, with the goal of enhancing the quality of coffee produced by local farmers. The region also has rice cultivars. Toraja coffee is famous in the world. Indonesia, the third largest coffee-producing country, produces the most famous high-quality coffee from the Tana Toraja region. Toraja Coffee's value chain includes farmers, collector traders for the domestic sale of coffee, collector traders for the sale of coffee across, and coffee processing firms(Salam et al., 2021). The coffee value can undergo drying, sorting, and processing into ground coffee. The ripe coffee cherries are picked, and the outer skin is removed. The beans are fermented in water for flavor development. The beans are washed and spread to dry in the sun or beds. The farmers strip coffee beans into beans and continue with the drying process. The dried beans are passed through a hulling machine to remove the parchment layer. The beans are sorted, graded, and polished to improve their appearance. The green coffee beans are packed and stored in a cool place. The coffee processing leads to byproducts of coffee, like coffee pulp, which can be composed to enhance soil nutrients. The coffee husk can be used as an energy source.

Orange Value Chain in South Africa

South Africa is located at the southernmost tip of the African continent, characterized by a diverse range of ecosystems. South Africa ranks among the world's leading producers in several agricultural sectors. It is the fourth-largest producer of chicory roots and grapefruit, the fifth-largest producer of cereals, and ranks seventh in the production of both green maize and maize. Additionally, the country is the ninth-largest producer of castor oil seeds and pears, and the tenth-largest producer of sisal and fiber crops. The dairy industry in South Africa is comprised supports the livelihoods of individuals. South Africa is one of the leading citrus-producing countries in the world. The agro-climatic zone has led the country to produce quality citrus produce. The citrus supply chain requires maintenance of optimum temperature to increase the shelf life. Citrus Research International in South Africa is undergoing research on ambient or warm loading of citrus produce. Cold chain protocol has been observed as an alternative to forced air cooling. The citrus produce is exported to the United States of America through a cold chain. The cold chain exports help maintain significantly low temperatures. The export of citrus is done through cold sterilization treatment protocols. Cold sterilization treatment helps to control pests and diseases by maintaining a temperature of -0.6 degrees Celsius for twenty-two days—low-temperature results in controlling the growth of insects, pests, and diseases among the fruits. The cold sterilization treatment varies across countries during export. The cold chain helps to stop the growth of insects and pests but may result in chilling injury. The citrus produce suffers from chilling injury due to extreme low temperatures. The injury results in the discoloration of the fruits and peel disorder (Goedhals-Gerber & Khumalo, 2020). The citrus supply chain of South Africa is compared with Spain since Spain is the leading producer and exporter of citrus produce. Spain is the primary producer of oranges in the European Union. A significant amount of investment in the supply chain in South Africa goes into cold sterilization, while the European Union supports Spain's citrus industry, providing compensation for fruit processing and, therefore, gains a competitive advantage.

The orange processing is a significant step for reducing wastage. Orange, once harvested, requires proper sorting and grading. They are then washed and cleaned. The fruits are peeled off; the commonly processed produce is juice from the slices. The fruit juice is pasteurized to destroy microorganisms and is concentrated by removing the water. The juice is then blended and flavored. The blended juice is packed. Byproducts like orange peels, pulp, and seeds are generated. Orange peel in the Xinhui district of China promotes the industrialization of orange peel to promote deep-processed products. The orange peel has a geographical indication of Xinhui (Jing et al., 2022). Orange peels make essential oils, perfumes, cosmetics, flavors, and animal feed, such as dried peels. Pulp and seeds are used in dietary supplements and as additives in food products. Pectin extracted from the pulp is used as a gelling agent to prepare jams, jellies, etc.

Avocado Value Chain in India

India is the leading producer of different agricultural commodities like milk, fruits, and vegetables. Avocado (Pesea Americana Mill) is an essential nutritional commodity globally. There are three distinct races of Avocado (Mexican, Guatemalan, and West Indian), which vary in shape, Size, and fat content. It can produced in semitropical, subtropical, and tropical climates. Earlier, avocados were grown mainly as a backyard crop in Kerala, Karnataka, Tamil Nadu, and Coastal Maharashtra. In the southern and western parts of India, the climate is most suitable, and the value chain can be created, making India an exporter of avocados in a few years and bringing fortune to the farmers. Innovation is lacking among the 55 Clusters the Ministry of Agriculture and Farmers' Welfare identified. ICAR-IIHR has already released two distinct varieties, viz. Arka Supreme and Arka Coorg Ravi for commercial cultivation in India. The changing dietary habits and a growing awareness of the health benefits are in huge demand; this will reverse the supply chain of this exotic fruit. Here, structural changes in the title of the State and Union government are required, and this cluster can only be developed in the form of a Cooperative Avocado zone.

Aloe Vera Value Chain in Greece

Greece is located in southeastern Europe, is composed of numerous islands scattered across the Aegean and Ionian seas. The agriculture includes a diverse range of crops, including olives, grapes, citrus fruits, cereals, and vegetables. The country's vineyards yield significant quantities of grapes, which contribute to the production of internationally recognized wines. In addition to crop cultivation, Greece also produces a variety of livestock products. Fisheries play a prominent role in the economy, while forestry, although important, holds a secondary position in comparison to other agricultural sectors. Aloe vera is adaptable to climate change and provides gel beneficial for health, skin, weight loss, and hair, and, is used in pharmaceutical and cosmetic industries (Solomou et al., 2020). Aloe vera suits Greece's climatic conditions, which are in high demand in the health sector. There is a need to increase Aloevera's leaf processing units. Aloe vera provides a range of byproducts in the form of latex and lead rind. The pulp from the leaf is used in the food, pharmaceutical, and cosmetic sectors. Aloe vera juice is also drained and concentrated by evaporation. The concentrated form is known as glossy aloe.

A PARADIGM SHIFT FOR AGRICULTURE

The pricing of food grains in a food-surplus country has to be very different from that in a food-deficient situation. We must look for trading opportunities in the global market. Agricultural Policy needs changes with a focus from cereals to fruits and vegetables. Sectoral diversification is necessary; workers should be shifted from agriculture to other sectors. Value chains can be created through agro-processing activities. Infrastructure, tech (seed, fertilizer, etc), and marketing support through massive investment by public and private players is required. The way forward is to abandon the present system and create a new one with a roadmap.

Creating Value Chain for Agro Processing

A globally linked network will provide the farmers with quality seeds and train them to improve output quality. Also, procure the output at remunerative prices for the national and global markets, such as Walmart. Land remains with the farmers, but they carry production under a contractual arrangement with assured prices for 5 to 7 years.

Structural Changes in the Agriculture Value Chain

Robust market linkages will help in better price realization and processing of agricultural commodities. This will also help in reducing the waste losses. Agriculture policies have also led to augmented production. Agriculture Production should be linked with agriculture processing to reduce the substantial losses. The strategies for linkage need to be extended to global business. Strict norms on the movement of commodities across borders are barriers to the supply chain. Sometimes, the truck drivers do not get a permit to cross the borders, resulting in the shift of jobs from truck drivers. This results in delays in the supply chain. As a result, an invisible break can be seen in the supply chains. Countries want to establish themselves as manufacturing hubs. Manufacturing-oriented businesses may have problems with logistic drivers. The goods may suffer from finding the right place at the right time. Sustaining the logistic carriers or drivers is required. The effect may be seen in the coming years. There may be a shortage of good container drivers. Countries may go for a strategic partnership to boost their trade alliance. The countries are moving towards friend sourcing for trade. Disruption in the global payment system from US dollars to the countries' currency may be a big challenge.

Competitive Advantage of Value Chains

Big players expect suppliers to produce the product efficiently and revolve around designing and testing the product. Walmart proposed the Collaborating Forecast Planning Replenish (CPFR) approach, which may be phenomenal in demand management of the buyers from the suppliers' side.

Technology

Investment can be driven by modern science and technology to improve yields. Transforming farming to value-added activity export clusters for improved margins is the need of the hour. Farmers should be equipped to discover better product prices through modern digital technology. Leveraging these skills and capabilities has led to increased investment and exports worldwide. Research incentives for research institutions to develop new technologies that promote cultivation with less chemical usage can be beneficial.

Indirect Controls Like Movement Restrictions

Changes in the labor union and contract system also impact the movement of containers. The creation of employment outside agriculture through a new value chain must be part of the action plan for addressing the apprehensions of agitating farmers.

Financial Inclusion for Agriculture Value Chain

Large-scale promotion of value chains and value-added exports are required. Special zones based on clusters and economic corridors that can capture identity-preserved supply chain systems based on private and PPP investments need to be implemented. Apprehensions about corporate capital dismantling public institutions, exploiting farmers, and causing irreparable environmental damage must be addressed. The government must take measures for equity and sustainability by regulating private investment in agriculture and trading.

Conservation of Natural Resources

Since the resources are scarce, the existing linear economy model will not be sustainable for a long time, and the world is shifting toward the Circular economy (CE) model. The CE model includes reverse logistics, design innovation, collaborative ecosystem, and business model innovation, concentrating on managing and integrating resource and value flows of the closed-loop chain. (Goyal et al., 2018). It is driven by several factors: creative brains, victorious business people, enhanced technologies, and friendly ecosystem management at zonal, regional, and country levels. However, CE has not been investigated to a great extent and is, in fact, untouched because of the overly emphasis on the linear model. (Goyal et al., 2021; Ungerman & Dědková, 2020).

Reduce Chemical Usage and Conserve the Environment

A policy that promotes cultivation systems that optimize the use of chemicals like fertilizers and pesticides should be implemented for safe food, minimum environmental impact, and Investment in modern technological tools that can reduce chemical usage.

Responsible Crop Portfolio

Diverse and Nutritious crop production systems are required to meet the changing demands of different sections of the population. Investment in cultivating crops demanded by consumers and moving away from traditional cropping patterns can help bring change. The cultivation of rice and wheat can be transformed to promote a diversified crop portfolio of oilseeds, millet, protein, nutrition, organic crops, and rural processing infrastructure with market linkages. Incentivize change in cropping patterns, produce fruits, vegetables, etc., through price incentives, infrastructural support, and other administrative measures.

Retailers Moving into Processing

Are more supermarkets getting involved in the processing operations of Own Label products? This change is due to the consumers' inclination to access cheaper, own-brand products instead of branded ones. By implementing this processing method, retailers can lower costs, gain more control over what is being done, and have more efficient control over promotions (Kumar et al., 2013).

Innovations for Structural Changes

Three interrelated kinds of innovation in reducing and preventing food waste within the food supply chain: technological (TI), organizational (OI), and marketing (MI) innovations (Aramyan et al., 2021).

Technological Innovation (TI)

There are smartphone applications targeting the fight against food waste, for instance, by encouraging the sale of products that will expire soon. While TI is a frequent type of innovation, it often needs to be a basis for another innovation level since changes in business practices and marketing strategies must be made for the new technological improvement to succeed. Technology responds mainly to problems concerning the agronomics and packaging of products with less focus on manufacturing, distribution, and consumption and how organizations can use technologies for tackling food waste by providing a classification of these technologies (Ciccullo et al., 2020).

Organizational Innovation (OI)

New ways of selling and distributing food must be implemented in the organization to maximize how the food is sold and the way it is distributed to eliminate wastage. For instance, changing the approach to retail sales and encouraging the sales of products that are near their sell-by dates via applications is an OI that supports TI. There are problems with waste by-products in the agri-food industry where by-products are mostly considered waste and, hence, have no value. This problem could be fixed by applying the circular economy principle, which means reusing these wastes as resources. However, problems are encountered, especially in producing organic fertilizers (Ciulli et al., 2019; Nattassha et al., 2020; Kang et al., 2020).

Marketing Innovation (MI)

Marketing innovations concern how food products are marketed and sold; for instance, using applications to sell almost expired foodstuffs or altering the marketing techniques (for instance, eliminating the "two-for-the-price-of-one" scheme). Coordination among the stakeholders results in the recovery of the supply chain (Zhao et al., 2023).

Business Enabled Environment

ASC requires strategic and structural phase transformation. The strategic phase drives the ASC toward leadership orientation, sustainable development goals, digitalization, sustainability principles, and a culture of responsibility. The structural phase drives the ASC towards leadership, technology sharing, environment management, specialized economic sustainability, food security, and community well-being for social sustainability and green energy, focusing on sustainable consumption for environmental sustainability. Strategic and structural transformation can be brought through institutional structures. Firms should increase positive environmental impact through collaborations. ASC activities have shifted to circular readiness, green procurement, waste management, collaborative inventory management, green behavior, and green benchmarks. Introducing solar system panels will reduce energy consumption in the storage rooms. Private branding for agricultural commodities will improve food safety assurance by establishing contract farming. Customer education by introducing a food waste program will reduce the usage of single plastic bags. Quality assurance through USDA certifications can improve the way stakeholders handle products. Building infrastructure through environment preservation can be introduced through tree plantation. Digitalization will reduce paperwork. Better material handling will reduce the contribution of carbon footprint through compressed packaging. Shop refurbishment leads to upgrading shops to provide a better customer experience. Infrastructure, Pest control, Coordination, Food safety, Private sector association, and Investment are a few measures for the agriculture value chain.

Strategic Interventions for Value Chain of Agriculture Commodities

The strategic interventions for the value chain in agriculture commodities are tabulated in Table 2.

Table 2. Strategic interventions for value chain of agriculture commodities

Strategic Interventions	Objective	Constraints	Proposed Solutions for Expected Outcomes
Increase Production for Exports			
Diversification of Production areas	Crop Diversification	Convincing farmers to introduce high-value-added crops	Consistent supply
Farmers' access to extension services	To build awareness and capacity building of farmers	Farmer's interests and their perception	Capacity Building of Farmers
Good Agricultural Practices	Good Agriculture Practice (GAP) certification for quality production	Awareness and expansion of good agricultural practices by all farmers	Expansion of extension services and agriculture companies to create awareness of good agriculture practice
Increase Standards Compliance and Traceability			
Modern Technology	Residue monitoring	Lack of Technical skills	High yield
Artificial Intelligence and Machine learning	Better standards	Lack of expertise	Competitive advantage
Blockchain technology	Better traceability of agricultural commodities	Technical expertise	Awareness of technology and its spread
Data Analytics and Visualization	Better visualization of the flow of agricultural commodities	Expansion of usage of data analytic techniques Technical literacy	Technical Education spread Adoption of data analytic and visualization techniques
Reduction in Post Harvest Losses			
New Investment Partners	Cold storage	Distance, cost	Quality management

continued on following page

Table 2. Continued

Strategic Interventions	Objective	Constraints	Proposed Solutions for Expected Outcomes
Cluster improvement	logistics	Price fluctuations	Increase in export
Cold Chain	Reefer Vans	Optimum facilities like cold storage, reefer vans, etc.	Logistic support for perishable commodities

CONCLUSION

The study focused on different theories and their application in ASC. The study further discusses the value chain of different agricultural commodities across countries. It has been observed that there is a dire need to create commodity-specific multistate cooperatives. All planned clusters need multi-state cooperatives, which will pass the profit of the trade business back to members and make the value chain more democratic, transparent, and farmer-centric. The New Agriculture transformation required Coop 2.0, and the role of the registrars needs to be revisited. The Ease of doing Business in the cooperative sector needs substantial structural changes, especially in the role of Registrars, who are the ultimate regulators. The value chain of commodities brings forth the strategic interventions required for the global agriculture supply chain.

REFERENCES

Aldrich, D. P. (2012). *Building resilience: Social Capital in post-disaster recovery*. University of Chicago Press. DOI: 10.7208/chicago/9780226012896.001.0001

Anastasiadis, F., Apostolidou, I., & Michailidis, A. (2020). Mapping sustainable tomato supply chain in Greece: A framework for research. *Foods*, 9(5), 539. DOI: 10.3390/foods9050539 PMID: 32357573

Aramyan, L., Grainger, M., Logatcheva, K., Piras, S., Setti, M., Stewart, G., & Vittuari, M. (2021). Food waste reduction in supply chains through innovations: A review. *Measuring Business Excellence*, 25(4), 475–492. DOI: 10.1108/MBE-11-2019-0105

Ayyaz, S., Bonney, L. B., & Adhikari, R. (2023). Consumers' insight and behaviour towards mango quality and safety attributes: Implications for integrated value chain development. *British Food Journal*, 125(12), 4518–4538. DOI: 10.1108/BFJ-01-2023-0022

Balié, J., Del Prete, D., Magrini, E., Montalbano, P., & Nenci, S. (2019). Does trade policy impact food and agriculture global value chain participation of sub-Saharan African countries? *American Journal of Agricultural Economics*, 101(3), 773–789. DOI: 10.1093/ajae/aay091

Bastein, A. G. T. M., Roelofs, E., Rietveld, E., & Hoogendoorn, A. (2013). *Opportunities for a circular economy in the Netherlands*. TNO.

Chamlee-Wright, E., & Storr, V. H. (2011). Social Capital as collective narratives and post-disaster community recovery. *The Sociological Review*, 59(2), 266–282. DOI: 10.1111/j.1467-954X.2011.02008.x

Chenhall, R. H., Hall, M., & Smith, D. (2010). Social capital and management control systems: A study of a non-government organization. *Accounting, Organizations and Society*, 35(8), 737–756. DOI: 10.1016/j.aos.2010.09.006

Ciccullo, F., Cagliano, R., Bartezzaghi, G., & Perego, A. (2021). Implementing the circular economy paradigm in the agri-food supply chain: The role of food waste prevention technologies. *Resources, Conservation and Recycling*, 164, 105114. DOI: 10.1016/j.resconrec.2020.105114

Ciulli, F., Kolk, A., & Boe-Lillegraven, S. (2020). Circularity brokers: Digital platform organizations and waste recovery in food supply chains. *Journal of Business Ethics*, 167(2), 299–331. DOI: 10.1007/s10551-019-04160-5

Co, H. C., & Barro, F. (2009). Stakeholder theory and dynamics in supply chain collaboration. *International Journal of Operations & Production Management*, 29(6), 591–611. DOI: 10.1108/01443570910957573

Coffé, H., & Geys, B. (2007). Toward an empirical characterization of bridging and bonding social capital. *Nonprofit and Voluntary Sector Quarterly*, 36(1), 121–139. DOI: 10.1177/0899764006293181

Coral, C., & Mithöfer, D. (2023). The backbone of agrifood value chain resilience: Innovation in the Ecuadorian banana value chain from a historical perspective. *World Development Perspectives*, 29, 100476. DOI: 10.1016/j.wdp.2022.100476

Daqiq, M. H. (2021). Analysis of Pomegranate Value Chain in Kandahar Province of Afghanistan: Issues and Prospects, https://repository.lahoreschool.edu.pk/xmlui/bitstream/handle/123456789/17448/05%20Hasham.pdf?sequence=1&isAllowed=y

Day, J. M., Melnyk, S. A., Larson, P. D., Davis, E. W., & Whybark, D. C. (2012). Humanitarian and disaster relief supply chains: A matter of life and death. *The Journal of Supply Chain Management*, 48(2), 21–36. DOI: 10.1111/j.1745-493X.2012.03267.x

de Sousa, M. A., Addison, T., & Ekman, B. (2001). From humanitarian assistance to poverty reduction in Angola.

Diezma, B., Barreiro, P., Baltazar, P., & Correa, E. C. (2021). A general procedure for predicting the remaining shelf life of nectarines and peaches for virtualization of the value chain. *Postharvest Biology and Technology*, 181, 1–7. DOI: 10.1016/j.postharvbio.2021.111677

Domínguez, I. P., Fellmann, T., Chatzopoulos, T., Pieralli, S., Jensen, H., Barreiro-Hurle, J., & Micale, F. (2018). EU commodity market development: Medium-term agricultural outlook. In *Proceedings of the October 2017 workshop* (p. 133).

El Bilali, H. (2020). Transition heuristic frameworks in research on agro-food sustainability transitions. *Environment, Development and Sustainability*, 22(3), 1693–1728. DOI: 10.1007/s10668-018-0290-0

Elgar, F. J., Davis, C. G., Wohl, M. J., Trites, S. J., Zelenski, J. M., & Martin, M. S. (2011). Social capital, health and life satisfaction in 50 countries. *Health & Place*, 17(5), 1044–1053. DOI: 10.1016/j.healthplace.2011.06.010 PMID: 21784694

European Commission. (2017). *EU Agricultural Outlook for the Agricultural Markets and Income 2017-2030*. European Commission.

Fiksel, J., Sanjay, P., & Raman, K. (2021). Steps toward a resilient circular economy in India. *Clean Technologies and Environmental Policy*, 23(1), 203–218. DOI: 10.1007/s10098-020-01982-0 PMID: 33199980

Food and Agriculture Organization of United Nations. (2021). https://www.fao.org/faostat/en/#data/FBS, as retrieved on 20th September 2024.

Franklin-Johnson, E., Figge, F., & Canning, L. (2016). Resource duration as a managerial indicator for Circular Economy performance. *Journal of Cleaner Production*, 133, 589–598. DOI: 10.1016/j.jclepro.2016.05.023

Gebre, G. G., Rik, E., & Kijne, A. (2020). Analysis of banana value chain in Ethiopia: Approaches to sustainable value chain development. *Cogent Food & Agriculture*, 6(1), 1742516. DOI: 10.1080/23311932.2020.1742516

Getahun, W., Tesfaye, A., Mamo, T., & Ferede, S. (2018). Apple value chain analysis in the Central Highlands of Ethiopia. *International Journal of Agriculture Innovation and Research*, 7(1), 2319–1473.

Giampietri, E., Verneau, F., Del Giudice, T., Carfora, V., & Finco, A. (2018). A Theory of Planned behaviour perspective for investigating the role of trust in consumer purchasing decision related to short food supply chains. *Food Quality and Preference*, 64, 160–166. DOI: 10.1016/j.foodqual.2017.09.012

Goedhals-Gerber, L. L., & Khumalo, G. (2020). Identifying temperature breaks in the export cold chain of navel oranges: A Western Cape case. *Food Control*, 110, 107013. DOI: 10.1016/j.foodcont.2019.107013

Goyal, S., Esposito, M., & Kapoor, A. (2018). Circular economy business models in developing economies: Lessons from India on reduce, recycle, and reuse paradigms. *Thunderbird International Business Review*, 60(5), 729–740. DOI: 10.1002/tie.21883

Goyal, S., Garg, D., & Luthra, S. (2021). Analyzing critical success factors to adopt sustainable consumption and production linked with circular economy. *Environment, Development and Sustainability*, •••, 1–30.

Grootaert, C., Narayan, D., Jones, V. N., & Woolcock, M. (2004). Measuring social Capital: An integrated questionnaire. The World Bank., Retrieved February 2019, from. DOI: 10.1596/0-8213-5661-5

Hazen, B. T., Overstreet, R. E., & Cegielski, C. G. (2012). Supply chain innovation diffusion: Going beyond adoption. *International Journal of Logistics Management*, 23(1), 119–134. DOI: 10.1108/09574091211226957

Hillman, A. J., & Dalziel, T. (2003). Boards of directors and firm performance: Integrating agency and resource dependence perspectives. *Academy of Management Review*, 28(3), 383–396. DOI: 10.5465/amr.2003.10196729

Hitt, M.A., L. Bierman, K. Uhlenbruck, and K. Shimizu. "The Importance of Resources in the Internationalization of Professional Service Firms: The Good, the Bad, and the Ugly," Academy of Management Journal, (49:1), 2006, pp. 1137- 1157

Honja, T. (2014). Review of mango value chain in Ethiopia. *Journal of Biology, Agriculture and Healthcare*, 4(25), 230–239.

Horton, D., Devaux, A., Bernet, T., Mayanja, S., Ordinola, M., & Thiele, G. (2023). Inclusive innovation in agricultural value chains: Lessons from use of a systems approach in diverse settings. *Innovation and Development*, 13(3), 517–539. DOI: 10.1080/2157930X.2022.2070587

India Environment Portal. (2016). Solid waste management rules. https://www.indiaenvironmentportal.org.in/content/427824/solid-waste-management-rules-2016/

Jeanneaux, P. (2023). Toward a more profitable value chain for New York state onions. *Journal of Agriculture, Food Systems, and Community Development*, 13(1), 1–7. DOI: 10.5304/jafscd.2023.131.022

Kang, Z., Zhao, Y., Chen, L., Guo, Y., Mu, Q., & Wang, S. (2022). Advances in machine learning and hyperspectral imaging in the food supply chain. *Food Engineering Reviews*, 14(4), 596–616. DOI: 10.1007/s12393-022-09322-2

Kayikci, Y., Kazancoglu, Y., Gozacan-Chase, N., & Lafci, C. (2022). Analyzing the drivers of smart sustainable circular supply chain for sustainable development goals through stakeholder theory. *Business Strategy and the Environment*, 31(7), 3335–3353. DOI: 10.1002/bse.3087

Kenis, P., & Raab, J. (2020). Back to the future: Using organization design theory for effective organizational networks. *Perspectives on Public Management and Governance*, 3(2), 109–123. DOI: 10.1093/ppmgov/gvaa005

Kherallah, M., & Kirsten, J. F. (2002). The New Institutional Economics: Applications for Agricultural Policy Research in Developing Countries: "New institutional economists are the blue-collar guys with a hearty appetite for reality." Oliver Williamson, 2000a. *Agrekon*, 41(2), 110–133.

Klugman, J. (1999). Social and economic policies to prevent complex humanitarian emergencies: lessons from experience.

Krishna, A. (2002). Enhancing Political Participation in Democracies: What is the Role of Social Capital? *Comparative Political Studies*, 35(4), 437–460. DOI: 10.1177/001041400203500400З

Kulwijila, M. (2021). Socio-economic determinants of postharvest losses in the grape value chain in Dodoma Municipality and Chamwino District, Tanzania. *African Journal of Economic Review*, 9(2), 288–305.

Kulwijila, M., Makindara, J., & Laswai, H. (2018). Grape value chain mapping in Dodoma region, Tanzania. *Journal of Economics and Sustainable Development*, 9(2), 171–182.

Kumar, M., Srai, J., Pattinson, L., & Gregory, M. (2013). Mapping of the UK food supply chains: Capturing trends and structural changes. *Journal of Advances in Management Research*, 10(2), 299–326. DOI: 10.1108/JAMR-05-2013-0034

Li, J., Liu, Y., & Chen, Z. (2022, July). Role of Industry Associations in the GI Value Chain: The Case of Xinhui Orange Peel. In *Worldwide Perspectives on Geographical Indications*.

Louwaars, N. P. (2023). Potato in the Netherlands–a remarkable story. In *Impact of hybrid potato* (pp. 45–58). Academic. DOI: 10.3920/978-90-8686-946-6_4

McMichael, C., & Manderson, L. (2004). Somali women and well-being: Social networks and social Capital among immigrant women in Australia. *Human Organization*, 63(1), 88–99. DOI: 10.17730/humo.63.1.nwlpjdj4d4l97561

Montalbano, P., & Nenci, S. (2020). The effects of global value chain (GVC) participation on the economic growth of the agricultural and food sectors.

Morgan, R. M., & Hunt, S. D. (1994). The commitment-trust theory of relationship marketing. *Journal of Marketing*, 58(3), 20–38. DOI: 10.1177/002224299405800302

Mvumi, B., Matsikira, L. T., & Mutambara, J. (2016). The banana postharvest value chain analysis in Zimbabwe. *British Food Journal*, 118(2), 272–285. DOI: 10.1108/BFJ-08-2014-0293

Nakagawa, Y., & Shaw, R. (2004). Social Capital: A missing link to disaster recovery. *International Journal of Mass Emergencies and Disasters*, 22(1), 5–34. DOI: 10.1177/028072700402200101

Nattassha, R., Handayati, Y., Simatupang, T. M., & Siallagan, M. (2020). Understanding circular economy implementation in the agri-food supply chain: The case of an Indonesian organic fertiliser producer. *Agriculture & Food Security*, 9(1), 1–16. DOI: 10.1186/s40066-020-00264-8

Pacheco, C. and Paul, B. (2023), Applying complexity theory perspective to knowledge management in the innovation context, VINE Journal of Information and Knowledge Management Systems, Vol. ahead-of-print No. ahead-of-print.

Patil, K., Ojha, D., Struckell, E. M., & Patel, P. C. (2023). Behavioral drivers of blockchain assimilation in supply chains–A social network theory perspective. *Technological Forecasting and Social Change*, 192, 1–16.

Piekkari, R., & Westney, D. E. (2017), Language as a Meeting Ground for Research on the MNC and Organization Theory, Multinational Corporations and Organization Theory: Post Millennium Perspectives (Research in the Sociology of Organizations, Vol. 49), Emerald Publishing Limited, Bingley, pp. 193-232.

Pienaar, L., & Barends-Jones, V. (2021). The economic contribution of South Africa's pomegranate industry. *Agriprobe*, 18(4), 57–64.

Porter, M. E. (1979). *Decision Support Tools: Porter's Value Chain*. Cambridge University, Institute for Manufacturing.

Priyadarshini, P., & Abhilash, P. C. (2020). Circular economy practices within energy and waste management sectors of India: A meta-analysis. *Bioresource Technology*, 304, 123018. DOI: 10.1016/j.biortech.2020.123018 PMID: 32087547

Qader, G., Shahid, Z. A., Junaid, M., Shaikh, I. M., & Qureshi, M. A. (2023). The role of diffusion of innovation theory towards the adoption of halal meat supply chain. *Journal of Islamic Marketing*, 14(5), 1211–1228. DOI: 10.1108/JIMA-01-2021-0032

Reinecke, J., & Donaghey, J. (2017). Collective action and social partnership in global supply chains: Lessons from the Bangladesh accord. *Annual Review of Social Partnerships*, 12, 32–37.

Salam, M., Viantika, N. M., Amiruddin, A., Pinontoan, F. M., & Rahmatullah, R. A. (2021, March). Value chain analysis of Toraja coffee. [). IOP Publishing.]. *IOP Conference Series. Earth and Environmental Science*, 681(1), 012115. DOI: 10.1088/1755-1315/681/1/012115

Solomou, A. D., Germani, R., & Georgakopoulou, M. (2020). Ecological value, cultivation, utilization and commercialization of Aloe vera in Greece. *JAPS: Journal of Animal & Plant Sciences*, 30(4), 1013–1023.

UNCTAD. (n.d.), United Nations Conference on Trade and Development. Published Reports, https://unctad.org/, as retrieved on 10th May 2024.

Ungerman, O., & Dědková, J. (2020). Model of the circular economy and its application in business practice. *Environment, Development and Sustainability*, 22(4), 3407–3432. DOI: 10.1007/s10668-019-00351-2

USAID-KAVES. (2015). French Bean value chain Analysis: An online publication prepared by Fintrac Inc. for review by the United States Agency for International Development, https://erepository.uonbi.ac.ke/bitstream/handle/11295/164474/Kabugua%20L_Assessment%20of%20Household%20Expenditure%20Patterns%20and%20Their%20Effects%20on%20Food%20Security-%20the%20Case%20of%20French%20Beans%20Producing%20Households%20in%20Mbooni%20Sub-county,%20Kenya.pdf?sequence=1, as retrieved on 10th May 2024.

Vachon, S., & Klassen, R. D. (2006). Green project partnership in the supply chain: The case of the package printing industry. *Journal of Cleaner Production*, 14(6/7), 661–671. DOI: 10.1016/j.jclepro.2005.07.014

Wachtendorf, T., & Kendra, J. M. "Considering Convergence, Coordination and Social Capital in Disasters," 2004 Presentation to the *Canadian Risk and Hazards Networks 1st Annual Symposium*. Winnipeg, Manitoba, November 19.

Wang, G., Zhang, Z., Li, S., & Shin, C. (2023). Research on the influencing factors of sustainable supply chain development of agri-food products based on cross-border live-streaming e-commerce in China. *Foods*, 12(17), 3323.

Wilson, H., & Daniel, E. (2007). The multi-channel challenge: A dynamic capability approach. *Industrial Marketing Management*, 36(1), 10–20. DOI: 10.1016/j.indmarman.2006.06.015

Woolcock, M. (1998). Social Capital and Economic Development: Toward a Theoretical Synthesis and Policy Framework. *Theory and Society*, 27(2), 151–208. DOI: 10.1023/A:1006884930135

World Bank. 2003. Social Capital for Development. Available at: http://www.worldbank.org/ poverty/scapital/. Accessed in February 2019.

Xue, L., Cao, Z., Scherhaufer, S., Östergren, K., Cheng, S., & Liu, G. (2021). Mapping the EU tomato supply chain from farm to fork for greenhouse gas emission mitigation strategies. *Journal of Industrial Ecology*, 25(2), 377–389. DOI: 10.1111/jiec.13080

Zaridis, A., Vlachos, I., & Bourlakis, M. (2021). SMEs strategy and scale constraints impact on agri-food supply chain collaboration and firm performance. *Production Planning and Control*, 32(14), 1165–1178.

Zhao, G., Vazquez-Noguerol, M., Liu, S., & Prado-Prado, J. C. (2024). Agri-food supply chain resilience strategies for preparing, responding, recovering, and adapting in relation to unexpected crisis: A cross-country comparative analysis from the COVID-19 pandemic. *Journal of Business Logistics*, 45(1), e12361. DOI: 10.1111/jbl.12361

Chapter 2
Enhancing Agricultural Competitiveness:
Harnessing Value Chains and Agro-Processing for Sustainable Growth

Vaishnavi Singh
Banaras Hindu University, India

Dharmik Borisagar
https://orcid.org/0009-0003-4807-9643
Banaras Hindu University, India

O. P. Singh
https://orcid.org/0000-0001-6119-8020
Banaras Hindu University, India

Shubham Kumar Verma
https://orcid.org/0009-0007-4793-8919
Banaras Hindu University, India

ABSTRACT

This chapter talks about how agro-processing and value chains are significant for making agriculture a successful and competitive sector.We will learn how these linkages help advancing the quality of agricultural produce and make them more tempting to buyers around the world. We will discuss how transforming agriculture is essential, and value chains play a big role in this by taking everyone work collectively and for better. We will also talk about the importance of supporting local supply chain. These are the networks which are important for producers as well

DOI: 10.4018/979-8-3693-4330-2.ch002

as consumers because they ensure there's enough food and create new economic opportunities in agricultural sector. With the highest population in the world and the ability to export goods, India plays a major role in the global agricultural sector. By understanding and supporting agro-processing and value chains, we can make sure farming continues to flourish, promoting both local communities and the global market.

INTRODUCTION

Agri-food systems worldwide face many challenges, including insufficient food access, nutritional insecurity, disturbances in global trade routes due to regional conflicts, and agricultural productivity issues caused by climate change (Piñeiro *et al.* 2020). Processing of agricultural produce provides significant opportunities for the application of blockchain in the agri-food industry can advance transparency and efficiency, transforming raw agricultural produce into consumer goods. This conversion reduces waste, extends shelf life of the produce, and adds value to it, leading to higher income transfers to farmers from various consumer groups, as processed commodities can reach a broader market. (Chengappa 2004). Agro processing is a vital link between agriculture and industry, as it plays a crucial role in the economic development of agrarian societies by increasing the value of agricultural products and providing diverse and sustainable employment opportunities. This sector helps in ensuring the economic growth and development of a country (Geyer 2019), as well as the food safety to the citizens across the globe. But as the agriculture commodities supply in market is on rise but with the increased production the farmers are not getting the optimum price of their products and the consumers are not getting the better-quality products. So, with increase in the production there must be some innovations which should be applied for making the agriculture a profitable business and the solution for this problem could be the agro-processing industries and advancement in value chain techniques. Agro Processing could help the farmers to increase the profit margin of their products by converting the primary raw materials into a more desirable form so that the consumer can use the product on a more regular basis and it is widely accepted across the society and nation, benefitting the farmers as well, as they will get now better price of their products through value addition to it.(Black *et al* 2020) Apart from agro-processing, market linkages and service sector should also be improvised which could be enhanced by adopting the newer methodologies like improvised value chain techniques keeping in mind the maximum satisfaction of the end consumers. So, this chapter aims to explore the diverse range of traditional and modern agro-processing techniques that not only enhance the value of agricultural products but also prioritize consumer satisfaction through efficient value chains. Understanding these methods not only ensures profitable prices for producers but also enhances consumer satisfaction. Agro-processing has become a vital component of modern agriculture, as it helps make agricultural products more competitive in both domestic and international markets. In this chapter, we will explore the various government policies which are aimed at promoting and supporting the agro-processing industry in India. Additionally, we will provide suggestions to tap into the full potential for expanding this industry further.

What is Agro-Processing?

Agro-processing involves transforming raw agricultural produce into finished or semi-finished products, either for further production or direct consumption. It encompasses a range of activities such as cleaning, grading, packaging, and preserving the final product. It can be further classified into primary and secondary processes:

1. **Primary Processing:** It involves essential processes such as cleaning, grading, and packaging of the produce, making it suitable for further processing.
2. **Secondary Processing:** In this stage, the produce is transformed into higher-value products through more intricate processes.

Another way to categorize agro-processing techniques is based on the technological advancements within the industry over time. This classification includes traditional methods used by farmers as well as modern techniques developed with the advent of industrialization

Traditional Techniques

1. **Sun Drying:** This traditional technique is still widely used by farmers, where produce is dried under the sun to reduce moisture content and prevent spoilage.
2. **Smoking:** Commonly applied to meat and seafood, this method uses smoke to impart a distinct flavour to the product.
3. **Fermentation:** This process involves using bacterial cultures to transform milk into a variety of dairy products, enhancing shelf life and providing a unique taste.
4. **Milling:** An age-old technique used to grind cereals into flour.
5. **Salting:** Salt is used to inhibit the growth of microorganisms, particularly in products like meat and fish.

Modern Techniques

1. **Freezing:** Produce is stored at extremely low temperatures, where only a few microorganisms can survive, while preserving its freshness.
2. **Vacuum packing:** This method involves removing all air from the packaging, preventing oxidation and halting microbial growth
3. **Irradiation:** It involves exposing food products to ionizing radiation to eliminate pathogens.
4. **High-Pressure Processing:** It utilizes high pressure to preserve food products, maintaining freshness and nutritional value.

These techniques not only enhance the value of agricultural products by preserving their quality and extending shelf life (In case of Sun Drying, Smoking, Freezing, Vacuum packing, Irradiation, Packaging Innovations. High-Pressure Processing, but also prioritize consumer satisfaction (Fermentation, milling, grinding, salting are techniques that improve the taste and quality of products, resulting in increased consumer satisfaction by ensuring food safety, maintaining nutritional content, and improving overall product quality. (Shankar 2023). Efficient value chains are established through the integration of these traditional and modern agro-processing techniques, ensuring that agricultural products reach consumers in optimal condition.

Examples of Agro Processed Products

- Flour, bread, and pasta from grains
- Canned vegetables and fruits
- Dairy products like cheese, butter, and yogurt
- Processed meats and seafood
- Jams, jellies, and sauces
- Beverages such as juices, wine, and beer
- Snacks and ready-to-eat foods

Challenges in Agro Processing

According to the FICCI survey 'Bottlenecks in Indian Food Processing Industry, 2010', several critical challenges require immediate attention to stimulate sectoral growth. (Sakina M. 2019). These includes-

1. **Consumer Preference for Fresh Produce:** Indian consumers generally prefer fresh agricultural products over processed foods, which makes it challenging to promote and sell processed food items. (Gupta *et al.*, 2022)

2. **Market Dominance of Large Companies:** Large companies have a competitive advantage due to their ability to invest significantly in marketing, brand building, and distribution networks. This dominance marginalizes smaller manufacturers who struggle with marketing expenditures and brand establishment.
3. **Inadequate Infrastructure:** More than 30% of farm produce is lost due to inadequate cold chain infrastructure and logistics. Issues include long and fragmented supply chains, insufficient cold storage and warehousing facilities, lack of modern logistics, and last-mile connectivity problems. (Maheshwar & Chanakwa, 2006),
4. **Complex Regulatory Environment:** Multiple and sometimes conflicting laws govern food safety, quality standards, and packaging requirements. Compliance with these diverse regulations, managed by various ministries and authorities at both central and state levels, poses a significant challenge for food processors.
5. **Multiplicity of Authorities:** The involvement of multiple ministries and administrative authorities creates a complex regulatory landscape that burdens the industry with compliance challenges and administrative complexities.
6. **Shortage of Skilled Manpower:** There is a significant shortage of skilled and semi-skilled workers in the agro-processing sector. This deficit limits the industry's capacity to adopt modern technologies and equipment effectively. (Dixit & Ravichandran, 2022)
7. **Other Factors:** Constraints in raw material procurement, issues related to taxation, limited access to credit for small and medium enterprises, and insufficient applied research and development further hinder the growth and competitiveness of the food processing sector in India.

Addressing these challenges through coordinated efforts in infrastructure development, regulatory simplification, skill development initiatives, and supportive policies is essential to fostering growth and sustainability in India's agro-processing industry.

What is a Value Chain?

According to NABARD, a value chain consists of a series of actors and activities involved in managing inputs, production, integrating technology, post-harvest management, adding value through processing, marketing, financing, exporting, and risk mitigation. This chain aims to deliver commodities, goods, and services in the desired form from their primary production location to end consumers through a series of sequential value addition stages and related services. In practice, value chains are often intricate, involving numerous interconnected activities and industries that span different regions within a country or across various countries globally. For example, agro-food value chains encompass activities from farming in rural and urban areas to the supply of inputs like seeds, fertilizers, and pesticides, as well as agricultural machinery and irrigation equipment. These chains also include handling, storage, processing, packaging, and distribution activities. Additional factors such as power generation and logistics play crucial roles in shaping the overall environment and performance of these value chains. Generally, an agricultural value chain includes three main activities:

1. supply/production,
2. processing or manufacturing,
3. marketing/distribution and consumption.

ROLE OF DIGITAL TECHNOLOGIES IN AGRO-PROCESSING AND EFFICIENT VC

The integration of digital technologies within supply chains, particularly by large and leading firms, holds substantial potential to elevate the participation of small and medium-sized enterprises (SMEs) in food value chains. At the foundational level, enhanced traceability enabled by digital tools allows farmers to add more value to their produce by linking it directly to individual farms. (Charlebois *et al.*, 2024). This capability can help create niche markets and facilitate the sale of higher-value crops. Additionally, straightforward digital tools can be used to monitor and manage disease outbreaks through early detection and warning systems communicated to farming communities. Digital technology also provides farmers with increased visibility when selling to supermarket chains. Research by Chisoro-Dube, das Nair, and Landani (2018); Barnes and Higginson (2020); das Nair and Landani (2020); and Krishnan and das Nair (2021) underscores how digital technology can tackle challenges at the primary production level, boosting productivity, spreading essential information, enabling payments, and improving access to inputs, finance, and end markets within food value chains.

However, digital technology can also pose challenges. Its adoption may create barriers to entry for SMEs, potentially excluding firms that are not digitally enabled over time. This could exacerbate inequalities in value chains if it results in the exclusion of SMEs. Furthermore, digital technologies generate valuable data that can be monetized, conferring significant market power to firms that control this data. Such power could raise competition law concerns if used to exclude competitors. Encouraging digital adoption also necessitates access to resources and skill development, which entail additional costs.

Value Addition

It is a result of Agro-processing, as we add value to the already existing agricultural products therefore value addition involves the transformation of raw agricultural products into forms that have greater economic worth, longer shelf life, improved quality, or increased utility (Anderson & Hanselka, 2009). In general value addition refers to the process of enhancing the economic value of a product by changing its current place, time, form, or characteristics to create a more valuable product for the consumer.

Key Aspects of Value Addition

1. **Form Utility:** Changing the physical form of a product to increase its usability and appeal. For example, conversion of raw tomatoes into tomato sauce. It's like changing the form of the original produced into something new and more valuable than before.
2. **Place Utility:** Making a product available at a location where it is more valued. For example, transporting fresh apples from temperate areas to tropical areas. It's like providing products to a place where it is scarce and also helps in clearing market glut in the areas where problem of overproduction avails so transfers of products from point of origin (where it is found abundantly) to the point of consumption (where it was scarcely found)
3. **Time Utility:** Making a product available at a time when it is more valuable. For example, storage of grains after the harvesting and selling them during off-season when prices are higher. As agricultural is season based so this utility has wide Importance, we humans have to survive throughout the year and certain things like potato, wheat etc grows in certain season so storing them, using cold storage, processing them allows us to keep them safe and sound for consumption in all the seasons.
4. **Information Utility:** Providing information that adds more monetary value, such as labelling organic products or indicating nutritional content. It will help consumers as they can make their buying decisions based on the information served to them, and it also make consumer aware about the products that have been sold in the market.

Benefits of Value Addition

1. **Increased Profitability:** By enhancing the value of raw produce, farmers and producers can achieve higher prices and so their income will increase. Earlier they will sell their produce in raw form that will fetch them lesser prices, so value addition is a tool that will help to get remunerative prices.
2. **Market Diversification:** Value-added products can open up new consumer segments, both locally and internationally. Consumers can have more options to buy, this will also create competition between suppliers and due to that quality of product will be better.
3. **Employment Creation:** The processes involved in value addition can create employment opportunities for the producers as well as processors and middlemen. Several new factories, warehouses and other infrastructure facilities will be developed for the value chain activities and agro-processing so it will engage the labour force of the country providing them consistent employment over a period of time.
4. **Reduced Post-Harvest Losses:** A very large quantity of agricultural produces goes in waste due to lack of processing facilities. Processing and better storage can reduce wastage of agricultural produce to a larger extent that will help in utilization of the resources properly and effectively.

Value addition is an activity for increasing the profitability and sustainability of agricultural produce that can provide significant benefits to farmers, processors, and consumers.

Recent Innovations in Value Chain

Recent innovations in value chains, particularly in agriculture and food sectors, have been driven by developments in technology and changing consumer preferences. Some of them are discussed below:

1. Digital and Smart Technologies

- **Blockchain**: It guarantees transparency and traceability in supply chains. Consumers and stakeholders can track the journey of food products from farm to table, enhancing food safety and trust. Blockchain technology can enhance food traceability in both new and existing supply chains. Many current systems are isolated, with traceability limited to internal records for specific stakeholders within the supply chain. (Antonuccin *et al.*,2021, Papa, 2017, Saurabh 2021).

Figure 1. Traceability in agri supply chain

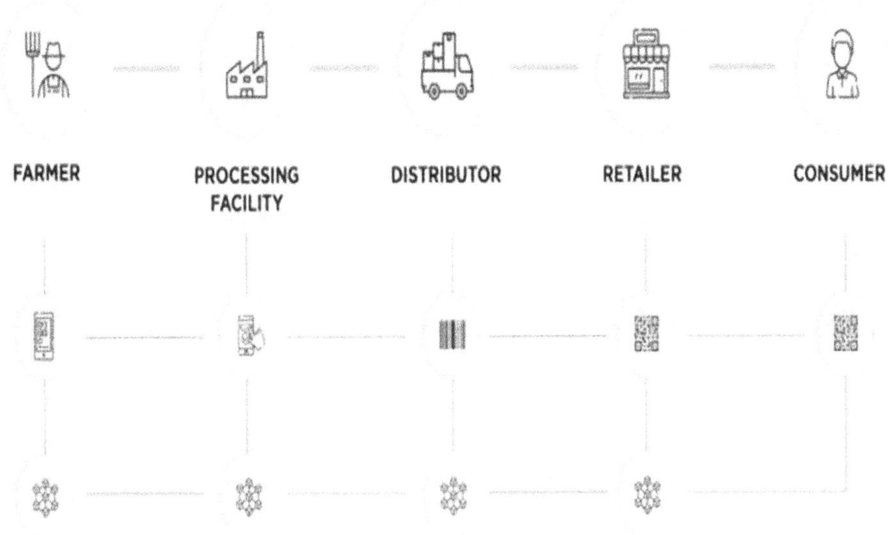

- **Internet of Things (IoT)**: Utilizes sensors and devices to monitor crops, livestock, and supply chain conditions in real-time. This helps optimize resource use, improve yield, and reduce waste. The Internet of Things contributes to enhancing competitiveness, efficiency, and safety in agricultural production. IoT platforms and mobile applications enable comprehensive monitoring and control throughout the agricultural value chain, facilitating informed decision-making, particularly in high-risk scenarios. Despite the abundance of agricultural data, much of it remains underutilized. Digitalization in agriculture enhances industry efficiency and productivity, marking IoT as a pivotal technology in advancing resource-efficient Agriculture 4.0 (A.I. Solodovnik *et al.*, 2021).
- **Artificial Intelligence (AI)**: AI adoption within agricultural value chains (AVC) has the potential to boost agricultural income, improve competitiveness, and lower costs (Ganesh Kumar 2023). However, research into AI applications in agricultural processing and the consumer sector remains underdeveloped compared to input procurement, production processes, and quality testing within AVC. Deep learning algorithms, particularly artificial neural networks, are widely utilized by various stakeholders in AVC. These algorithms are employed in tasks such as water resource management, yield prediction, price and demand forecasting, energy efficiency enhancement, optimization of fertilizer and pesticide usage, crop planning, personalized

advisory services, and predicting consumer behaviour (Ganeshkumar *et al.*,2023).

2. Sustainability Innovations

- **Vertical Farming and Hydroponics**: vertical farming involves growing crops in stacked layers, often in controlled indoor environments, allowing for year-round production in urban areas. It not only allows for more efficient use of limited land resources but also reduces costs by eliminating the need for transportation from distant markets (Singh et al., 2024). This cultivation method regulates the plants' microenvironment while also utilizing water and other scare inputs in the most sustainable way. (Riccaboni *et al.*, 2021).
- **Biodegradable Packaging**: Conventional plastic materials take an extremely long time to degrade in the environment. Biodegradable materials, developed in recent years, address this issue while appealing to consumers who have become more conscious of their carbon footprint. (Foschi, E. 2020).
- **Renewable Energy Integration**: Various types of renewable energy sources are being explored in Agro-processing to conserve and utilize the energy they produce, aiming to reduce dependence on fossil fuels and to curtail greenhouse gases emission. (Gorjian, S., *et al.*, 2022)

3. Other Innovations

Several other alternative innovations are currently gaining attention in the industry, including the following:

1. Circular Economy Models: Developing systems that reuse, reuse, and recycle materials in continuous cycles, minimizing waste and resource use (Joensuu *et al.*, 2020).
2. Carbon Capture and Utilization (CCU): Technologies that capture carbon dioxide emissions from industrial processes and reuse them into making of useful products like building materials or synthetic fuels (Ekemezie 2024 *et al.*,).
3. Algae-Based Products: Utilizing algae to produce biofuels and biodegradable plastics. Algae can absorb large amounts of CO_2 and grow in environments that don't compete with agriculture (Olabi 2023 *et al.*,).
4. Sustainable Urban Mobility Solutions: Innovations like electric vehicle (EV), smart grids for EVs, and improved public transportation networks that reduce the carbon footprint (Paalosmaa 2021 *et al.*,).
5. Waterless Dyeing Technologies: Processes that eliminate the need for water in textile dyeing, help in reducing water consumption (Samanta 2019 *et al.*,).

6. Aquaponics: A system in which aquaculture (raising fish) is combined with hydroponics. The fish excreta provide nutrients for plants, and the plants help filter and purify the water for the fish, creating a mutualism (Mintz 2019*et al.*,).
7. Regenerative Agriculture: Farming practices that restore and enhance soil health, biodiversity, and ecosystems while producing food. Practices like cover cropping, no-till farming, and agroforestry comes under this (Rehberger 2023 *et al.*,).
8. Edible Food Packaging: Packets made from food-grade materials that can be eaten along with the product, reducing waste and help in creating a new form of sustainable packaging (Katiyar 2021 *et al.*,).
9. Green Hydrogen: Production of hydrogen fuel through electrolysis that is powered by renewable energy. Green hydrogen can be used in various sectors, as a zero-emission energy source (Saha 2024 *et al.*,).
10. Enhanced Logistics and Distribution
    - Cold Chain Solutions: The cold chain is a distinguished method for reducing food losses and food waste, and has long been promoted by established industry focused organizations. Improved refrigeration technologies and cold chain logistics extend the shelf life of perishable agricultural produce, reduce food waste, and ensure the freshness of products during transportation.
    - Automated Warehousing: Warehousing is an important part of the supply chain and plays a key role in aiding efficiency of the production process (De Koster et al., 2007). The increased accessibility of the internet has expanded the scope of e-commerce, allowing businesses to sell directly to consumers and resulting in higher profit margins. (Hao *et al.*, 2020). The integration of technology in warehousing, through systems like Automated Storage and Retrieval Systems (ASRS) and Warehouse Management Systems (WMS), has significantly improved order completion and optimized inventory management.

4. Consumer-Centric Innovations.

- **Direct-to-Consumer (D2C) Models**: The increase of e-commerce platforms helps all the producers to sell directly to consumers, that will increase their margin of profit.

5. Circular Economy Practices

- **Waste-to-Value Technologies**: Converting agricultural and food waste into valuable products like biofuels, fertilizers, and animal feed. For example, anaerobic digestion processes turn organic waste into biogas. By employing innovative scientific and technological methods, we can transform by-products from the food industry into products that hold higher economic value than the primary products. The utilization of these by-products can be categorized into five main areas: as sources of food and feed ingredients, as carbon sources for cultivating beneficial microorganisms, as fertilizers through composting, for direct energy generation or biogas production, and as sources for creating high value-added products. (Ajila *et al.*, 2012).

- **Upcycling Food Waste**: Transforming food by-products and surplus into new, value-added products. Companies are creating snacks, beverages, and other goods from what would otherwise be waste.

6. Collaborative Platforms and Networks

- **Agri-Food Hubs**: Platforms that connect small farmers with markets, resources, and services. These hubs facilitate access to technology, training, and finance, improving productivity and income. (Sgroi *et al.*, 2022).
- **Crowdsourcing and Crowdfunding**: usage of collective intelligence and funding to support agricultural innovations and startups, help small-scale farmers to adopt new technologies. (Troise *et al.*, 2021). These innovations are transforming value chains by making them more efficient, sustainable, transparent, and responsive to consumer needs.

MARKET LINKAGE AND VARIOUS DISTRIBUTION CHANNELS

The market linkage and various distribution channels is crucial for ensuring that products reach consumers from producers in a proper manner. This linkage involves several intermediaries that facilitate the movement, handling, and sale of goods. Understanding these linkages helps in optimizing supply chain efficiency and meeting consumer demand effectively. (Shiferaw *et al.*, 2007).

- **Market Linkage**: means connection between producers and consumers in the market. It involves mechanisms that safeguard products reach to the end users. (Shepherd, 2007)
- **Distribution Channels**: Channels through which goods move from producers to consumers, involving multiple intermediaries such as wholesalers, retailers, distributors etc. (Satheeshkumar *et al.*, 2022)

Types of Distribution Channels

1. **Direct Channels**: Producers will sell its products directly to consumers without intermediaries.
    - o **Examples**: owned stores, local Mandi etc.
    - o **Benefits**: Greater control over the produce, higher profit margins.
2. **Indirect Channels**: Involve one or more intermediaries between the producer and consumer.
    - o **Examples**:
        - ▪ **One-level Channel**: Producer → Retailer → Consumer.
        - ▪ **Two-level Channel**: Producer → Wholesaler → Retailer → Consumer.
        - ▪ **Three-level Channel**: Producer → Agent/Broker → Wholesaler → Retailer → Consumer.
    - o **Benefits**: Wider market reach, reduced burden on producers for logistics and distribution, access to established networks.

Functions and Linkages

1. **Producer to Wholesaler**:
    - o Wholesalers purchase large quantities from producers and sell them in smaller quantities to retailers or other intermediaries.
    - o **Benefits**: Producers can focus on production without worrying about distribution logistics.
2. **Wholesaler to Retailer**:
    - o Retailers buy from wholesalers and sell directly to consumers.
    - o **Benefits**: Retailers offer a wide range of products to consumers and provide locations for convenient shopping.
3. **Retailer to Consumer**:
    - o Retailers are the final link in the distribution chain, offering products directly to consumers.
    - o **Benefits**: Direct interaction with consumers, providing opportunities for personalized services and feedback.
4. **Producer to Agent/Broker**:
    - o Agents/brokers facilitate the sale of products without taking ownership, earning a commission for their services.
    - o **Benefits**: Useful for producers to access distant or specialized markets without extensive investment.

Market Linkage Strategies

1. **Aggregation Centres**:
   - o Collect products from various small producers to create larger, more marketable quantities.
   - o **Benefits**: Helps small-scale farmers access larger markets and reduce transportation costs.
2. **Cooperatives**:
   - o Groups of producers who collaborate to market and sell their products collectively.
   - o **Benefits**: Improved bargaining power, shared resources, and better access to markets.
3. **Digital Platforms**:
   - o Online marketplaces that connect producers directly with consumers or retailers.
   - o **Benefits**: Expanded market reach, reduced transaction costs, real-time market information.
4. **Contract Farming**:
   - o Agreements between producers and buyers (often large retailers or processors) where terms of production and sale are pre-agreed.
   - o **Benefits**: Guaranteed market for producers, quality control for buyers.
5. **Third-Party Logistics (3PL)**:
   - o Outsourcing logistics and supply chain management to specialized service providers.
   - o **Benefits**: Efficient logistics management, cost reduction, and focus on core business activities.

Trends in Market Linkage and Distribution Channels

1. **Omnichannel Retailing**:
   - o Integrating online and offline channels to provide a seamless shopping experience. (Sharma et al 2023).
   - o **Example**: Buy online, pick up in-store (BOPIS).
2. **E-commerce and Direct-to-Consumer (D2C)**:
   - o Brands selling directly to consumers via their own websites or online marketplaces.
   - o **Example**: Subscription boxes for regular delivery of products.
3. **Sustainability and Ethical Practices**:
   - o Focus on eco-friendly distribution practices and ethical sourcing.

- **Example**: Using electric vehicles for delivery, sourcing from fair-trade certified producers.

4. **Technology Integration**:
   - Using technologies like blockchain for traceability, IoT for real-time tracking, and AI for demand forecasting.
   - **Example**: Blockchain to ensure transparency in the supply chain.

The linkage of markets and various distribution channels is integral to the efficient flow of goods from producers to consumers. It involves a network of intermediaries and strategies designed to optimize the supply chain, meet consumer demand, and enhance market reach. Understanding these linkages can lead to improved profitability, customer satisfaction, and sustainable business practices.

Compliances regarding the value chain policies

Compliance with value chain policies involves obeying various regulations, standards, and best practices across different stages of the supply chain. (GS1. 2013) These policies are essential for ensuring quality, safety, ethical practices, in the production and distribution of goods.

1. Quality and Safety Standards
   - **Food Safety Modernization Act (FSMA)**: Requires food producers to follow stringent safety protocols to prevent contamination and ensure the safety of food products.
   - **ISO Standards:** ISO 9001 (Quality Management) and ISO 22000 (Food Safety Management) standards established by the International Organization for Standardization are mandatory for manufacturers of food products.
   - **Good Manufacturing Practices (GMP)**: Regulations that ensure products are consistently produced and controlled according to quality standards.

Components of food safety regulations

1. Hazard Analysis and Critical Control Points (HACCP)
2. Food Facility Registration
3. Good Manufacturing Practices (GMPs)
4. Labelling Requirements
5. Allergen Management
6. Food Additives and Ingredients

7. Microbiological Safety
8. Pesticide and Chemical Residue Limits
9. Import and Export Regulations
10. Inspections and Audits
11. Traceability and Recall
12. Temperature Control

Environmental Compliance

- **Environmental Protection Agency (EPA) Regulations**. The government has implemented various laws for the protection of environment and natural resources which companies must comply with to obtain the necessary certifications.
- **Sustainability Reporting Standards**: Global Reporting Initiative (GRI) and Sustainability Accounting Standards Board (SASB) have established well-structured frameworks for companies to report on their environmental impact.

Labor and Ethical Standards

- **Fair Labor Standards Act (FLSA)**: U.S. law governing minimum wage, overtime pay, and child labour.
- **International Labour Organization (ILO) Conventions**: Standards set by the ILO to ensure fair working conditions globally.
- **Ethical Trading Initiative (ETI) Base Code**: A set of labour standards based on ILO conventions to promote fair labour practices.

Supply Chain Transparency and Traceability

- **Blockchain Technology**: A new technology with exclusive combination of features such as decentralized structure which help in traceability and transparency in supply chains, helping to fulfil the set standards which related to product origin and handling. (Aung *et al.*, 2014, Dutta *et al.*, 2020)

Trade and Customs Compliance

- **Incoterms**: International commercial terms published by the International Chamber of Commerce (ICC) that define the responsibilities of buyers and sellers in international transactions.

- **Tariff and Trade Regulations**: Compliance with import/export regulations, including tariffs, trade agreements, and documentation requirements.

Health and Safety Regulations

- **Occupational Safety and Health Administration:** It outlines safety and health guidelines for employers to implement in their workplaces.
- **REACH (Registration, Evaluation, Authorisation, and Restriction of Chemicals)**: The European Union has implemented regulations to manage hazardous chemicals and safeguard human health.

Product-Specific Regulations

- **Pharmaceutical Regulations**: Drug manufacturers are required to adhere to the standards set by the Central Drug Standard Control Organisation (CDSCO).

Strategies for Ensuring Compliance

1. **Internal Audits and Inspections**: Internal Mechanism put in place for ensuring adherence to all the regulation.
2. **Third-Party Certifications**: To secure certification from accredited bodies.
3. **Training and Education**: To offer training to employees for raising awareness on safety-related compliances.
4. **Technology Integration**: Use of technology to monitor the compliance process.

Compliances Regarding the Value Chain Policies in India and Other Developing Nations

1. Quality and Safety Standards

- **Food Safety and Standards Authority of India (FSSAI)**: It pertains to food safety standards that ensure the quality and safety of food products in the country.
- **Bureau of Indian Standards (BIS)**: It establishes product reliability through labelling, which is granted after rigorous testing.
- **Good Manufacturing Practices (GMP)**: In industries like food processing and pharmaceuticals, it ensures that consistency is maintained in the production of each unit of output.

2. Environmental Compliance

- **Extended Producer Responsibility (EPR)**: Entails producers of electronic waste and plastic waste to manage the disposal and recycling of their products.
- **National Green Tribunal (NGT)**: Focuses in handling environmental disputes and enforcing legal rights related to the environment.

3. Labor and Ethical Standards

- **Minimum Wages Act, 1948**: Sets the minimum wage rates for different types of employment to ensure fair wages.
- **Factories Act, 1948**: Regulates labour conditions in factories, including working hours, safety standards etc
- **Child Labour (Prohibition and Regulation) Act, 1986**: Prohibits the employment of children below a certain age in specific hazardous occupations.

4. Supply Chain Transparency and Traceability

- **Goods and Services Tax (GST)**: Requires companies to comply with GST regulations, ensuring transparency in financial transactions throughout the supply chain.
- **Aadhaar and PAN Integration**: Ensures traceability and verification of entities involved in the supply chain.

5. Trade and Customs Compliance

- **Customs Act, 1962**: It addresses the processes, documentation, and formulation of customs duties for exported and imported goods.
- **Directorate General of Foreign Trade (DGFT)**: It is the governing agency responsible for external trade in all products. It regulates licensing of firms in trade sector and also provide measures for trade facilitation.

6. Data Protection and Privacy

- **Information Technology Act, 2000**: Governs data protection and privacy, including amendments that address cybersecurity and digital data protection.
- **Personal Data Protection Bill (pending)**: Once enacted, it will provide comprehensive data protection regulations, similar to GDPR in the European Union.

7. Product-Specific Regulations

- **Pharmaceuticals**: Compliance with the Central Drugs Standard Control Organization (CDSCO) for drug approval and manufacturing standards.
- **Automotive**: Adhering to safety and environmental standards set by the Automotive Research Association of India (ARAI).

8. Sustainability Initiatives

- **National Action Plan on Climate Change (NAPCC)**: Outlines India's strategy for sustainable development and climate change mitigation.
- **Corporate Social Responsibility (CSR)**: Companies are required to spend a portion of their profits on social, environmental, and community development projects.

Inspiring Stories Through Case Study

The agro-processing industry has demonstrated its role as a catalyst for positive change in various regions around the world. Below are some examples presented as case studies:

1. The Transformation of Ethiopia's Coffee Industry

Ethiopia, despite being the birthplace of coffee, has never held a leading position in trade due to an unregulated value chain, which has led to a loss of market potential. (Taye, B. M. 2020).

Intervention:

- **Establishment of the Ethiopian Commodity Exchange (ECX)**: The establishment of the ECX has introduced essential reforms regarding pricing and quality, as well as creating a market for coffee producers.
- **Training and Certification**: Farmers and cooperatives received training on quality standards, traceability, and certification (such as Fair Trade and Organic certifications).

Impact:

- **Economic Growth**: The ECX facilitated better pricing and market access, leading to increased incomes for over 2.5 million coffee farmers.
- **Food Safety and Quality**: Improved quality control and certification processes ensured that Ethiopian coffee met international safety standards, boosting its global reputation.

- **Exports**: Coffee exports surged, making it one of Ethiopia's top foreign exchange earners, which significantly contributed to the country's economic growth.

2. India's Dairy Revolution (Operation Flood)

In the 1960s, India faced a severe milk shortage, which affected nutritional security and economic development. (Khemka, M. 2022).

Intervention:

- **Operation Flood**: This program was launched in 1970, which aims to create a nationwide milk grid by linking rural dairy farmers with urban markets through cooperatives.
- **Amul Cooperative Model**: Farmers were organized into cooperatives, which provided them with inputs, and definite market access.
- **Processing and Value Addition**: Investment in dairy processing plants to produce a variety of dairy products like milk powder, butter, and cheese.

Impact:

- **Economic Growth**: The dairy sector became a significant contributor to India's GDP, and millions of rural households saw their incomes rise.
- **Food Safety and Nutrition**: Improved processing standards and quality control ensured the safety of dairy products, enhancing nutritional security across the country.
- **Self-Sufficiency and Exports**: India became the world's largest milk producer, achieving self-sufficiency and exporting dairy products globally.

3. Kenya's Horticulture Success

Kenya had potential in horticulture, but smallholder farmers faced challenges in accessing markets and maintaining product quality. (Neven *et al.*, 2004).

Intervention:

- **Export Promotion Programs**: Initiatives like the USAID-Kenya Horticulture Competitiveness Project provided training and support to farmers.
- **Certification and Standards**: Farmers were helped to achieve certifications like Global GAP (Good Agricultural Practices) to meet international market standards.

- **Value Addition**: Investment in packaging, cold storage, and processing facilities for fruits and vegetables.

Impact:

- **Economic Growth**: Horticulture became one of Kenya's top foreign exchange earners, creating jobs and increasing incomes for thousands of smallholder farmers.
- **Food Safety**: Adherence to international standards ensured that Kenyan produce was safe and of high quality.
- **Global Market Access**: Kenyan fruits and vegetables gained access to lucrative markets in Europe, Asia, and the Middle East.

4. Cocoa Processing in Ghana

Ghana is one of the world's largest cocoa producers, but traditionally, most cocoa was exported in raw form, yielding lower profits for farmers. (Monastyrnaya, *et al.*, 2016)

Intervention:

- **Farmer Cooperatives**: Formation of cooperatives to pool resources and invest in small-scale cocoa processing facilities.
- **Product Diversification**: Production of cocoa butter, cocoa powder, and chocolate bars.
- **Certification Programs**: Participation in Fair Trade and Organic certification programs to access premium markets.

Impact:

- **Higher Incomes**: Farmers earned more by selling processed cocoa products and accessing premium markets.
- **Community Development**: Increased revenues were invested in local infrastructure, such as schools and healthcare facilities.
- **Market Expansion**: Ghanaian cocoa products gained international recognition, boosting export opportunities.

5. Cashew Nut Processing in Vietnam

Vietnam is a major producer of raw cashew nuts, but processing capabilities were initially limited, reducing farmers' potential earnings. (Costa *et al.*, 2019).

Intervention:

- **Investment in Processing**: Government and private sector investments in cashew processing plants.

- **Skill Development**: Training programs for farmers and workers on processing techniques and quality control.
- **Export Promotion**: Efforts to meet international standards and promote Vietnamese cashew products abroad.

Impact:

- **Value Addition**: Processed cashew nuts fetched higher prices than raw nuts, increasing farmers' incomes.
- **Economic Growth**: The cashew processing industry created jobs and contributed significantly to Vietnam's export revenues.
- **Global Market Access**: Vietnamese cashew nuts became a preferred choice in international markets due to improved quality and competitive pricing.

6. Rice Processing in Thailand

Thailand is a one of major rice producer, but small farmers struggled with fluctuating prices and limited market access for raw rice. (Ritthaisong *et al.*, 2014).

Intervention:

- **Cooperative Milling**: Establishment of cooperative rice mills to produce high-quality milled rice, rice flour, and rice-based snacks.
- **Branding and Certification**: Development of brands and obtaining certifications like organic and non-GMO to attract premium markets.
- **Value Chain Integration**: Enhancing the entire value chain from production to marketing, including better storage and packaging solutions.

Impact:

- **Increased Profit Margins**: Farmers benefited from higher prices for processed and branded rice products.
- **Market Stability**: Cooperative milling provided more stable income streams and reduced dependency on raw rice prices.
- **Product Diversification**: Introduction of value-added rice products diversified farmers' income sources.

These case studies highlight the changing impact of agro-processing and value addition on farmers' incomes and livelihoods. By converting primary agricultural produce which are raw in nature into more desirable, higher-value products, farmers can get much higher prices than before along with a stable market with new consumer demands. This not only improves their incomes but also contributes to broader economic development. These examples confirm the potential of value chain advances which can bring prosperity in the agricultural sector.

Specific Indian Case Studies

1. Amul Cooperative Model

Amul, was founded in 1946 in Gujarat, is now one of India's most successful dairy cooperatives. (Ghosh 2015).

Intervention:

- **Farmer Cooperatives**: Formation of village-level cooperatives for milk collection.
- **Value Addition**: Processing milk into a variety of dairy products, including butter, cheese, ice cream, yoghurt etc
- **Branding and Marketing**: Development of the Amul brand, known for its quality and affordability.

Impact:

- **Economic Empowerment**: Over 3.6 million milk producers benefit from stable and higher incomes.
- **Innovation and Expansion**: Amul introduced several innovative dairy products, expanding its market reach.
- **National and Global Presence**: Amul products are widely available across India and exported to several countries, showcasing the potential of cooperative models.

3. ITC e-Choupal

An initiative launched by ITC Limited in 2000, e-Choupal aimed to enhance the efficiency of the agricultural value chain by the use of technology. (Anupindi *et al.*, 2007)

Intervention:

- **Digital Platforms**: Setting up internet kiosks in villages to provide farmers with real-time information on weather, market prices, etc.
- **Direct Procurement**: Facilitating direct procurement of agricultural produce from farmers, avoiding traditional intermediaries.
- **Training and Support**: Providing farmers with training on sustainable agricultural practices and quality standards.

Impact:

- **Increased Farmer Incomes**: Farmers received better prices for their produce by eliminating middlemen and accessing real-time market information. Farmers secured better prices for their produce by cutting out middlemen and gaining the benefit of real-time market information.

- **Empowerment and Knowledge**: Access to information and training empowered farmers to make informed decisions and improve productivity.
- **Supply Chain Efficiency**: ITC benefited from a more efficient supply chain, ensuring timely procurement of high-quality produce.

5. Punjab's Basmati Rice Value Chain Enhancement

Punjab is a major producer of Basmati rice, but farmers faced issues with fluctuating prices and market access. (Pandey *et al.*, 2018).

Intervention:

- **Cooperative Models**: Formation of cooperatives to pool resources and invest in modern rice milling and processing facilities.
- **Quality Standards**: Implementation of strict quality control measures to ensure high standards of Basmati rice.
- **Export Promotion**: Efforts to access international markets, particularly in the Middle East and Europe.

Impact:

- **Higher Incomes**: Farmers benefited from higher prices for processed and well-packaged Basmati rice.
- **Market Stability**: Cooperative models provided more stable income streams and reduced dependency on raw rice prices.
- **Global Market Access**: Punjab's Basmati rice gained international recognition, boosting export revenues.

SWOT ANALYSIS OF AGRO-PROCESSING AND VALUE CHAIN PROCESSES OF INDIA (Manjunath *et al.*, 2021)

Strengths: The primary strength of India's food processing sector is the year-round availability of raw materials. With a vast workforce and extensive geographical area dedicated to agriculture and related activities, there is no shortage of raw materials. This steady supply ensures continuous production of final products for both domestic and international markets, making the agro-processing sector more promising than other agricultural and industrial activities. Development in this sector directly impacts income levels. Rapid transportation is essential for food processing due to the short shelf-life of most food products, especially fruits and vegetables. To reduce post-harvest losses, a widespread network of manufacturing hubs is necessary. The government has supported this by improving transport facilities that connect fields to processing plants. Additionally, even minor changes in consumer preferences can lead to significant increases in demand due to the large population.

Weaknesses: A major obstacle in the sector's development is the high initial working capital requirement. The capital needed from the initial processing stage to marketing is substantial. Available technology in the country is often not suitable for small and medium entrepreneurs without government assistance. A significant portion of processed food comes from the unorganized sector, which is characterized by low skill levels, insufficient funds for modernization, and a lack of technical knowledge in food processing. Many processing plants use outdated technologies, resulting in low efficiency and high per-unit operational costs. The PMFME scheme, introduced by the Indian government in September 2020 alongside the ODOP concept, aims to address these weaknesses. Despite considerable budget expenditure on research and development, the gap between laboratory research and industry application remains. Information dissemination is sometimes slow, causing technologies to become outdated by the time they reach the industry. Additionally, the high cost of available technologies makes them inaccessible to small and medium entrepreneurs. This gap between research and practical application limits sector growth to about 2%.

Opportunities: Despite having advantages in raw material availability, agricultural production, and increasing demand for processed food, India's food processing level is still low compared to other countries. The lack of processing and storage facilities for fruits and vegetables leads to significant wastage, estimated at 20-35%, valued at approximately INR 330 billion annually. (Hussain, 2016.) Improper storage methods contribute to this wastage. However, the agro-processing sector remains a sunrise industry due to future opportunities. The average Indian spends almost half of their income on food, and this is expected to increase by 20%. An anticipated 8% increase in per capita disposable income by the end of the decade presents a significant opportunity. Demographic and economic changes, along with lifestyle shifts, will lead to changes in consumer consumption patterns. Factors include a growing number of nuclear working families with limited cooking time, increased incidence of lifestyle diseases driving demand for healthy products, greater international exposure raising awareness of new concepts, and increased consumer awareness fuelled by media penetration, which is not limited to urban areas.

Threats: The major threats faced by food processing industries include infrastructural gaps in the supply chain (such as inadequate primary processing, storage, and distribution facilities), lack of product development and innovation, insufficient focus on quality and safety standards, seasonality of operations, low-capacity utilization, and poor linkage between production and processing. As a WTO signatory, India's market is open to both imports and exports. Increased equity for multinational corporations (MNCs) through FDI allows them to enter the domestic market, raising competition for local producers in terms of price, quality, and standards. The intensive promotion strategies adopted by MNCs can overshadow domestic products.

FUTURE OF AGROPROCESSING AND VALUE CHAIN SECTOR IN UPCOMING YEARS

Looking forward, the agro-processing and value chain sector is poised for significant advancements and transformations driven by several key factors. Technological inventions, such as precision agriculture using IoT, drones, and AI, promise to revolutionize how crops are monitored and managed (Alazzai *et al.*, 2024). These technologies allow producers to optimize input use, enhance productivity, and reduce waste, thereby improving overall efficiency in agricultural production. Blockchain technology is set to play a key role in ensuring transparency and traceability throughout the food supply chain. By providing immutable records of transactions and product origins, blockchain enhances food safety measures and builds consumer trust in the genuineness of agricultural products. Automation and robotics are gradually being adopted in processing plants, offering benefits such as increased output, reduced labour costs. These advancements not only reorganize operations but also contribute to higher-quality end products that meet rigorous market demands. (Ganeshkumar *et al.*, 2023) Sustainability is becoming a cornerstone of agro-processing practices, with a growing importance of eco-friendly farming methods like organic farming, water conservation, and sustainable packaging solutions. Consumers are ever more demanding healthier food options, as well as plant-based and alternative protein products, driving innovation across the value chain. (Haggblade, S. 2011). Government support through favourable policies, subsidies, and infrastructure development is important for the sector's growth.

Smallholder farmers are being combined into larger value chains through cooperatives and digital platforms, which improve market access and reduce transaction costs. Continuous capacity-building efforts are essential to prepare farmers with modern agricultural practices and quality standards, ensuring their participation in and benefit from value-added processes. To adapt to climate change and changing customer demands, India's food processing sector must use new technology and procedures. This will help the sector remain adaptable and profitable. (Rayhan *et al.*, 2024).

The future of India's food processing sector is hopeful. To thrive, it must employ modern technology, be sustainable, understand client preferences, and have favourable government regulations. (Sharma *et al.,* 2022). By implementing these things, the sector may become more competitive and efficient, therefore improving agriculture worldwide.

India's Potential in Supply Chain and Agro-Processing Industries

The food processing sector in India is extremely significant and quickly growing. It will help India's economy expand, generate employment, and increase exports to other countries. It has lifted people out of poverty by providing jobs, increasing the value of agricultural goods, and reducing food waste. Because it requires a large workforce, this business has aided rural areas with high poverty rates by providing employment, raising income, and lowering poverty. (Malik, *et al.*,2020). By enhancing the value of raw agricultural products through processing, the sector has delivered higher returns for farmers and rural communities, supporting the government's goal of doubling farmers' incomes. The Ministry of Food Processing Industries vigorously engaged in organizing promotional activities related to millets as part of (International year of Millets) IYoM 2023. It utilized scheme like (Production linked incentive) PLI to enhance millet processing infrastructure nationwide and coordinated millet conferences and exhibitions in 30 districts throughout the year. Additionally, the Ministry hosted World Food India 2023 from November 3rd to 5th in New Delhi, following the example set by the Ministry's earlier organization of World Food India (WFI) in 2017 (MOFPI, 2023).

The food processing industry has significantly boosted employment opportunities for women, employing 2.27 lakh women workers in the registered sector during 2019-20. The Ministry of Food Processing Industries has supported numerous mega food parks and integrated cold chain projects, seven of which have female promoters. The increasing involvement of women entrepreneurs in this sector, coupled with efforts to enhance their participation and growth, reflects a positive trend (MOFPI, 2023). Empowering women in the food processing sector via entrepreneurship and job possibilities can lead to a more equitable and sustainable food system for everybody. Women's success in this area requires consistent assistance, such as specific policies and initiatives.

Overall, the future of India's food processing sector is optimistic, supported by technological innovations, government initiatives, and a strong foundation of agricultural richness. This sector has the potential to drive economic growth, promote sustainability, serving farmers, rural communities, and the nation as a whole.

CONCLUSION

Food processing can be sustainable only when it takes into account the practices which mainly focuses on reducing food waste, efficient use of resources (raw materials), carbon emissions reduction, effective waste utilization etc, within the entire food supply chain, these principles apply to final food products prepared from the processing method as well as all the intermediate products and raw materials throughout their production processes. There is also a growing emphasis on ethical practices and the creation of "guilt-free" products which can draw more consumers who were earlier sceptical of processed food products. While many of these technological improvements are motivated by consumer demand and to earn profit, the main aim of the agro processing sector is to provide safe and nutritious food for everyone.

REFERENCES

Adhikari, K. B., Regmi, P. P., Gautam, D. M., Thapa, R. B., & Joshi, G. R. (2017). Value chain analysis of orthodox tea: Evidence from Ilam district of Nepal. *Journal of Agriculture and Forestry University*, 1(1), 61–68.

Ajila, C., Brar, S. K., Verma, M., & Prasada Rao, U. (2012). Sustainable Solutions for Agro Processing Waste Management: An Overview. In Malik, A., & Grohmann, E. (Eds.), *Environmental Protection Strategies for Sustainable Development. Strategies for Sustainability*. Springer., DOI: 10.1007/978-94-007-1591-2_3

Alazzai, W. K., Abood, B. S. Z., Al-Jawahry, H. M., & Obaid, M. K. (2024). Precision Farming: The Power of AI and IoT Technologies. In *E3S Web of Conferences* (Vol. 491, p. 04006). EDP Sciences.

Anderson, D. P., & Hanselka, D. (2009). *Adding value to agricultural products*. Texas FARMER Collection.

Antonucci, F., Figorilli, S., Costa, C., Pallottino, F., Raso, L., & Menesatti, P. (2019). A review on blockchain applications in the agri-food sector. *Journal of the Science of Food and Agriculture*, 99(14), 6129–6138. DOI: 10.1002/jsfa.9912 PMID: 31273793

Anupindi, R., & Sivakumar, S. (2007). Supply chain reengineering in agri-business: A case study of ITC's e-Choupal. In *Building supply chain excellence in emerging economies* (pp. 265–307). Springer US. DOI: 10.1007/978-0-387-38429-0_12

Aung, M. M., & Chang, Y. S. (2014). Traceability in a food supply chain: Safety and quality perspectives. *Food Control*, 39, 172–184. DOI: 10.1016/j.foodcont.2013.11.007

Ba, M. N. (2016). Strategic agricultural commodity value chains in Africa for increased food: the regional approach for food security.

Barnes, J., & Higginson, L. (2020). *Practical implications for the South African food processing industry. Digital Industrial Policy Brief 4*. Industrial Development Think Tank.

Bammann, H. (2019). Participatory value chain analysis for improved farmer incomes, employment opportunities and food security.

Black, A., Edwards, L., Gorven, R., & Mapulanga, W. (2020). Agro-processing, value chains, and regional integration in Southern Africa. WIDER Working Paper 2020/36. Helsinki: UNU-WIDER. https://doi.org/DOI: 10.35188/UNU-WIDER/2020/793-4

Charlebois, S., Latif, N., Ilahi, I., Sarker, B., Music, J., & Vezeau, J. (2024). Digital traceability in Agri-Food Supply Chains: A Comparative analysis of OECD member countries. *Foods*, 13(7), 1075. DOI: 10.3390/foods13071075 PMID: 38611378

Chengappa, P. G. (2004). Emerging trends in agro-processing in India. *Indian Journal of Agricultural Economics*, 59(1).

Chisoro-Dube, S., Das Nair, R., Nkhonjera, M., & Tempia, N. 2018. Structural transformation in agriculture and agro-processing value chains. Centre for Competition, Regulation and Economic Development. Working Paper 8/2018

Costa, C., & Delgado, C. (2019). The Cashew Value Chain in Mozambique.

Das Nair, R., & Nontenja, N. (2023). Digital Technology Adoption in Agro-processing Value Chains.

Dixit, P., & Ravichandran, R. (2022). Bridging the skill gap in the Indian food processing sector: A review. *FoodSci Indian Journal of Research in Food Science and Nutrition*, 9(2), 85. DOI: 10.15613/fijrfn/2022/v9i2/217902

De Koster, R., Le-Duc, T., & Roodbergen, K. J. (2007). Design and control of warehouse order picking: A literature review. *European Journal of Operational Research*, 182(2), 481–501. DOI: 10.1016/j.ejor.2006.07.009

Dube, S., Das Nair, R., Nkhonjera, M., & Tempia, N. (2018). Structural transformation in agriculture and agro-processing value chains.

Dutta, P., Choi, T. M., Somani, S., & Butala, R. (2020). Blockchain technology in supply chain operations: Applications, challenges and research opportunities. *Transportation research part e: Logistics and transportation review, 142*, 102067.

Ekemezie, I. O., & Digitemie, W. N.Ifeanyi Onyedika EkemezieWags Numoipiri Digitemie. (2024). Carbon Capture and Utilization (CCU): A review of emerging applications and challenges. *Engineering Science & Technology Journal*, 5(3), 949–961. DOI: 10.51594/estj.v5i3.949

Foschi, E. (2020). System innovation and life cycle thinking in packaging value chain: the circularity of plastics.

Ganeshkumar, C., Jena, S. K., Sivakumar, A., & Nambirajan, T. (2023). Artificial intelligence in agricultural value chain: Review and future directions. *Journal of Agribusiness in Developing and Emerging Economies*, 13(3), 379–398. DOI: 10.1108/JADEE-07-2020-0140

Hao, J., Shi, H., Shi, V., & Yang, C. (2020). Adoption of automatic warehousing systems in logistics firms: A technology–organization–environment framework. *Sustainability (Basel)*, 12(12), 5185. DOI: 10.3390/su12125185

Ganeshkumar, C., Jena, S. K., Sivakumar, A., & Nambirajan, T. (2023). Artificial intelligence in agricultural value chain: Review and future directions. *Journal of Agribusiness in Developing and Emerging Economies*, 13(3), 379–398. DOI: 10.1108/JADEE-07-2020-0140

Geyer, H. S. (2019). Global Value Chain Participation and Trade Barriers in Sub-Saharan Africa. In Scholvin, S., Black, A., Diez, J. R., & Turok, I. (Eds.), *Value Chains in Sub-Saharan Africa: Challenges 21 of Integration Into the Global Economy* (pp. 13–27). Springer., DOI: 10.1007/978-3-030-06206-4_2

Ghosh, A. (2015). 16| EMBEDDEDNESS AND THE DYNAMICS OF GROWTH: THE CASE OF THE AMUL COOPERATIVE, INDIA. *Social and Solidarity Economy: Beyond the Fringe*, 284.

Gorjian, S., Fakhraei, O., Gorjian, A., Sharafkhani, A., & Aziznejad, A. (2022). Sustainable food and agriculture: Employment of renewable energy technologies. *Current Robotics Reports*, 3(3), 153–163. DOI: 10.1007/s43154-022-00080-x PMID: 35600253

GS1. 2013. White Paper: The need for global standards and solutions to combat counterfeiting. Online. Available: https://bit.ly/3ZdTojX

GS1. n. d. -a. How traceability standards work. Online. Available: https://www.gs1.org/howtraceability-standards-work. (Accessed: 11 March 2023)

Gupta, N., Bhattacharjee, M., & Saha, A. R. (2022). SUSTAINABILITY AND DIETARY CHANGE: AN ANALYSIS OF INDIAN FOOD CONSUMPTION PATTERNS. *INDIAN JOURNAL OF APPLIED ECONOMICS AND BUSINESS*, 5(1), 85–105. DOI: 10.47509/IJAEB.2023.v05i01.05

Haggblade, S. (2011). Modernizing African agribusiness: Reflections for the future. *Journal of Agribusiness in Developing and Emerging Economies*, 1(1), 10–30. DOI: 10.1108/20440831111131532

Humphrey, J., & Schmitz, H. (2001). Governance in global value chains. *IDS Bulletin*, 32(3), 19–29. DOI: 10.1111/j.1759-5436.2001.mp32003003.x

Hussain, A. (2016). *A STUDY ON INDIAN FOOD PROCESSING INDUSTRIES (FPI)*. EPRA Journals. https://eprajournals.com/IJES/article/8721/

Joensuu, T., Edelman, H., & Saari, A. (2020). Circular economy practices in the built environment. *Journal of Cleaner Production*, 276, 124215. DOI: 10.1016/j.jclepro.2020.124215

Katiyar, V., & Ghosh, T. (2021). *Nanotechnology in edible food packaging*. Springer., DOI: 10.1007/978-981-33-6169-0

Khemka, M. (2022). Operation Flood: The Successful Rural Development Vision. *Indian Business@ 75—Different Aspects*, 65.

Maheshwar, C., & Chanakwa, T. (2006). Postharvest Losses Due to Gaps in Cold Chain In India-A Solution. *Acta Horticulturae*, (712), 777–784. DOI: 10.17660/ActaHortic.2006.712.100

Manjunath, B. L., & Narayana, C. K. (2021). Boosting agro-processing industry in India. *Indian Farming*, 71(9).

Malik, R. M. M. K. V. K. (2020). *Indian food processing: an untapped growth opportunity*. Invest India. https://www.investindia.gov.in/siru/indian-food-processing-sector-untapped-growth-opportunity

Mintz, A. (2019). *Integrating Sustainable Practices: Compost Tea as a Nutrient Supplement for Aquaponic Plant Production* (Doctoral dissertation, Evergreen State College).

Monastyrnaya, E., Joerin, J., Dawoe, E., & Six, J. (2016). Assessing the resilience of the cocoa value chain in Ghana. *Case study report*, 1-137.

Neven, D., & Reardon, T. (2004). The rise of Kenyan supermarkets and the evolution of their horticulture product procurement systems. *Development Policy Review*, 22(6), 669–699. DOI: 10.1111/j.1467-7679.2004.00271.x

Olabi, A. G., Shehata, N., Sayed, E. T., Rodriguez, C., Anyanwu, R. C., Russell, C., & Abdelkareem, M. A. (2023). Role of microalgae in achieving sustainable development goals and circular economy. *The Science of the Total Environment*, 854, 158689. DOI: 10.1016/j.scitotenv.2022.158689 PMID: 36108848

Papa, S. F. (2017, June). Use of blockchain technology in agribusiness: Transparency and monitoring in agricultural trade. In *2017 International Conference on Management Science and Management Innovation (MSMI 2017)* (pp. 38-40). Atlantis Press. DOI: 10.2991/msmi-17.2017.9

Paalosmaa, T., & Shafie-khah, M. (2021). Feasibility of innovative smart mobility solutions: A case study for vaasa. *World Electric Vehicle Journal*, 12(4), 188. DOI: 10.3390/wevj12040188

Pandey, P. R., & Bhandari, H. (2018). *Rice Technological Innovation and Value Chain Development in South Asia: Current Status and Future Directions: SAARC Agriculture Centre* (p. 164). Bangladesh.

Piñeiro, V., Arias, J., Dürr, J., Elverdin, P., Ibáñez, A. M., Kinengyere, A., Opazo, C. M., Owoo, N., Page, J. R., Prager, S. D., & Torero, M. (2020). A scoping review on incentives for adoption of sustainable agricultural practices and their outcomes. *Nature Sustainability*, 3(10), 809–820. DOI: 10.1038/s41893-020-00617-y

Rehberger, E., West, P. C., Spillane, C., & McKeown, P. C. (2023). What climate and environmental benefits of regenerative agriculture practices? an evidence review. *Environmental Research Communications*, 5(5), 052001. DOI: 10.1088/2515-7620/acd6dc

Riccaboni, A., Neri, E., Trovarelli, F., & Pulselli, R. M. (2021). Sustainability-oriented research and innovation in 'farm to fork'value chains. *Current Opinion in Food Science*, 42, 102–112. DOI: 10.1016/j.cofs.2021.04.006

Ritthaisong, Y. M., Johri, L., & Speece, M. (2014). Sources of sustainable competitive advantage: The case of rice-milling firms in Thailand. *British Food Journal*, 116(2), 272–291. DOI: 10.1108/BFJ-01-2012-0003

Rayhan, M. J., Rahman, S. M., Mamun, A. A., Saif, A. N. M., Islam, K. A., Alom, M. M., & Hafiz, N. (2024). FinTech solutions for sustainable agricultural value chains: A perspective from smallholder farmers. *Business Strategy & Development*, 7(2), e358. DOI: 10.1002/bsd2.358

Saha, P., Akash, F. A., Shovon, S. M., Monir, M. U., Ahmed, M. T., Khan, M. F. H., Sarkar, S. M., Islam, M. K., Hasan, M. M., Vo, D.-V. N., Aziz, A. A., Hossain, M. J., & Akter, R. (2024). Grey, blue, and green hydrogen: A comprehensive review of production methods and prospects for zero-emission energy. *International Journal of Green Energy*, 21(6), 1383–1397. DOI: 10.1080/15435075.2023.2244583

Samanta, K. K., Pandit, P., Samanta, P., & Basak, S. (2019). Water consumption in textile processing and sustainable approaches for its conservation. In *Water in textiles and fashion* (pp. 41–59). Woodhead Publishing. DOI: 10.1016/B978-0-08-102633-5.00003-8

Satheeshkumar, D. R., Krishnan, D. S. G., & Rawath, S. S., & S J, A. K. (2022). A Study on Distribution Channel Strategy: Retailers' Perspective. *Rangasamy, S., Krishnan, SG, Rawath, S., Rakshith, M., & Ajith Kumar, S. (2022, September 27). A STUDY ON DISTRIBUTION CHANNEL STRATEGY: RETAILERS'PERSPECTIVE.International Journal for Innovative Engineering and Management Research*, 11(9), 240–252.

Saurabh, S., & Dey, K. (2021). Blockchain technology adoption, architecture, and sustainable agri-food supply chains. *Journal of Cleaner Production*, 284, 124731. DOI: 10.1016/j.jclepro.2020.124731

Sgroi, F., & Marino, G. (2022). Environmental and digital innovation in food: The role of digital food hubs in the creation of sustainable local agri-food systems. *The Science of the Total Environment*, 810, 152257. DOI: 10.1016/j.scitotenv.2021.152257 PMID: 34902402

Singh, S., Singh, P., Kumar, A., Baheliya, A. K., & Patel, K. K. (2024). Promoting environmental Sustainability through Vertical Farming: A review. *Journal of Advances in Biology & Biotechnology*, 27(5), 210–219. DOI: 10.9734/jabb/2024/v27i5780

Sharma, H. K., & Kumar, N. (2022). Agro processing: scope and importance. In *Agro-Processing and Food Engineering: Operational and Application Aspects* (pp. 1–22). Springer Singapore. DOI: 10.1007/978-981-16-7289-7_1

Sharma, N., & Dutta, N. (2023). Omnichannel retailing: exploring future research avenues in retail marketing and distribution management. *International Journal of Retail & Distribution*

Shankar Tumuluru, J. (2023). *Introductory Chapter: Food Processing, Preservation, and Packaging – A Brief Overview*. IntechOpen., DOI: 10.5772/intechopen.110229

Sakina, M. (2019). The food processing industry in India: challenges and prospects. *Gap Interdisciplinaries, 2*(3).

Shepherd, A. W., & Farina, E. M. (2013). Value chain finance: Taking stock of recent learning and looking ahead. *World Development*, 44, 181–193.

Shepherd, A. (2007). *Approaches to linking producers to markets*. https://vtechworks.lib.vt.edu/handle/10919/68275

Shafi, U., Mumtaz, R., García-Nieto, J., Hassan, S. A., Zaidi, S. A. R., & Iqbal, N. (2019). Precision agriculture techniques and practices: From considerations to applications. *Sensors (Basel)*, 19(17), 3796. DOI: 10.3390/s19173796 PMID: 31480709

Shiferaw, B., & Teklewold, H. (2007). Structure and functioning of chickpea markets in Ethiopia: Evidence based on analyses of value chains linking smallholders and markets. *IPMS Working Paper*.

Smith, G. (2009). *Interaction of Public and Private Standards in the Food Chain. OECD Food, Agriculture and Fisheries Papers, No. 15*. OECD Publishing., DOI: 10.1787/18156797

Solodovnik, A. I., Savkin, V. I., & Amelina, A. V. (2021, September). The role of the internet of things as direction for the development of agriculture 4.0 for rural areas. [). IOP Publishing.]. *IOP Conference Series. Earth and Environmental Science*, 839(3), 032040. DOI: 10.1088/1755-1315/839/3/032040

Tallontire, A., Opondo, M., Nelson, V., & Martin, A. (2011). Beyond the vertical? Using value chains and governance as a framework to analyse private standards initiatives in agri-food chains. *Agriculture and Human Values*, 28(3), 427–441. DOI: 10.1007/s10460-009-9237-2

Taye, B. M. (2020). Coffee: The Driver of Economic Development and Socio-Cultural Integration in Ethiopia.

Troise, C., Tani, M., Dinsmore, J.Jr, & Schiuma, G. (2021). Understanding the implications of equity crowdfunding on sustainability-oriented innovation and changes in agri-food systems: Insights into an open innovation approach. *Technological Forecasting and Social Change*, 171, 120959. www.iifiir.org. DOI: 10.1016/j.techfore.2021.120959

KEY TERMS AND DEFINITIONS

Agro-Processing: Agro processing refers to the activities and processes involved in transforming raw agricultural products into finished or semi-finished goods.

Blockchain Technology: Used for ensuring traceability and transparency in supply chains, helping to comply with regulations and standards related to product origin and handling.

Distribution Channels: Pathways through which goods move from producers to consumers, involving multiple intermediaries such as wholesalers, retailers, and distributors.

Market Linkage: Refers to the connection between producers and consumers in the market. It involves strategies and mechanisms that ensure products are accessible to the end users.

Traceability: Traceability refers to the completeness of the information about every step in a supply chain.

Value Chain: Value chain consists of a series of actors and activities involved in managing inputs, production, integrating technology, post-harvest management, adding value through processing, marketing, financing, exporting, and risk mitigation.

Value Addition: Value addition involves the transformation of raw agricultural products into forms that have greater economic worth, longer shelf life, improved quality, or increased utility.

Chapter 3
A Collectivization Approach to Agricultural Supply and Value Chains

Ashok Dalwai
Ministry of Agriculture and Farmers Welfare, Government of India, India

Vishita Khanna
International Agribusiness Management Institute, India

Soumyadeep Das
https://orcid.org/0000-0001-7441-2508
Uttar Banga Krishi Vishwavidyalaya, India

Ritambhara Singh
https://orcid.org/0000-0001-9857-1415
Dr. Rajendra Prasad Central Agricultural University, India

ABSTRACT

Since agriculture as an enterprise can sustain, only if it can yield net positive returns in favour of the farmer-producers, robustness of agricultural value chains assumes critical importance and this is feasible when efficient scales of operations support the system. In the given context of predominance of small & marginal farmers, the solution lies in promoting collectives of farmers and enabling aggregation of produce. It is in this context that the Government has consciously adopted the approach of farmers' mobilisation and aggregation of agri-produce, which will bring in efficiency of higher scale and facilitate the nation's commitment to establishing robust supply chains

DOI: 10.4018/979-8-3693-4330-2.ch003

1. INTRODUCTION

Following the October 2021 supply chain summit convened by the United States (US) President Joe Biden, in 2022, a group of 18 economies, including India, the US, and the European Union (EU) came together to unveil a four-point roadmap to build collective long-term supply chains. The roadmap is guided by the global principles of transparency, diversification, sustainability, and security. In June 2024, India took another step forward by signing the Indo-Pacific Economic Framework for Prosperity Agreement (IPEFPA) relating to Supply Chain (an initiative of 14 member-countries). Further, India is also part of the Supply Chain Resilience Initiative (SCRI) with Japan and Australia to promote collective, long-term resilient supply chains based on international partnerships. The Commerce Secretary to the Government of India has reiterated that "India, with its skilled manpower, natural resources, and policy support aims to become a major player in the global supply chain."

Now that India has envisioned to establish itself as a developed economy by 2047, and has been taking firm steps towards its realisation, one critical initiative that will be integral to the new paradigm is the strengthening of the diverse supply chains across a basket of commodities to create its own niche in the global trade. Considering the continued predominance of agriculture from the perspective of high dependence of the working population on it for livelihood at more than 45 percent, enhancing the efficiency of the primary sector assumes critical importance in the nation's developmental strategy. Monetization of the growing magnitude of agricultural produce optimally, with a view to rendering the agriculture sector profitable, and simultaneously enabling the farmer-producers to benefiting from optimal share in the value captured thereof, calls for building robust domestic and global supply chains. This is a pre-requisite to upgrading of the current low share (around 17 percent) of agriculture in the nation's Gross Value Added (GVA).

Since independence, and with the commencement of the first five-year plan in 1951, the country has focussed on agriculture. With the adoption of the Green Revolution strategy in 1965, the country has seen more wholesome interventions to promote an integrated food system, but is yet to blossom completely. Further, over the last 75 years, the country's agricultural production basket has seen diversification across all the sub-sectors, even while some produce-anomalies exist, either as surplus or deficit vis-à-vis the domestic demand. Additionally, there is a recognition that the robust agricultural growth has not yielded proportionate incomes for the farmer-producers on account of inadequacies relating to post-harvest management, causing sub-par monetization. Responding to the new paradox and keeping in mind the welfare of the farming community, the government has adopted income approach to agriculture. The recommendations of the Inter-Ministerial Committee on Doubling Farmers Income constitute the government's strategy in this direction.

Amongst several recommendations made by the Committee in its 2018 Report based on analysis of the situation, the challenges to the agricultural sector in becoming a profitable avenue have been identified and put together as 'structural issue' and appropriate solutions suggested. As per the DFI Committee, the structural challenges emanate from the agricultural structure of the country dominated by more than 86 percent of small & marginal farmers who suffer from several constraints including operational inefficiency, poor financial capacity, etc. Since agriculture as an enterprise can sustain, only if it can yield net positive returns in favour of the farmer-producers, robustness of agricultural value chains assumes critical importance and this is feasible when efficient scales of operations support the system. In the given context of predominance of small & marginal farmers, the solution lies in promoting collectives of farmers and enabling aggregation of produce. It is in this context that the Government has consciously adopted the approach of farmers' mobilisation and aggregation of agri-produce, which will bring in efficiency of higher scale and facilitate the nation's commitment to establishing robust supply chains.

The market share of very first farmer aggregates- the cooperatives, however, varies across sectors and countries. Cooperatives have successfully made investments in establishing rural infrastructure and have aided in strengthening the supply chains. Examples in this regard exist in respect of various commodities in both India and across the world. They can therefore be assumed to play a key role in establishing sustainability. Cooperatives can help farmers adopt better and sustainable agricultural practices, commodities, and marketing strategies, which are more environmentally, ecologically, socially, and economically sustainable. The cooperatives that stand on the principles of equality and have a democratic way of participation of memberships are unique enterprises that value democratic decision-making and have very distinct organizational characteristics. Indian cooperatives have evolved too.

However, depending upon the challenges and complexities that are location-specific, apart from cooperatives alternate forms of collectives are also feasible, and India provides scope for adopting from the multiple options. Since 2011, there has been concerted effort by the union government with support from various state governments to promote farmer Producers Organisations (FPOs) in different legal formats. These include societies, cooperatives, and companies. In order to facilitate registration of the FPO as a company, the Companies Act was amended in the year 2012 based on Alagh Committee report. Further, there have been evaluations undertaken from time to time and policies amended to make it easy for the formation and sustenance of the FPOs. The last Guidelines issued by the Union Ministry of Agriculture and Farmers Welfare in 2020 are being looked at once again. This is reflective of the flexible policy approach of the government to make FPOs successful in the country. Many Farmers Producers Organization (FPOs) covering large number of small & marginal landholders of India, are seen to have successfully mobilised them into efficient & effective members of a robust organisation, resulting in profit-centric input and output management. There certainly exists vast scope for improvement and better performance.

In addition to cooperatives and FPOs as collectives of farmers, there are other interventions that can be made to promote mobilisation of farmers & aggregation of produce to serve as the starting points of healthy supply chains. These include contract farming (to negotiate price risks), legalisation of land lease (to negotiate land division & fragmentation), establishment of decentralised village aggregation platforms (GrAMs), and promotion of alternate market and trade channels. Supply chains will also need to be supported by offering to the stakeholders operational, technical, financial and support services. The cooperatives, for example, provide for agricultural inputs and loans to farmers, as also support production, processing, transportation, storage and marketing of agricultural produce.

This chapter links the activities of co-operatives and other forms of collectives with the agricultural supply chain models and offers managerial implications to cooperative societies/other collectives, policymakers and agricultural supply chain actors.

2. UNDERSTANDING THE CHALLENGES IN AGRICULTURAL SUPPLY CHAINS- A REVIEW WITH SPECIFIC REFERENCE TO PERISHABLES

For industries operating in a variety of areas and domains, sustainable development is an international imperative. Globally, organizations are making great efforts to reduce their environmental impact, but they struggle to do so without balancing the needs of the environment, society, and economics. In food supply chains (FSCs), sustainability presents an even bigger problem. The biggest issue facing humanity by 2050 will be feeding a planet that will have ten billion people. Approximately one-third of the world's food supply is lost or squandered at this time (Food and Agriculture Organisation of the United Nations, 2019). According to Gardas et al. (2018), the quantity of food being wasted worldwide would need the production of 300 million barrels of oil and 230 km^3 of water. Like vegetables and fruits, perishable food products or PFPs account for over 70% of all food waste (Gardas et al., 2018). Large quantities of resources used in the manufacture, marketing, and transportation of those food commodities are squandered because of PFP waste, which results in the loss of natural resources in significant amounts (Sgarbossa and Russo, 2017).

The unsustainability of perishable food supply chains (PFSCs) is exacerbated by several product attributes, including product perishability and sensitivity to the environment, demand fluctuations, and unpredictability in harvest and production yield (Van Der Vorst & Beulens, 2002). There are several challenges with managing PFSCs sustainably, including waste and spoiling, inefficiencies in these supply chains (SC), environmental impacts of the operations, different safety concerns because the product is organic, etc. PFSC encompasses a diverse range of stakeholders, including farmers, organizations, cold-chain service providers, wholesale markets, local shops, end users, and second-tier suppliers of fertilizers, seeds, and farm gear. Each of these players works in a different industry, such as farming, harvesting, logistics, packaging and processing, wholesale, and retail. Therefore, to comprehend the complexity, and the obstacles to sustainable management of the PFSCs, a system view is necessary. According to Joshi and Visvanathan (2019), the problem of reducing food waste in emerging nations has not gotten the attention it deserves.

Management of these issues from the perspective lens of developing countries is essential, as 90% of the world's population will reside in developing countries by the year 2050. Compared to industrialized nations, developing nations waste food in quite different ways. A report of the United Nations' High-Level Panel of Experts on Food Security and Nutrition (HLPE) commissioned by the Food and Agriculture Organisation states that in developed countries, food wastage and loss occur more often at the consumer's end, with nearly 40% of food wastage or lost only at the consumer's side in North America (Pinstrup-Andersen et al., 2014). PFP waste is more common in low-income as well as in developing nations at the beginning of the food value chain, on farms, and during transportation and storage, less at the customers' end (Pinstrup-Andersen et al., 2014).

The variation in the patterns might be caused by structural inefficiencies prevalent in the food supply chains, technological, managerial, and financial limitations, and deficiencies in infrastructure. Furthermore, in less developed nations, organisations have been sluggish to adopt sustainable practices. Poor supply chain architecture and poor logistical performance causes 370 million tonnes of vegetables and fruits to spoil in China itself (Shabani et al., 2012). The magnitude of the issue is observed in the fact that 40% of India's agricultural production gets wasted (Shabani et al., 2012; Gardas et al., 2018). With 17% of the world's population living in India, and over half of them dependent on farming and agriculture for a living, the sustainability of the PFSC presents a significant challenge to the country.

According to Jonkman et al. (2019), including perishability in the design of the FSCs is crucial. Transport, storage, weather, and harvesting circumstances all have a big impact on the product-quality of PFSCs. Thus, SC design is essential to controlling PFSCs (Balaji and Arshinder 2016). A number of research (Blackburn and Scudder, 2009; Rijpkema et al., 2014; Ali et al., 2017) have examined PFSCs. After studying the existing literature on how perishability affected distribution and planning issues, it was concluded that a thorough examination of the perishability issue in the food industry is necessary (Amorim et al., 2013). The three most important difficulties were identified in the supply chain of agri-fresh goods as being reducing Post-Harvest Loss (PHL), maximising income, and ensuring customer happiness (Shukla and Jharkharia, 2013).

Agri-supply chains, agri-retail chains, cold chains, and the food supply chain have been identified as the main arenas for efficient management and sustainability in food supply chains (Singh and Shabani, 2016; Sgarbossa and Russo, 2017; Naik and Suresh, 2018; Chauhan et al., 2018; Sharma et al., 2019). The old financial viewpoint may give way to a TBL-focused strategy, according to recent research (Chkanikova, 2016). Nevertheless, judgments about the risks and difficulties associated with this new approach have not received as much attention as they should. Numerous researches have emphasized different dangers, difficulties, and obstructions in PFSCs. According to Prakash et al. (2017), there are four types of risks associated with PFSCs: supply, demand, environment, and process risk. According to Raut et al. (2018), the primary cause of PHL in developing countries like India is a lack of ties between government, business, and institutions. Food attributes such as perishability, quality variances, bulkiness, seasonality, regulatory regulations, and infrastructure were recognized by Gokarn and Kuthambalayan (2017) as the main obstacles preventing the decrease of waste in FSCs.

One major obstacle to the sustainability of PSFCs is the failure to take food quality and characteristics into account when designing and planning SCs (Deng et al., 2019). In another study, Siddh et al. (2018) investigated the connection between the performances of sustainability-based organizations and PFSC quality. Important problems in PFSCs include product deterioration, quality degradation, and wastage owing to loss of quality and absence of standardised quality control (Osvald and Stirn, 2008; Rijpkema et al., 2014; Balaji and Arshinder, 2016; Gligor et al., 2018). To reduce PFP food waste, strict measures for controlling and managing temperature are crucial (Ali et al., 2018). Monitoring of the temperature data over time in a cold chain showed that a prominent source of temperature abuse was from the consumer end (Derens-Bertheau et al., 2015). One of the biggest obstacles to managing PFSCs in Indian context is the absence of cold chain-related infrastructure (Negi and Anand, 2015, 2016; Joshi et al., 2009). PHL in the case of fresh food is mostly caused by farms lacking ripening chambers, cold storage, and other necessary infrastructure. In addition to identifying critical success elements, the obstacles to sustainable cold chain management were also noted by Singh and Shabani (2016). In papers that focused on developing countries like India, Vietnam, etc., market availability and infrastructure-related challenges such as road, electricity, have been particularly addressed (Srivastava et al., 2015; Gokarn and Kuthambalayan, 2017; Gligor et al., 2018; Naik and Suresh, 2018). By lowering supply and demand uncertainty, sharing of information and traceability can increase the effectiveness of PFSCs (Sharma et al., 2019; Lusiantoro et al., 2018). Suboptimal information exploitation was noted by Solér et al. (2010) as a hindrance to long-term competitive advantage. Information exchange, traceability, and visibility concerns in PFSCs are also given a lot of attention in the literature (Bosona and Gebresenbet, 2013; Aung and Chang, 2014a; Balaji and Arshinder, 2016). Bosona and Gebresenbet (2013) discovered that the main motivators for traceability systems of food in PFSCs were quality, safety, and regulatory, as well as economic, social, and technological considerations. Farmers have a lack of knowledge and it is important to provide information to them as well as integrate them with consumers and vendors (Zhong et al., 2015). Food waste may be reduced by implementing the circular economy and knowledge exchange (Halloran et al., 2014). Food waste in case of the food supply chains is mostly caused by inefficient management of demand (Balaji and Arshinder, 2016). Management of supply and infrastructure requirements needs assessment aided by demand management (NCCD, 2015). FSCs typically exhibit traits such as inflexible supply and demand needs, mismatched output, and unmatched upstream-downstream lead times (Adebanjo, 2009).

The literature has examined other issues such as inadequate pre-harvest management, high operating costs, and want of government assistance (Prakash et al., 2017; Gokarn and Kuthambalayan, 2017; Gligor et al., 2018). Effective legislation and support from the government are needed to expand sustainability across the supply chains (Sajjad et al., 2019). In addition to these difficulties, PFSCs also confront difficulties that are already present in non-perishable supply chains, such as inadequate integration among the supply chain actors (Gardas et al., 2019), inadequate logistical performances (Gokarn and Kuthambalayan, 2017), and a lack training and professional expertise (Gligor et al., 2018). Table 1 effectively summarizes major challenges in perishable food supply chains and implications

Table 1. Major identified challenges in perishable food supply chains and implications

Major challenges	Implications
Lack of product characteristics and perishability	Diverse food items require diverse handling, storage, and temperature ranges with logistical skills. Facilities in all PFSCs have not been created with product-specific design factors in mind.
Poor management	Production is not demand-based, which frequently leads to overstocking or stock-outs, wasting resources and causing spoiling.
Inadequate performance in logistics	When compared to non-refrigerated vehicles, vehicles that transport perishable goods use a great deal more energy. Logistics performance is a huge problem due to poor service quality and inadequate logistics infrastructure, as well as due to the failure to take their environmental effect into account when designing networks.
Insufficient quality control	Inadequate quality control in farming, handling, storage, and transportation results in the wastage of goods and other resources.
Ineffective temperature control	Inadequate temperature control restricts the product's geographic reach and enhances the risk of spoiling, which makes the product unsustainable from an environmental and economic standpoint.
Insufficient supply chain visibility and traceability	PFSCs with no traceability carry a higher risk of spoiling, recalls, returns, and inaccurate batch detection. Furthermore, the consumer can receive false information about the items' shelf life, place of manufacture, etc.
Farmers' lack of horizontal integration	The supply is fragmented and disaggregated due to the lack of horizontal integration among farmers. Unsustainability on both an socio-economic level results from farmers' poor integration with wholesalers and merchants.
Absence of means for communication and information	Facilitating information exchange via ICTs enhances PSFC sustainability overall, farmer income, and productivity.
Inadequate pre-harvest handling	Pre-harvest factors, including high-quality seeds, irrigation methods, and harvest timing, have a big impact on the future sustainability, quality, and holding life of PFSC.
Infrastructural problems	Power outages, inadequate agricultural infrastructure connection, and a dearth of roadways all contribute to unforeseen supply delays, lost shelf life, and a reduction in economic activity across the PFSC.
Uncertain pricing in the market	Prices that fluctuate have a direct impact on an organization's capacity to make money, maintain itself financially, and lose customers.
Inadequate infrastructure for the cold chain	In PFSCs, cold chain assistance is essential. PHL and the unsustainability of the supply chains in PFSC are mostly caused by unavailability and insufficient cold chain assistance.

3. COOPERATIVES AND PERISHABLE FOOD SUPPLY AND VALUE CHAINS: EXPERIENCES FROM THE WEST

In Western countries, agricultural cooperatives contribute substantially to agri-food supply chains. For example, in the European Union, they contributed as high as 40 percent in the agri-food sector in the year 2010. Within the European Union, the share of cooperatives in the market is more than 50 percent in several countries including Austria, Denmark, Finland, France, Ireland, the Netherlands, and Sweden, In the case of perishable produce, the farmers in the European Union, prefer to engage with the markets via cooperatives or collectives. Another fine example relating to the perishables comes from the United States of America, wherein around 75 percent of milk produced in the country is routed into the market through collectives or dairy cooperatives.

3.1 European Parliamentary Research Service Elaborates on Status of Cooperatives in the European Union

As per the European Parliamentary Research Services, cooperatives are significant players in the global economy, according to a 2018 research on Exploring the Cooperative Economy published by the ICA and the European Research Institute on Cooperative and Social Enterprises (EURICSE). Economic and social data were gathered from 2 575 cooperative firms and mutual organizations worldwide for the report's aims. In terms of the financial elements, the survey found that the 300 largest cooperatives and mutuals had a combined turnover of US$2 018 trillion in the reference year 2016 (up from US$2 164 trillion in 2015). The paper states that there are 3 million cooperatives globally, which collectively employ 280 million people, or 10% of the global labour force. Cooperatives engage in a variety of economic endeavours.

Of all the cooperatives that took part in the survey in 2016, for example, 33% worked in the agriculture and food industry; 19% provided banking and financial services; 16% provided other services (such as business services, transportation, and communications); 7% dealt with industry and utilities; 6% operated as insurance cooperatives or mutuals; and 4% worked in the fields of health, education, and social care. Approximately half of the 2,575 cooperatives that were polled fit the definition of small and medium-sized firms (SMEs) as defined by the European Commission, provided that their annual revenue is up to €50 million. Europe and Japan produced two of the top five cooperatives in the world in terms of turnover in 2016.

Examining Europe and the EU more closely, we may conclude that cooperative businesses significantly contribute to the social economy of Europe. The ICA's regional office, Cooperatives Europe, released a report in 2016 titled "The power of cooperation," which used 2015 as its reference year. The report stated that there were approximately 176 500 cooperative enterprises with over 4.7 million employees in Europe overall, with 131 000 of those cooperatives being located in EU Member States. Over €992 billion of the €1 004 trillion total revenue of European cooperatives was produced within EU member states. Over 141 million people, or 17% of the total population of Europe, were associated with cooperative (in the EU: more than 127 million). In Europe, there were 14% more cooperative firms and

14% more members between 2009 and 2015.9. Comparing by industry, workforce size, and employee turnover uncovers some fascinating data. For example, the retail sector, which comprises fewer than 1% of all businesses, accounts for about 30% of the total revenue of cooperatives, while the industrial and services sector, which comprises the majority of businesses and employees, accounts for less than 10%. This difference in turnover is likely due to the fact that the industry and services sector has a higher proportion of SMEs.

When nations are compared, Italy (39 600), Turkey (33 857), France (22 517) and Spain (20 050) had the highest percentage of cooperatives in Europe in 2015. The countries with the highest percentage of cooperative members in Europe were France (26 106 829), Germany (22 200 000), the Netherlands (16 912 900), the United Kingdom (14 919 093), and Italy (12 620 000). In Europe, the countries with the most cooperative employees were France (1 217 466), Italy (1 150 200), Germany (860 000), Poland (300 000), and Spain (290 221). The biggest yearly turnover in Europe was recorded by cooperatives in France (€307 billion), Germany (€195 billion), Italy (€150 billion), and the Netherlands (€81 billion).

Cooperatives have significant market shares in a number of industries, according to studies by the Commission: - agriculture: 83% in the Netherlands, 79% in Finland, 55% in Italy, and 50% in France; - forestry: 60% in Sweden and 31% in Finland. Given the significance of agriculture to cooperatives, it merits more examination. Cooperatives are widely distributed across the European food supply chain. The General Confederation of Agricultural Cooperatives in the European Union (COGECA) released a study in 2014 titled Development of Agricultural Cooperatives in the EU, which states that the combined revenue of all agricultural cooperatives in 2013 was €347 billion. There were about 22,000 cooperatives in the same year, with about 6.2 million members. The top 100 farming cooperatives showed a noteworthy upward trend.

3.2 The United States Department of Agriculture Elaborates on the Cooperatives and Their Role in Value Chains

As per the United States Department of Agriculture, since early 1900s, agricultural cooperatives have been a staple in rural America. Around 306 co-ops (17.5 percent of all agricultural co-ops) were 100 years of age or older in 2020; 53.6 percent were 75 years of age or older, and 76.5 percent were 50 years of age or older. Even though they are becoming fewer in number, cooperatives are still a reliable and effective way for members to sell their goods; many even enhance the products, which benefits member-owners even more. In addition, cooperatives still offer members the goods and services they require to run farms and ranches in the demanding and ever-changing agricultural landscape.

In the USA, the agricultural co-operatives are generally well-run, effective, and financially stable, which contributes to building a solid basis for the stability of the rural communities that are home to the majority of producer-members and many of the thousands of co-op facilities. Numerous less-rural communities where agricultural cooperatives may establish offices, factories, or other infrastructure also benefit economically from co-ops. The overall success of agricultural cooperatives in 2020 demonstrates that the tried-and-true model of a member-owned, member-governed co-op business structure is still vital to American farmers and the country's food and fibre system.

In 2020, net income of cooperatives (co-ops) reached a record $8.4 billion, despite overall gross revenue declining for the second consecutive year, according to the USDA's annual survey of the 1,744 farmer, rancher, and fishing cooperatives throughout the country. The following are some particular conclusions drawn from the 2020 aggregated income statement:

- Co-ops' marketing sales climbed by 1.0 percent to $126.3 billion in 2019, but their supply sales fell by 5.6 percent, from $71.0 billion to $67.1 billion. While service revenues and other operating income saw a tiny gain of $22,000, total operating revenue as a whole decreased by 1.3 percent to $198.6 billion.
- To the tune of $8.4 billion, net income after taxes increased by 7.1 percent. This net income is important as it allows co-ops to return earnings to members in patronage and equity redemption, and/or to invest further in operational assets for continued prosperity and service to members.
- Both the return on member equity and the return on assets increased from 2019 to 29.7 percent and 9.3 percent, respectively.
- The nation's agricultural cooperatives maintained a robust combined balance sheet in 2020, with record assets of $102.4 billion and record member equity of $47.5 billion.
- Members' equity rose by $1.6 billion (3.4 percent) and assets by $2.2 billion (2.2 percent). The amount that agricultural cooperatives invested in property, plant, and equipment (fixed assets) increased to a record $31.4 billion, up $932 million from 2019. These fixed assets included major food and beverage processing plants, grain elevators, warehouses, farm supply stores, fertiliser and feed plants, and petroleum/convenience stores.

4. RISE OF FARMER PRODUCER ORGANIZATIONS (FPOs) IN INDIA

To develop and bolster the promotion of 10,000 new Farmer Producer Organisations (FPOs), in 2020, the Indian government rolled out the policy of "Formation and Promotion of 10,000 FPOs" initiative into action in the form of a Central Sector Scheme (CSS). As per the Press Information Bureau of India, as of June 2024, 8875 FPOs have been registered nationwide. The number of FPOs registered in each state is listed in table 2.

The entire paid-up capital to the Farmer under the Central Sector Scheme (CSS) for the "Formation and Promotion of 10,000 FPOs" is Rs. 630.3 Crore. Under the mentioned 10,000 FPOs initiative, matching equity grants totalling 210.1 Cr. were granted to qualifying FPOs as of June 30, 2024.

Table 2. Number of registered FPOs in India

State	Number of Producer Companies
Andaman & Nicobar	7
Andhra Pradesh	449
Arunachal Pradesh	150
Assam	427
Bihar	580
Chattisgarh	213
Dadra Nagar Haveli	2
Goa	8
Gujarat	419
Haryana	172
Himachal Pradesh	168
Jammu & Kashmir	318
Jharkhand	355
Karnataka	336
Kerala	137
Ladakh	3
Lakshadweep	1
Madhya Pradesh	622
Maharashtra	579
Manipur	76

continued on following page

Table 2. Continued

State	Number of Producer Companies
Meghalaya	57
Mizoram	49
Nagaland	87
Odisha	448
Puducherry	6
Punjab	136
Rajasthan	533
Sikkim	13
Tamil Nadu	414
Telangana	302
Tripura	59
Uttar Pradesh	1,246
Uttarakhand	150
West Bengal	353
Grand Total	8,875

Source: Press Information Bureau, June 2024

A credit guarantee fund, supported by the plan, has allowed loans totalling Rs. 50.4 crores to be given to these FPOs. Around 19,68,868 shareholder farmers were enrolled in FPOs established and promoted under the 10,000 FPOs initiative as of June 30, 2024 and approximately 6,374 CEOs were appointed in 8,875 registered FPOs under the program as of June 30, 2024 (Press Information Bureau, August 2024). The FPOs in India have been promoted on four distinct grounds.

4.1 Value-Chain-Based Approach to Promote FPOs

The largest initiative in the nation for reducing poverty, Deendayal Antyodaya Yojana - National Rural Livelihood Mission (DAY-NRLM), has been assisting low-income and vulnerable women by organising them into Self Help Groups (SHGs). The majority of SHG members rely on agriculture, especially raising cattle and Non-Timber Forest Produce (NTFP), to support themselves. Village Organisations (VOs) and Cluster Level Federations (CLF) are created from Self Help Groups (SHGs) that are founded at the village level. Small and marginalised female farmers are formed into institutions under Mahila Kisan Sashktikaran Pariyojana (MKSP), a DAY-NRLM subcomponent. Producer Groups (PGs) are founded at the cluster or village level by women farmers engaged in similar activities such as livestock, agriculture, or non-traditional farming practices.

The program's concept called for value chain-based interventions to be established in areas with sufficient numbers of PGs, SHGs, and social capital. Market-linked Producer Enterprises (PEs), with strong business models for improved price realisation for small and marginal farmers, were supported by the value chain development plan. The first step in the program that SRLMs with technical assistance agencies carry out is to identify clusters that have the potential to produce commodities. A thorough commodity-wise value chain development exercise might be conducted for the listed commodities (about 10–20 per state) to determine the necessary intervention. Post-farm gate value addition process and marketing are the main areas of intervention. The establishment of producer groups in the project region is preceded by a planned value chain intervention. Following their federation, the producer groups become Producer Enterprises, which are officially recognised as either Producers' Companies or Cooperatives. Under the program, the value chain-based interventions made a substantial contribution to the growth and development of producers' collectives, or producers' groups (PGs), and producers' enterprises (PEs), which give small-scale and marginalised women farmers access to markets. NRLM has approved development plans for value chain including 81.82 million SHG families as of March 2022. Promoter Enterprise included more than 575,000 Mahila Kisan (women farmers). The initiatives are concentrated on value addition activities and market connections regarding agricultural crops like paddy, wheat, maize, mustard, groundnut, pigeon pea, green gram, black gram; horticultural crops like vegetables, ginger, tamarind, mango, cashew, aonla (Indian gooseberry), bael (stone apple), floriculture, hill broom, salai gum (Indian Olibanum); dairy, goatery, fishery, backyard poultry, piggery, duck farming, apiculture, and other Non-Timber Forest Products (NTFPs), etc., through Producer Enterprises. DAY-NRLM has used a value-chain strategy to promote more than 1,000 FPOs. To implement a methodical approach based on value chain intervention, DAY-NRLM partnered with Tata Trust to form the Foundation for Development of Rural Value Chain (FDRVC). Farmer income improvement is facilitated by FDRVC through the organisation of farmers into big Producer Enterprises

4.2 One District One Product (ODOP) Approach to FPOs

As per clause 4.7 of operational guidelines, in addition to using a cluster-based approach for produce or produce mixes, the creation of FPOs needs to be concentrated on the "One District One Product" (ODOP) strategy for the advancement of product specialisation, provided that the district has declared focused agriculture produce. In this scenario, FPOs are encouraged to support the processing, branding, marketing, and export of the product in order to maximise value realisation. The ODOP plan calls for making use of economies of scale when it comes to purchasing inputs, using shared services, and promoting goods. The plan offers a platform for developing value chains and coordinating support infrastructure. In a single district, there could be many ODOP product clusters. A state may have a cluster of ODOP products (table 3) made up of many neighbouring districts. Nonetheless, the FPOs, as per the guidelines have to also offer a broader range of products and services to ensure that there are adequate events and opportunities for member interaction throughout the business year. Additionally, FPOs can federate at the district, state, and federal levels for the product specified based on their branding, processing, and marketing needs.

Nine implementing agencies, designated by the government as the task force, work with promotional institutes known as Cluster-Based Business Organisations (CBBOs) to gather, register, and offer expert handholding support to all the FPOs for five years. All facets of FPO management, inputs, production, processing, and value addition, as well as market and credit connections and technology utilisation, are covered by the handholding support. Long-term support is to strengthen the FPOs' entrepreneurial skills and competencies so they may become financially independent and self-sufficient after receiving government assistance. It is anticipated that the cluster-based, ODOP strategy would result in setting up specialist FPOs with forward connections to markets and agricultural value chains. Government-sponsored FPO marketing initiatives have grown over time, starting as a small-scale project component and then expanding to a large-scale pan-Indian initiative. The strategy shifted from a numerical and business-oriented concentration to a minutely watched cluster-based strategy designed to attain the development of commodity value chains with scale. Promotion's goal placed a strong emphasis on supporting production clusters of certain commodities as well as on FPO creation. However, the procedure and administrative delays makes it difficult to allocate cash and provide matching grant supports to leverage finances.

Table 3. ODOP districts in the Indian states and UTs

State	Number of ODOP districts based on single product cluster						Total
	Agriculture crop-based product	**Horticulture crop-based product**	**Marine and Fish products**	**Animal based products (milk, meat, animal feed)**	**Other processed products (Bakery, snacks, pickle, jaggery, etc.)**	**Products from other allied sectors (honey, minor forest produce, mushrooms, etc.**	
Andaman & Nicobar	0	2	1	0	0	0	3
Andhra Pradesh	10	9	2	2	3	1	27
Arunachal Pradesh	2	22	0	0	2	0	26
Assam	12	15	0	2	2	2	33
Bihar	10	22	0	0	3	2	37
Chandigarh	0	0	0	0	1	0	1
Chattisgarh	8	9	0	1	1	9	28
Dadra Nagar Haveli	0	0	2	0	0	1	3
Delhi	0	0	0	2	9	0	11
Gujarat	14	19	0	0	0	0	33
Goa	0	2	0	0	0	0	2
Haryana	1	7	0	9	4	1	22
Himachal Pradesh	1	9	0	1	0	1	12
Jammu & Kashmir	1	8	2	5	2	1	19
Jharkhand	2	15	0	0	3	4	24
Karnataka	4	20	2	1	3	0	30
Kerala	2	11	0	1	0	0	14
Ladakh	0	2	0	0	0	0	2
Lakshadweep	0	1	0	0	0	0	1
Madhya Pradesh	10	42	0	0	0	0	52
Maharashtra	13	19	3	1	0	1	37
Manipur	1	13	2	0	0	0	16
Meghalaya	0	10	0	0	0	1	11
Mizoram	0	11	0	0	0	0	11
Nagaland	3	6	1	0	1	0	11
Odisha	8	9	5	6	0	2	30
Puducherry	0	0	1	1	0	0	2
Punjab	4	8	0	5	4	2	23
Rajasthan	9	20	0	1	2	2	34
Sikkim	0	4	0	1	1	0	6

continued on following page

Table 3. Continued

State	Number of ODOP districts based on single product cluster						Total
	Agriculture crop-based product	Horticulture crop-based product	Marine and Fish products	Animal based products (milk, meat, animal feed)	Other processed products (Bakery, snacks, pickle, jaggery, etc.)	Products from other allied sectors (honey, minor forest produce, mushrooms, etc.	
Tamil Nadu	11	15	4	5	1	0	36
Telangana	11	14	1	3	4	0	33
Tripura	1	4	0	1	2	0	8
Uttar Pradesh	13	35	1	8	14	4	75
Uttarakhand	1	9	1	0	1	1	13
Grand Total	152	392	28	54	63	37	726

Source: Authors' analysis from list of ODOP districts available on the website of the Ministry of Food Processing Industries of the Government of India.

4.3 FPOs Promoted Through Sector-Focused Organisational Competencies: Commodity Approach

Certain major organisations having high skills and abundant resources are fostering high-achieving FPOs, in stark contrast to the promoting institutions that were resource-constrained and depended on project-based financing.

One such organization with years of expertise handling the commodity and a thorough grasp of the milk industry is the NDDB Dairy Services (NDS) . It can obtain funds from outside sources in addition to using its internal resources. Between the years 2012 and 2021, NDS financed for establishing sixteen Milk Producer Companies (MPCs) in ten Indian states (table 4). Six of these MPCs in the states of Gujarat, Uttar Pradesh, Bihar, Rajasthan, Punjab, Andhra Pradesh, and Tamil Nadu were promoted under National Dairy Plan I. For Tata Trusts in Punjab, Haryana, Uttar Pradesh, Maharashtra, and Rajasthan, five greenfield MPCs were established; five more MPCs were established for NRLM/SRLMs in Uttar Pradesh, Bihar, Rajasthan, and Madhya Pradesh. These businesses attained size in terms of market reach, sales, and membership. As of January 2022, these sixteen milk producer companies together registered around 0.697 million milk producers as members. When combined, the enterprises purchase 2.9 million litres of milk each day with a share capital of Rs 1,670 million. For the financial year 2020-21, these firms had a total revenue of Rs 47,920 million. In the foreseeable future, NDS intends to support FPCs working in the non-dairy industry, namely horticulture.

Table 4. Milk Producer Companies (MPCs) Facilitated by NDDB Dairy Services (NDS) Incorporated/ Operationalised till March 2021

Sl. No.	Category	No. of MPCs	No. of States Covered (Operational Area)	Status as on Jan 31, 2022			Avg. Daily Milk Procurement ('000 LPD)	Total Revenue (Rs in Million)
				No. of Villages Covered	No. of Members	Share Capital (Rs. in Million)		
1	National Dairy Plan	6	7	12469	509762	1450	2566	42550
2	Greenfield-Tata Trusts	5	5	2126	89435	140	235	3,700
3	Greenfield NRLM/ SRLM	5	4	2410	97949	80	97	1,670
	Total	16	10	17005	697146	1670	2898	47920

Source: NDDB Dairy Services

Currently, India has 210 operational milk producer firms, of which 81% are under 5 years old. Dairy products account for nearly 50% of the top 20 farmer producer companies (Neti et al., 2019).

The poultry industry is another example of how sector expertise aided in the promotion and development of FPO. Through the support of a small-holder community poultry model, the National Smallholder Poultry Development Trust (NSPDT) empowers impoverished women in rural areas of India to launch and manage prosperous chicken businesses. In the scenario, a small-town woman farmer located in a far-off village is given industrial poultry. It groups women into formalised collectives and develops procedures and systems to enable them to achieve scale efficiencies and production levels competitive with the industry. From impoverished homes, women form cooperatives and work together to raise broilers in contemporary backyard chicken farms. Typically, a mature cooperative has between 300 and 500 members. The cooperatives throughout an area or a state form a typical federation. This gives members access to professional and technical assistance, enhanced compliance, economies of scale in input procurement, and a platform for sharing knowledge, and process exchange across member co-ops, all of which foster unity. By adjusting and reacting to technological and economic developments in the sector, the federation makes sure co-ops maintain their competitiveness. across 27 poultry producers' groups (Co-operative & Producers company) across Madhya Pradesh, Maharashtra, Assam, Jharkhand, and Odisha, NSPDT has so far supported 15,000 women who raise chickens. Over Rs 5,240 million was turned over in total, with members earning over Rs. 400 million in earnings.

Commodity-specific FPO development and promotion is a component in the objective of several state-supported programs, such as the Spices Board, and the Coconut Development Board, among others. Producer Companies and Producer

Societies are among the 445 spices that the Spices Board is promoting in 21 states. Nearly 10,000 coconut farmers' societies and 60 coconut FPCs have been assisted by the Coconut Development Board in four states.

Large quantities, mentorship from a skilled mentor, increasing market demand, and sufficient finance support are necessary components for the development of commodity-specific collectives. These FPOs are authorised to grow their paid-up capital as well as their ability to use credit for expanding commercial transactions if they mobilise a sizable membership base. Small sums are routinely withheld by the Milk Producer Companies (MPC) from member payments, increasing the members' share capital in their business over time. The paid-up capital is increasing quickly as a result.

4.4 Fodder Centric FPOs

After requests from Ministry of Fisheries, Animal Husbandry and Dairying proposed the establishment of fodder-centric FPOs in 2020 to look after the fodder deficit situation in India., the Department of Agriculture and Farmers Welfare's competent authority authorised the designation of NDDB as the implementing agency for the 10,000 FPOs formation and promotion scheme. The primary focus of the FPOs is animal husbandry, with fodder-centric activities serving as a secondary activity (fodder plus model). NDDB was tasked with creating 100 FPOs between 2020 and 2023 while adhering to the scheme's parameters.

In August 2022, the fodder inflation based on the Wholesale Price Index reached a nine-year high of 25.5%. When comparing the costs of dry fodder in 2022 to those of the previous year, they are still high. Rural households that depend on the cattle industry for their livelihoods have suffered as a result. The National Statistical Office (NSO) released a report titled "Situation Assessment of Agricultural Households and Land and Livestock Holdings of Households in Rural India, 2019" that states that in 2018, 48.5% of the total 17.24 crore rural households owned cattle in the categories of "in milk," young cattle, and cattle in "others." Forty-eight percent of the 9.3 crore agricultural families utilised green fodder, while fifty-four percent used dry fodder, 30.4 per cent used concentrated fodder and 12.5 per cent other animal feed over the same time period.

In India, the current fodder shortfall is 12–15% for green fodder, 25–26% for dry fodder, and 36% for concentrated fodder. Both geographical and seasonal variables contribute to the shortage. The drop in wheat yields and the increase in input costs, such as diesel, are the main causes of the high WPI-based fodder inflation.

5. FPOS AND VALUE CHAIN: SUCCESSFUL CASES FROM INDIA AND ABROAD

5.1 Anand Milk Union Limited (AMUL)

It is worth citing the case of India's largest dairy milk cooperative – Anand Milk Union Ltd. (AMUL) as a successful cooperative. It has successfully introduced its dairy products in the US markets this year, thanks to 3.6 million farmers associated with it in the form of cooperatives. The dairy products are marketed through the Gujarat Cooperative Milk Marketing Federation which has 18 district cooperative milk unions under its umbrella, pouring 30 million litres of milk every day. In 2023-24, the federation clocked a turnover of Rs 80,000 crore. Today, the Federation has 98 dairy plants with a daily milk handling capacity of 50 million litres of milk and is equipped with modern facilities to process, pack, and store the milk and milk products. The brand, namely 'AMUL' is celebrating 2024 as its 50th year, and it has every reason to celebrate given high success it has achieved as a dairy cooperative. The expansion of the brand continues in terms of memberships, procurement, production, product development, and distribution, courtesy its highly efficient and sustainable supply chains.

5.2 Amrapali Farmer Producers Organization

Amrapali FPO is based in a small district of Jharkhand called as Lohardaga District that is famous for Bauxite supply and is called as Bauxite town. This region has seven blocks, out of which the Kairo block is renowned for quality fruits and vegetables. Earlier this block was majorly covered with forests but with deforestation, it reduced to 32 to 35 percent of the total area.

Table 5. Livelihood in Kairo District Block

Sr. No.	Category	Percentage of Population in Block
1	Cultivators	46.57
2	Agricultural Labours	42.26
3	Household Industries	1.07
4	Other Workers	10.09

Source: Registrar General of India, Census 2011

Looking at the background information of the block (table 5) and its dependence on rains for agriculture purpose, a non-profit organisation was formed in 1985 under the Trust Act which was called Lohardaga Gram Swarajya Sansthan (LGSS). Established for rural upliftment and sustainable development of the community.

The LGSS trust analysed that inspite of having natural resources and agricultural land, their farmers were still in a miserable condition. The reason was found to be inadequate working capital availability which made the farmers borrow from informal institutions like money lenders. In 2014-15, Producers' Organization Development and Upliftment Corpus' (PRODUCE) Fund granted support worth Rs. 9.06 Lakh to LGSS. This PRODUCE fund worth Rs. 200 Crores was created by the Government of India in NABARD for creation of 2000 Farmer Producer Organisation in India. Thc PRODUCE Fund aims to help new FPOs and their early financial needs, fostering credibility, economic viability, and sustainability for farmers' businesses. With this view, Amrapali FPO was formed at Utka Village of Kairo Block in Lohardaga District with the help of NABARD in 2016. This FPO is registered under Company's Act with a coverage 9 villages along with 5 Gram Panchayats. This FPO basically focussed on Fruits and Vegetables. They achieved a revenue of 1,18,046 in 2019.

By 2022, Amrapali FPO had 551 members out of which 535 are active members. They also have started a retail outlet for the sale of seeds and other inputs which is only operational in sowing season and as and when required. Farmers have realised the power of collectivism as they are reaping much better prices of their produce. Member farmers are also given training on primary value addition like cleaning, grading, sorting and other manual operations as well as the basics of inter cropping are being taught to them.

Economic and Social Impact

Being a tribal zone, the land is organic by default. This gives an additional advantage to the farmers that the produce is paid higher if marketed efficiently. The FPO team members have organised vermi compost kits, agro spray equipment and power tillers by partnering with the operational departments at various levels.

The FPO also conducts training programs to create awareness on health, literacy, education programs and other services for upliftment of the society. Overall, we can see that through collectivism approach even small farmers living in remote locations can be benefitted. (Nayak, 2022a)

5.3 SEEDS Farmer Producer Company Limited

Background Information

The FPCL was formed on 18th December, 2014 in Virudhnagar, Tamil Nadu which is Agro climatically a Semi-Arid region. Seeds Farmer producer Company Limited covers 117 villages and has over 4700 farmer members. Their Annual Turnover was approximately 4.7 Crore Rupees in FY 2019-20. (Nayak, 2022b). This FPO has 95 percent female members with an aim to promote women empowerment and capacity building.

This area was then adopted by NABARD under the Watershed development scheme, "Pisindi 'Azhagiya' Nallur Watershed Project" in 2011. The construction of agricultural ponds, check dams, trenches, bunds, well recharge pits, sunken ponds and other watershed development features resulted in the production of long-term water resources which made the harvest almost double in quantity and much better in quality but their income levels did not change much. They were basically facing the issue of marketing for their produce. An NGO named SEEDS NGO (Social Education Environment Development Society) developed this FPC in 2014 with an aim to help the farmers get the remunerative price of their produce. They believed that the farmers do not think beyond production and in such scenarios, they have to sell the produce in a very nominal price.

Mechanism to Increase the Value for Farmers

The FPO helps the farmers of over 117 villages by collectivising the farmers, providing capacity building, supplying quality inputs to them like seeds, fertilizer, cattle feed and many other inputs. They also procure and add value to their produce through sorting, grading and packaging of the produce. Major crops grown by the farmers are black gram, green gram, red gram, chillies, coriander and millets.

The FPO procures these commodities from the members and carries out primary processing like drying, cleaning, destoning, polishing, sorting, grading and packing in the value addition centre of the FPO established with the support of state Government and NABKISAN. They employ around 29 employees from the village including 10 women, 18 men and 1 transgender. The FPO has further started secondary processing of the produce including coriander powder, Pulse milling as well as masala powders along with ready to cook dosa batter. All the value-added products are sold under the brand name of Seeds Agro and marketed to the bulk buyers, business centres and local markets

Farmers associated with this FPC claim that they receive 500 additional for 100kg seeds if they sell through the FPC compared to open market through third-party agent. He also claims that they receive payments immediately at the farm gate and the pricing mechanism is very fair and transparent. They also motivate farmers to try something different and take the produce in distant market as per their market intelligence where the price is higher. These strategies can be only effective through collectivism and unity amongst farmers

Unique Approach

The have 8 directors in the board of director who hold up meetings every month to discuss the issues prevalent amongst farmers and resolve them immediately with out any delays. The FPo also holds a weekly market for the sale of fresh fruits and vegetables as a direct marketing model where in approximately 1500 people visit the market to purchase direct from farmers. They also started an F- Mart for all the grocery needs of the locals

Seeds Farmers Produce Company Limited has obtained Equity Grant support from SFAC, established credit linkage with various financial institutions. Besides the initial support for collectivizing the farmers as SEEDS FPCL over a period of three years, NABARD has extended its financial assistance for 2 more years for nurturing the FPO and provided Business development grant for expanding the business.

With the support of NABARD, SEEDS FPCL is also acting as Joint Liability Groups promoting institution, wherein NABARD has sanctioned a grant assistance of Rs. 20 lakhs for credit linking 1,000 JLGs. The FPO has signed an MoU with Indian Bank, to identify eligible famers for obtaining Kisan Credit Card with an aim to provide KCC to all its members.

All these measures have resulted in increase in average income level of farmers by 75%.The FPO also performed well in the most adverse situations of the pandemic when trucks transporting pulses got stuck in midway. Seeds FPCL delivered grocery to the farmers at their doorstep placing orders on a phone call as an effort to support the community and livelihood. This FPCL protects and uplifts its community not only economically but socially also.

5.4 Samcholing Green Tea Cooperative, Bhutan

This Cooperative is a woman led cooperative that received the award of "Best Green Tea Producer" in 2018 by Government of India. There are 28 female members in this cooperative who were not schooled formally but they received trainings in various countries like Nepal, Japan, China and Sikkim which was sponsored by Bhutan Government. Each farmer has one to two-acre land and cumulatively they grow green tea on 43 acres of land producing 600kg of fresh leaves annually. (Tashi et al., 2022)

Uniqueness of This Tea

Samcholing tea is one of the rarest in the world and was discovered in 1990s. The seeds of this tea were originally gifted to the King in 1950s which he planted in his winter palace but only for ornamental usage. In 1900s, a tea expert from New York visited this place and revived the potential of Camelia Sinesis. Later, an agricultural project was setup there with the help of Ministry of Agriculture as an initiative to uplift the population of this region socially and economically. The tea plantation takes around three to four years to grow and then can be harvested for about 20 years. The tea leaf harvesting season usually starts from April and ends in September. This tea is prepared from unfermented leaves of Camelia Sinesis which are then handpicked, pan fired and rolled to give a unique tea with buttery notes of vegetables like coriander and lemon peel. Unlike other green tea leaves, these teas can be reused upto multiple steeping. This tea is grown at an altitude of 1800-2000m which offers ideal conditions to maximize the aromatic compounds and polyphenols resulting in the unique taste of this tea. This tea is grown with a spacing higher then any of the commercial tea gardens that favors the nutrition retention in the fight against soil depletion. This makes the tea rare and unique in its own way and provides additional nutritional benefits. (Beckwith, 2019)

Progress of the Co-Operative

In a remote place like this, people usually work together for their economic well-being. There are many cooperatives in the region growing potatoes, corn and other commodities but they were unable to get competitive prices. In such scenario, a product like green tea led to a path for self-sufficiency and sustainable future. The market for this tea was assured since the begin as the tea production was very limited and the local demand was huge. In 2012, Ministry of Agriculture developed a small processing facility to promote standardisation in processing and improve the quality. In 2016, this factory started working full-fledged and they also received food safety license in 2021 (Beckwith, 2019).

Inspired from Thailand's best practices of OTOP (One Tambon One Product), Bhutan also launched One Gewog One Product (OGOP) where Gewog stands for group of villages. Under this scheme, farmers are given trainings on raw material sourcing, product creation, testing, quality assurance, packaging, branding as well as marketing. This initiative links Bhutan's products with rest of the world and enlarges their market to get better price realisation. (www.ogop.bt). The cooperative currently faces challenges in connecting with the international markets and collaborating with tea corporates due to lack of linkages. (Tashi et al., 2022)

5.5 Bangladesh Milk Producers' Cooperative Union Limited (BMPCUL)

Dairy Industry in Bangladesh originated in 1946. Small Dairy Farmers in rural areas of Bangladesh were not getting remunerative prices of milk (Shamsuddoha *et al.*, 2000). Later they saw Kolkata as a promising market round the year due to rail connectivity but processing was required to supply the milk to distant markets. With the aim to raise the income of poor and landless farmers by ensuring a guaranteed market, they established a dairy facility in Lahiri Mohanpur which is now known as Sirajganj (MilkVita, 2024). In 1965, the first cooperative for milk producers was established, known as the Eastern Milk Producers' Cooperative Union Limited (EMPCUL). In 1977, the organization's name was changed to Bangladesh Milk Producers' Co-operative Union Ltd., while retaining its product brand, MILK-VITA. (Nedacadmin, 2022)

Action Plan

The project proposed establishing dairy plants in milk-abundant regions like Pabna, Tangail, Manikganj, and Faridpur, and the government included it in its First Five-Year Plan (1973-78). These plants would operate through milk collection networks of co-operative societies in milk-producing regions, collecting milk from local farmers twice daily. After initial processing at rural plants in Tangail, Manikganj, and Faridpur, the milk would be transported to Dhaka in insulated tankers for further processing, packaging, and marketing. The Baghabarighat Plant, meanwhile, would convert its collected milk into products such as butter, milk powder, and ghee, all marketed through Dhaka's Marketing Unit. The organization's headquarters, Dugdha Bhaban, oversees operations. (MilkVita, 2024).

In 1991, major reforms were made in its administration by appointing a professionally qualified chief executive instead of personnel being deputed from the Government. They also implemented 195 policies for the smooth running of the Dairy. BMPCUL created a managing committee consisting of nine members out of which six were elected by the milk producers' primary societies (including chairman and vice- chairman) and the rest three were nominated members from the Government. The responsibility of implementing the decision of the committee relies solely on the executive head who is the General Manager of the organisation. Under his leadership there are professionals who are the experts in individual fields like mechanical engineers, electrical engineers, animal husbandry experts, veterinary experts along with accounts, marketing and finance experts. BMPCUL has a very limited coverage in Bangladesh in terms of area but still runs 7 dairy plants and shares 60% of total marketed milk in Bangladesh. (Khan *et al.*, 2014)

Benefits of MilkVita Delivered to the Farmers

Current population of Bangladesh is 174 million out of which around 65 percent is dependent on Agriculture directly or indirectly. In Bangladesh dairy farming acts as an important source of income as well as a vital means of achieving nutritional requirements (BBS, 2019). Through this cooperative model, the farmers were able to (Shamsuddoha *et al.*, 2000):

a. Achieve better price of the raw milk and stabilize the incomes of the small dairy farmers
b. Increase processing of the surplus milk and reduce the losses
c. Received trainings required to produce better quality of the produce
d. Received veterinary services on time. BMPCUL offers free veterinary services like vaccinations and medical check-up for the
e. Reduce dependency on the middlemen
f. Increase the shelf life of milk by packaging and value addition which led to higher income realization for the farmers
g. Increase the access to distant markets
h. Supports gender equity as dairy farming is generally carried out by female farmers in Bangladesh

Products

i. Pasteurized Liquid Milk
ii. Toned milk
iii. Skimmed milk
iv. Flavoured Milk (Chocolate, Mango)
v. Ghee
vi. Butter
vii. Full Cream milk Powder
viii. Skimmed Milk Powder
ix. Rasagolla (MilkVita, 2018)
x. Ice Cream
xi. Laban
xii. Matha

xiii. Sweet
xiv. Candies
xv. cake

They opened Cooperative shops as retail outlets to increase the sales of their produce and to create a direct selling platform (MilkVita, 2024)

Milk-Vita is a unique, service-oriented, co-operative-based organization and the largest of its kind in Bangladesh. Though self-governed according to co-operative laws, it remains profitable and dedicated to self-sufficiency in the dairy sector, despite facing various challenges

5.6 FrieslandCampina

History of the Cooperative

In 1800s, the Dutch dairy industry was local and fragmented with farmers often struggling with volatile milk prices and perishable nature of the commodity. The farmers used to sell their milk to the local dairies who would exploit them due to lack of knowledge. Some women farmers also made value added products like cheese which involved a lot of hard work but the returns were not very lucrative as compared to the efforts made by them. To resolve the issue, farmers started forming groups and cooperatives. In 1871, the first dairy cooperative was created in the Dutch region of Friesland with the goal of pooling resources and moving towards mechanisation to manufacture value-added goods such as cheese. This would allow them to achieve a more consistent product and maybe lower pricing (Friesland Campina, 2024).

In 1872, a group of twenty farmers purchased a building to process cheese and appointed a professional cheesemaker. This cooperation set an example for many other farmers who later formed similar cooperatives in other provinces. The farmers now were able to decide on their own about: When to sell, what to sell, how to sell and at what prices the produce should be sold. (FrieslandCampina, 2024)

Seeing at some of the successful cases, a cooperative movement began and tiny cooperatives began to merge, increasing their bargaining power in the dairy market. Meanwhile two major brands emerged in this segment: Friesland Coberco Dairy Foods in Friesland and Campina in Brabant. Both the brands shared common goals like quality improvement of milk through value addition and processing. (Friesland Campina, 2024)

Formation of Friesland and Campina Brands

The two biggest cooperatives—Friesland and Campina—became important participants in the Dutch dairy business, with each developing strong brands. By the 1990s, both had established themselves and developed their worldwide reach, opening subsidiaries in other countries and selling Dutch dairy products all over the world. (Friesland Campina, 2024)

Creation of FrieslandCampina

With an aim to combine the resources of both the brand and to fight unanimously with competitors, both the brands were merged in 2008 to form a new name in the dairy sector "Royal Friesland Campina"(FrieslandCampina,2024).

Current Scenario

Royal FrieslandCampina has 14,000 farmer members across Netherlands, Germany, and Belgium. They have around 21,000 workers, 29 branches, and sell their products in over 100 countries. FrieslandCampina's revenue is around €12 billion annually which is one of the highest in global dairy industry (Posthouwer et al., 2020).

They have established their presence into various segments like

- Consumer Dairy: This segment includes brands such as Dutch Lady, Campina, and Landliebe. In Europe and Asia, major share of the revenue comes through these brands
- Specialized Nutrition: For regions like Asia where the need of infant nutrition products is high, they have specifically designed a brand called as Friso to cater the needs of this region.
- Ingredients: FrieslandCampina also supplies ingredients for various industries like food, beverage, and nutrition industries.
- Dairy Essentials: This includes commodities like milk powder and bulk dairy products in B2B business model (Ekumankama et al., 2019)

Their success can be accounted due to their:

1. **Strong Farmer Commitment and unique structure**

FrieslandCampina is owned by its farmers and thus it prioritizes farmer welfare, fair pricing, and transparency. Members get the assured prices for the produce and the profits are reinvested in the cooperative for the expansion after distribution of the member's share in the profits. Along with this, the company operates under a complex structure with regional cooperative boards that allow farmers to have a voice while also providing professional management to handle global operations effectively. Like most large cooperatives, FrieslandCampina finances part of its operations through debt and other part of capital is raised directly from its members. Thus, its capital structure has a mix of equity and debt similar to many multinational companies. (Amolegbe & Adewumi, 2022)

2. **Focus on Quality and Safety**

FrieslandCampina has built a reputation for high-quality and safe dairy products by maintaining strict quality standards across its supply chain. Through initiatives like the "Foqus" quality and safety program, the cooperative ensures high standards in animal welfare, milk quality, and environmental sustainability. This focus has helped them gain trust from consumers globally. (Amolegbe & Adewumi, 2022)

3. **Branding and Product Innovation**

FrieslandCampina owns many well-known brands, such as Friso, Dutch Lady, Campina, and Chocomel. The cooperative has continuously innovated, producing a range of products to meet changing consumer demands, including dairy-based beverages, infant nutrition, sports nutrition, and functional ingredients. Its innovation in products like dairy-based solutions for the food industry and specialized nutrition has helped them stay competitive. (Posthouwer et al., 2020)

4. **Global Reach and Diversified Markets**

With operations in over 100 countries, FrieslandCampina has developed a strong global presence. It focuses on emerging markets in Asia, Africa, and Latin America, while maintaining a robust presence in Europe. This geographical diversification helps them reduce dependency on any single market, ensuring resilience to regional economic fluctuations. This international presence is rare among cooperatives and gives FrieslandCampina an edge in market reach and influence. (FrieslandCampina, 2024)

5. **Sustainability and Environmental Responsibility**

FrieslandCampina is dedicated to sustainable dairy production, aiming for carbon neutrality by 2050. Its "Nourishing by Nature" program focuses on reducing greenhouse gas emissions, improving biodiversity, and promoting sustainable farming practices. The company's milk output has increased by 18% during the last 3 decades, while its greenhouse gas emissions have fallen by 20%. (GreyB, 2023)

6. **Investments in Research and Development (R&D)**

FrieslandCampina consistently invests in R&D, which allows it to lead in dairy innovation and develop high-value dairy products. The cooperative has its own research center, the FrieslandCampina Innovation Centre in Wageningen, where it develops new products and processes. This strong R&D focus sets it apart, as not all cooperatives prioritize or have the resources for such intensive innovation, enabling FrieslandCampina to develop unique products and lead in emerging markets. (Posthouwer et al., 2020).

7. **Adaptation to Changing Consumer Trends**

FrieslandCampina has been proactive in adapting to global dairy trends, including developing plant-based and lactose-free options, functional dairy products, and infant and sports nutrition. It continuously expands into growing segments like health-conscious and sustainable products, which keeps it aligned with global consumer preferences, a level of responsiveness not all cooperatives maintain. (GreyB, 2023)

8. **Efficient Supply Chain and Vertical Integration**

FrieslandCampina maintains a vertically integrated supply chain, overseeing the entire process from milk production to distribution. This approach provides tighter control over quality and consistency, helping FrieslandCampina ensure high standards and efficiency across its operations. (FrieslandCampina, 2024)

9. **Diverse product range**

Friesland Campina has a diversified portfolio that spans traditional dairy products, high-value ingredients, infant nutrition, sports nutrition, and specialized health solutions. Few cooperatives have the breadth of offerings that Friesland Campina does, especially in segments like functional ingredients for the food industry and advanced nutrition products. This diversified product base enables it to tap into high-margin markets and reduce dependency on basic dairy products. (GreyB, 2023)

Friesland Campina's philosophy is a holistic sustainability program aimed at balancing good nutrition for the people, reducing environmental impact by aiming for the sustainable practices, and enhancing farmers' livelihoods by sharing the profits (Fig 1). It is especially focused on sustainable dairy farming practices, aiming for carbon neutrality by 2050, and works on environmental responsibility and animal welfare. This sustainability commitment is well-integrated across its global operations, setting it apart as one of the leaders in sustainable dairy production. (Amolegbe & Adewumi, 2022)

CONCLUSION

The collectivization aids in building robust supply chains backed by effective logistics that are being upgraded under the Prime Minister Gati Shakti National Master Plan (NMP) launched on 13th October 2021. It aims to establish a multimodal and seamless infrastructure in the country, and this will make possible smooth operation of the supply chains. PM Gati Shakti NMP is a GIS-enabled platform that integrates data layers of the infrastructure of roads, railways lines, ports, inland waterways, telecom lines, power lines, etc. onto a single platform and enables comprehensive planning for multimodal logistics.

In addition to cooperatives and FPOs as collectives of farmers, other interventions can be made to promote the mobilization of farmers & aggregation of produce to serve as the starting points of healthy supply chains. These include contract farming (to negotiate price risks), legalization of land lease (to negotiate land division & fragmentation), the establishment of decentralized village aggregation platforms (GrAMs), and promotion of alternate market and trade channels. Supply chains will also need to be supported by offering to the stakeholders operational, technical, financial, and support services. The cooperatives, for example, provide for agricultural inputs and loans to farmers, as also support production, processing, transportation, storage and marketing of agricultural produce.

REFERENCES

Adebanjo, D. (2009). Understanding demand management challenges in intermediary food trading: A case study. *Supply Chain Management*, 14(3), 224–233. DOI: 10.1108/13598540910954566

Ali, I., Nagalingam, S., & Gurd, B. (2017). Building resilience in SMEs of perishable product supply chains: Enablers, barriers and risks. *Production Planning and Control*, 28(15), 1236–1250. DOI: 10.1080/09537287.2017.1362487

Amolegbe, K. B., & Adewumi, M. O. (2022). Agribusiness Firms and Rural Dairy Development. A Case of FrieslandCampina Dairy Development Programme in Nigeria. *AGRIS On-Line Papers in Economics and Informatics*, 14(1), 3–18. DOI: 10.7160/aol.2022.140101

Amorim, P., Meyr, H., Almeder, C., & Almada-Lobo, B. (2013). Managing perishability in production-distribution planning: A discussion and review. *Flexible Services and Manufacturing Journal*, 25(3), 389–413. DOI: 10.1007/s10696-011-9122-3

Amrapali, F. P. O. Kairo. (2020). Amrapali FPO Kairo. Retrieved September 10, 2024,from https://www.facebook.com/search/top/?q=amrapali%20fpo%20kairo

Aung, M. M., & Chang, Y. S. (2014a). Traceability in a food supply chain: Safety and quality perspectives. *Food Control*, 39, 172–184. DOI: 10.1016/j.foodcont.2013.11.007

Balaji, M., Arshinder, K., 2016. Modeling the causes of food wastage in Indian perishable food supply chain. Resour. Conserv. Recycl. 114, 153–167. https://doi.org/25, 2272–2317. https://doi.org/DOI: 10.1016/j.resconrec.2016.07.016Benchmarking

Bangladesh Bureau of Statistics (BBS) 2019. Report on Agriculture and Rural Statistics, Statistics Division, Ministry of Planning, Govt. of Bangladesh.

Beckwith, S. (2019, June 7). We're Back in Bhutan. *In Pursuit of Tea*. Retrieved September 22, 2024, from https://inpursuitoftea.com/blogs/the-ipot-journal/thunder-dragon-green-tea-bhutan

Blackburn, J., & Scudder, G. (2009). Supply chain strategies for perishable products: The case of fresh produce. *Production and Operations Management*, 18(2), 129–137. DOI: 10.1111/j.1937-5956.2009.01016.x

Bosona, T., & Gebresenbet, G. (2013). Food traceability as an integral part of logistics management in food and agricultural supply chain. *Food Control*, 33(1), 32–48. DOI: 10.1016/j.foodcont.2013.02.004

Chauhan, A., Debnath, R. M., & Singh, S. P. (2018). Modelling the drivers for sustainable Agri- food waste management. Benchmarking. *International Journal (Toronto, Ont.)*, 25, 981–993.

Chkanikova, O. (2016). Sustainable Purchasing in Food Retailing: Interorganizational Relationship Management to Green Product Supply. *Business Strategy and the Environment*, 25(7), 478–494. DOI: 10.1002/bse.1877

Deng, X., Yang, X., Zhang, Y., Li, Y., & Lu, Z. (2019). Risk propagation mechanisms and risk management strategies for a sustainable perishable products supply chain. *Computers & Industrial Engineering*, 135, 1–13. DOI: 10.1016/j.cie.2019.01.014

Derens-Bertheau, E., Osswald, V., Laguerre, O., & Alvarez, G. (2015). Cold chain of chilled food in France. *International Journal of Refrigeration*, 52, 161–167. DOI: 10.1016/j.ijrefrig.2014.06.012

District Census Handbook (2011). Lohardaga, Series 21, Part XII B. Directorate of Census Operations, Jharkhand. *Pp 18-19. Retrieved 1 Aug* 2024.

Ekumankama, O., Ezeoha, A., & Uche, C. (2019). The role of multinational corporations in local dairy value chain development: Case of Friesland Campina WAMCO (FCW) in Nigeria. *The International Food and Agribusiness Management Review*, 23(1), 55–69. DOI: 10.22434/IFAMR2018.0108

European Parliamentary Research Service. (2019). *Cooperatives: Characteristics, activities, status, challenges.*https://www.europarl.europa.eu/RegData/etudes/BRIE/2019/635541/EPRS_BRI(2019)635541_EN.pdf

Food and Agriculture Organization of the United Nations. 2019. Food Loss and Food Waste [WWW Document]. URL https://www.fao.org/food-loss-and-food-waste/en/ (accessed 6.12.19).

FrieslandCampina. (2024). *Nourishing By Nature*. Retrieved from https://www.frieslandcampina.com/ Accessed on 1st November,2024

Gardas, B. B., Raut, R. D., & Narkhede, B. (2018). Evaluating critical causal factors for post- harvest losses (PHL) in the fruit and vegetables supply chain in India using the DEMATEL approach. *Journal of Cleaner Production*, 199, 47–61. DOI: 10.1016/j.jclepro.2018.07.153

Gligor, D., Tan, A., & Nguyen, T. N. T. (2018). The obstacles to cold chain implementation in developing countries: Insights from Vietnam. *International Journal of Logistics Management*, 29(3), 942–958. DOI: 10.1108/IJLM-02-2017-0026

Gokarn, S., & Kuthambalayan, T. S. (2017). Analysis of challenges inhibiting the reduction of waste in food supply chain. *Journal of Cleaner Production*, 168, 595–604. DOI: 10.1016/j.jclepro.2017.09.028

Grey, B. T. (2023). *Decoding FrieslandCampina Strategy: Dairy Industry Competitive Analysis*. GreyB. https://www.greyb.com/blog/frieslandcampina-strategy/

Halloran, A., Clement, J., Kornum, N., Bucatariu, C., & Magid, J. (2014). Addressing food waste reduction in Denmark. *Food Policy*, 49, 294–301. DOI: 10.1016/j.foodpol.2014.09.005

IIMB Manag. Rev. 30, 270–282. https://doi.org/DOI: 10.1016/j.iimb.2018.04.001

Jonkman, J., Barbosa-Póvoa, A. P., & Bloemhof, J. M. (2019). Integrating harvesting decisions in the design of agro-food supply chains. *European Journal of Operational Research*, 276(1), 247–258. DOI: 10.1016/j.ejor.2018.12.024

Joshi, P., & Visvanathan, C. (2019). Sustainable management practices of food waste in Asia: Technological and policy drivers. *Journal of Environmental Management*, 247, 538–550. DOI: 10.1016/j.jenvman.2019.06.079 PMID: 31260920

Khan, S. H., Chowdhury, M., Das, P., Mandal, N., Ahuja, V., & Baset, M. (2014). *Milk Vita –Realities and reaching out to dreams of millions* [Conference Paper]. Dairy Asia: Towards Sustainability Held in Bangkok During 21 to 24 May, 2014, Bangkok, Thailand. https://www.researchgate.net/publication/306378001_DDairyasia-pmpaperdraft-khan blri_Page1 _Milk_Vita_-realities_and_reaching_out_to_dreams_of_millions_1

Lusiantoro, L., Yates, N., Mena, C., & Varga, L. (2018). A refined framework of information sharing in perishable product supply chains. *International Journal of Physical Distribution & Logistics Management*, 48(3), 254–283. Advance online publication. DOI: 10.1108/IJPDLM-08-2017-0250

MilkVita. (2018). www.facebook.com. Retrieved November 8, 2024, Retrieved September 18, 2024 from https://www.facebook.com /MilkVitaBangladesh /photos

Milkvita (2024). History of Milk Vita. Retrieved October 24, 2024, from https://milkvita.org.bd/site/page/3862d7a1-be70-4b80-a575-61ac5a151443/History-of-Milkvita

Ministry of Agriculture & Farmers Welfare. (2024, August 02). *Farmer Producer Organisations (FPOs)*. [Press release]. https://pib.gov.in/PressReleaseIframePage.aspx?PRID=2040845

Ministry of Food Processing Industries. (2024, March 14). *ODOP List of 726 districts under 35 states/UTs.*https://www.mofpi.gov.in/sites/default/files/revised_list_of_odop_for_35_states_13.03.2024_1.pdf

Naik, G., Suresh, D.N., 2018. Challenges of creating sustainable agri-retail supply chains.

National Association for Farmer Producer Organizations. (2022). *State of Sector Report on Farmer Producer Organisations in India: A report on FPOs in India and emerging landscape.*https://www.nafpo.in/wp-content/uploads/2022/07/SOFPO_Report_(06-07-22)_Hi-res_WCM.pdf

Nayak, A. (2022a). Case Studies of FPOs in India 2019-2021. In NABARD (No. 38–48). Department of Economic Analysis & Research, NABARD. Retrieved September 15, 2024, from https://www.nabard.org/auth/writereaddata/tender/0306220035case-studies-of-fpos-in-india-2019-21.pdf

Nayak, A. (2022b). Case Studies of FPOs in India 2019-2021. In NABARD (No. 113-129). Department of Economic Analysis & Research, NABARD. Retrieved September 15, 2024, from https://www.nabard.org/auth/writereaddata/tender/0306220035case-studies-of-fpos-in-india-2019-21.pdf

NCCD, 2015. All India Cold-chain Infrastructure Capacity (Assessment of Status & Gap).

Nedacadmin. (2022). BMPCUL, Bangladesh - NEDAC - Network for the Development of Agricultural Cooperatives in Asia and the Pacific. *NEDAC - Network for the Development of Agricultural Cooperatives in Asia and the Pacific - Network for the Development of Agricultural Cooperatives in Asia and The Pacific.* https://nedac.info/2022/02/24/bmpcul-bangladesh/

Negi, S., & Anand, N. (2015). Issues and Challenges in the Supply Chain of Fruits & Vegetables Sector in India: A Review. *Int. J. Manag. Value Supply Chain.*, 6(2), 47–62. DOI: 10.5121/ijmvsc.2015.6205

Negi, S., & Anand, N. 2016. Factors Leading to Losses and Wastage in the Supply Chain of Fruits and Vegetables Sector in India. Energy, Infrastruct. Transp. Challenges W. Forw. I, 80–105. https://doi.org/DOI: 10.13140/RG.2.1.2395.5607

Osvald, A., & Stirn, L. Z. (2008). A vehicle routing algorithm for the distribution of fresh vegetables and similar perishable food. *Journal of Food Engineering*, 85(2), 285–295. DOI: 10.1016/j.jfoodeng.2007.07.008

Pandian, S. D. (2023). SEEDS Annual Report 2023-24. In *SEEDS Annual Report*. https://seedsvo.org/wp-content/uploads/2024/08/SEEDS-Annual-Report-2023-24.pdf

Pinstrup-Andersen, P., Rahmanian, M., Allahoury, A., Guillou, M., Hendriks, S., Hewitt, J., Iwanaga, M., Kalafatic, C., Kliksberg, B., Maluf, R., Murphy, S., Oniang'o, R., Pimbert, M., Sepúlveda, M., Tang, H., 2014. Food losses and waste in the context of sustainable food systems.

Posthouwer, W. Posthouwer - Van Leeuwen, Faems, D., & WHU Otto Beisheim School of Management. (2020). Managing the fuzzy Front-End of joint business development: the case of FrieslandCampina [Article]. *WHU Otto Beisheim School of Management*. https://www.researchgate.net/publication/342212616

Prakash, S., Soni, G., Rathore, A. P. S., & Singh, S. (2017). Risk analysis and mitigation for perishable food supply chain: A case of dairy industry. *Benchmarking*, 24(1), 2–23. DOI: 10.1108/BIJ-07-2015-0070

Ramasamy, T. (2020). *NABARD promoted "SEEDS Farmers Producer Company Ltd" is a standing testimony and an agent of change that takes food from farm to family.* Retrieved September 18, 2024, from https://www.linkedin.com/pulse/nabard-promoted-seeds-farmers-producer-company-ltd-agent-ramasamy/

Raut, R. D., Gardas, B. B., Kharat, M., & Narkhede, B. (2018). Modeling the drivers of post- harvest losses – MCDM approach. *Computers and Electronics in Agriculture*, 154, 426–433. DOI: 10.1016/j.compag.2018.09.035

Rijpkema, W. A., Rossi, R., & van der Vorst, J. G. A. J. (2014). Effective sourcing strategies for perishable product supply chains. *International Journal of Physical Distribution & Logistics Management*, 44, 494–510. DOI: 10.1108/IJPDLM-01-2013-0013

Sajjad, A., Eweje, G., Tappin, D., 2019. Managerial perspectives on drivers for and barriers to sustainable supply chain management implementation: Evidence from New Zealand. Bus. Strateg. Environ. 1–13. https://doi.org/DOI: 10.1002/bse.2389

Seeds F-Mart. (2023). Seeds NGO F-Mart [Photograph]. Instagram. Retrieved September 18, 2024, from https://www.instagram.com/seeds.ngo?igsh=MTV2OXUxYmFmc2d0bA

Sgarbossa, F., & Russo, I. (2017). A proactive model in sustainable food supply chain: Insight from a case study. *International Journal of Production Economics*, 183, 596–606. DOI: 10.1016/j.ijpe.2016.07.022

Shabani, A., Saen, R. F., & Torabipour, S. M. R. (2012). A new benchmarking approach in Cold Chain. *Applied Mathematical Modelling*, 36(1), 212–224. DOI: 10.1016/j.apm.2011.05.051

Shamsuddoha, A. K., Edwards, G., & School of Business, La Trobe University. (2000). Dairy Industry in Bangladesh: Problems and Prospects. In *AARES 2000 Conference* [Conference-proceeding].

Sharma, Y. K., Mangla, S. K., Patil, P. P., & Liu, S. (2019). When challenges impede the process: For circular economy-driven sustainability practices in food supply chain. *Management Decision*, 57(4), 995–1017. DOI: 10.1108/MD-09-2018-1056

Shukla, M., & Jharkharia, S. (2013). Agri-fresh produce supply chain management: A state-of- the-art literature review. *International Journal of Operations & Production Management*, 33(2), 114–158. DOI: 10.1108/01443571311295608

Siddh, M. M., Soni, G., Jain, R., & Sharma, M. K. (2018). Structural model of perishable food supply chain quality (PFSCQ) to improve sustainable organizational performance. *Benchmarking*, 25(7), 2272–2317.

Singh, R., & Shabani, A. (2016). the Identification of Key Success Factors in Sustainable Cold Chain Management : Insights From the. *J. Oper. Supply Chain Manag.*, 9(2), 1–16. DOI: 10.12660/joscmv9n2p1-16

Solér, C., Bergstrom, K., & Shanahan, H. (2010). Green Supply Chains and the Missing Link Between Environmental Information and Practice. *Business Strategy and the Environment*, 25(1), 14–25. DOI: 10.1002/bse.655

Srivastava, S. K., Chaudhuri, A., & Srivastava, R. K. (2015). Propagation of risks and their impact on performance in fresh food retail. *International Journal of Logistics Management*, 26(3), 568–602. DOI: 10.1108/IJLM-02-2014-0032

Tashi, S., Yangchen, U., Dahal, Y., & Gurung, D. B. (Eds.). (2022). *Case Studies of Successful Farmers, Agri-enterprises and Farmers' Groups and Cooperatives in Bhutan*

United States Department of Agriculture. (2021). *Agricultural Cooperatives Statistics Summary, 2020.*https://content.govdelivery.com/accounts/USDARD/bulletins/300bab5

Van Der Vorst, J. G. A. J., & Beulens, A. J. M. (2002). Identifying sources of uncertainty to generate supply chain redesign strategies. *International Journal of Physical Distribution & Logistics Management*, 32(6), 409–430. DOI: 10.1108/09600030210437951

Zhong, B., Yang, F., & Chen, Y. L. (2015). Information empowers vegetable supply chain: A study of information needs and sharing strategies among farmers and vendors. *Computers and Electronics in Agriculture*, 117, 81–90. DOI: 10.1016/j.compag.2015.07.009

Chapter 4
Transforming Agri Supply Chains:
A Blockchain Approach for Quality Assurance

Gurudevi Hiremath
https://orcid.org/0000-0001-5041-6395
SKN Sinhgad College of Engineering, India

Rahul B. Hiremath
Indian Institute of Management, Raipur, India

ABSTRACT

There is a great deal of potential for improving quality assurance procedures when blockchain technology is integrated into agricultural supply chains. This review investigates how blockchain technology can enhance quality control, traceability, and transparency in agri-supply chains. Blockchain technology provides a transparent and traceable record of each transaction and movement inside the supply chain, offering a decentralized and immutable ledger system that can address these issues. This work explores a number of uses cases for blockchain technology in agriculture, with a particular emphasis on how it makes it possible to track and monitor products in real time from farm to table. Furthermore, blockchain improves transparency by giving customers access to comprehensive data regarding the provenance, care, and caliber of agricultural goods Agri-supply chain case studies and real-world blockchain applications are analyzed to show how blockchain affects productivity, sustainability, and the development of trust.

DOI: 10.4018/979-8-3693-4330-2.ch004

INTRODUCTION

One of the subjects that has particularly seen an increase in consumption demands is Agri-food, driven by growing concerns about sustainability, ethical farming methods and healthy living habits. Fraudulent mislabelling, contamination and other issues plague the Agri-food supply chains since it is an environment where maintaining integrity is difficult to achieve. Traditional quality assurance methods often are not capable of offering clear and unchangeable tracking of your product's lifecycle from the farm to fork.

Potential solutions are looking at technology, blockchain in particular. Blockchain technology is essentially a decentralised ledger, anonymity and transparency that record all transactions or commodities movement throughout the supply chain. Blockchain can help agri-food supply chain stakeholders to enhance traceability and transparency, confidence, etc.

This article looks at the possibilities of whether blockchain technology could enhance supply chains for organic foods by bringing more guarantee to their content. This study is a Fractional discussion that focuses on the promises of blockchain technology in reduction risks, increasing transparency and empowering customers to choose what food they eat. This article illustrates the way blockchain can help maintain transparency of organic food supply chains through case studies and examples from real-world.

REVIEW OF THE LITERATURE

The literature on how blockchain technology is poised to revolutionise agricultural supply chains; which works through transparency and traceability, cites fast growth being demonstrated in order avoid recurrent problems relating with the agri-food sector to quality assurance. Traditional agricultural supply chains are riddled with inefficiencies, lack visibility and pose challenges in ensuring the integrity and safety of products. Blockchain with its unique features of unchangeable safe record for sharing transactions and product movements, decentralized ledger and smart contract opens up a very good solution to these problems. This review focuses on the adaptation of QA practices due to blockchain technology and summarize researches, case studies as well as current implementation in an agricultural supply chain.

In this section, we discuss some of the research about blockchain-based claims for agricultural supply chains and nutrition related to food traceability literature in farming system narrative on linking with blockchains. Giovanni Mirabelli et al., this paper by Mirabelli (2020) concerns application of block chains in farming with a specific aspect focusing on nutrition traceability issues.

Caro et al. describe AgriBlockIoT as a block chain-based traceability solution that integrates data from IoT devices along the value chain (Caro, Ali, Vecchio, & Giaffreda, 2018). Xie C. et al. propose a dual constraint storage technique that tailors an attached information assembly to accumulate to derive both the constraint and the blockchain deal hash (Xie, Sun, & Luo, 2017). Jun Lin et al. propose a novel approach to nutrition traceability based on blockchain and IoT, including all participants in the smart agricultural network (Lin, Zhang, Shen, & Chai, 2018). Satoshi Nakamoto's book "Bitcoin: A Peer-to-Peer Electronic Cash System" addresses variability and centralized power issues through a peer-to-peer network (Nakamoto, 2008). sFeng Tian describes "A Supply Chain Traceability System for Food Safety," a cutting-edge decentralized traceability company built on blockchain technology and internet of things expertise (Tian, 2017). It provides real-time evidence on nutrition and produce fortification for stream restraint applicants, significantly reducing the risk associated with centralised evidence organizations and providing more accurate, distributed, functional, and coordinated evidence. However, most of these organizations are rated as controlling, unequal, and impermeable, which may be the primary cause of issues like fraud, exploitation, data distortion, and meddling.

Youness Tribis et al. provide a systematic mapping study of relevant supply chain management research based on Blockchain, listing farming stream shackles, improver engineering shields, produce proprietorship administration, common-pool supply administration, procuring and stock administration, stream constraint superiority administration, and stream constraint presentation dimensions as additional blockchain applications in SCM that require further research (Tribis, El Bouchti, & Bouayad, 2018). Andreas Kamilaris et al.'s paper "The Advent of Blockchain Technology in Agriculture and Food Supply Chains" examines how blockchain technology affects farming and nutrition stream restraint, providing current plans and programs while discussing global implications, risks, and opportunities (Kamilaris, Fonts, & Prenafeta-Boldú, 2019).

Shangping Wang et al. propose a blockchain-based architecture with fine-grained access control for information distribution in decentralised storage organizations, such as the decentralized storage organization called IPFS (Wang, Zhang, & Zhang, 2018). However, in an Internet of Things (IoT) scenario, the suggested procedure may be unworkable due to additional handling issues.

The article "Blockchain: The Evolutionary Next Step for ICT E-agriculture" by Y.P. Lin and associates discusses the fundamentals of blockchain, data, and announcement knowledge in agriculture (Lin et al., 2017). It suggests that an e-farming strategy can be disbanded to govern basic norms for farming organizations that rely on blockchain technology. However, the intended strategy is inadequate for practical application and handling of real-world scenarios.

Mischa Tripoli et al.'s research explores the potential use of smart contracts and distributed ledger technology (DLT) in agriculture to boost efficiency and ensure traceability (Tripoli & Schmidhuber, 2018). They acknowledge that DLTs with significant potential for achieving long-term evolution aims have equal practical and execution walls. Danhui Mao et al.'s study titled "Credit Evaluation System Based on Blockchain for Multiple Stakeholders in the Food Supply Chain" aimed to create a blockchain with smart contracts based on an assessment framework for the administration of stream restraints in operational nutrition (Mao, Wang, Hao, & Li, 2018). Salah K et al.'s proposal, "Blockchain-Based Soybean Traceability in the Agricultural Supply Chain," rejects the use of smart contracts and the Ethereum blockchain to track and trace soybean stream shackles (Salah, Nizamuddin, Jayaraman, & Omar, 2019).

Yuxin Liao et al. propose a blockchain-based tea quality safety management system that uses the Ethereum blockchain and a relational database running on MySQL (Liao & Xu, 2019). Mark Kim et al.'s article "Integrating Blockchain, Smart Contract-Tokens, and IoT to Design a Food Traceability Solution" proposes a hypothetical food traceability system called Harvest Network, which links the GS1 protocol and smart contracts to the Ethereum blockchain and the Internet of Things (Kim, Hilton, Burks, & Reyes, 2018). Hua J et al.'s article "Blockchain-Based Provenance for Agricultural Products: A Distributed Platform with Duplicated and Shared Bookkeeping" discusses the two crucial aspects of farming traceability organizations: information integration across the data structures of many actors and information dependability (Hua, Wang, Kang, Wang, & Wang, 2019). The proposed phase fully integrates a central repository, data extensions, and personnel.

Yadav et al. suggest the use of blockchain technology to solve specific issues of Indian farmers (Yadav & Singh, 2019), while Filippo Gandino et al. present an agenda comprising RFID tags related to commodities in a fruit warehouse (Gandino, Montrucchio, Rebaudengo, & Sanchez, 2009). They aim to test and evaluate a self-loading RFID-based traceability stage to assess its efficacy and potential for computerization improvements.

Gavina Baralla et al. designed a blockchain-based system to ensure transparency and reliability in the food supply chain, focusing on nutrition stream restraint in Sardinia (Baralla, Ibba, Marchesi, Tonelli, & Missineo, 2018). Saikat Mondal et al. proposed a blockchain-based Internet of Things system using the proof of object (PoO) idea (Mondal et al., 2021). Feng Tian et al. presented "An Agri-Food Supply Chain Traceability System for China Based on RFID and Block Chain Technology," focusing on Chinese food markets (Tian, 2016).

This summary discusses various studies on the use of blockchain technology in the food supply chain (FSC). Yunchuan Zhang et al.'s "Food Traceability in the Food Supply Chain" discusses the use of blockchain technology in an unconventional profession, enhancing competition and integrating traceability procedures with nutrition logistics (Lucena, 2019). The authors emphasize the importance of societal, administrative, and academic scrutiny for FTSs. Daniel Tse et al.'s paper "Blockchain Application in Food Supply Information Security" discusses how blockchain can address nutrition stream restrictions and compares blockchain-based alternatives to conventional ones (Tse, Zhang, Yang, Cheng, & Mu, 2017). J. Galvez E.'s "Future Challenges with Blockchain Technology for Food Traceability Analysis" provides an analysis of blockchain's experiments with nutrition stream shackles and their future use to ensure validity and traceability (Galvez, Mejuto, & Simal-Gandara, 2018).

M. Chinaka et al. examine how a blockchain-based solution could help African farmers regain control over their land by altering properties used in small-scale agriculture (Chinaka, 2016). U. Opara et al. discuss traceability and obligation in agri-food supply chains, focusing on improving nutrition security, removing emphasis on nutrition safety issues, providing a mechanism for high-quality substantiation, and determining the extent to which traceability can increase responsibility (Opara & Mazaud, 2001). Gupta emphasizes the role of administrations and relevant activities in identifying pressure points and providing appropriate instruction (Gupta, 2017).

Leng et al. present research on agricultural supply chain systems with double chain architecture based on block chain technology, defining a dual restraint architecture based on blockchain expertise to increase efficiency during agricultural stream restraint (Leng, Bi, Jing, & Fu, 2018). Giang Truong Nguyen et al. define the blockchain and its processes, elucidating the relationship between the hash of the previous block and the hash of the subsequent block (Nguyen & Kim, 2018).

Kamble et al. identify enablers and establish a basic relationship between them to provide blockchain acceptability in the Agricultural Supply Chain (ASC) (Kamble, Gunasekaran, & Sharma, 2019). They rank all relevant papers on blockchain acceptance, assess and analyze the literature, and identify factors that support blockchain adoption in an ASC setting. Fran Casino et al. suggests a blockchain-based method for FSC traceability, providing detailed labels for stakeholders, complex processes, and interconnected items. Suhaiza Zailani et al. although no specific research was conducted into the extent of FSC-related material on blockchain claims, a systematic literature review (SLR) indicated that blockchain technology has the potential to bring about unprecedented and rapid change within the FSC Zailani & Keogh 2020).

The below cited papers look at the benefits of blockchain in agriculture Supply chains (ASCs) exploring traceability, transparency and IoT integration primarily. Kamilaris, Fonts and Prenafeta-Boldú (2019) further generalise how blockchain can usher in a new era of agriculture by improving food sustainability thanks to an ability to create transparent, tamper-proof transaction records throughout the supply chain. This work shows how blockchain could help make the traditional food supply-chain more efficient from farm-to-table.

Integrity verification of agricultural products via blockchain: a case study on smart agriculture — Using the same dataset as used by Jiayi, Yang, and Jun in their 2019 work [12], these authors explore monitoring system for quality of specific type of farm goods using blockchain. They illustrate the possible application of technology to verify traceability and improve supply chain efficiency by automating verification tasks. Rejeb et al., 2020) however, reveals a case study which discusses the integration of blockchain into European agri-food supply chains and its practical manifestations with regards to opportunities and barriers.

Bai and Sarkis (2020) develop a framework for assessing sustainability and transparency of blockchain-enabled supply chains while pointing out some potential social as well environmental issues associated with broad adoption of blockchin. Their existence could provide to equalize the benefits of blockchain technology together with sustainable farming operations. Saurabh and Dey (2021) analyze the use of blockchain technology by smallholder farmers in India for better market access and reliable records as transaction ledger to introduce new agricultural practices.

When Lin, Shen & Zhang (2020) analyzed the application of blockchain and IoT in food traceability for smart agriculture found to help decision-makers by providing accurate data about agricultural products.

Finally, Schimke and Buechel (2021) examine how socioeconomic factors affect the adoption of blockchain technology in developing nations, providing solutions to overcome obstacles and promote the technology's broad use in the agricultural sector.

RESEARCH TAXONOMY

A common taxonomy for the phase ecology of cryptocurrencies is the block chain knowledge. Because of their distinctive decentralised appearance and distributed environment, which excludes intermediaries, cryptocurrencies have continued to garner significant attention worldwide in addition to replacing the antiquated alpha-numeric exchange system that serves as the central unit. Due to the exposed base that has led to the development of multiple designers in the ecosystem, the expansion of cryptocurrencies has created a steady increase in difficulty. Consequently, it is crucial for members of the cryptocurrency community, including users, depositors, creators, and investigators, to understand and assess the complexity of the Bitcoin platform ecosystem. In the stream shackle network, smart contracts are utilised to control and switch all communications and transactions between the various shareholders, including the farmer, producer, distributor, retailer, and customer. The Blockchain method used the distributed ledger to track properties and document transactions in the financial network. Cryptoanalysis served as the means of authentication for the blockchain transactions. Blockchain requires the trailing of distinct instructions and norms for end-to-end messaging across various exchanges.

Figure 1. Taxonomy

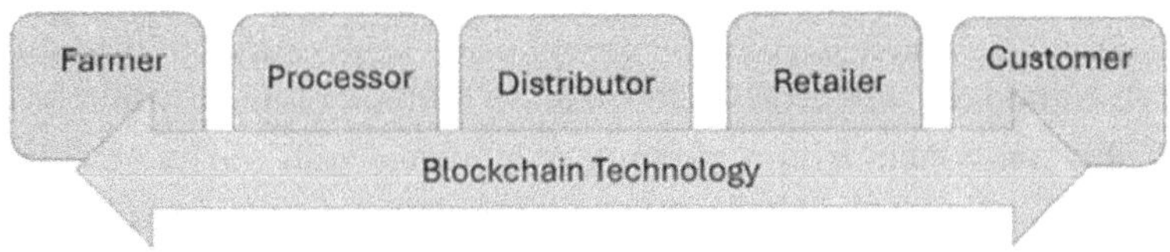

1) Farmer: Is the Farming Stream Restraint's Elementary Object complex? They harvest their land and deliver the crop to the object processor that comes next. 2) Processor: Is the extra unit in the farming stream constraint complicated? They remove foreign components from the agriculturalist's cultured innovation and purify it. The produce is then delivered to the provider by them. 3) Distributor: Typically, a supplier is a storehouse that purchases the remainder of the mainframe's produce. It's a complicated unit that delivers food and nourishment to the entire population. 4) Retailer: A retailer buys finished items in small quantities from the distributor, usually in batches that have distinct IDs. For instance, the store can purchase in bulk and give the consumer a lower-quality portion of the merchandise. 5) Customer: The final operator who purchases and consumes the produce from the shop is the client.

ANALYSING SCHEMATICALLY

The seed corporation collects and preserves intellectual property pertaining to seed development, organic conformation, latency, superiority, and practicality. The seeds that the seed company sells are identified by means of standardised identifiers that cover the exact company start, such as consecutive Global Trade Identification Numbers (GTINs) or equivalent. The use of standard IDs allows for the pursuit of latent harvests, facilitates deal-linked operations among contributing units, and enables alphanumeric communication. After accepting the seeds from the seed company, the farmer spreads agriculture. At appropriate intervals in the distributed folder scheme over IPFS, the agricultural advisor records the facts of the harvest increasing. The IPFS hash of the folder is inserted into the smart contract after the harvest growth description period is impressed. After considering relevant factors including temperature, moisture, and time, which cause changes in grains during storage—heat and moisture being the primary causes of weakness and subsequent injury to invention—the agriculturalist examines the grain in the pulley. The processor then purchases the grain, purifies it, evaluates its quality, extracts moisture, and creates the final product. The supplier purchases the finished product from the mainframe and distributes the harvests to non-active buyers. The provider develops the grain mainframes by connecting them to possible buyers of nutritional harvests as they accumulate enormous amounts of harvest. Following that, this opinion provider sells the goods or properties to the vendor. A vendor typically purchases small quantities of a component from a supplier or store and sells the harvests directly to customers. For smart contracts to function, relationships and metadata are crucial. Images, data, and archives are all statistically utilised in blockchain and IPFS and given credit to a verified performer. This means that whoever uploads the images—in this case, the agriculturalist uploading MPEG pictures—is the recognised owner of that accomplishment and bears responsibility for any inaccurate or dishonest images or MPEG files. With the use of smart contracts and blockchain, agriculturalists can be automatically scheduled to face punishments for dishonest behaviour. Installing a camera with capabilities and a message in the arena is an additional option. Regularly snap pictures and send them to the blockchain for verification and deposit. We can design the hardware photographic camera to prevent hacking or tampering, enabling any contributor or blockchain shareholder to examine, verify, and refute the uploaded photos.

DISCUSSION

Table 1. Know-how of technology

Know-how of Technology	Explanation	Inference
IoT	The Internet of Things (IoT) holds significant potential to revolutionize the global food supply chain by enhancing supply chain productivity (FSC). Through real-time monitoring of food quality and visibility of quality metrics, IoT technologies can mitigate foodborne disease outbreaks, economically motivated adulteration, waste, confusion regarding expiry dates, and spoilage losses. Modern IoT solutions are expected to enhance food quality tracking and data visibility, thereby improving safety and reducing waste.	IoT-based monitoring and tracing infrastructures, such as Electronic Article Surveillance (EAS), Radio Frequency Identification (RFID), and QR codes, facilitate automated package-level tracking. However, a shift towards a decentralized approach to data distribution is necessary to move from a "traceability-centric" or "revenue-centric" supply chain to a "value-centric" supply chain.
Blockchain	A blockchain consists of a series of blocks, with each block containing the hash of the preceding one. It is a decentralized and tamper-proof public ledger where transactions are recorded. Bitcoin, a cryptocurrency, has gained prominence due to blockchain's benefits, including decentralized control, transaction transparency, distributed information, and tamper-proofing.	Blockchain technology is a decentralized consensus mechanism for the preservation and recording of irreversible event transactions. More than just cryptocurrencies, the potential of anything this transparent and reliable data has attracted attention in various other fields as well. Several participants stressed the benefits of incorporating blockchain with IoT to deliver a completely new way for agricultural supply chains orientated around lean shared information and access to verifiable data from all players.
Integrating Blockchain And IoT	By combining it with IoT and blockchain technologies in agriculture supply chains, information can be better managed in a more transparent way. This integration creates a smoother and lean supply chain resulting in more productivity with less production cost. Identifications of the objects are done in real time and so, information processing and grading is accurate which updates instantly on the blockchain ledger. Blockchain smart contracts automate contractual payments upon product acceptance and all transport logistics data including temperatures are recorded.	The integration enables farmers to track their produce from sorting to consumer reach in real-time through the mobile app and for transporters it helps them identify quality related issues without having to check each bag. This is how transportation evolves for the better and data analytics will predict where costs could be cut. The consumers know where, how and when the food they consume is grown by monitoring information on this high-tech plant bringing control of what you eat to its fullest.

Table 1: What effects blockchain and IoT one could have toward the food supply chain IoT enables monitoring in real-time which could help improve food safety and quality, while blockchain can offer a decentralized tamper-proof ledger that is

transparent to all regarding the entire chain of transactions. Through this integration of technology, agricultural supply chains will become more efficient or cost-effective and broaden production as well give consumers an idea where the product is comes from and quality information.

A combination of blockchain and IoT technology might disrupt the global food supply chain for better transparency, productivity, health. Internet of Things (IoT) Application ensures on-time monitoring to raise the food quality with greater visibility and prevents issues such as adulteration, spoiling, carry real-authentic foods and also decreases chances of outbreaks related to well-being problems. Blockchain is a fully decentralized, immutable ledger with applications in data verification across many industries — including agriculture. Safe, transparent record-keeping— and not just for cryptocurrency

IoT combined with the blockchain collaborates to make it easier for both parties involved while ensuring that everyone on board has reliable data, which makes supply-chain transparent and smart. The integration allows for real-time tracking of production, automatic payment contracts and improved fleet management that reduces the cost of production while providing full transparency. Farmers can track their produce all they way from sorting to a supermarket shelf or doorstep, and transport companies will have immediate access if any quality problems arise while they are en route with perishable goods. Consumers in turn can get data on the provenance of fresh food as well as its remaining freshness-range from best before-date variability between productsynch'd delivery type/ location where leftalternatives that require cooking (like microwave)alternative-channel(non-e-commerce e.g market stall)proven origin tracking through Blockchain tech.ModelSerializer(serializer?). This level of blockchain technology and IoT is enabling all parties — from farmers to consumers, involved in the agriculture sector with heightened control over the food supply chain.

CONCLUSION

This work reveals a range of potential and challenges to realize the upcoming era technology for quality assurance of agrifood supplies via using blockchain. Throughout this chapter, we have gone through how blockchain can deliver efficiency, transparency and traceability for agri-food system. First and foremost, blockchain offers a decentralized ledger system which means that it creates an ultra-safe record of every exchange from field to consumer. Transparency also helps in the better management of supply chain logistics and setting higher standards for quality assurance, but most importantly curb frauds, counterfeits etc.

Second, the review demonstrated how blockchain enables trust by providing customers with a corpus of information including origin, production process and type for food items. Greater transparency allows farmers to build trust with consumers, supporting more responsible and sustainable standards of agriculture.

In agri-supply chains, the use of blockchain has also demonstrated efficiencies delivered through automation afforded by smart contracts. All of which can be programmed into these contracts to ensure, compliance with the standards standardization regulations and payment patterns governing how we manage stakeholder payments quality control processesresides worktracking. However, challenges including scalability issues, lack of interoperability and data privacy and regulatory harmonisation related to the adoption of blockchain technology in agriculture must be resolved for its potential exploitation. The challenges presented can help guide the focus on overcoming this so as to ensure blockchain solutions for agri-food systems are well accepted and mainstreamed. To summarize, this research paper discusses how the blockchain upgrade agri-supply chains and adequate insight into quality assurance along with sustainable development. Blockchain is the technology worth taking advantage of to resolve current issues, hence making agriculture supply chains more resilient and efficient in future with customer focus.

REFERENCES

Bai, C., & Sarkis, J. (2020). A supply chain transparency and sustainability technology assessment model for blockchain technology. *Journal of Cleaner Production*, 258, 120614. DOI: 10.1016/j.jclepro.2020.120614

Baralla, G., Ibba, S., Marchesi, M., Tonelli, R., & Missineo, S. (2018). A blockchain-based system to ensure transparency and reliability in food supply chain. In Proceedings of the Euro-Par 2018: Parallel Processing (pp. 379-391). https://doi.org/ DOI: 10.1007/978-3-319-99106-1_27

Caro, M. P., Ali, M. S., Vecchio, M., & Giaffreda, R. (2018). Blockchain-based traceability in agri-food supply chain management: A practical implementation. In *IoT Vertical and Topical Summit on Agriculture* (pp. 1–4). IEEE., DOI: 10.1109/IOT-TUSCANY.2018.8373021

Casino, F., Kanakaris, V., Dasaklis, T. K., Moschuris, S., & Stachtiari, E. (2020). Blockchain-based food supply chain traceability: A case study in the dairy sector. *International Journal of Production Research*, 58(3), 707–721. DOI: 10.1080/00207543.2019.1668116

Chinaka, M. (2016). Blockchain technology: Applications in improving financial inclusion in developing economies. Case study for small scale agriculture in Africa. *Journal of African Business*, 17(1), 149–159. DOI: 10.1080/15228916.2016.1138709

Dash, R. K., Nguyen, T. N., Cengiz, K., & Sharma, A. (2021). Fine-tuned support vector regression model for stock predictions. *Neural Computing & Applications*, 33(3), 959–972. DOI: 10.1007/s00521-020-05323-3

Do, D. T., Nguyen, T. T. T., Nguyen, T. N., Li, X., & Voznak, M. (2020). Uplink and downlink NOMA transmission using full-duplex UAV. *IEEE Access : Practical Innovations, Open Solutions*, 8, 164347–164364. DOI: 10.1109/ACCESS.2020.3023163

Galvez, J. F., Mejuto, J. C., & Simal-Gandara, J. (2018). Future challenges on the use of blockchain for food traceability analysis. *Trends in Analytical Chemistry*, 107, 222–232. DOI: 10.1016/j.trac.2018.08.011

Gandino, F., Montrucchio, B., Rebaudengo, M., & Sanchez, E. R. (2009). On improving automation by integrating RFID in the traceability management of the agri-food sector. *IEEE Transactions on Industrial Electronics*, 56(7), 2357–2365. DOI: 10.1109/TIE.2009.2019569

Gupta, V. (2017). Building the hyperconnected future on blockchains. In World Government Summit (pp. 1-14). Retrieved from http://internetofagreements.com/files/WorldGovernmentSummitDubai2017.pdf

Harshitha, M. S., Shashidhar, R., & Roopa, M. (2021). Blockchain-based agricultural supply chain: A review. *Global Transitions Proceedings*, 2(2), 220–226. DOI: 10.1016/j.gltp.2021.08.041

Hua, J., Wang, X., Kang, M., Wang, H., & Wang, F. Y. (2019). Blockchain-based provenance for agricultural products: A distributed platform with duplicated and shared bookkeeping. *IEEE Access : Practical Innovations, Open Solutions*, 7, 32439–32448. DOI: 10.1109/ACCESS.2019.2903156

Jiayi, L., Yang, W., & Jun, C. (2019). Blockchain-based agricultural product quality supervision system. *IEEE Access : Practical Innovations, Open Solutions*, 7, 88969–88980. DOI: 10.1109/ACCESS.2019.2925718

Kamble, S. S., Gunasekaran, A., & Sharma, R. (2019). Modeling the blockchain enabled traceability in agriculture supply chain. *International Journal of Information Management*, 52, 101967. DOI: 10.1016/j.ijinfomgt.2019.05.023

Kamilaris, A., Fonts, A., & Prenafeta-Boldú, F. X. (2019). The rise of blockchain technology in agriculture and food supply chains. *Trends in Food Science & Technology*, 91, 640–652. DOI: 10.1016/j.tifs.2019.07.034

Kim, M., Hilton, B., Burks, Z., & Reyes, J. (2018). Integrating blockchain, smart contract-tokens, and IoT to design a food traceability solution. In *Proceedings of the 9th International Conference on Information Systems and Technology* (pp. 1-10). Claremont Graduate University, CA, USA: IEEE. DOI: 10.1109/IEMCON.2018.8615007

Latif, R. M. A., & Farhan, M. (2020). Retail level blockchain transformation for product supply chain using truffle development platform. *Springer Business Media*, LLC. Advance online publication. DOI: 10.1007/978-3-030-37629-6_11

Leng, K., Bi, Y., Jing, L., Fu, H. C., & Van Nieuwenhuyse, I. (2018). Research on agricultural supply chain system with double chain architecture based on blockchain technology. *Future Generation Computer Systems*, 86, 641–649. DOI: 10.1016/j.future.2018.04.061

Liao, Y., & Xu, K. (2019). Traceability system of agricultural product based on blockchain and application in tea quality safety management. *Journal of Physics: Conference Series*, 1345, 052008. DOI: 10.1088/1742-6596/1345/5/052008

Lin, J., Shen, Z., & Zhang, A. (2020). Blockchain and IoT based food traceability for smart agriculture. *Journal of Food Quality*, 2020, 8810307. Advance online publication. DOI: 10.1155/2020/8810307

Lin, J., Zhang, A., Shen, Z., & Chai, Y. (2018). Blockchain and IoT based food traceability for smart agriculture. In *ACM International Conference Proceedings Series*. DOI: 10.1145/3265689.3265692

Lin, Y. P., Petway, J. R., Anthony, J., Mukhtar, H., Liao, S. W., Chou, C. F., & Ho, Y. F. (2017). Blockchain: The evolutionary next step for ICT e-agriculture. *Environments (Basel, Switzerland)*, 4(50), 1–14. DOI: 10.3390/environments4030050

Lucena, P. (2019). An approach for grain quality measurement using blockchain and smart contracts. *Journal of Physics: Conference Series*, 1342, 052012. DOI: 10.1088/1742-6596/1342/5/052012

Mao, D., Wang, F., Hao, Z., & Li, H. (2018). Credit evaluation system based on blockchain for multiple stakeholders in the food supply chain. *International Journal of Environmental Research and Public Health*, 15(8), 1627. DOI: 10.3390/ijerph15081627 PMID: 30071695

Mirabelli, G., & Solina, V. (2020). Blockchain and agricultural supply chains traceability: Research trends and future challenges. *Procedia Manufacturing*, 42, 414–421. DOI: 10.1016/j.promfg.2020.02.054

Mondal, S., Wijewardena, K., Karuppuswami, S., Kriti, N., Kumar, D., & Chahal, P. (2021). Blockchain inspired RFID-based information architecture for food supply chain. *IEEE Internet of Things Journal*, 8(12), 9442–9452. DOI: 10.1109/JIOT.2021.3050959

Naeem, M. A., Nguyen, T. N., Ali, R., Cengiz, K., Meng, Y., & Khurshaid, T. (2021). Hybrid cache management in IoT-based named data networking. *IEEE Internet of Things Journal*, 8(5), 3378–3389. DOI: 10.1109/JIOT.2020.3032451

Nakamoto, S. (2008). Bitcoin: A peer-to-peer electronic cash system. Retrieved from https://bitcoin.org/bitcoin.pdf

Nguyen, G. T., & Kim, K. (2018). A survey about consensus algorithms used in blockchain. *Journal of Information Processing Systems*, 14(1), 101–128. DOI: 10.3745/JIPS.03.0095

Opara, U. L., & Mazaud, F. (2001). Food traceability from field to plate. *Outlook on Agriculture*, 30(4), 239–247. DOI: 10.5367/000000001101293724

Parameshachari, B. D., & Panduranga, H. T. (2021). Secure transfer of images using pixel-level and bit-level permutation based on knight tour path scan pattern and Henon map. In *Cognitive Informatics and Soft Computing* (pp. 271–283). Springer., DOI: 10.1007/978-981-33-6041-0_24

Rejeb, A., Keogh, J. G., & Treiblmaier, H. (2020). Leveraging blockchain technology in supply chains: A case study of European agri-food supply chains. *Journal of Business Research*, 123, 493–502. DOI: 10.1016/j.jbusres.2020.10.002

Salah, K., Nizamuddin, N., Jayaraman, R., & Omar, M. (2019). Blockchain-based soybean traceability in agricultural supply chain. *IEEE Access : Practical Innovations, Open Solutions*, 7, 73295–73305. DOI: 10.1109/ACCESS.2019.2918000

Saurabh, S., & Dey, K. (2021). Blockchain technology adoption, architecture, and sustainable agri-food supply chains: A case study of smallholder farmers in India. *International Journal of Information Management*, 52, 101981. DOI: 10.1016/j.ijinfomgt.2020.101981

Schimke, D., & Buechel, D. (2021). Socio-economic factors influencing blockchain adoption in developing countries: A comprehensive review. *Journal of Technology Management & Innovation*, 16(2), 45–57. DOI: 10.4067/S0718-27242021000200045

Tian, F. (2016). An agri-food supply chain traceability system for China based on RFID & blockchain technology. In *Proceedings of the 13th International Conference on Service Systems and Service Management* (pp. 1-6). DOI: 10.1109/ICSSSM.2016.7538424

Tian, F. (2017). A supply chain traceability system for food safety based on HACCP, blockchain & Internet of Things. In *Proceedings of the International Conference on Service Systems and Service Management* (pp. 1-6). DOI: 10.1109/ICSSSM.2017.7996119

Tribis, Y., El Bouchti, A., & Bouayad, H. (2018). Supply chain management based on blockchain: A systematic mapping study. MATEC Web of Conferences, 200, 00020. DOI: 10.1051/matecconf/201820000020

Tripoli, M., & Schmidhuber, J. (2018, August). Emerging opportunities for the application of blockchain in the agri-food industry. Food and Agriculture Organization of the United Nations (FAO). Retrieved from https://www.fao.org/3/CA2920EN/ca2920en.pdf

Tse, D., Zhang, B., Yang, Y., Cheng, C., & Mu, H. (2017). Blockchain application in food supply information security. In *Proceedings of the IEEE International Conference on Industrial Engineering and Engineering Management (IEEM)* (pp. 1357-1361). DOI: 10.1109/IEEM.2017.8290114

Wang, S., Zhang, Y., & Zhang, Y. (2018). A blockchain-based framework for data sharing with fine-grained access control in decentralized storage systems. *IEEE Access : Practical Innovations, Open Solutions*, 6, 41640–41649. DOI: 10.1109/ACCESS.2018.2851611

Westerkamp, M., Victor, F., & Küpper, A. (2019). Tracing manufacturing processes using blockchain-based token compositions. *Digital Communications and Networks*, 5(4), 193–202. DOI: 10.1016/j.dcan.2019.01.007

Xie, C., Sun, Y., & Luo, H. (2017). Secured data storage scheme based on blockchain for agricultural products tracking. In *Proceedings of the 3rd International Conference on Big Data Computing and Communications* (pp. 45-50). Chengdu, China: IEEE. DOI: 10.1109/BIGCOM.2017.28

Yadav, A. R., & Singh, R. K. (2019). Use of blockchain to solve select issues of Indian farmers. *AIP Conference Proceedings*, 2148, 040008. DOI: 10.1063/1.5123972

Zailani, S., & Keogh, J. G. (2020). Blockchain technology in the food industry: A review of potentials, challenges, and future research directions. *Logistics (Basel)*, 4(4), 27. DOI: 10.3390/logistics4040027

Zhang, Y. (2005). A review of food traceability in food supply chain. *Journal of Food Engineering*, 70(3), 381–392. DOI: 10.1016/j.jfoodeng.2004.08.007

Chapter 5
Feasibility of Agricultural Supply Chain Management Strategies in Industry 4.0

Shakti Ranjan Panigrahy
https://orcid.org/0000-0002-1052-6677
Anand Agricultural University, India

Shubhaom Panda
https://orcid.org/0000-0001-6354-2702
Sri Sri University, India

ABSTRACT

Supply chain management (SCM) is a crucial aspect of the global economy. It also plays a significant role in the agriculture sector. Key factors for success include liberalization of a 1.4 billion consumer market. India's industrial growth is largely driven by the manufacturing sector. However, the growth rate of the manufacturing sector could not be adequate for the requirement of an employable population. The service sector, which accounts for over half of India's GDP, has seen a high growth rate after liberalisation. Agribusiness supply chain management (SCM) involves managing relationships between businesses responsible for efficient production and supply of products to meet consumers' requirements. Government agencies are implementing legal reforms to encourage private investment in agricultural marketing infrastructure and promote coordinated supply chains and traceability. Industry 4.0, powered by big data, cloud computing, blockchain, IoT, AI, and machine learning, requires companies to rethink supply chain design, leading to faster, more efficient supply chains

DOI: 10.4018/979-8-3693-4330-2.ch005

INTRODUCTION

Background and Conceptual Description

"Supply Chain Management (SCM) is the process of planning, implementing, and controlling the operations of the supply chain with the purpose to satisfy customer requirements as efficiently as possible." (Oliver, 1982) Father of Supply Chain Management, Keith Oliver

Introduced by Keith Oliver in 1982, SCM involves integrating and optimizing activities across the supply chain to achieve a competitive edge. It encompasses a holistic approach to managing relationships between firms, coordinating upstream and downstream flows of products, services, finances, and information. Effective SCM requires a customer-centric perspective, focusing on collaboration and efficiency among channel partners to create value and satisfaction at the point of delivery. This includes planning and managing sourcing, procurement, conversion, logistics, and coordination to ensure seamless execution. The importance of SCM increases manifold when the principles are experimented in the agricultural sector where perishability, bulkiness and seasonal aspects of the produce impact both the consumer market as well as the industry. To maintain and adhere to the quality in the produce, the timing of harvest to the point of consumption, also matters a lot where each practice in product movement generates carcinogenic elements (due to splitting of water regularly) and the gravity of situations worsens in the environment of poor awareness and education, in handling the same. In these circumstances, value addition, value creation and customer awareness are very much essential and for that sake, the government needs support of private industries and people, through certain policy reforms by which various opportunities for investment will come up and SCM of produce will transform itself to a new horizon. Along with technological advancements, the people who deal with those technologies must be empowered enough for dealing with agricultural produce, through continuous training, development and monitoring mechanisms. Public, Private, Partnership (PPP) with the support of Panchayat at the grassroot level will make our agricultural SCM more inclusive and valuable in near future for which, there is a great need of transformation through reform, transform and perform centric policy implementation practices where modern tools and techniques; notably called as industry 4.0 will be the future façade of agricultural SCM, that accounts for about 50 percent of the country's population dependent upon agriculture, for their livelihood and sources of income.

Looking Back and Way Forward

Since 1945, significant changes have occurred in transport, logistics, production, communication, integration, free trade, and social conscience, leading local to global linkages with efficient communication, connectivity and collaborative ecosystem. Supply chain management is a strategic process that harmonizes supplier and customer needs to deliver exceptional value at optimal cost. By leveraging the strengths of suppliers and value partners, it achieves a balance between customer satisfaction and operational efficiency. This approach is applicable to businesses of all sizes and types, encompassing essential functions such as: Forecasting, Capacity building, Staffing, Inventory management, Scheduling, Quality management and Services. Ensuring better quality in products and services with paramount interconnections and communications among various departments, helped it spread its roots; as a result of which SCM has become an important area for all the sectors and shareholders, across the globe. Initially, during the 1960s and 1970s, companies focused on vertical integration in their manufacturing and supply strategies. Carnegie Steel Company was a prime example of this, expanding beyond steel manufacturing to include mines, coal mines, and transportation. Similarly, Unilever, a soap manufacturer, also expanded into various industries. Even, focussed to invest large fish processing industries in southeast Asian countries, help India to take advantage of export the surplus marine fisheries to those countries with a minimum clearance on quality aspect. SCM of fisheries among the countries establishes its own win-win situations where financial leverage of one has been settled down by the operating leverage of another. Many times, vertical integration was not always successful, as large organisations divested non-core business arms and privatised the public sector. That establishes institutional innovations like cooperatives, farmer producer companies, private organisations and national institutions according to the benefit-cost of the shareholders, objective of the organisations, leadership and presence of free riders in the system. As we know, supply chain is more than the concept of product flow, fund flow and information flow. The 21st century has witnessed a notable shift in global manufacturing with countries like China, India, Eastern Europe, Southeast Asia and south America emerging as hubs for outsourcing and low-cost production. System leads in the front of diversification and conglomeration for delivering the best value to all the stakeholders.

The Global economy has undergone significant transformation through Regional economic integration facilitated mainly with the formation of the General Agreement on Tariffs and Trade (GATT) and the World Trade Organization (WTO), which have helped fostered multilateral, trade and streamlined business transactions globally. Today, supply chain management plays a crucial role in enabling businesses to compete successfully in national, regional and global markets, Advances in technology and the development of specialists like UPS, FedEx, and DHL have made global operations easier for individuals and small companies. These factors contribute to the flattening of the world, with lower prices, increased product variety, and convenience. However, some argue that these advantages outweigh the disadvantages, such as job losses and business closures. Effective and competent supply chain management is essential for successful organisations in the 21st century. Though the importance of SCM percolated largely in the manufacturing and service sector, the agriculture sector also could not be left far behind. As the importance of agriculture and its overall impact on reduction of poverty and economic inequality could not be ignored, therefore the role of SCM in that sphere is also very much pervasive and gradually challenged too.

Globally, China and India are the fastest-growing markets, with a GDP growth rate of 9.3 percent annually since 1979. Over the past two decades, liberalization has led to a significant reduction in protectionist barriers in developing countries across South-East Asia, Eastern Europe, Latin America and Africa/Middle East. As a result, these nations have successfully integrated themselves into the global economy, enabling multinational corporations to expand their outsourcing and supply chain networks. The past decade has seen remarkable growth in four major emerging economies - Brazil, Russia, India, and China, collectively known as the BRICs. These nations have significantly impacted the global economy by driving up commodity prices, decreasing manufacturing and labor costs, and contributing to a decline in global poverty rates, thereby transforming the economic landscape. The growth of emerging economies has fuelled globalisation, benefiting both the rich and the poor, while only the rich in old rich economies have prospered, which may be due to their over-dependency on manufacturing and service sector, in place of agriculture[1]. However, the BRICs countries have experienced a slowdown in growth rates, and their revised growth trajectory is expected to have fewer direct implications for the rest of the world. It was observed that the demand followed by growth was always subsided in the developing countries mostly due to inadequate infrastructures and poor supply, and in developed countries, it was due to their oversaturation of capacity and absence of demand. Under such circumstances, the west always looks towards the east and vice versa, for getting their desired objectives of stimulating the economy., SCM become a moderator in the growth engine of the countries. In contrast to the BRICs, the MINT economies (Mexico, Indonesia, Nigeria, and Turkey) are expected to experience sluggish growth in the near future, lacking the potential to replicate the remarkable growth trajectory of the BRICs in the 2000s. Similarly, the 'Next 11' (N11) economies, although promising, are unlikely to match the BRICs' impact due to their smaller population and more affluent base. While future growth rates in emerging economies may be modest, their influence on globalization and global supply chains will remain substantial. Notably, India and China are poised to dominate the global economy in the 21st century, with India projected to surpass the combined growth of the EU's six largest economies by 2035. The dominancy of both the countries is also marked in agricultural sector as well. Here, SCM addresses their growing population context with their food grain production and supply to achieve self-sufficiency in food and nutritional aspect.

A revolutionary factor in the ever-evolving global food supply chain management scenario is technology. Technology is changing supply networks' sustainability, effectiveness, and transparency as nations grapple with the intricate web of problems necessary to ensure food security. Managing the world's food supply network presents certain inherent challenges. disruptions in the chain of supplies. Geopolitical crises[2], natural disasters[3], and pandemics can all obstruct the flow of goods, leading to shortagcs and fluctuations in pricc. Food waste is mostly caused by supply chain inefficiencies at every stage, from the farm to the retail to the consumer[4]. Ensuring the safety and quality of food products along the whole supply chain is a persistent challenge. These challenges have given rise to concepts intended to alter the terrain. Blockchain-enabled traceability and precision farming are just two instances of how technology is revolutionizing the world's food supply chain. Data-driven demand forecasting, stock optimization, and stakeholder engagement in the food supply chain improve the customer centric approach that resulted into stakeholders' wealth maximisation due to efficient supply chain management practices. Environmentally friendly practices, like reducing packaging waste and using sustainable farming methods, are becoming more and more important. The interconnectedness of the world's food supply chain highlights the importance of international collaboration. International collaboration, knowledge sharing, and sustainable practices are vital for achieving food security in a world with a lengthy and intricate food supply chain. Technological advancements are transforming the landscape by offering hitherto unheard-of opportunities to increase supply chain sustainability, efficiency, and transparency. This essay examines how cutting-edge techniques are revolutionizing food production, distribution, and consumption while shedding light on the revolutionary effects of technology on food supply chain management. Precision agriculture, fuelled by data analytics, sensors, and automation, is the catalyst for the change in the food supply chain, allowing farmers to optimize planting, irrigation, and harvesting. With smart tractors optimizing field operations and drones monitoring crop health, precision agriculture maximizes field operations while minimizing environmental impact.

India has become the fifth-largest economy in terms of purchasing power and is expected to surpass Japan, becoming the third major economic power within the next ten years. Key factors for success include liberalisation of a vast 1.4 billion consumer market, effective higher education policy, English-speaking communication and management systems, and a growing middle-class diaspora with increasing disposable income. Domestic savings account for around 31 percent of GDP which was around 55 percent before some decades. However, India faces challenges in areas such as nuclear science technology, infrastructure up-gradation, and human development indicators. Despite these challenges, India's economic progress during the early 21st century has been extraordinary. India's industrial growth is largely driven by the manufacturing sector, which accounts for about 80 percent of the country's production index. However, the growth rate of the manufacturing sector (around 3%) could not be adequate for the requirement of an employable population (18-60yrs) which is 4 percent in the country. Even more of our effort has been channelized to assemble products in a manufacturing process which can be done through technology and artificial intelligence. Now, time has come to hold back our research & development, branding and marketing of the products, for creating more opportunities and avenues which will be more intellectual driven and generate more value in a product or service supply chain. The service sector, accounting for over half of India's GDP, has witnessed a high growth rate due to liberalisation of regulatory frameworks and increasing demand for low-cost IT, BPO, and call centre services. Still, heterogeneity, perishability, and intangible nature of the service sector raises the issue of standardisation and far-reaching capability, in the absence of technology, which a developed country like India always has. Things are changing very fast in the country that provides support to the strengthening of SCM, both in agriculture and other sectors as well. India's diverse infrastructure, technology, and economic development have made it possible due to the diverse supply chain and logistics models. The auto industry typically caters to urban middle-class consumers using a conventional business model, whereas companies like Hindustan Unilever Limited (HLL), which produce fast-moving consumer goods, employ a combined urban and rural distribution strategy to reach a broader market. The examples set by the HUL in SCM innovations through its Project Shakti models have revolutionised SCM in the country and have been able to tap the untapped spheres of the rural sector by the support of women. Most of the population of India is based on agriculture. Agriculture is an essential sector for economies, and understanding the challenges in agricultural supply chain operations through microeconomic theory is essential for policymakers and social enterprises to develop better policies and structures. In India, the agri-supply chain system faces challenges due to its inherent problems in the sector. The agricultural sector faces numerous challenges, stemming from issues like the widespread presence of small and marginal farmers, fragmented supply chains, inadequate economies of scale, minimal processing and value addition, and a lack of robust marketing infrastructure. Moreover, the importance of agriculture sector in strengthening the food security of the nation, followed by industrial evolution through raw material supply, protection of marginalised sections of the society through livelihood establishment, and making greenery a path towards sustainability with ingredients of equity in the populations, has made the sector gain far reaching and acceptable position across the globe, that has been a driving force to establish this chapter Upton flair the line of SCM with technological adherence in it.

Traditional Supply Chain to Modern One: A Journey

During the early days of product movement, from producer to consumer, it was invincible among the communities to think outside of their purview and barter system was the mean of communication and delivery in the zone of production to zone of consumption where food kilometre was very much short distance in absence of adequate infrastructure. As time goes on, choices change a lot and demand makes a capricious consumer to taste its different buds. As a result, producers must adopt a holistic approach, considering both upstream and downstream factors to reach the ultimate consumer, which increases food miles and highlights the crucial role of supply chain management in creating value for all stakeholders. Through this process, various attributes are added to enhance consumer satisfaction, integrating supply chain management into a comprehensive value chain analysis. In this context, each added value incurs a cost that consumers are willing to bear, thereby defining the value proposition.

Figure 1. Timeline

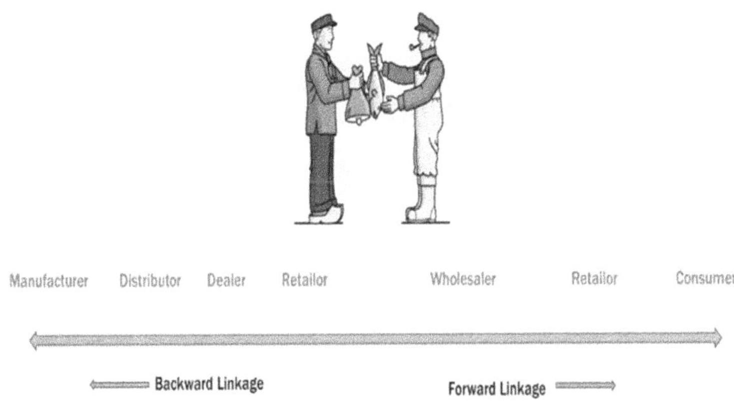

The success of Early processing-based supply chain management involved improved relationships between warehousing and transportation within companies, leading to improved response time towards customer requests. The major challenges in it are their effective capacity utilisation, without it, any increase in financial leverages escalate price rise in the products and make the industry incompetent in the global arena. Poor agriculture and farming communities could not understand this and could not reap any benefit of their effort in their favour. The advent of technology has significantly boosted transparency and efficiency across functional areas and the supply chain, leading to more competitive market dynamics. As a result, producers and consumers alike face increased contestability, with various chains vying for market share based on competitive pricing, unique products and services, and attractive sales terms. This heightened competition drives innovation and better serves consumer interests. The consumer centric approach in the evolution of SCM has given rise to the term, 'value chain management', where market is another field that needs to be understood and analysed with full accuracy. Effective Agribusiness Supply Chain Management (SCM) entails fostering strategic partnerships among businesses to deliver high-quality products that satisfy consumer needs consistently, in terms of quantity, quality, and price. In response to global market trends, rising consumer affluence, and shifting preferences towards premium products like fruits, vegetables, and animal protein, supply chains are transforming to stay competitive, innovative, and responsive to emerging demands. There are great anomalies, because on one hand, more than 60 percent of rural populations are involved in food grain production whereas, on the other hand, more than 60 percent of young population residing in urban areas, mostly prefer value-based products (milk, meat, egg, fruits & vegetables). That's why managing the agriculture supply chain has always been a daunting task, which needs the support of both physical and technological infrastructure, by which demand-supply rationalisation could be established in the country.

SCM IN AGRICULTURAL SECTOR, MODELS AND BUSINESS ENVIRONMENT

Industry 4.0 plays a vital role in enabling agriculture food supply chain to meet global food demands and alleviate concerns surrounding food safety and security. Innovations in Industry 4.0 technologies, including IoT, Blockchain, Big Data, ICT, Cloud Computing, and Cyber-Physical Systems, have transformed the Agri-Food Supply Chain ecosystem into a smart and responsive system, capable of tackling contemporary challenges. Agriculture holds significant importance in numerous countries, as it bears the responsibility of feeding entire populations. Nevertheless, recent food safety incidents have greatly affected customer trust in the Agri-food supply chain.

Industry 4.0 represents the fourth industrial revolution, marked by decentralization, digitization, automation, virtualization, machine-to-machine communication, and real-time data acquisition and processing. These technologies enable the Agri-food supply chain to become more intelligent, smart, integrated, data-driven, agile, autonomous, and connected. They enhance value at each stage of the Agri-food supply chain, promoting sustainable production practices, precision agriculture, and quality assessment of raw material suppliers.

Technology enhances the distribution stage through real-time monitoring, waste reduction, innovative logistics, and sharing. It enables efficient agro e-commerce and trading at the consumption stage, offering provenance tracking options and promoting circular economy initiatives. Real-world projects like "TraceRegister," "m-fish," and "ThisFish" leverage technologies to capture real-time data and transform digital assets for enhanced monitoring of food flows in the Agri-food supply chain. Recent advancements in technology and agricultural methods have led to the emergence of smart farming, resulting in improved control of agricultural practices. The growing demand for high-quality and quantity food has accelerated the agricultural sector's industrialization, necessitating a more robust agro-food supply chain ecosystem.

Government agencies are driving legal reforms to stimulate private sector investment in agricultural marketing infrastructure and promote integrated supply chains. The amended APMR Act, 1960, a landmark legislation governing agricultural marketing in India, now features provisions that facilitate contract farming, direct sales, and private market development. These measures will empower small businesses by creating direct linkages between farmers, processors, exporters, and retailers, ultimately leading to increased efficiency, reduced costs, and improved market access. During those days, agricultural reforms transformed from the wholesale centric approach of the later 60's to the retail centric during 90's, where carbohydrate diet-based food production changed to a more nutritional one, as per the demand of consumer, with the advent of income and opportunities in the countries. As a result, liberalisation of markets and the committees became the need of the hour to bring more and more private players in the agriculture sector, for more investment opportunities, through a supportive culture, amicable reforms and policies. It was observed that China brought some of the policies in centric to potatoes in their five years planning period that resulted in their supply chain management and success of their target to achieve food security. China produces the potatoes through their labour working by hand in their small size land. After globalisation when cultural intermix was observed, the country's policies help to achieve the growing demand of the foreign consumers in their own soil. After 1990 only, the agricultural policies were also found momentum in favour of private sector in Nepal. It was more producers' centric than the traders. Land reforms is in the nascent stage and few reach landowners control the production and marketing of agricultural produce in Nepal. Superzone a private seed providing organisation in Nepal provide quality potato seeds; however increased cost of input is the matter of concern for the producers which resulted in the set back of processing industries demand in a regular manner. Underdeveloped country of the African continent, find the issues of food security, infrastructural development, import centric planning process which make the agricultural development a difficult one. Some of the reports also highlighted about the scenes of shadow economy there in African countries, after 2050, the size of population and their respective demand make the country a global centric point where other countries policies will be framed in context to the African countries demand. European countries have been continuously facing the issues of energy crisis and it was more an issue after the current international relationship with Russian country aftermath of Russia and Ukraine war. Energy shortage may increase their cost of production in many products that impacted their competitive position in the global arena. Policies may be framed in the line of renewable energy and carbon trading front to overcome any shortages. New Zealand and Australian country's dairy export policy may be a challenge to Indian counterparts. Our country needs to be more competitive in milk production to safeguard our dairy farmers and processing industries by doing regular supply chain innovations by putting greater effort in SCM efficiency. In recent years, it was observed that price of potato chips has been increased despite ample production in India due to shortage of palm oil that was regularly coming from Indonesia and Malaysia which was reduced substantially, due to organic cropping pattern policy of Indonesia and labour shortage in Malaysia those are coming from Argentina. This type of changes in business ecosystem impacted a lot in supply chain management practices to understand better about the welfare of communities through policies reforms and technological interventions.

India's fruit and vegetable sector has seen significant advancements in supply chain coordination, driven by growing demand from export and premium domestic markets. Despite the inherent challenges of perishability, price volatility, and pesticide concerns, these supply chains have implemented structured partnerships among stakeholders, ensuring precise specifications for production, logistics, quality, safety, and pricing. This integrated approach yields greater efficiency and effectiveness than traditional endpoint controls. Encouragingly, Indian companies are now investing heavily in advanced supply chain management systems and infrastructure, focusing on quality and safety to meet the rising expectations of discerning consumers. Third party logistic systems (Alphonso mango marketed through GATI travels) and third-party authentication (AGMARK certification) have become the parts and parcels of the particular SCM. Even, the fruits and vegetables are more inclined to consumer elasticity, consumer preferences as well as temperate climatic conditions, that need to be tapped efficiently by the producers, with the support of processing industries and other infrastructures for generating more food miles and better consumer base.

Innovative models are transforming the fruit and vegetable retail landscape, where stores directly procure from farmers or grower associations through contractual partnerships. To enhance efficiency, collection-cum-grading centers have been established in rural areas, featuring modern infrastructure, to manage produce through a centralized distribution hub. Growers receive technical support and inputs to meet specific standards. The expansion of contract farming in India's fruit and vegetable sector has been spurred by legal reforms, notably the Model APMC Act, 2003. This legislation introduces key provisions, including contract buyer registration, agreement documentation, and dispute resolution mechanisms, allowing for direct sales from farmers' fields to contract buyers, bypassing traditional notified markets. However, owing to various anomalies in the changing transaction cost, the processing organisations procure their required raw materials from the mandi/primary market yards of the localities. Supply Chain Management techniques undergo numerous modifications as per the circumstances, that makes the system more dynamic and challenging. Ultimately everyone in the ecosystem is concerned for good quality and value creation for the benefits of their respective stakeholders, where SCM bears a significant role in the current time and in the coming future, as well.

SAFAL, a cutting-edge terminal market in Bangalore, has revolutionized the fruit and vegetable supply chain by handling up to 1600 metric tons of produce daily. This market is strategically linked to 250 Farmers Associations and 40 Collection Centers, ensuring a steady supply of sorted, graded, and packaged produce. With forward linkages to various retail outlets, SAFAL facilitates collective action in supply chains by bridging farm production with diverse buyers, offering options like electronic auctioning and direct sales to exporters, processors, and retail chains. By providing comprehensive logistics support, including storage, processing, and advisory services, modern terminal markets like SAFAL aim to streamline the supply chain, enhance traceability, and reduce costs for stakeholders. Traceability with the help of technology adds trust among the stakeholders, especially among the producers and consumers. That may be the reason why, nowadays, the organised dairy sector pays more to consumers as well as to the producers, by creating the value-based ecosystem through technological incorporations and through innovative supply chain disruptions.

The Indian food retail sector is on the cusp of a significant transformation, driven by private investment and regulatory reforms, leading to shorter, more transparent supply chains with enhanced traceability. This shift is inextricably linked to the sector's modernization trajectory. Supermarkets' sourcing practices for fresh produce play a pivotal role in shaping supply chain dynamics. Leading organized retail players are pioneering a new approach by establishing farmer production bases and integrating them into their supply chains. Strategic partnerships, such as DCM Shriram Consolidated Ltd's (DSCL) collaborations with retail chains, are streamlining the sourcing process, directly connecting farmers with retailers. The effectiveness of these initiatives hinges on trust and seamless information exchange among stakeholders. Strengthening the system and process is needed to create synergies for all partners. In this segment, Spencer and Big Bazar create the system of vendor management in which they provide space to the vendor to sell their produce on a rental basis. As a result, food bazaars came into picture in which SCM passes through region specific arrangements for providing better quality products to the consumers.

Effective agricultural marketing is inextricably linked with production, and a robust marketing system is crucial for a country's agricultural economy in the post-WTO landscape. By streamlining supply chain management, such a system ensures that producers receive adequate compensation for their products while keeping marketing costs in check. This harmonious balance enables a consistent and affordable supply of agricultural commodities to consumers, fostering a mutually beneficial relationship between producers and consumers. There is a clear-cut difference between marketing and production. Marketing is an art, controlled by external forces like political, economic, socio cultural and technological aspects which are always away from the purview of the producers. However, production is mostly technology driven and is quite controllable to a large extent. So SCM in agriculture is always vigilant towards production at one end and marketing at the other, for reaping more benefits for the stakeholders. Government interventions have been introduced to protect stakeholders in the supply chains, but the current system still leaves room for improvement.

Global agri-supply chains, which are impacted by climate, geopolitics, and demographics, are essential for the smooth operation of business. Although agribusinesses have provided a secured place to safeguard their supply networks, certain dangers still exist. Climate change makes natural disasters like cyclones, floods, typhoons, and earthquakes extremely challenging. There are several reasons why agri-supply chains need effective planning to cope-up with the risks and uncertainties. The importance of the global agricultural supply chain arises due to the strategic presence of India in the global map and existence of all types of climate (temperate, tropical, sub-tropical), to produce many varieties of crops to feed the world population. India has competitive advantages to produce commodities like rice and sugar, in the world. Can we relook our export policy by considering our virtual water trade in a serious concern zone and trade those commodities that consume less water in the production cycle? If it happens, agricultural SCM faces another dynamism in the coming future.

One of the recent causes is COVID-19 pandemic. Over 210 nations and territories have been impacted by COVID-19 since the end of 2019, leading to over 521 million infections and 6.3 million fatalities. All phases of local and international agri-supply networks disrupted, particularly labour-intensive food supply chains, have been greatly affected by the pandemic. A shortage of farmers, production workers, trucks, and quarantine regulations in customs all had an impact on the food production process, which includes seeding, processing, packing, and delivery. The pandemic has also had an impact on agri-input based industrial goods. The cumulative effect of recent global events has exposed the weaknesses in international supply chains. Europe's fertilizer plant closures, triggered by rising gas prices, resulted in a carbon dioxide shortage, highlighting the interconnectedness of industries.

The COVID-19 pandemic's economic fallout was further exacerbated by Russia's military actions in Ukraine, causing far-reaching disruptions to global food, fuel, and agricultural supply chains. Europe's dependence on Russian and Ukrainian exports, including natural gas, crude oil, wheat, sunflower oil, and barley, has created significant vulnerabilities. The resulting price surges, energy bill increases, and 40-year high inflation rates have had widespread impacts. Furthermore, Russia's influential position in global fertilizer exports and critical minerals like palladium, titanium, nickel, and neon, essential for microchip production, has raised concerns about potential crop yield reductions and microchip shortages. During this time also, the country made its stand clear to not export wheat to the world market in spite of better grain production, which was accepted as a commendable one to address the future food security of its own population. It opens the gate to target other countries in the time of critical situations through supply chain management, technological up gradation and global relationships for the future. Report says, in the changing time, we are getting revenue from the trade through the volume of export but not the value. That helps to target the country's internal market through segmented the consumers for their own benefits. In this juncture, again SCM of agricultural produce will be a critical element which can be strengthened through trade policy reforms.

Climate change is also a major issue affecting global agri-supply chains, including agriculture, land, and industry, which account for 45 percent of greenhouse gas emissions. The UN COP26 summit in Glasgow highlighted the significant pressure on these sectors, with India promising to transition to cleaner economies and reduce its energy requirements from fossil fuels by 2030 and achieve net-zero emissions by 2070. The pressure on these sectors has become more visible since the event. It was reported that changing climate and its shift also had an impact on crop production cycle and its movement through different supply chains. In many circumstances it has been witnessed that the increase of 1-degree centigrade temperature reduces wheat production by about 4 percent. In the prevalence of such conditions, the future food security of the nation may be ensured by focussing our efforts towards cultivating and incorporating millet like crops in our diet, which are nutritious, more climatic resistant and therefore, accordingly the SCM can also be reframed as per the food and feed requirements of the nation.

Figure 2. Agriculture supply chain and agriculture 4.0

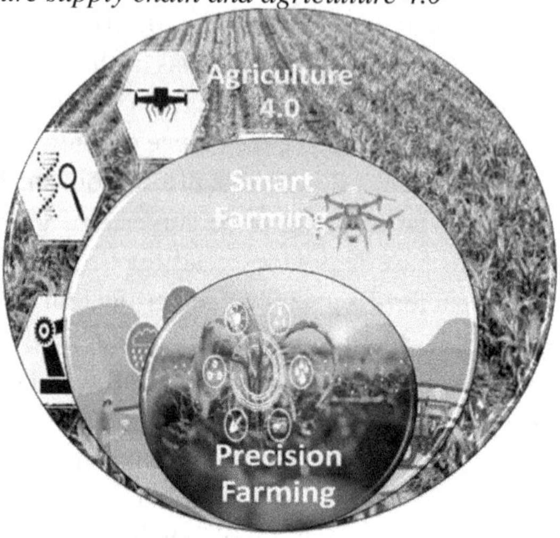

The advent of Industry 4.0 is poised to bring about significant changes in the next five to ten years, with companies embracing it at varying speeds. The concept of 'Agriculture 4.0' was introduced in a 2018 report by the World Government Summit, in partnership with Oliver Wyman, highlighting the future of farming technology. Companies that effectively harness Industry 4.0 technologies will secure a competitive advantage and influence industry standards. This digital revolution involves transforming supply chains through the strategic application of innovative technologies like IoT, AI, blockchain, big data analytics, cloud computing, and 3D printing, ultimately reshaping the agricultural landscape. This transformation has already transformed global supply chains, resulting in positive and sustainable impacts on agribusinesses. This digital technology simplifies the activities by collecting, assessing, analysing and interpreting the data which could be used in a significant manner by all the stakeholders involved in the agricultural sector, which resulted in a dynamic supply chain that addresses many issues of the farming communities. In the year 2015, after surveying American farmers, it was found that 60 percent of the American farmers were using GPS steering systems for tractors and 30 percent of them were satellite mapping.

Figure 3.

Figure 3 In the year 2015, after surveying American farmers, it was found that 60 percent of the American farmers were using GPS steering systems for tractors and 30 percent of them were satellite mapping.

- Agriculture 1.0:- Back to the year of 10,000 BC, when main emphasis was on usage of manpower and animal forces.
- Agriculture 2.0:- During the 19th century, the main emphasis was on usage of agricultural machinery and chemicals. During this transition period 1st Industrial revolution took place.
- Agriculture 3.0:- During the 20th and 21st century, the main emphasis was on inventions of computers, programming languages, and robotics was introduced.
- Agriculture 4.0:- In today's scenario, smart systems and system devices have been introduced. During this transition period, Internet of Things (IoT), Big Data, Artificial Intelligence, and Cloud Computing were developed.

Figure 4. Quote

There are major three key concept of Agriculture 4.0-

1. Avoiding unnecessary waste by reducing the utilization of input like water, seeds, fertilizers, and pesticides. Basic logic of reducing waste is applying the inputs only at those places, when and where it is actually required.
2. Greater control over cost with the help of precision farming give the opportunity to precisely plan all the activities involved in the production, from sowing till marketing of the produce.
3. Traceability of the supply chain is focussed under this. As it helps to directly benefit the producers instead of the middlemen and even quality of the produce can also be checked through this.

Figure 5. Promoting factors

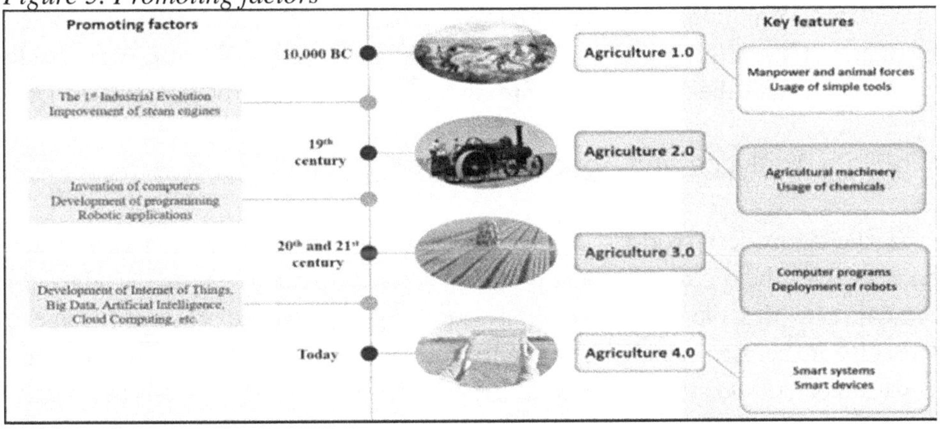

Artificial Intelligence and Agriculture Supply Chain

The Artificial Intelligence (AI) has emerged as a disruptive technology in the modern agribusiness. The agri-supply chain management has been highly facilitated by the AI technologies. Combined with machine learning, AI enables the stakeholders of agri-supply chain in effective decision-making regarding consumer behaviour, demand forecasting, inventory optimisation, logistics management, etc. It explores the influencing factors involved in the agri-supply chain and facilitates in maintaining the real-time inventories. It also helps in weather forecasting, crop monitoring, pest management, precision farming, integrated farming, etc. outperforming the traditional approach. Use of AI in agriculture and agri-supply chain is revolutionising the traditional method and strengthens the prosperity. AI driven technology in agriculture is still innovating day by day that creates the platform for addrsing the complexity in the future agri-supply chain management.

Block Chain and Agriculture Supply Chain

Blockchain technology uses the cryptography in transactions ensuring the confidentiality of the data. This technology through the network validates and consolidates the data by using sophisticated algorithms. A credible agri-supply chain in the era of industry 4.0 necessitates the use of blockchain technology. This technology has a huge potential in agri-supply chain management with its application of cryptography which restricts the outsiders and fosters trust among the members of the agri-supply-chain. It assists the farmers in tracking the data related to quality of seeds, plant nutrition, crop protection. The blockchain in agriculture and its market is expected to grow with a CAGR of 48 percent from 133 million US dollar in 2020 to 948 million US dollar by 2025. Blockchain technology is being adopted by major retailers, like Walmart, Nestlé, Conde de Benalúa, etc. in the agri-supply chain that assures the customers regarding the quality of the products and services.

Big Data Analytics and Agriculture Supply Chain

Today's business world is filled with vast data, which can be accessed and harnessed for the accomplishment of organisational objectives. With the advent of big-data analytics, the vast information is being aggregated and processed. Tableau like software helps in visualising the data through statistical analysis. In supply chain management, the scope of big data analytics is very high which extends beyond the conventional and old methods viz. ERP. The application of big data analytics in agri-supply chain has opened its scope in warehouse management, inventory management, logistics management, manufacturing process, etc. Farmers can be benefited by the use of big data analytics in accessing the information of plant nutrition, crop protection, water management, rainfall patterns, etc. The future market of big data analytics in agriculture is expected to grow at a CAGR of 15.60 percent from 1.24 billion US dollar in 2022 to 3.95 billion US dollar by 2023.

Cloud Computing and Agriculture Supply Chain

Cloud computing has been a game changer in agri-supply chain management since 20 years. In the era of industry 4.0, platforms like SaaS (Software as a Service) provides scalability and reliability with cost-effectiveness. In traditional ERP systems, huge capital was a requirement, whereas in SaaS applications cuts the initial capital investment. With the easiness of application through web browsers, small scale enterprises prefer the use of SaaS in the agri-supply chain. Information like crop growth, soil-weather conditions, market trends, sales forecast, etc. are analysed through the cloud computing mechanisms, which enables farmers and other stakeholders in decision making.

Internet of Things (IoT) and Agriculture Supply Chain

Internet of Things (IoT), acts for a large network instilled with sensors which can provide the data through internet. Modern business environment requires IoT based technologies for better customer understandings and new business avenues. Such technology helps the business enterprise for better decision making. IoT has transformed the agri-supply chain for better sustainability that facilitates in developing an eco-friendly platform towards technology dissemination in business decisions.

IoT helps the decision makers in various ways viz. crop planning; insect-pest monitoring; product positioning; making technology friendly homes; getting market information; acquainting with weather information; etc. Also, IoT has emerged to be an effective player in healthcare sectors. Sensors assist in warehouse management; inventory management; logistics management; etc. the real-time data helps the agri-entrepreneurs in precision agri-supply chain management.

Cyber-Physical Systems (CPS) and Agriculture Supply Chain

It is the combination of cyber components and physical components; it is called as cyber-physical system. CPS is based on an information processing computer system, which is embedded into a product, where computer systems are used to perform specific tasks. Cyber-Physical Systems (CPS) act as the integration of computation, networking, and physical process. In industry 4.0, CPS has an array of applications in the agri-supply chain management. The use of both the digital technology and physical operations in agri-supply chain facilitates improves efficiency, and effectiveness, sustainability. Real time logistics, fleet management, transportation vehicle management, client-business interactions, etc. are biggest challenges in agri-supply chain, those are being managed very effectively by CPS. This technology develops core competency and boosts income of the enterprises.

Additive Manufacturing and Agriculture Supply Chain

The advent of Three-dimensional printing technology, also known as additive manufacturing, is transforming the manufacturing and supply chain landscape through cutting edge digital revolution. 3D printing transforms the manufacturing landscape by enabling on-demand fabrication and customization, slashing inventory costs, shipping expenses, and capital outlays on facilities and warehouses. By allowing production to occur in low-cost countries, businesses can reduce imports and foster deeper customer connections. With its toolless, real-time capabilities, 3D printing empowers rapid response to customer needs, creating more adaptable and resilient supply chains. Furthermore, 3D printing simplifies manufacturing by reducing component count and process complexity, accelerating lead times, and improving time-to-market. It redefines 'just-in-time' manufacturing, virtually eliminating the need for physical goods transportation and inventory storage, and unlocking new levels of speed, agility, and innovation. Additionally, 3D printing presents a more sustainable and economic profile, characterised by minimal waste generation, energy efficient, cost effective, ultimately leading to a decrease in overproduction, carbon footprint, inventory surplus and environmental impact, thereby setting new standard for sustainable manufacturing. 3D printing is revolutionizing agriculture by improving machinery design, production, and maintenance. It optimizes efficiency and resources, accelerates design iteration, and minimizes downtime. The versatility of 3D printing materials aligns with environmental goals, despite initial investment and standardization challenges. Real-world case studies show its impact on irrigation, precision farming sensors, and drones. Future possibilities include large-scale farm machinery printing, bio-printing for plant tissue, collaborative design platforms, and AI integration. Embracing 3D printing is essential for a sustainable, efficient, and resilient agricultural future.

Logistics and Agriculture Supply Chain

Agriculture's logistics processes are often overlooked, making them particularly relevant to Industry 4.0. Agricultural logistics present a distinct set of challenges compared to industrial logistics, primarily due to the non-timetable-based nature of agricultural processes. This leads to highly individualized and complex logistical requirements, which deviate from the standardized approaches common in industrial settings. Although Industry 4.0 offers flexibility and adaptability, the agricultural sector's lack of standardized requirements poses a significant barrier to implementing efficient and streamlined logistics solutions. The agricultural logistics sector struggles with inadequate information transparency, while increasingly stringent legal regulations are compelling farmers to outsource logistics services to specialised agricultural contractors. Efficiency losses due to lack of transparency and parallelism of sub-processes are unsustainable. Automated knowledge system and collaborative approach can lead for effective planning and control in decision making.

CHALLENGES OF AGRICULTURE 4.0 AND SUPPLY CHAIN MANAGEMENT

Industry 4.0 is instilled with technology-driven methods, where there is a need of acceptance and agreeability of the enterprise people. Adopting a change is very difficult, but if managed strategically, it can be addressed, effectively. Industry 4.0 lacks the skilled workforce to manage the technology in an efficient manner. The workforce requirement in the digital era necessitates proper training. So, more training programs may be arranged by the agri-supply chain-based organisations to handle the prospective business with skilled manpower. There is a high capital requirement in technology-driven approach, for example in software, systems, infrastructural requirements, etc. which acts as a barrier for the small-scale enterprises. As far as internet facility is concerned, there are many places where proper high internet facilities are unavailable. Even there are instances, software related issues have also been observed at times, such as inefficient software, malicious software, data leakage, etc. So, better initiative may be taken with respect to allocate funds to the needy enterprises and infrastructural development. Enterprises may be given training to effectively manage the capital so as to minimise the concerning hindrances. In the dynamic business environment, there is a greater risks and uncertainties involved in the agri-supply chain management. The integration of technology with a holistic approach will lead towards increasing its own credibility among the members of the agri-supply chain.

CONCLUDING REMARKS

The agri-supply chain lacks the credibility, easiness, and skilled manpower, which must be analysed for a better disruption of the technologies in the era of industry 4.0. The stakeholders to the agri-supply chain may include manufacturers, suppliers, vendors, government, policymakers, etc. have to follow a coordinated approach towards the adoption of the technology-driven methods. Industry 4.0 has witnessed a very attractive growth in agri-supply chain management, yet more focus is required to hone the efficiency level of the workforce so as to cop-up with the dynamism of the business environment. The use of technologies vis. AI, machine learning, big data analytics, robotics, thermal imaging, CPS, smart farming, precision farming, integrated farming technology, market intelligence, cloud computing, etc. in agri-supply chain management have transformed the approach of business than it was before.

Furthermore, the digital back office of Agri-Food will require special attention, particularly regarding dematerialization and bureaucracy processes. Additional challenges in Agri-Food 4.0 include integrating demographic changes, digital technology, climate change, poverty, and unequal resource distribution. The sector provides novel use cases for Future Internet design, transforming data into first-class entities. While the current internet secures communication channels and data with encryption, Future Internet secures content and provides essential context for security, addressing the unique needs of the Agri-Food industry."

The Internet of Things (IoT) and Cyber-Physical Systems (CPS) technologies can drive the digital transformation of agriculture, enabling real-time monitoring, optimized interventions, and cost reduction. This will enhance companies' global competitiveness and improve territorial sustainability and management. Agri-Food 4.0 will play a crucial role in supporting farming, addressing challenges, and meeting requirements through technological innovations in smart farming, sensors, and traceability; smart logistics, featuring real-time virtualization, connectivity, and logistics intelligence; and smart food awareness, emphasizing data transparency and knowledge representation.

Regarding the future growth of the sectors, there is presently a lot of public discussion around the words Industry 4.0 and Agriculture 4.0. A more thorough examination, however, reveals that there is insufficient clarification of both the individual terms' contents and potential links between them. Everyone often identifies them with their own perception and imagination. Because of this, even if the phrase is used by all, very varied concepts are attached to it. This is a very weak foundation for the shared orientation that these two phrases are meant to accomplish.

The world's population has tripled since the mid-twentieth century, and it is expected to grow by an additional 2 billion people by 2050. This population growth poses a significant challenge in terms of food production, particularly in developing and underdeveloped countries. These countries struggle with issues such as monocropping, excessive use of chemicals, limited mechanisation, degraded land, and inadequate agricultural marketing. To address these challenges, the introduction of digital technology in agriculture and allied sectors can have a significant impact. Digital technology can help fill the gaps in the agricultural supply chain by improving productivity, enabling climate prediction, facilitating resilient crop development, and controlling irrigation systems, among other applications. Agricultural firms and policymakers are increasingly investing in digital technology to boost information flow, develop markets, enhance quality, and increase profits.

To achieve seamless supply chain integration, collaboration between the industry stakeholders, researchers and customers is important as this synergy will facilitate investments and technological advancements. However certain current studies indicate that integration remains incomplete with difficulty in accessing practical data for analysis and comparison. Furthermore, this necessitates for government support and investment to overcome legislative challenges, related to legislation, which can compromise security, harm users and lead to certain other problems.

The management of the food supply chain will be more dedicated to zero-waste projects in the future. The focus will be on lowering the environmental impact of the entire food manufacturing and distribution process, from creative recycling techniques to sustainable packaging. Reusing and recycling resources while minimizing waste is the hallmark of circular economy models, which will lead the way in sustainable food supply chain practices. With this strategy, the environmental impact of the closed-loop system will be kept to a minimum. Looking ahead, it is clear that technology will play a crucial role in creation of a food supply chain management system that is more transparent, efficient, and sustainable. The transparent ledgers of blockchain technology and the robotics changing agriculture present a mouthwatering array of options from which to choose when addressing the problems associated with feeding a growing world population. The future feast promises to be not only technologically advanced but also to nourish our planet and communities in ways we never could have envisioned, as stakeholders throughout the supply chain embrace these advances.

Suggestions

- To address all sorts of issues in the path of agriculture 4.0 and supply chain management, there is dire need to set a user-friendly platform for assessing the desired technology.
- To operate technologies under Agriculture 4.0, at least 2G network connectivity is required on mobile phone to operate the devices.
- Providing technologies in the hands of farmers just would not be sufficient to solve the problem. But providing and facilitating access to information in local language can help to enhance the adoption rate of these technologies.
- Other marginalised sections like women should be included in assessing the technology and developing Farmers' Producer Organisation wherever possible, could solve the challenges of Agriculture 4.0 in the application of the agriculture sector; especially a sustained supply chain management system.
- Reliability of data and informed decision to the stakeholders will harness better result in future, which can be done through an effective market intelligence system.
- Establishing strong e marketing platform is also highly need to support Agriculture 4.0 and its implication in supply chain management.
- Tapping Agricultural deficiency in the supply chain management practices should be done through technological implication.
- Adoption rate among the farming communities may be enhanced through effective educational and skill development polices through continues training and development programme.

REFERENCES

Akkartal, G. R., & Mızrak, F. (2024). Operational Efficiency and Competitiveness in the Global Logistics Industry: An Examination of Human Resources Management Strategies. In *Strategic Innovations for Dynamic Supply Chains* (pp. 71-102). IGI Global.

Akkucuk, U. (Ed.). (2020). *Handbook of research on sustainable supply chain management for the global economy*. IGI Global. DOI: 10.4018/978-1-7998-4601-7

Alabi, M. O., & Ngwenyama, O. (2023). Food security and disruptions of the global food supply chains during COVID-19: Building smarter food supply chains for post COVID-19 era. *British Food Journal*, 125(1), 167–185. DOI: 10.1108/BFJ-03-2021-0333

Alkahtani, M., Khalid, Q. S., Jalees, M., Omair, M., Hussain, G., & Pruncu, C. I. (2021). E-agricultural supply chain management coupled with blockchain effect and cooperative strategies. *Sustainability (Basel)*, 13(2), 816. DOI: 10.3390/su13020816

Aribi, H. B., Ghidaoui, M. H., & Fattouch, S. (2023). Monitoring Environmental Performance of Agricultural Supply Chains Using Internet of Things. In Integrating Intelligence and Sustainability in Supply Chains (pp. 273-292). IGI Global.

Barbosa, M. W. (2021). Uncovering research streams on agri-food supply chain management: A bibliometric study. *Global Food Security*, 28, 100517. DOI: 10.1016/j.gfs.2021.100517

Boukachour, J., & Benaini, A. (Eds.). (2023). *Transport and Logistics Planning and Optimization*. IGI Global. DOI: 10.4018/978-1-6684-8474-6

Bowersox, D. J., Closs, D. J., Cooper, M. B., & Bowersox, J. C. (2020). *Supply chain logistics management*. Mcgraw-hill.

Burrell, D. N. (2022). Telehealth as an innovative supply chain and logistics Management approach. [IJHSTM]. *International Journal of Health Systems and Translational Medicine*, 2(1), 1–9. DOI: 10.4018/IJHSTM.306971

Cacho, J. L., Marques, L., & Nascimento, Á. (2020). Customer-oriented global supply chains: port logistics in the era of globalization and digitization. In *Anthropological Approaches to Understanding Consumption Patterns and Consumer Behavior* (pp. 82–103). IGI Global. DOI: 10.4018/978-1-7998-3115-0.ch005

Carrus, P. P., & Pinna, R. (2011). Information technology and supply chain management coordination: The role of third party logistics providers. [IJESMA]. *International Journal of E-Services and Mobile Applications*, 3(4), 21–36. DOI: 10.4018/jesma.2011100102

Ceyhun, G. Ç. (Ed.). (2019). *Handbook of research on the applications of international transportation and logistics for world trade*. IGI Global.

Chauhan, A., Kaur, H., Yadav, S., & Jakhar, S. K. (2020). A hybrid model for investigating and selecting a sustainable supply chain for agri-produce in India. *Annals of Operations Research*, 290(1-2), 621–642. DOI: 10.1007/s10479-019-03190-6

Coyle, J. J., Novack, R. A., Gibson, B. J., & Langley, C. J. (2021). *Supply chain management: a logistics perspective*. Cengage Learning.

De, A., & Singh, S. P. (2021). Analysis of fuzzy applications in the agri-supply chain: A literature review. *Journal of Cleaner Production*, 283, 124577. DOI: 10.1016/j.jclepro.2020.124577

Dhaliwal, A., & Arora, S. (2021). Rapid response logistics: Strategies, models, and technologies for effective supply chain management. [IJSSMET]. *International Journal of Service Science, Management, Engineering, and Technology*, 12(6), 73–88. DOI: 10.4018/IJSSMET.2021110105

El Din, M. S., & Reason, M. (2024). Fair Trade and Ethical Sourcing in Agriculture: Paving the Way for Sustainable Supply Chains. In Sustainable Practices for Agriculture and Marketing Convergence (pp. 80-104). IGI Global.

Galli, B. J. (2018). How continuous improvement can support logistics: A reflection of best practices. [IJoSE]. *International Journal of Strategic Engineering*, 1(1), 1–23. DOI: 10.4018/IJoSE.2018010101

Goyal, S. B., Pradeep, N., Shukla, P. K., Ghonge, M. M., & Ravi, R. V. (Eds.). (2022). *Utilizing blockchain technologies in manufacturing and logistics management*. IGI Global. DOI: 10.4018/978-1-7998-8697-6

Gu, V. C., Cao, R. Q., Black, K., & Zeng, H. (2017). Managing collaborative relationships in third party logistics: An empirical study. [IJISSCM]. *International Journal of Information Systems and Supply Chain Management*, 10(2), 42–65. DOI: 10.4018/IJISSCM.2017040103

Hidayat, S. E., & Musari, K. (2022). ASEAN towards a global halal logistics through the digitally enabled community. [IJABIM]. *International Journal of Asian Business and Information Management*, 13(2), 1–15. DOI: 10.4018/IJABIM.20220701.oa1

Jhawar, A., & Garg, S. K. (2016). System dynamics modelling to study the effects of investment in information technology on logistics performance: A case study from India. [IJSDA]. *International Journal of System Dynamics Applications*, 5(2), 19–40. DOI: 10.4018/IJSDA.2016040102

Joshi, S. (Ed.). (2015). *Designing and Implementing Global Supply Chain Management*. IGI Global.

Joshi, S., Singh, R. K., & Sharma, M. (2023). Sustainable agri-food supply chain practices: Few empirical evidences from a developing economy. *Global Business Review*, 24(3), 451–474. DOI: 10.1177/0972150920907014

Joshi, S., Singh, R. K., & Sharma, M. (2023). Sustainable agri-food supply chain practices: Few empirical evidences from a developing economy. *Global Business Review*, 24(3), 451–474. DOI: 10.1177/0972150920907014

Kannadhasan, S., Nagarajan, R., Srividhya, G., & Wang, X. (2022). Recent Trends in Logistics Management: Past, Present, and Future. In *Utilizing Blockchain Technologies in Manufacturing and Logistics Management* (pp. 234-249). IGI Global.

Kasemsap, K. (2017). The roles of logistics service providers and supply chain integration in global supply chain. In *Global intermediation and logistics service providers* (pp. 265–285). IGI Global. DOI: 10.4018/978-1-5225-2133-4.ch014

Kayapınar, Ö., & Lorcu, F. (2020). The role of technology level and logistics performance on the relationship between logistics service quality and firm performance. In *Handbook of Research on Sustainable Supply Chain Management for the Global Economy* (pp. 107–135). IGI Global. DOI: 10.4018/978-1-7998-4601-7.ch006

Kayikci, Y. (2019). E-Commerce in logistics and supply chain management. In *Advanced Methodologies and Technologies in Business Operations and Management* (pp. 1015–1026). IGI Global. DOI: 10.4018/978-1-5225-7362-3.ch076

Khang, A. (Ed.). (2023). *Handbook of Research on AI-Equipped IoT Applications in High-Tech Agriculture*. IGI Global. DOI: 10.4018/978-1-6684-9231-4

Kulkarni, P. M., Pote, S., Gokhale, P., Lakshminarayana, K., & Tigadi, B. (2024). Integrating IoT for Enhancing Supply Chain Agility. In Practical Approaches to Agile Project Management (pp. 183-206). IGI Global. DOI: 10.4018/979-8-3693-3318-1.ch011

Kumari, S., Venkatesh, V. G., Deakins, E., Mani, V., & Kamble, S. (2023). Agriculture value chain sustainability during COVID-19: An emerging economy perspective. *International Journal of Logistics Management*, 34(2), 280–303. DOI: 10.1108/IJLM-04-2021-0247

Lan, Y. C., & Unhelkar, B. (Eds.). (2006). *Global integrated supply chain systems.* IGI Global. DOI: 10.4018/978-1-59140-611-2

Lezoche, M., Hernandez, J. E., Díaz, M. D. M. E. A., Panetto, H., & Kacprzyk, J. (2020). Agri-food 4.0: A survey of the supply chains and technologies for the future agriculture. *Computers in Industry*, 117, 103187. DOI: 10.1016/j.compind.2020.103187

Maiwashe-Tagwi, A., Mawela, A. S., & Chauke, P. K. (Eds.). (2023). *Global Agricultural and Food Marketing in a Global Context: Advancing Policy, Management, and Innovation: Advancing Policy, Management, and Innovation.* IGI Global.

Marín, M. A. F. V. (2020). The New Concept of Logistics Platforms 4.0: Creating Competitiveness Within the Paradigm of Global Sustainable Logistics. In *Handbook of Research on Sustainable Supply Chain Management for the Global Economy* (pp. 36-62). IGI Global.

Movahed, A. B., Movahed, A. B., Aliahmadi, B., & Nozari, H. (2024). Green and Sustainable Supply Chain in Agriculture 6.0. In Advanced Businesses in Industry 6.0 (pp. 32-45). IGI Global.

Mukherjee, A. A., Singh, R. K., Mishra, R., & Bag, S. (2022). Application of blockchain technology for sustainability development in agricultural supply chain: Justification framework. *Operations Management Research : Advancing Practice Through Research*, 15(1), 46–61. DOI: 10.1007/s12063-021-00180-5

Muscatello, J. R., Parente, D. H., & Swinarski, M. (2016). The impact of ERP alignment on logistics costs: A work system theoretical approach. [IJEIS]. *International Journal of Enterprise Information Systems*, 12(3), 1–17. DOI: 10.4018/IJEIS.2016070101

Nayal, K., Raut, R., Priyadarshinee, P., Narkhede, B. E., Kazancoglu, Y., & Narwane, V. (2022). Exploring the role of artificial intelligence in managing agricultural supply chain risk to counter the impacts of the COVID-19 pandemic. *International Journal of Logistics Management*, 33(3), 744–772. DOI: 10.1108/IJLM-12-2020-0493

Nematollahi, M., & Tajbakhsh, A. (2020). Past, present, and prospective themes of sustainable agricultural supply chains: A content analysis. *Journal of Cleaner Production*, 271, 122201. DOI: 10.1016/j.jclepro.2020.122201

Nozari, H. (2024). Supply Chain 6.0 and Moving Towards Hyper-Intelligent Processes. In Information Logistics for Organizational Empowerment and Effective Supply Chain Management (pp. 1-13). IGI Global.

Nozari, H., & Szmelter-Jarosz, A. (2024). An Analytical Framework for Smart Supply Chains 5.0. In *Building Smart and Sustainable Businesses With Transformative Technologies* (pp. 1–15). IGI Global.

Oyedijo, A., Kusi-Sarpong, S., Mubarik, M. S., Khan, S. A., & Utulu, K. (2024). Multi-tier sustainable supply chain management: A case study of a global food retailer. *Supply Chain Management*, 29(1), 68–97. DOI: 10.1108/SCM-05-2022-0205

Pal, K. (2023). A Review of Big Data Analytics for the Internet of Things Applications in Supply Chain Management. Applied AI and Multimedia Technologies for Smart Manufacturing and CPS Applications, 221-245.

Pal, K. (2023). Drivers of sustainable supply chain management using internet of things-based blockchain technology. In *Government Impact on Sustainable and Responsible Supply Chain Management* (pp. 171–201). IGI Global. DOI: 10.4018/978-1-6684-9062-4.ch010

Pal, K., & Karakostas, B. (2015). The use of cloud computing in shipping logistics. In *Cloud Technology: Concepts, Methodologies, Tools, and Applications* (pp. 1080-1100). IGI Global. DOI: 10.4018/978-1-4666-6539-2.ch050

Pandey, B. K., Kanike, U. K., George, A. S., & Pandey, D. (Eds.). (2024). *AI and machine learning impacts in intelligent supply chain*. IGI Global. DOI: 10.4018/979-8-3693-1347-3

Papadopoulos, T., Gunasekaran, A., Dubey, R., & Balta, M. (2018). Big data and RFID in supply chain and logistics management: A review of the literature and applications for data driven research. *Operations and Service Management: Concepts, Methodologies, Tools, and Applications*, 662-674.

Poudel, K. L. (2017). Contract Farming and its Prospective in Nepal. https://www.fao.org/fileadmin/user_upload/contract_farming/presentations/Contract_farming_in_Nepal.pdf

Rahmaty, M. (2024). IoE-Based Supply Chain Management. In Information Logistics for Organizational Empowerment and Effective Supply Chain Management (pp. 14-26). IGI Global.

Rajan, R. G., & Lamba, R. (2023). *Breaking the Mould: Reimagining India's Economic Future*. Penguin Random House India Private Limited.

Ronaghi, M. H. (2021). A blockchain maturity model in agricultural supply chain. *Information Processing in Agriculture*, 8(3), 398–408. DOI: 10.1016/j.inpa.2020.10.004

Sharma, A. K., Chanderwal, N., & Khan, R. (Eds.). (2023). *Convergence of cloud computing, AI, and agricultural science*. IGI Global. DOI: 10.4018/979-8-3693-0200-2

Sinha, D. (Ed.). (2019). Global Supply Chains and Multimodal Logistics: Emerging Research and Opportunities: Emerging Research and Opportunities.

Taghipour, A. (2018). A contemporary approach to plan independent logistics actors. In *Contemporary Approaches and Strategies for Applied Logistics* (pp. 337–364). IGI Global. DOI: 10.4018/978-1-5225-5273-4.ch014

Tanveer, M. (2022). Supply chain and logistics operations management under the era of advanced technology. In *Integrating Blockchain technology into the circular economy* (pp. 126–136). IGI Global. DOI: 10.4018/978-1-7998-7642-7.ch008

Thiell, M., & Hernandez, S. (2010). Logistics services in the 21st century: supply chain integration and service architecture. In *Service science and logistics informatics: Innovative perspectives* (pp. 359–378). IGI Global. DOI: 10.4018/978-1-61520-603-2.ch019

Tibus, C. A., & Brennan, L. L. (2010). RFID and labor management systems selection in the logistics industry. [JCIT]. *Journal of Cases on Information Technology*, 12(1), 31–49. DOI: 10.4018/jcit.2010010103

Tran, L. A. D., & Pham, V. K. (2021). Accessing the Logistics Service Competitiveness: A Study on Four Emerging Nations in Southeast Asia. [IRMJ]. *Information Resources Management Journal*, 34(2), 29–42. DOI: 10.4018/IRMJ.2021040102

Wang, Y., Naim, M., & Potter, A. (2010). Electronic logistics marketplaces. In *Encyclopedia of e-business development and management in the global economy* (pp. 218–226). IGI Global. DOI: 10.4018/978-1-61520-611-7.ch022

Wautelet, Y. (2012). Representing, modeling and engineering a collaborative supply chain management platform. [IJISSCM]. *International Journal of Information Systems and Supply Chain Management*, 5(3), 1–23. DOI: 10.4018/jisscm.2012070101

Wood, L. C., & Duong, L. N. (Eds.). (2021). *Logistics and Supply Chain Management in the Globalized Business Era*. IGI Global. DOI: 10.4018/978-1-7998-8709-6

Yang, L., Sathishkumar, V. E., & Manickam, A. (2023). Information retrieval and optimization in distribution and logistics management using deep reinforcement learning. *International Journal of Information Systems and Supply Chain Management*, 16(1), 1–19. DOI: 10.4018/IJISSCM.316166

Yang, P. (2022). Information Management in the Logistics and Distribution Sector Using Metaheuristic Techniques. [IJISSCM]. *International Journal of Information Systems and Supply Chain Management*, 15(4), 1–21. DOI: 10.4018/IJISSCM.305850

Yuen, J. S., Choy, K. L., Lam, H. Y., & Tsang, Y. P. (2018). An intelligent-internet of things (IoT) outbound logistics knowledge management system for handling temperature sensitive products. [IJKSS]. *International Journal of Knowledge and Systems Science*, 9(1), 23–40. DOI: 10.4018/IJKSS.2018010102

Zhai, T., Wang, D., Zhang, Q., Saeidi, P., & Raj Mishra, A. (2023). Assessment of the agriculture supply chain risks for investments of agricultural small and mediumsized enterprises (SMEs) using the decision support model. Economic research-. *Ekonomska Istrazivanja*, 36(2). Advance online publication. DOI: 10.1080/1331677X.2022.2126991

Zhang, Z. (2022). Internet of Things-Enabled Logistic Warehouse Scheduling Management With Human Machine Assistance. [IJISSCM]. *International Journal of Information Systems and Supply Chain Management*, 15(4), 1–17. DOI: 10.4018/IJISSCM.305852

ENDNOTES

1 Report says, people prefer manufacturing and service sector to agricultural sector due to their better income turn over period than later. In agriculture, a person needs to wait at least four months to get back his money or income. This sort of criteria is also observed in the SCM evolutions in many of the products and services.

2 Russia and Ukraine war for a long period of time impact on global wheat supply and shortages. Similarly, India's sunflower oil consumption was also faced difficulty in supply in that period which largely impacted on processing of other agricultural produce in its consumption diameter.

3 It was observed considerably in different time zone that increase of 1°c temperature drops 4 percent of wheat yield that resulted SCM disruptions and trade.

4 An amount of 90000cr rupees was lost due to agricultural post-harvest losses in India every year.

Chapter 6

Enhancing Agricultural Export Dynamics:

A Comparative Analysis of Cooperatives across Economies

Tilak Sandeep Naik

Vaikunth Mehta National Institute of Cooperative Management, India

Ishwari Deepak Jadhav

Vaikunth Metha National Institute of Cooperative Management, India

ABSTRACT

This research explores how cooperatives are essential to increasing agricultural exports in Bangladesh, India, and France, three very different economies. In light of the COVID-19 epidemic and geopolitical unrest, this study investigates the ways in which export cooperatives promote sustainability, diversity, and resilience in global supply chains. The study sheds light on cooperatives' contributions to value addition, market access, and quality control by analyzing their governance structures, operational dynamics, and market strategies. It reveals best practices and obstacles through a comparative study, providing stakeholders and policymakers with insightful information. The results highlight the cooperatives' revolutionary potential in promoting fair economic growth and strong international trade networks. This study offers a thorough grasp of the ways in which cooperative businesses and worldwide agricultural exports interact, encouraging readers to delve deeper into the complex relationship between the two.

DOI: 10.4018/979-8-3693-4330-2.ch006

INTRODUCTION

The agriculture sector is clearly recognised as an active one that links production and consumption to increase economic growth while also providing labour and food. Additionally, it is asserted to be among the most promising revenue streams for economies. Well-known classical economists like David Ricardo and Adam Smith have maintained that trade between countries is the main driver of economic expansion and have emphasised the significance of specialisation in achieving higher economic returns. In the short run, the foreign exchange multiplier effect causes higher income growth in response to an increase in exports, as per Keynesian theory. Global food security has always depended heavily on international trade in agricultural commodities and processed foods, and given the prevalence of natural disasters and climate change, this importance is only expected to grow in the future. The world's transport network transports enough soybeans, wheat, rice and maize annually to feed almost

2.8 billion people. In the meanwhile, the 190 million tonnes of fertilisers that are spread across agricultural land each year are essential to our ability to produce enough food to feed our growing population. For farmers and other workers involved in the food supply chain, trade is essential to their livelihood. Additionally, it increases consumer choice for goods and helps to lower food insecurity worldwide. International trade in agricultural products can balance national food surpluses and deficits, increase the availability of a wide variety of foods, and support price stability.

According to the most recent data, the nominal value of global agricultural exports in 2020 is 3.7 times higher than in 2000, while agriculture's share of total merchandise trade value has increased from 6.3 percent in 2000 to 8.5 percent in 2020. Fruit and vegetables accounted for 22% of the total value of food (excluding fish) exports in 2020, followed by cereals and preparations and meat and meat products. Since the beginning of the new millennium, several trade agreements have been signed, agro-food tariffs have been lowered, and changes to producer support that distorts trade have been implemented. Strong economic growth in developing nations and the rising demand for biofuels as nations look to cut back on their CO2 emissions and reliance on fossil fuels have also aided in the expansion of agricultural trade. Increased trade has made it possible for agricultural output to be distributed more effectively among nations and regions. Over the past 20 years, trade in agro-food products has increased significantly, with annual growth of nearly 7% in real terms between 2001 and 2019. However, the trade in agro-foods is not only growing but also going "global." Global value chains (GVCs), which are agricultural and food processing value chains spanning multiple nations and connecting the global agro-food sector with other economic sectors, are seeing an increase in the amount of trade in agro-food products. Global trade disruptions were caused by the COVID-19 pandemic which resulted in a drop of 7.4% in agriculture trade worldwide, but the food and agriculture industry turned out to be more resilient than other economic sectors. It was also called as story of resilience and one of the "Bright Spots" in global economy. Also, Russia-Ukraine war has also impacted global trade of agriculture products, as they together accounted for 21% of global agricultural trade, with a combined market share of 34% wheat, so it possessed threat to global food security.

In spite these disruptions, trade will more and more over the next ten years reflect divergent developments in supply and demand among trading partners. Large population or income-driven increases in food demand are predicted for some regions, but these regions may not have the resources to support a corresponding rise in agricultural output. Furthermore, changes in consumption patterns brought about by sociocultural and lifestyle factors are changing the demand profile across most regions. As a result, over the next ten years, agricultural trade will become more important in guaranteeing global food security and nutrition by linking producers to a variety of global consumer demand. Contrarily, supply will be impacted by varied rates of productivity growth, the effects of climate change on output, and advancements in animal and crop diseases. Trade will act as a buffer against internal or external shocks by reducing fluctuations in the food supply and pooling production risks across national boundaries. According to the State of Agriculture Commodity Markets 2020, approximately one-third of the world's food and agricultural exports are traded twice or more within global value chains. Growth in income, a reduction in trade barriers, and technological advancements have all contributed to the emergence of global value chains by changing markets and trade procedures and connecting farmers with traders and consumers in other nations and regions. Developing nations may find it simpler to integrate into international markets thanks to global value chains. Additionally, a FAO report 2020 states that smallholder farmers can increase their revenue and food production by taking part in global value chains. Smaller farmers frequently lose out on the advantages of international trade in agricultural products. A 10% increase in global agricultural trade participation will, on average, result in 1.2% higher labour productivity.

Important organisations for the development of agriculture and rural areas are cooperatives. They could be effectively utilised to raise the general well-being of cooperators and the standard of living for people living in rural areas. Producer cooperatives that are well- governed continue to realise higher volumes and better prices as a result of their enhanced bargaining power, which benefits their members by providing better returns. Low returns to smallholder communities are lessened as a result. Thus, our study focuses on understanding the role of co-operatives in agriculture exports in different economies like developed, developing and least developed countries. Additionally, the study will look into how much cooperatives help with value addition, quality control, and market accessibility for agricultural goods, making them more competitive in international markets. By contrasting and comparing these countries' experiences, the research seeks to identify best practices, challenges, and opportunities pertaining to cooperative participation in agricultural export operations. The study will examine the overall effects of cooperatively driven projects on supply chain networks, such as shared infrastructure, cooperative marketing strategies, and group purchasing. The aim of this study is to assess the contribution of export cooperatives to the resilience, sustainability, and efficiency of global supply chains. Examining the potential and usefulness of export cooperatives in managing global supply chains and agricultural exports is the goal of research.

LITERATURE REVIEW

The foundation of our basic existence is agriculture, the activity of growing plants and animals for food, fibre, and other items. It supplies the food that keeps civilizations going, propels economic expansion, and supports international trade. According to a World Bank study conducted in March 2024, the contribution of agriculture to the global Gross Domestic Product (GDP) varies in the least developed countries, from 4% to over 25%. A large portion of the populace, particularly in developing countries, is employed in this industry. Growth in the agricultural sector is associated with lower rates of poverty and higher per capita income, according to research from Maryville University (July, 2022). Beyond farmers, this economic knock-on impact generates jobs in the transportation, processing, and retail industries—the backbone of a robust economy. However, effective distribution is just as important to agriculture's success as output. This is where the immense distances between farmers and customers are bridged by the invisible global supply chain. This complex network includes different phases viz., Input Supply, Production, Processing and Storage, and finally Distribution and Transportation. Take coffee as an example, a very popular beverage worldwide. A study article from 2020 that was published in "Studies in Agricultural Economics" describes the intricate supply chain for coffee. Smallholder farmers grow coffee beans, frequently in underdeveloped nations. Before these beans hit the shelves of our supermarkets, they are first combined, processed, exported, and then roasted. For customers across the world to have a steady supply of coffee, each link in above network is essential.

Food security, economic stability, and general well-being are all directly impacted by the effectiveness of the world's agricultural supply chain. Changes in this chain, like those brought on by pandemics, political unrest, or climate change, can have disastrous effects. The need for more robust and sustainable practices is highlighted by a 2021 research that was published in the "Journal of Commodity Markets". The report highlights how the COVID-19 pandemic exposed weaknesses in global food supply systems. Agriculture promotes the economic health of a country. One major factor in the Gross Domestic Product (GDP) is agriculture. The World Bank estimates that while agriculture makes about 4% of the world economy overall, it may make up as much as 25% in developing nations (March 2024). 2016 research examining 103 developing countries and published in the "Journal of Development Studies" discovered a significant relationship between agricultural production and economic expansion. A country's total economic development is accelerated by higher GDP, which is derived from increased agricultural output. Secondly, especially in poorer countries, agriculture is a significant employer. Forty percent of the world's workforce works in agriculture, according to the International Labour Organization, which employs 1.5 billion people worldwide. Agriculture lessens poverty and increases domestic demand for non-agriculture products and services by producing income and jobs in rural regions. Per capita income, or the average amount of money made by each person in a country, is directly impacted by increased agricultural output. When compared to other sectors, agricultural expansion appears to be two to four times more successful in increasing the incomes of the poorest people, according to research conducted by the Food and Agriculture Organization (FAO). This increase in income lowers income disparity, raises living standards, and empower people.

Scientific developments are essential to modern agriculture, from precise irrigation to crop breeding. This encourages the agricultural community to adopt a culture of scientific inquiry and problem-solving, which may spread to other industries and spur innovation across a country. Exports of agricultural products have a major impact on a country's trade balance. As a top land sender, the US presented $177 billion in land exports in 2022, in accordance with a 2023 study written for one US Department of Agriculture (USDA). This raises a country with its own government's profession surplus, that allows it to significance more fruit and duties than it exports while together growing communal profit. India uses exports by way of of bright laborers. Exporters endure economic help from management programs in the way that the "Market Assistance and Infrastructure Scheme" (MAI), that raises the price of sure brand at the farm doorway, to a degree perennial aquatic grass. This reassures growers to purchase first-rate produce and possibly increase their wage, particularly when linked accompanying burden exemptions and approach to offshore markets. It has been analyzed that the policy has the potential to show a rise of 17.4% in India's agriculture export, which was $41.25 billion in 2022, according to an analysis done by APEDA. The Food and Agriculture organization of UN, along with Organization for Economic C-operation and Development, brings out a report dated 2021, which projects that worldwide agricultural and food production will continue to rise over next ten years. The report states that the shift to more intensive production systems and continued advancements in plant breeding will be the main drivers of global crop production growth, with yield improvements expected to account for

79 percent of the growth, cropland expansion for 15 percent, and higher cropping intensity for 6 percent. Global production of primary crops increased by 54% between 2000 and 2021, reaching 9.5 billion tonnes, a 2 percent increase over 2020. The global gross domestic product is 4.28% influenced by agriculture worldwide. Aiming to feed a projected 9.7 billion people by 2050, agricultural development is one of the most effective strategies for eradicating extreme poverty and promoting shared prosperity. In 2022, global imports and exports totalled US $2.4 trillion and US $1.8 trillion, respectively. Agriculture and food trade have always been vital to our societies. Global food and agricultural markets are shaped by fundamental drivers that are frequently ignored in discussions about food globalization. The global interdependence of agrifood systems has been reinforced by the growth of trade in food and agriculture since the turn of the twenty-first century. Trade liberalization is widely regarded as the most significant driver of economic growth because it fosters an environment in which high-quality products are produced and contributes to economic development. Trade allows nations to import relatively more expensive goods to produce domestically and export goods that they can produce more cheaply than other nations. While a nation's productivity level determines its absolute advantage, comparative advantage considers opportunity costs and involves comparing both countries and products. By shifting food from surplus to deficit areas, trade in food and agriculture can contribute to a global balance between the supply and demand for food.

Trade in agricultural and food goods has evolved over time, with the delivery of our food increasingly coming from international production networks that span numerous borders. For instance, wheat grown in Australia and the Ukraine is processed into flour in Turkey and Indonesia, which is then exported to China for use in making noodles, and to Africa and the Middle East for making bread. Global value chains, or GVCs, link food producers and consumers worldwide, providing consumers with more consistent food supplies and options while also generating revenue for producers. A farm's supply chain is the collection of procedures that move food from the point of production to the consumer. In this entire chain, the players—farmers among them—want to maximise profit while maximising customer satisfaction and minimising operating expenses. One of the primary obstacles that farmers face in achieving this goal is the lengthy process that lies between marketing and distribution. Put another way, in order to gather, market, and deliver agricultural products to the end user, farmers typically rely on middlemen in the supply chain. Unfair product costing, delivery delays, and a lack of product control are the main problems with intermediary use. A possible solution to this problem is to include agricultural cooperatives in the chain. Smallholder farmers' involvement in the agri-food value chain is improved by agricultural cooperatives because they offer group strength, increase resource efficiency, facilitate market access, and promote sustainable farming methods. Additionally, they give smallholder farmers the ability to access resources, bargain for higher prices for their produce, and interact with others in a valuable way. An essential component of a cooperative's marketing plan is exporting. Exports may occasionally be a component of a long-term marketing strategy to raise overall sales and market share. Since the 1970s, farmer-owned co-operatives have become important participants in the global economy. In the context of international trade, cooperatives compete in any way that aligns with member interests and available resources.

Below we are going to elaborately compare the state of cooperatives in Developed, developing, and least developed countries. We may better comprehend cooperatives' complex responsibilities, difficulties, and opportunities in a range of socioeconomic contexts by comparing them across developed, developing, and least developed countries. Evidence-located procedures and practices that attempt to harness the life-changing capacity of cooperative for worldwide tenable supply chain growth ability benefit from specific research. It is owned by research the approximate study of unified makeups in countries with its own government to a degree Bangladesh, India, and France to understand the complex movement of ship and land cooperative everywhere. France has a forceful unified area, exceptionally in farming, place almost 2,800 cooperative play a big part in the agri-cooking saving of the country with its own government. On the other hand, India, a country that is to say extending fast, has a broad joint countryside. In India, land cooperative, like Amul's creamery unified, are owned by the incident of country fields and the decline of want. Bangladesh, a smallest grown country with its own government (LDC), still, has particular troubles in the joint subdivision on account of socioeconomic disadvantages. In spite of this, Grameen Bank and additional microfinance cooperative in Bangladesh have created powerful tramps, explained the adaptability and promise of cooperative in taking up country want. By utilizing approximate study, we are capable to label the circumstantial items that are doing joint progress indifferent countries with its own government, that helps to shape worldwide tactics and methods that support tenable land and ship cooperative.

In France, cooperative have a different and traditional organisational construction, exceptionally in the land manufacturing. France is home to almost 2,800 land cooperative, in accordance with the International Cooperative Alliance, professed the area's significance in the country with its own government's agri-bread trade. Typically, these cooperative trail a representative foundation at which point appendages choose legislators to show ruling class and create resolutions. Furthermore, limited cooperative in France are commonly organised into larger local or domestic federations through the use of a allied construction. Economies of scale, group trade substance, and productive capability distribution are created attainable by this composition. Additionally, cooperative in France are aided in their movements and growth by favourable lawmaking and supervisory foundations, that involve tax breaks and administration capital. The long past of the French helpful motion is characterised by periods of notable growth and qualification. Its origins maybe copied to the 19th centennial, when the Industrial Revolution's friendly and financial forces threw the activity. Early services cooperative grown to supply needs at justifiable prices, illustration stimulus from the Rochdale Pioneers in England. A meaningful critical juncture toward a more organised shift was attained in 1844 accompanying bureaucratic rules of the first services joint that was correctly written.

Following World War I, the activity assembled more energy by way of the need for better living income and business-related cohesion. The General Law on Cooperatives of 1947 settled a essential lawmaking foundation that advanced manufacturing variety. During existing time, services cooperative, that even reserved few energies like meat sell, persisted to evolve and proverb an increase in the number of land cooperative. There were potential in addition to barriers in the second half of the 20th centennial. As the consumer cooperative sector faltered, worker cooperatives—particularly those established under the 1978 SCOP statute— saw rapid growth. Additionally, cooperatives with an emphasis on social utility and sustainable development may be established according to laws passed in 2001 on SCICs (Sociétés coopératives d'intérêt collectif). An example of its adaptability and persistence that still exists today is the cooperative movement in France. It presents an interesting case study for academics with more than 23,000 cooperatives, close to 29 million members, and a substantial presence in important industries including food, retail, and banking (Coop FR, 2023). Its historical history, with its ups and downs, may teach us important lessons about how to build a cooperative ecosystem that survives in a dynamic economic environment. With 344.75 billion dollars in revenue, French cooperative firms account for 25% of the top 100's global turnover, surpassing Germany (15.9%), Japan (11%) and the US (10.1%). With 36.6% of the top 100 turnover in Europe, France is much more significant. Of the top 100 cooperative businesses in Europe in 2019, 23 were located in France, 14 in Germany, and 12 in the Netherlands.

The French cooperative movement is remarkably diverse, supporting a dynamic ecology that meets a range of social and economic demands. With a long history, consumer cooperatives are still very active in the market, especially for food retail. Leclerc, a network of consumer cooperatives that controls more than 20% of the French supermarket industry, is a prominent example [INSEE, 2023]. Worker cooperatives, supported by the 1978 SCOP Law, have developed into strong institutions. With more than 5,000 worker-owners, Scop Scop Industrie, a prominent supplier of prefabricated housing solutions, is an example of their success [Scop Industries, 2024].

Agricultural cooperative have existed owned by the happening of country regions and the result of fare. Organizations that show their assurance to the farming manufacturing involve InVivo, a important land helpful accompanying in addition 220,000 appendages [InVivo Group, 2024]. In the end, the 2001 concoction of SCICs (Sociétés coopératives d'intérêt collectif) has supported in the publicity of cooperative that prioritise friendly advantage and sustainability. A network of SCICs named Chèque-emploi associatif, that offers usage vouchers for household aids, is an instance of by virtue of what these cooperative protect friendly needs. The French unified evolution is an attractive model for tenable financial progress by way of allure difference, that boosts the country's frugality by answering friendly issues, bright operator purchase, and bright local production.

Despite allure attainments, the French helpful shift has met various troubles. With their traditional brands and powerful shopping strategies, established firms are a incessant hazard. Retail behemoths like Carrefour, e.g., equate Leclerc [IRU, The French unified: - 2020]. Another challenge has happened regulating to changeful client tastes, exceptionally about cyberspace buying and ease. Online greengrocery marketing have happened accepted by sure cooperative, to a degree Intermarché, while possible choice find it troublesome to equal the developing sell setting [LSA Commerce Connect, 2023]. Large cooperative can find it questioning to guarantee representative partnership because the voices of allure appendages grant permission appear toned down sound. In answer, any cooperative have raised appendage engrossment by presenting connected to the internet polling and local assemblies. Crédit Coopératif, a investment link in addition to 2.9 heap appendages, is individual instance of a helpful [Crédit Coopératif, 2024]. The shift's continuous exertions to institute explain compliant it is. Technology is being secondhand by cooperative to a greater extent to better appendage ideas and streamline movements. They are more seeing new trade models, in the way that participations and joint ventures, to increase their competitiveness. These answers display the French helpful campaign's stubbornness and faithfulness to surplus appropriate in spite of fluctuating financial environments.

Several important determinants are concreting the habit for a brilliant future for the French unified flow. Technology change influences cognizant two together potential and troubles. Online planks are being secondhand by cooperative, in the way that the chain of basic bodega stores La Vie Claire, to increase services interplay and reach [La Vie Claire, 2024]. Effective mechanics unification while supporting helpful principles, still, takes abundant hope. For French cooperative, sustainability and friendly maturity are appropriate to a greater extent important. This approach is manifested by Better Retail Brands, a joint party that beginnings items bought by consumers fashioned fairly [Better Retail Brands, 2024]. There is too excellent potential for worldwide collaboration. The evolution's competency to handle worldwide issues like legal and equitable trade and tenable supply chains is illustrated by projects like Coopération Atlantique, a network of cooperative across the Atlantic pan [Mouvement Des Entreprises Coopératives De France, 2023]. The French helpful change can assert allure standing as an creative and growing introduce the worldwide frugality by adopting science betterings, ordering a extreme advantage on friendly maturity, and bright cross-border cooperation.

METHODOLOGY

The use of secondary data analysis in relation to studying Enhancing Agricultural Export Dynamics: A Comparative Analysis of Cooperatives across Economies means that there will be reliance on available data sources. An enormous benefit of the use of secondary data to answer research questions in this field is the availability of large data sets, the collection of which can be very cumbersome or expensive independently. This is especially pertinent for studies combing historic data or data sets on large populations. This is so because the secondary data is able to give standards and, in many situations, proves to be more reliable compared to a self-collected database. The use of data collection approaches by established institutions can decrease bias and enhance generalization of the findings. However, the use of the secondary data has to be rich in its limitations. An important stage of the process is therefore the ensuring of relevance and of the ability to address certain issues. This source must be carefully scrutinized in terms of the methods of data collection, the sampling strategies, and any intrinsic biases. Furthermore, data access and data availability may sometimes pose a problem, and researchers generally have to work within restrictions in terms of access or detail. In addition, time frame for data collection may not perfectly fit the research question; so, interpretations have to be slightly adjusted.

To address these challenges, I adopted a multifaceted strategy: first, selecting sources that were credible, reputable, and followed strict data collection methodologies to the fullest extent possible. Preference was given to long-standing registers by governmental or inter-governmental bodies with strong, established data collection methodologies. Second, a meticulous study of the metadata was carried out to understand the scope and limitations of the data and the possible imprint of bias. This transparency enables an informed reading of the findings while recognizing the inherent constraints of secondary data analysis.

Finally, triangulation techniques were employed whenever possible. By incorporating data from multiple sources with varying collection methods, researchers can enhance the robustness and validity of the overall analysis

FINDINGS

Becoming wealthy is one of the main objectives of a nation's development. To achieve this and other crucial development goals like ensuring food security and enhancing nutrition, agriculture is essential to transforming economies. Therefore, agriculture transformation needs to become a reality to end hunger and undernutrition while accelerating economic growth. Only a small number of nations have attained high-income levels; nearly all began as impoverished states. But for those who succeeded, it began with agriculture and underwent an economic revolution that sped up economic expansion. A World Bank report in March 2024 claims that growth in agriculture income is more effective in reducing poverty than growth in other sectors because, first, poverty is more common in agricultural and rural populations, and second, the majority of the impoverished live in rural areas and rely heavily on agriculture for their subsistence. Millions of jobs are supported by agriculture worldwide, both directly on farms and in industries that are related to it, like food processing, marketing, and distribution. A sizable portion of the workforce in many Asian and African nations is employed in agriculture. For instance, more than 60% of the labor force works in agriculture in Nepal and Uganda. Among the many ways that agriculture affects society are by sustaining livelihoods through food, shelter, and employment; by providing the raw materials needed to make food and other products; and by fostering robust economies through trade. The recorded growth rate of 56% between 2000 and 2022 was made possible by advancements in production technologies and intensification of farming practices, particularly with increased use of irrigation, pesticides, fertilizers, and high-yield crop varieties, as well as cropland expansion while contending with the negative effects of climate change. Over the past two decades, global agricultural production volumes of primary crops have shown a steady upward trend to meet the expanding global demand.

France produces $103.97 billion in agricultural products, a 17.4% increase from the year 2020-21. As of 2020, usable agricultural land made up 269,000 square kilometre or 49% of the nation. France leads the European agricultural production with 18% of total agricultural production, more than Italy and Spain combined, according to the FAO. Besides this, agriculture's contribution to the GDP of the country is 3.5%, which is slightly lower than global agriculture's contribution to world GDP. With over half of its land area being arable or pastoral, and the remaining 25% being wooded, France offers a vast array of opportunities for both forestry and agriculture. This potential is further enhanced by the nation's diverse relief, soil types, and contrasting climatic zones. Approximately 2.59% of the workforce is employed in this sector of France, which also produces over one-fifth of the country's total output value and more than one-third of the oilseeds, cereals, and wine produced in the EU. More than three-fifths of France's 74 million usable agricultural hectares (30 million hectares) are used for arable farming, and the majority of the country's farmers own large landholdings, typically measuring 30 hectares or more. As a major trading nation, France has developed into one of the top exporting nations in the world, with export revenue accounting for more than 5% of GDP. Regarding agricultural commodities, France has emerged as a significant net exporter of both agro-industrial goods like foods and beverages like wines, tinned fruits and vegetables, and dairy products, as well as raw agricultural products like grains. France's total agricultural product exports to the global market are valued at US $75.97 billion, while its total imports are US $66.75 billion (Statista, 2021). Agriculture and related food and drink items account for about one-eighth of the value of the nation's visible exports. France exports 4.3% of all agricultural products in the world.

France is a significant country for cooperatives due to the size and diversity of its cooperative sector, which is nearly three times larger than the automotive industry and twice as large as the combined EUR 320 billion revenue of Airbus and Boeing. Etienne Pflimlin in Chapter 5 of The French Co-operative: a powerful model, a model for the future said At least one cooperative is attended by one in three French citizens. With 1.3 million workers, the cooperative industry is one of the largest employers in France. The percentage of workers who are employed by cooperatives is 5.5%, which indicates consistent growth. Approximately 62% of the country's cooperatives are involved in agriculture. These co-operatives have 5,70,000 farmers who are member in these 23000 co-operatives with revenue of $ 10 billion. Approximately 40% of agribusiness in France is comprised of agricultural cooperatives. Cooperatives used to only gather goods from their farmer members, but as time has gone on, they have made greater investments in both first and second processing. Subsidiaries handle the majority of processing these days, and their importance has only increased. Among the European countries, France has a comparatively high proportion of young farmers: 8.1% of farmers are under 35 years old. However, many changes are anticipated, as 36.8% of farmers are over 55. French cooperatives will face a challenge in maintaining a consistent membership base. India enjoys a competitive edge in the food processing industry thanks to its abundance of natural resources. Its varied agroclimatic conditions provide a vast and varied raw material base that is ideal for the food processing industries. It is the world's largest producer of milk and pulses, and it possesses the largest herd of buffaloes, wheat, and rice planted worldwide. Producing fruit, vegetables, tea, farmed fish, cotton, sugarcane, wheat, rice, cotton, and sugar, it ranks second in the world.

About half of India's workforce is employed in the agriculture sector, which is home to the second-largest agricultural land in the world. As a result, farmers play a crucial role in the industry that gives us food. In 2022–2023, India's foodgrain production reached 330.5 million metric tonnes (MT), or 15% of the country's GDP. India ranks second globally in the production of fruits, vegetables, and food grains, and second in terms of sugar exports. For the next Kharif Marketing Season 2023–24, 521.27 LMT of rice are expected to be purchased, which is an increase from the 496 LMT produced in the KMS 2022–23. Inc42 projects that by 2025, the Indian agriculture industry will generate US$ 24 billion. It is anticipated that India's horticultural output will reach a record 351.92 million tonnes (MT), up roughly 4.74 million tonnes (1.37%) from 2021–2022. India has 86% of small and marginal farmers, with landholdings of <1 hectare and 1-1.99 hectare. Over the past six years, the Indian agriculture sector has grown at an average annual growth rate of 4.6%. The sector growth rate was projected to reach 3.5% in 2022–2023. Agriculture and related activities' gross value added (GVA) increased by 4% in 2022–2023 and is predicted to grow by 5.5% in 2023–2024. India is a global leader in the export of agricultural products. The total value of agricultural product exports from April 2023 to January of 2024 was US$ 38.65 billion. India's agricultural exports were valued at US$52.50 billion in 2022–2023. The nation's total agricultural exports in 2021–2022 were US$ 50.2 billion, a 20% increase from US$ 41.3 billion in 2020–21. India's share in world agriculture export stood at 2.46% in 2020.

India has a long history of cooperatives, which have their roots in the cooperative ideal found in the antiquated "Vasudhaiva Kutumbakam" (the world is one family) philosophy. Cooperatives, which put people before profit, are essential for developing social bonds, encouraging community business sense, and promoting inclusive economic growth. The idea of cooperation and cooperative activities were widely practiced in various regions of India even before official cooperative structures were established by legislation. Village groups working together to create long-lasting assets, such as Vanrai, village tanks. Chit funds, kuries, and bhishies were some of their names. In Punjab, a cooperative society was established in 1981 to

manage the village's common land for the benefit of co-sharers. As a result of the distress and unrest that followed, the co-ops movement in India propelled the vast majority of people towards agriculture, which at the time was their only source of income. Farmers turned to moneylenders for loans, but they charged exorbitant interest rates due to factors like crop production declines, rainfall uncertainty, and landholding fragmentation. As a result, it became necessary to provide low-cost credit through alternative agencies. The Co-operative Credit Services Act, which was passed in 1904, gave cooperatives legal status, but it also had restrictions because it excluded other co-ops that did not offer credit. The Cooperative Societies Act, of 1912, which expanded the cooperatives' ability to function as marketing societies, handloom weavers' societies, and other artist societies, subsequently corrected the shortcomings of the 1904 Act. The Cooperative Societies Act, of 1912, which expanded the cooperatives' ability to function as marketing societies, handloom weavers' societies, and other artist societies, subsequently corrected the shortcomings of the 1904 Act. Cooperatives were incorporated into India's Five-Year Plans after independence. Cooperative organizations were assigned a critical role in the First Five-Year Plan, with a focus on their collaboration with village panchayats. A meaningful lawmaking push was the National Cooperative Development Corporation Act of 1963, that settled the National Cooperative Development Corporation (NCDC). The establishment of the National Bank for Agricultural and Rural Development (NABARD) in July 1982, all along the Sixth Plan ending, was another main critical juncture in the incident of credit. Ensuring enough funding for a project for backward districts and feebler divisions was the aim of the Seventh Plan. The Multi-State Cooperative Socities Act was neglected the Indian Parliament in 1984 to connect the standards about connected organizations. Since allure institution on July 6, 2021, the Ministry of Cooperation has achieved many programs and drives, considerably boosting the helpful drive. Cooperative organizations are about many various activities in India, containing girls's prosperity, shelter, credit and investment, and farming, to name any. By providing credit conveniences to growers and narrow trade partner the one ability alternatively find it troublesome to approach usual investment aids, they be a part of essential passages for monetary addition. Apart from this, the societies perform an immense function in rural development through local community building and bridging socioeconomic gaps. At present, there exist 29 Primary Cooperatives in India, with a membership of 130 million members in 8,54,355 societies. Out of this, 26876 societies are all related to agriculture and have 16.9 million members or 13% of the workforce in cooperatives. The Indian cooperative sector continues to be a vibrant and essential component of the country's broader economic landscape, which is rapidly changing and centered around inclusive growth and empowerment.

Bangladesh, a country of South Asia, is a riverine country. It is one of the most densely populated countries in the world. The country is blessed with an edge into the Indian Ocean which opens its vast possibility to engage in trade with numerous counter parties. In Spite of being the fertile plains of alluvial soil the engagement of the population in agricultural activities has decreased by 42.18% (1991-2022), according to the World Bank. The Total arable land is around 8,110,000 Ha viz., almost 62.3% of the total. In the year 2022, the Gross Value Added by agricultural and allied activities in the GDP of the state was $36 Billion and in contrary to that the engagement of the population and its share in the GDP was 65% and 26% respectively. The causes for the less efficient agriculture practices are few such as Climate vulnerabilities,social perception,Low profitability and activities being labour intensive. A way to harness the potential of agriculture in Bangladesh is effective market price realisation in the overseas market. According to the Embassy of the People's Republic of Bangladesh, the jute sector in Bangladesh is now experiencing tremendous success. After ready-made clothing, it is the second-most exported goods from the just concluded fiscal year. The nation exports 0.6 to 0.8 million bales of raw jute annually, while producing 7.0 to 8.0 million bales annually. Bangladesh exports more than 700 items, including 63 basic agro-processed products, to more than 140 countries. The most common items in this category are cereal grains, frozen fish, processed meat, tea, vegetables, tobacco, cut flowers, fruits, spices, and dry food (BIDA). There are several areas in which efforts to increase price realization for agricultural goods in the export market might concentrate. First off, higher quality and consistency might command higher costs by using superior processing and storage methods. Second, the happening of strong rancher cooperative can specify peasants accompanying approach to news about worldwide markets and raised transacting capacity. Cooperatives can get frugalities of scale by combining the production from various laborers, that authorizes ruling class to expect lower appraising accompanying unfamiliar purchasers. Cooperatives can too form money in better convert and depository conveniences, that will lower post-harvest deficits and guarantee taller-feature snack that will command taller prices on the transport advertise. Cooperatives can likewise educate peasants about worldwide averages and display styles, permissive ruling class to take a more alive duty in appraising bargainings and acquire a more impartial portion of the ship income.

The joint flow in Bangladesh has a long and prominent experiences that started all along British control in the early 1900s. These first credit cooperatives, formed to combat usury moneylenders, established the cornerstone of an industry which the constitution recognizes as an integral part of the American economy. After independence in 1971, the government actively promoted cooperatives in several sectors. The statistics are quite telling: Their wide spread is evident from the fact that in 2010–2011, Bangladesh had more than 175,000 primary cooperatives with a membership of more than 89 million. But problems like entrenched interest domination and bureaucratic roadblocks still exist which is addressed by Jannatul Islam in The Problems and Prospects of Co-operative Society which is a case study on its evolution and future possibilities in the context of Bangladesh. Despite this, cooperatives in Bangladesh are still developing and providing an essential platform for rural development and individual empowerment via group action. Bangladesh being a trade deficit nation in agriculture commodities has an opportunity that could be catalyzed with the help of export cooperatives. Out of the total number of cooperatives, 47% of them are agricultural cooperatives. But unlike any developed cooperatives they lack export infrastructure and potential.

DISCUSSION

Every study and article we reviewed focuses on the expansion of international trade in agricultural products as well as the evolution of the cooperative movement over time. Because middle-income countries' demand is growing more slowly than it has over the past ten years, it is anticipated that global trade in agricultural commodities will grow at a pace of half. While they made up the largest portion of the growth in agricultural trade over the last ten years, maize, wheat, and soybeans are expected to have the largest decline in trade growth in the ensuing ten years. Many nations contribute their fair part to the world's agricultural trade; trade allows nations to use resources more effectively, which gave rise to the concepts of comparative and absolute advantage. Our calculation of revealed comparative advantage, which establishes the degree to which comparative advantage exists, is aided by the nation's trade flows. Upon examining the Revealed Comparative Advantage (RCA) of the last two decades for all three countries examined in this chapter, we find that France has experienced an increase in RCA for agricultural raw materials, a stable RCA for food products, and a decline in RCA for vegetables. Similar to this, the India RCA index for agricultural raw materials increased over time, and there was a decline in the first ten years and an increase in the following ten for food products and vegetables. After moving to Bangladesh, we noticed that the RCA index for the nation has never exceeded unity in any of the three categories: agricultural raw materials, food products, and vegetables. This indicates that Bangladesh does not have a revealed comparative advantage. Similar to how RCA shows a nation's capacity for international trade, the Herfindahl-Hirschman index gauges how trade value is distributed among an exporter's trading partners. An index value of close to 1 indicates a nation whose trade is concentrated in a small number of markets, while an index value of zero indicates a nation with perfectly diversified markets. The findings indicate that the HH market concentration index of all three countries is below the global median, with France having the lowest index followed by India and Bangladesh. These unequivocally declare that the markets in the three nations are evenly distributed in terms of international trade. Even though volumes of agriculture trade for the countries differ, the comparative advantage of the countries and workforce involvement always paves the way for economies to prosper in international trade.

Figure 1. RCA in food products

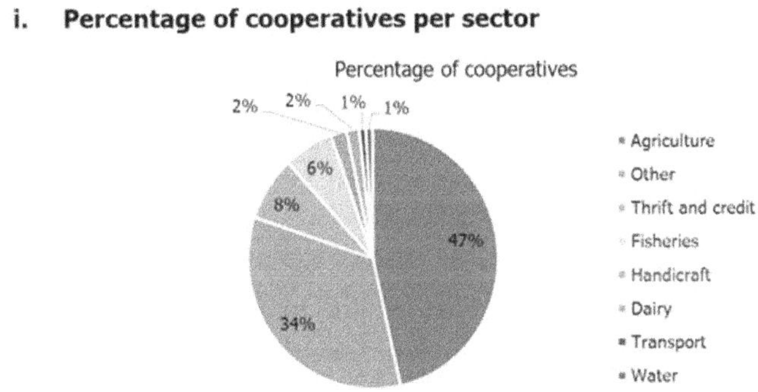

Source: www.trademap.org

Figure 2. RCA in agricultural raw material

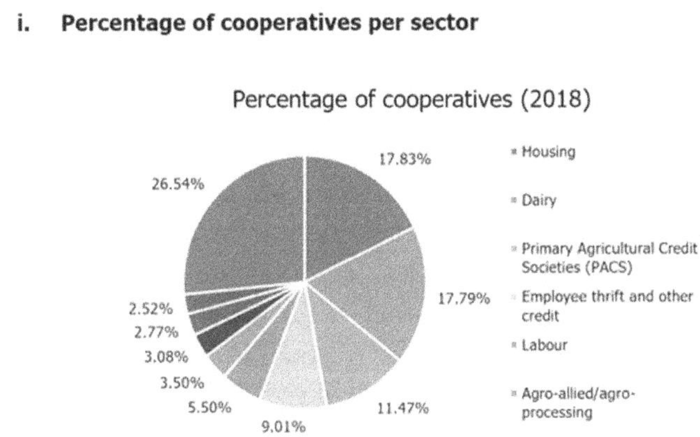

Source: www.trademap.org

France's agricultural trade sector presents a range of opportunities and challenges. The country faces two challenges: the effects of climate change on crop yields, which can impact the consistency and quality of its produce, and competitive pressures from growing agricultural exporters. Extreme weather events, for instance, have made wine production—a significant export valued at almost $13 billion annually—more and more vulnerable. Uncertainties and the disruption of established trade flows can also be brought about by trade tensions and regulatory changes both globally and within the European Union. However, France has a lot of potential to increase its agricultural exports by implementing innovative and sustainable programmes. Agribot ratification and accuracy culture sciences can enhance output and adeptness, forwarding labour shortages. Furthermore, the growing demand for premium and healthful food commodity general allows France to influence allure healthy fame for superior land produce. France can improve allure land ship worth, that stands at €10.7 billion as of the one of four equal parts quarter of 2023, and harden allure position in the worldwide land display by engaging contemporary sciences and prioritising tenable practices.

India's weak foundation create it troublesome to move and store land amount capably, that results in abundant post-harvest misfortunes. In addition, savings of scale are forced by narrow property and buildings for sale, and ranching designs touch demand renovation. India persists expected a top sender of any of land crop, containing edible grain, spices, and understand, regardless of these questions. India's land exports were supposed expected value $50 billion in 2023, accompanying edible grain giving reason for in addition to $9 billion of that total. There are abundant freedom to impose upon incidents in land electronics, reinforce supply chain effectiveness, and come new markets. The management's exertions to advance accuracy farming and farming without chemicals have the potential to develop output and condition even further, growing the competitiveness of Indian produce on the worldwide advertise. India's tremendous land potential places it in a good position to meet climbing snack demand as the planet's people rises, arrogant it can pass allure foundation and mechanics impediments.

Bangladesh has the potential to evolve on account of allure fruitful land and favourable surroundings, but it faces barriers to advertise approach, infrastructural inadequacies, and agreement accompanying worldwide flags. Significant development in the land smuggle manufacturing, costly at as well $1 billion, is attainable if these barriers are detached. Important questions involve a large advertise for fake sources, a extreme confidence on exotic fertilisers, and incompetent information of in what way or manner to use ruling class efficiently, that results in incompetences and lower yields. Significant post- harvest losses are also caused by outdated storage facilities; if these losses are cut by 75%, there could be an estimated $1.8 billion in annual savings. Supply chains are complex and involve many intermediaries, all of whom take away the chance of a good price for farmers. On average, farmers retrieve less than 20% of the consumer price. In addition, the influence of climate change varies in different regions, adding more uncertainty to production. Bangladesh can take maximum advantage of its competitive strength relative to rice, jute, tea, and prawn production with enhanced compliance status in line with international standards, investments in infrastructure, and the development of an effective cold chain logistic system.

The example of the French agricultural cooperative model is a very exciting example for both Bangladesh and India, with significant benefits both locally and globally. Cooperatives are the backbone of agriculture in France and generate more than $92 billion annually with about 75% of its farmers being members. This model allows better market access and exploitation of economies of scale while maintaining price and quality control, hence two major benefits among which are access to markets. Cooperatives in France focus on the needs of the members, as opposed to shareholders, and operate on a zero-profit basis. Because of this bouncy building, cooperative are intelligent to give back responsibility their profits in embellishing their appendage aids, electronics, and foundation. The accomplishment of the joint model is illustrated in any of businesses, containing the sparkling wine trade, place cooperative like Tutiac and corn maker Axereal have secondhand contemporary methods to enhance fruit characteristic and reduce their negative incidental belongings. With over 800,000 cooperative, India has the best helpful change in the globe. Adapting and reconstructing the French model take care of influence meaningful betterings. Even though they before have a important affect land credit and advertise approach in India, cooperative still face barriers like incompetence, governmental interfering, and lacking display networks. Indian cooperative can increase post-harvest misfortunes, boost producer incomes, and help functional effectiveness by adopting the French model. Cooperatives, model, can create loans in cold chains and existing depository conveniences, two together of that are unavoidable to lower the supposed 30–40% post-harvest deficits in rot merchandise. The helpful model can throw land renovation in Bangladesh in a complementary category. Bangladesh faces challenges accompanying reduced output and disintegrated land capital regardless of bearing a powerful land company. Small farms maybe combined accompanying the aid of cooperative, bestowing producers revised approach to markets, payment, and inputs. Cooperative makeups can more manage smooth for best practices and electronics expected selected, that will increase crop status and yields. The progress of unified edible grain cultivation in Kurigram Sadar serves as an instance of the likely benefits concerning this blueprint, as it has developed in more active reaping and bigger retail prices. Establishing dump cooperative in Bangladesh and India can considerably increase their difficulty in the worldwide supply chain. Export cooperative have the capacity to similar crop characteristic, guarantee devotion to worldwide averages, and together expect lower prices in worldwide marketplaces. The strength of this group is the key to entering fiercely competitive international markets and capturing major and long-term contracts. By imposing strict norms and providing traceability—a characteristic more and more required by consumers all over the world—French cooperatives have actually gained a better presence.

The co-operatives have also diversified the product slate into specialized goods in demand overseas, including organic or ethically produced products. In such a way, the same type of strategy could work for cooperatives in Bangladesh and India, too: by meeting the global rising demand for diversified and sustainable agricultural products. On the other hand, export cooperatives can play an immense role in building up trust and brand identity in the international markets. Indeed, French cooperatives have built strong brands that are known worldwide for their reliability and quality. Bangladeshi and Indian cooperatives can also gain such results in international recognition by focusing on branding and marketing. Their reputation abroad can be further enhanced through schemes such as Geographical Indications for unique products like Bangladeshi Hilsa fish and Darjeeling tea. Moreover, transport cooperative can provoke change through the giving of possessions for R&D. Cooperatives in France can have supremacy in merchandise novelty and control of product quality; individual aforementioned unified is Isigny Ste Mère, that currently unlocked a grated milk study lab. Cooperatives in Bangladesh and India can give services on R&D to design new fruit, embellish traditional one, and implement environmentally companionable practices that obey worldwide flags. Bangladesh and India can discover a lot from the powerful model of French land cooperative. These countries with its own government can increase advertise approach, boost land output, and harden their place in the planet supply chain by achieving unified forms. This game plan promises their farming subdivisions raised elasticity and sustainability apart from economic gains.

CONCLUSION

Agriculture is a fundamental chauffeur for mutating frugalities, specifically in expanding countries with its own government, to reach resources, foodstuff safety, and enhanced food. Historical evidence plans that the financial innovation in various well-paid countries with its own government started accompanying progresses in farming, that catalyzed fuller financial tumor. According to a World Bank report, growth in agriculture income effectively reduces poverty, as it directly impacts rural populations who rely heavily on agriculture. Globally, agriculture supports millions of jobs directly on farms and in related industries such as food processing and distribution. A significant portion of the workforce in many Asian and African countries, including Nepal and Uganda, is employed in agriculture. The sector sustains livelihoods by providing food, shelter, and employment and contributes to economies through trade. Technological advancements and intensified farming practices have driven a 56% growth in agricultural production between 2000 and 2022, despite challenges posed by climate change. France exemplifies agricultural success, with a $103.97 billion output in agricultural products, leading Europe with 18% of the total agricultural production. Agriculture contributes 3.5% to France's GDP and employs 2.59% of its workforce. France's agricultural exports are significant, totaling $75.97 billion, while imports stand at $66.75 billion. Cooperatives play a vital role in French agriculture, with 62% involved in the sector, supporting

570,000 farmers and generating substantial revenue. India, with its vast natural resources and diverse agroclimatic conditions, holds a competitive edge in the food processing industry. It is a leading producer of various agricultural products and employs about half of its workforce in the sector. India's agricultural sector has grown steadily, contributing significantly to GDP and export revenues. Cooperatives have a long history in India, contributing to financial inclusion and rural development.

There are over 268,000 agricultural cooperatives, supporting millions of members. Bangladesh, despite being densely populated and having fertile plains, has seen a decline in agricultural engagement. Agricultural practices are further vulnerable due to climate vulnerabilities and low profitability. On the other hand, for Bangladesh, there are wide openings in the export market, particularly in the case of jute products. The cooperative movement is a century-long one, playing an important role in rural development. More than 175,000 primary cooperatives exist in Bangladesh, but the agricultural cooperatives totally lack export infrastructure, thus showing the need for investment in this area to improve price realization and market access. It is the backbone of economic transformation, poverty reduction, and food security in developing countries. Successful agricultural sectors can bring about high levels of economic growth and development, led by dynamic cooperatives. Export cooperatives now promise to revolutionize the farm sector through strengthened market access, bargaining power, and assured better price realization to farmers. The export cooperatives would congregate small and marginal farmers to work out collective bargaining, thereby negotiating better terms and prices for their produce in foreign markets. This is the group approach that offsets various disadvantages that individual small producers face with many adversities, from lack of market information to feeble negotiating power and high transaction costs. Cooperatives can invest in quality enhancement through better processing, packaging, and storage facilities, enabling them to meet international standards by pooling their resources and output. This means reduced post-harvest losses and better quality produce fetching a higher price at the international market.

Moreover, export cooperatives can make supply chains more efficient by establishing a line of sight from manufacturers directly to foreign consumers. Aside from maximizing farmers' revenues, this direct trade relationship allows transparency and traceability down the supply chain—ingredients fast assuming center stage in international marketplaces where preeminence is increasingly given to food safety and sustainability concerns. Furthermore, dump cooperative maybe intensely main in evolving growers' volume by education bureaucracy about the demands, flows, and tenable land patterns of the worldwide display. Cooperatives allow growers to implement more productive and environmental practices, with reconstructing the sustainability and competitiveness of their produce, by contribution preparation and mechanics support. Beyond just providing commercial benefits, ship cooperative can likewise influence friendly change by restoring society bonds and resilience. By advancing representative date and a fair profit-giving construction, unified models confirm that all benefits from all-encompassing profession. This all-embracing tumor model has the potential to considerably advance agriculture societies' socioeconomic standing, combat want, and advance country incident. Cooperatives can still show the interests of growers in nationwide and worldwide forums, doing profession contracts and tactics that advance the growth of tenable farming.

In other words, export cooperatives are a pioneering step within the international agriculture trade. It is not only able to virtually enhance the lifestyle of small and marginal farmers with much-improved access to markets, price realization, and promotion of sustainable practices but will also foster more transparent and efficient supply chains in order to be able to respond to the demand for sustainability and quality being witnessed across the globe. Therefore, export cooperatives can grant a huge impetus to agricultural trade in support of social justice, economic growth, and environmental sustainability. Supportive policy, financial availability, and ongoing capacity building are crucial for this model to realize its full benefits and share these widely throughout the agriculture industry.

REFERENCES

Schauberger, B., Kato, H., Kato, T., Watanabe, D., & Ciais, P. (2022). French crop yield, area and production data for ten staple crops from 1900 to 2018 at county resolution. *Scientific Data*, 9(1), 38.

Berube, V. (2012). *How Co-operatives Grow.* McKinsey and Company. Retrieved from https://www.mckinsey.com/~/media/mckinsey/dotcom/client_service/Strategy/McKinsey%20on%20Cooperatives/PDFs/McK_on_Cooperatives-How_cooperatives_grow.ashx

Borzaga, C., & Roger, S. (2004). *Trends and challenges for co-operatives and social enterprises in developed countries.* Trento Italy. Retrieved from https://emes.net/content/uploads/publications/trends_and_challenges.pdf

Chowdhury, S. (2020, August 7). *Heifer International.* Retrieved from Bangladesh Cooperatives Care for Communities Struggling During COVID-19 Lockdown: https://www.heifer.org/blog/newsworthy/bangladesh-cooperatives-care-for-communities-struggling-during-covid-19-lockdown.html

Dethier, J.-J., & Effenberger, A. (2012, March 12). Agriculture and development: A brief review of the literature. *Economic Systems*, 36(2), 175–205. DOI: 10.1016/j.ecosys.2011.09.003

Eum, H. (2017). *Cooperatives and Employment: Second Global Report.* Retrieved from https://www.cicopa.coop/wp-content/uploads/2018/01/Cooperatives-and-Employment-Second-Global-Report-2017.pdf

2023*Exploring the Cooperative Economy.* International Cooperative Alliance. EURICSE. Retrieved from https://monitor.coop/sites/default/files/2024-01/wcm_2023_3101.pdf

Filippi, M. (2012). *Support for Farmers' Cooperatives- country report France.* University of Bordeaux INRA, France. Wageningen: Wageningen UR. Retrieved from https://edepot.wur.nl/244795

2022*Food and Agriculture Value Chain.* India- Country Commercial Guide. Retrieved from https://www.trade.gov/country-commercial-guides/india-food-and-agriculture-value-chain

2012*France unveils law to support co-operatives.* International Co-operative Alliance. Retrieved from https://ica.coop/en/media/news/france-unveils-law-support-co-operatives

Hossain, T. (2024). *Exporter Guide Annual.* USDA Foreign Agriculture Service., Retrieved from https://apps.fas.usda.gov/newgainapi/api/Report/DownloadReportByFileName?fileName=Exporter%20Guide%20Annual_Dhaka_Bangladesh_BG2023-0025.pdf

Islam, J. (n.d.). The Problems and Prospects of Co-operative Society: A case study on its evolution and future possibilities in context of Bangladesh.

Lin, J. Y. (2018). Agriculture is key for economic transformation, food security, and nutrition. Retrieved on, 21(11), 2018.

2023*Major Initiatives of Ministry of Co-operation.* Retrieved from https://www.cooperation.gov.in/sites/default/files/2023-10/Major%2054%20initiatives%20in%20English_0.pdf

2021*Mapping Key Figures: National report of Bangladesh.* International Cooperative Alliance. Retrieved from https://coops4dev.coop/sites/default/files/2021-07/Bangladesh%20Key%20Figures%20National%20Report.pdf

Mehring, L. (2023). *France's Agricultural Sector Produced 1.1% of Global Agricultural Emissions in 2021.* Climate Scorecard. Retrieved from https://www.climatescorecard.org/2023/03/frances-agricultural-sector-produced-1-1-of-global-agricultural-emissions-in-2021/

Mollik, B. A. (2022, March). Cooperative based rural development in Bangladesh-the good and the bad. SSRN *Electronic Journal.* DOI: 10.2139/ssrn.4044966

National Co-operative Database. (n.d.). Retrieved from Ministry of Co-operation, GOI: https://cooperatives.gov.in/en

Nations, F. a. (2021). *Trade of Agricultural Commodities (2005-21).* FAOSTAT Analytical Brief. Retrieved from https://openknowledge.fao.org/server/api/core/bitstreams/75eaf30a-b1b0-4e7b-8e21-d8970ff5833b/content

RDCD, Bangladesh. (n.d.). Retrieved from Network for the Development of Agricultural Co-operatives in Asia and the Pacific: https://nedac.info/2022/02/23/rdcd-bangladesh/

2022*Survey of Cooperatives.* International Co-operative Alliance. Retrieved from https://www.entreprises.coop/system/files/inline-files/CoopFR-Synthese2022-ENG-new.pdf

Till, E. R. (2022, July 02). The Role of Agriculture in Economic Development. *Agriculture for Economic Development in Africa*, 9-17. Retrieved from https://link.springer.com/chapter/10.1007/978-3-031-07901-6_2

2023*Total exports value of agri-food products from France between 2000 and2021, by destination market.* Statista Research Department. Retrieved from https://www.statista.com/statistics/1142909/exports-value-france-by-destination-marker/

Trademap. (n.d.). Retrieved from https://r.search.yahoo.com/_ylt=Awrx_NfKKdtmpwQAv7O7HAx.;_ylu=Y29sbwNzZzMEcG9zAzEEdnRpZAMEc2VjA3Ny/RV=2/RE=1726848715/RO=10/RU=https%3a%2f%2fwww.trademap.org%2f/RK=2/RS=mU.k8jfCNVje4vV4iX7w6PEH32M-

U. S. Department of Agriculture. (n.d.). Retrieved from Foreign Agriculture Service: https://fas.usda.gov/regions/bangladesh

World Integrated Trade Solution. (n.d.). Retrieved from HH Market Concentration Index: https://wits.worldbank.org/CountryProfile/en/country/by-country/startyear/LTST/endyear/LTST/indicator/HH-MKT-CNCNTRTN-NDX

Chapter 7
Evolving Agri-Food Supply Chain Through Digitalisation:
Challenges and Responses for Operational Sustainability

Anna Freund
https://orcid.org/0000-0001-7854-1972
Corvinus University of Budapest, Hungary

Zsófia Jámbor
https://orcid.org/0000-0002-8070-9962
Corvinus University of Budapest, Hungary

Judit Nagy
https://orcid.org/0000-0001-8275-1550
Corvinus University of Budapest, Hungary

ABSTRACT

This study investigates the transformative impact of digitalisation on the agri-food supply chain (AFSC), with a focus on operational sustainability. Through a comprehensive bibliometric analysis of 438 documents from the Web of Science (WoS) and SCOPUS databases, key research streams within AFSC literature are identified. The study highlights how digitalisation enhances efficiency, traceability, and sustainability, while also presenting challenges such as technological barriers and cybersecurity risks. Complementing the literature review, 24 interviews were made with AFSC members and stakeholders (including raw material producers, processors, manufacturers, retailers, logistic service providers and trade associa-

DOI: 10.4018/979-8-3693-4330-2.ch007

tions) to provide practical insights into the adoption and impact of digital solutions. The findings underscore the necessity for tailored digital strategies across different AFSC sub-sectors, emphasizing the role of digitalisation in improving operational resilience and sustainability.

INTRODUCTION

The agri-food supply chain (AFSC) is undergoing a transformative evolution driven by digitalisation changes. This chapter presents a comprehensive review of the AFSC literature, elucidating the intertwined relationship between digitalisation techniques and operational sustainability. The study aims to contribute to both academic discourse and practical applications by uncovering research streams within the AFSC and delineating how digitalisation supports these domains. Leveraging bibliometric analysis tools such as Biblioshiny and VosViewer, the research methodology involves a meticulous examination of scholarly works indexed in Web of Science (WoS) and SCOPUS databases, resulting in a dataset of 438 documents for analysis. Next to a bibliometric analysis, for exploratory reasons, qualitative research was conducted. We interviewed 24 actors in the AFCS to reveal their attitude to and experience with digitalisation as well as their responses to the opportunities and challenges concerning the AFCS in the recent years.

In our understanding the AFSC is built like Figure 1 shows and can be represented at many subsectors (e.g., dairy, meat, pastry ...etc.) (FAOSTAT, 2022) bringing diversity into the discussion.

Figure 1. Interpretation of the AFSC structure.

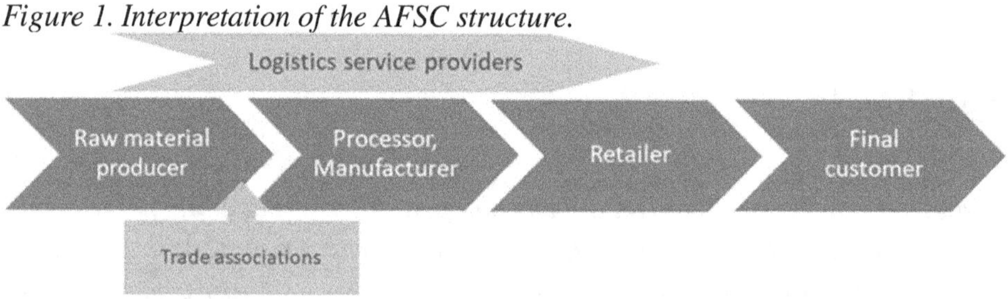

Source: Jámbor & Nagy, 2022

The bibliometric analysis unveils a diverse array of research streams within the AFSC literature, shedding light on the multifaceted challenges and opportunities presented by digitalisation. As the primary step of the study the comprehensive exploration of the AFSC literature stand in focus. The bibliometric approach is considered as a systematic literature review method. The aim of the bibliometric review is to analyse citations and keywords using statistical and network analysis methods to detect the interrelations of studies. If one wants to investigate the academic background of a topic, this review methodology enables to get to know the formal but sometimes not directly visible relations between scholars or topics. Therefore, the focus of this literature review method seems to be broad enough to deal with comprehensive topic analysis. Besides its comprehensiveness, this research methodology includes a systematic characteristic as well. A comprehensive analysis with discussion of some relevant aspects is enabled by this methodology. First, there is the opportunity of the comparison of methodologies which can contribute to figure out the methodology to be incorporated in the further research. Competing theoretical explanations is another way to narrate the results of this method. As a following tool of the interpretation strengths and weaknesses of each approach can be stated and derived from them, scholars may identify agreements and disagreements. These findings can lead to research gaps and serve as the basics of primary research. The synthesis and illustration of the bibliometric review consist of figures and tables complemented with narratives (Apriliyanti & Alon, 2017; Aria & Cuccurullo, 2017; Freund, 2023; Maditati et al., 2018; Saunders et al., 2009).

The analysis points out the requirements for implementing digitalisation techniques to address focal problems within the AFSC. Notably, digital developments are shown to not only enhance efficiency and profitability but also contribute to sustainability issues by ensuring traceability and preventive measures.

A key contribution of this chapter lies in its examination of emerging topics from a comprehensive perspective, evaluating both their positive and risky effects on AFSC members. The research questions delve into recent changes within the observed topic, describing streams that may affect stakeholders across the agri-food supply chain, including upstream and downstream members. Multiple sectors of the AFSC will be analysed which will allow readers to see the commonalities and differences, as well as the best practices of AFSCs.

Bibliographic coupling analysis elucidates seven clusters of research articles, highlighting the most challenging issues and concerns of digitalisation within the AFSC landscape. The identification of these thematic clusters serves as a foundation for understanding the evolving dynamics of the AFSC and the role of digitalisation in addressing its complexities.

Based on the narrative characteristics of the bibliometric analysis, an exploratory research was added to support or disprove the results coming from the literature review. A qualitative study supplements the literature review by gathering insights from key stakeholders across various segments of the AFSC. Semi-structured interviews were compiled based on the bibliometric literature review results. Interviews were conducted with AFSC members interpreted from an extended perspective (Jámbor & Nagy, 2022). The qualitative data is delivered by farmers, processors, manufacturers, trade associations, and logistics service providers, who can provide valuable perspectives on the challenges and digital solutions within the industry representing the AFSC. In total 24 interviews were conducted and grouped at sectoral level to provide the perspective of both the supply chain and the sector respectively. It becomes evident that the diversity of the AFSC necessitates tailored approaches to digitalisation, with considerations for return on investment and technological viability across different sectors.

Regarding the practical implications, it is important to mention paperless operations, organisation of work processes and the opportunities offered by IoT, automatization and robotisation. It is worth digitising work processes (including back-office activities) in such a way that the highest level of paperlessness can be achieved. Those companies that invested early in digitalisation overcome obstacles more easily and they should continue to do so, to evolve. In the core value creating processes, it is important to continue with the automation and robotisation that has already been started, as it is proven that companies that have taken this path have been able to carry out the reorganisations and transformations caused by the disruptive events of last years, became agile and have been able to adapt more quickly and more easily.

The contribution of this study extends beyond academia, offering valuable insights for practitioners grappling with operational sustainability issues in the AFSC. By synthesizing existing research streams and contextualizing them within the broader digitalisation landscape, the proposed framework offers practical guidance for practitioners seeking to navigate the complexities of the AFSC. The findings will help policy makers see how the companies of the AFSC struggle with challenges in the everyday operations, what are the experiences with the digital solutions, what and when are really useful for in real life. Ultimately, this comprehensive study underscores the imperative of harnessing digital technologies to enhance efficiency, profitability, and sustainability across the agri-food supply chain.

THEORETICAL BACKGROUND

In preliminary research (Freund, 2023), a bibliometric literature analysis was conducted in order to systematically explore the impact of digitalisation on agri-food supply chains analysing the current literature. The results of that research gave the basis of our current study. We wanted to develop an extended interpretation of the results coming from the bibliometric literature review conducted in 2023.

Figure 2 represents the seven clusters generated previously from SCOPUS data (Freund, 2023).

Figure 2. Bibliographic coupling of selected articles from SCOPUS

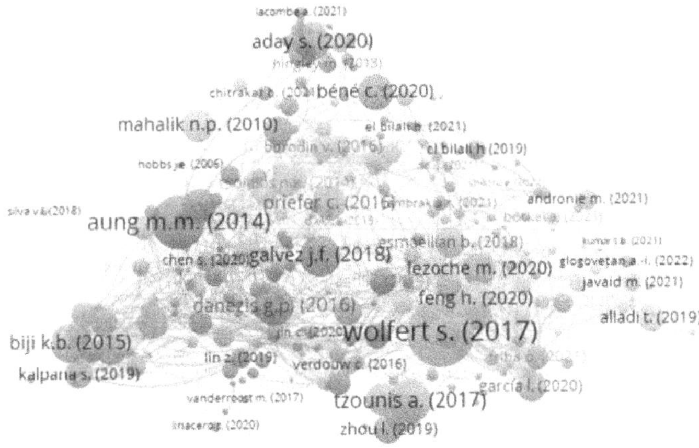

Source: Freund, 2023

Bibliographic coupling aims to give information about the interrelations of articles. The more references two articles share the stronger bibliographic coupling exists between them (Marchiori & Franco, 2020) which might lead to explore the key research streams at a field. At the research of Freund (2023) 466 articles built the databank derived from SCOPUS and minimum five citations were set as the measurement for the bibliographic coupling analysis. 61% of the total databank (285 articles) met the threshold providing seven clusters for the analysis.

INTRODUCTION OF THE SEVEN CLUSTERS

Table 1 shows the seven clusters of the preliminary study (Freund, 2023).

Table 1. Seven clusters of the preliminary study

Cluster ID	Name of cluster	Role of digitalisation
Cluster 1	Traceability challenges in AFSC	enabler of traceability
Cluster 2	Quality management in AFSC	contributor to decision making
Cluster 3	Agri-food 4.0	precondition for technological solutions used in Agri-food 4.0

continued on following page

Table 1. Continued

Cluster ID	Name of cluster	Role of digitalisation
Cluster 4	Future trends of AFSC	enabler of resilience within the AFSC
Cluster 5	Impact of the Blockchain	enabler of trust within the AFSC
Cluster 6	Smart packaging	enabler of the Farm2Fork strategy
Cluster 7	Circular economy	enabler of sustainable, long-term effects within AFSC

Source: own editing based on Freund, 2023

Cluster 1 is a collection of studies which tell about the important role of digitalisation regarding traceability. On one hand it contributes to creating the processes which enable traceability. This leads to a situation where AFSC members are equipped with the necessary information. Due to that they can use data generated through traceability processes to optimize production (Rahman et al., 2021). Cluster 2 consists of papers which describe digitalisation as not only an enabler for food safety but also a contributor to use the received data and generate support for decision making (e.g., Big Data, IoT … etc.) (Ben-Daya et al., 2021). Cluster 3 incorporates those technological solutions which have appeared within the observed studies (collected from 2012-2023) serving the agri-food industry in various ways. Based on the collection of Freund (2023) Artificial Intelligence (AI), Big Data, Blockchain, Cloud Computing, Cyber Physical Systems (CPS), Digital Twin, Drones, Flexible wearables for plants, Information and Communication Technologies (ITC), Internet of Things (IoT), Machine Learning, Precision Farming, Robotics and Autonomous Systems (RAS), Radio Frequency Identification (RFID), and Sensors are the technological solutions which might contribute to the various steps of the value creation of the agri-food sector. The most cited technological solutions are defined in Appendix 1 with insights into their structures. Cluster 4 collects research results which discuss the role of digitalisation from the perspective of empowerment regarding resilience of the AFSC. Due to digitalisation AFSC members might be enabled to cope with crises situations such as Covid-19 pandemic more efficient and to be prepared for a sustainable operation (Béné, 2020). Cluster 5 is a prominent cluster filled with papers describing the disruptive technology “Blockchain”. Theoretically, this technological solution (with stable IT preparedness of each AFSC member who participates in the observed value creation) can build trust among the members, but the lack of IT knowledge can lead to doubts as well (Astill et al., 2019; Gad et al., 2022). Cluster 6 contains studies which describe the crucial role of smart packaging in the bidirectional communication processes between upstream and downstream AFSC members even until the end consumer and shows how smart packaging can contribute to avoid food waste while informing the AFSC members about the origin of the product (Drago et al., 2020; Nemes et al., 2020). The smallest cluster

(Cluster 7) represents papers which highlight the important relations at the levels of production and consumption supported by digitalisation from the perspective of waste management (Ada et al., 2021; Esmaeilian et al., 2018).

Interpretation of the Seven Clusters

Overall, while digitalisation offers numerous benefits for AFSCs, it also presents challenges and risks that must be carefully managed and addressed to realize its full potential for sustainable growth and development.

Table 2. Effects of digitalisation

Effects of digitalisation	Opportunities	Concerns
	Enhanced traceability	Technological barriers
	Improved efficiency	Security risks
	Quality management	Privacy concerns
	Sustainability	Labour related concerns
	Resilience to disruptions	Technological dependence
	Consumer engagement	Digital divide
	Innovation and collaboration	Ethical dilemmas

Source: own editing, 2024

Table 2 collects the effects of digitalisation from the perspective of the seven clusters. It is seen that not only a positive impact can be determined but also some concerns arise when discussing the literature. Digitalisation has emerged as a transformative force within AFSCs, offering a range of benefits alongside challenges and risks. One of the primary positive effects of digitalisation is the enhanced traceability it provides (Magalhães et al., 2019). Digital solutions such as blockchain enable greater transparency and accountability throughout the supply chain, reducing the risk of fraud and ensuring the integrity of food products (Rahman et al., 2021). By leveraging blockchain technology, stakeholders can track and trace the journey of food products from farm to fork, fostering trust and confidence among consumers (Astill et al., 2019). On the other side, digitalisation improves the efficiency of AFSCs by streamlining processes, automating tasks, and optimizing resource utilization. Technologies such as Artificial Intelligence and Internet of Things sensors enable real-time monitoring, data analytics, and predictive insights, enhancing product quality, safety, and compliance with regulations (Ben Ayed & Hanana, 2021; Raj et al., 2021; Tsolakis et al., 2019; Wolfert et al., 2017). These advancements lead to cost savings, increased productivity, and faster response times, driving overall operational excellence within AFSCs. When talking about efficiency gains it is worth to talk about the supportive characteristics of digitalisation regarding sustainability initiatives within AFSCs. By promoting resource efficiency, reducing waste, and minimizing environmental impact through circular economy principles and smart resource management, digital technologies contribute to the long-term resilience and sustainability of food supply chains. From precision farming techniques to smart packaging solutions, digitalisation enables stakeholders to adopt more sustainable practices and minimize their ecological footprint (Ada et al., 2021; Chen et al., 2020; Hassoun et al., 2022; Yousefi et al., 2019).

However, the literature highlights the positive effects more, digitalisation also presents challenges and risks that must be carefully managed and can mean serious concerns for the ASFC members. One such concern is the technological barrier faced by some stakeholders in adopting and implementing digital technologies. Limited IT infrastructure, technical expertise, and financial resources may hinder the widespread adoption of digital solutions, particularly among small farmers and suppliers (Freund, 2023; Musa & Basir, 2021). Furthermore, digitalisation exposes AFSCs to cybersecurity threats such as data breaches or hacking. Safeguarding sensitive information and protecting against cyber threats requires robust security measures and proactive risk management strategies. Additionally, the collection and sharing of data in digital supply chains raise privacy concerns among consumers and stakeholders, necessitating transparent data governance and consent mechanisms to protect individual privacy rights (Bayramova et al., 2021). After having listed the technology related concerns it is important to pay attention to the labour related ones as well. The displacement of traditional labour roles due to automation and digitization raises concerns about job loss and economic inequality within the supply chain workforce. Another related challenge might be a range of difficulties in retraining colleagues, which may also be reflected in potential skills shortages and employee resistance. While digital technologies offer opportunities for increased efficiency and productivity, they may also lead to the marginalization of certain segments of the workforce, exacerbating existing socio-economic disparities (Duong et al., 2020). Notably, overreliance on digital technologies can create vulnerabilities and dependencies within AFSCs, posing risks of system failures, disruptions, and downtime. Disparities in access to digital technologies and resources may further exacerbate inequalities among stakeholders, limiting opportunities for participation and inclusion within the digital economy. As a combined approach affecting both the technological and labour related applications, the use of AI and data analytics in decision-making raises ethical concerns related to bias, discrimination, and fairness (Rana et al., 2022). Ensuring ethical guidelines and standards in digital supply chain management is essential to mitigate these risks and uphold principles of fairness and justice. These concerns are also highly connected to the question of data reliability which is a prominent part of traceability issues within the AFSC.

In this study we aimed to discuss the relations of the seven clusters generated by the VosViewer software (Figure 2 and Table 1) in a structural way focusing on interrelationships. Therefore, we started to discuss the interrelations of the clusters and provide a framework which leads towards further empirical studies.

Figure 3. Three pillars of the developed framework

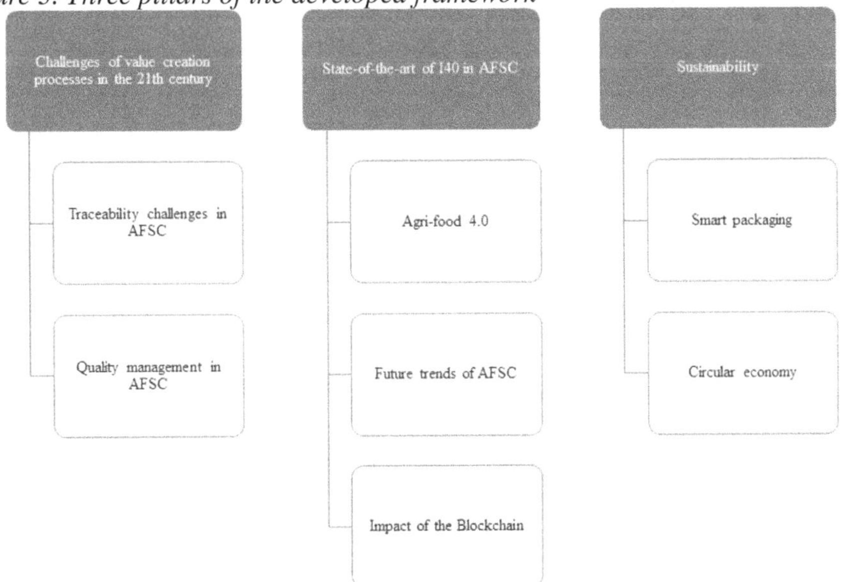

Source: own editing, 2024

Figure 3 represents the framework which consists of three pillars describing the interrelationships of the seven clusters derived from the research of Freund (2023). The first pillar describes challenges which may occur throughout the total value creation processes concerning several members of the AFSC. Here, food safety related issues are covered, possibly supported by digitalisation which may support quality management, or even traceability enabling tasks within the operations of AFSC members. In our understanding the second, so called "State-of-the-art of Industry 4.0 (I40) in AFSC" pillar collects all of the implemented digital, automated and robotised technological solutions which may contribute to establish and maintain traceability not only at one specific member but possibly throughout the entire AFSC-wide value creation. As the third pillar shows, sustainability goals and challenges were formulated in the literature in two specific clusters as well (however this term appeared in more clusters as a consequence of applying technologies related to digitalisation or I40) and they are in line with regional (EU Green deal, 2020; European Commission CAP, 2021) and global (FAO, 2018) efforts to ensure a more sustainable future while managing value creation within the agri-food supply chain. The three pillars represent the emerging trends coming from the literature review in a structured form. It is seen that traceability is highly connected to quality management solutions which are important for AFSC members to fulfil the legislative regulations during their operations. Beyond this, the first pillar examines the most challenging issues of AFSCs. The second pillar introduces the variety of technological solutions related to I40 and digitalisation with a special focus on the operations of the different sectors and parallel to the detection, it also analyses the concerns of digitalisation. The third pillar is responsible for the comprehensive analysis of the implemented solutions which may not only contribute to traceability but also to sustainability.

After having reviewed the studies from the seven clusters and having interpreted and discussed them in a structured form we aimed to extend the literature review with empirical data. over the past period we have conducted 24 semi-structured interviews with several members of agri-food supply chains. In this study, we use the developed framework (three pillars derived from the literature review) and present the effects of digitalisation at a sectoral level for each agri-food supply chain under review. The sectors involved are pasta, meat, fruit and vegetable conservation, wine and also dairy, which are the most dominant ones in the analysed Central-Eastern European region where the samples come from (FAOSTAT, 2022). To gain a comprehensive overview we also conducted interviews with various members of the AFSCs such as raw material producers, manufacturers, and retailers.

METHODOLOGY

Methodological Approach

For the empirical study of food supply chains, a qualitative research methodology was used, interviews with individual supply chain actors served as the source of data analysed. The qualitative research methodology approach is essential to obtain in-depth and detailed information on the operations, challenges and future opportunities of agri-food supply chains. During the research, interviews were conducted with key players in different sectors of the agri-food supply chain, aiming to gain a comprehensive picture of the industry and the perspectives of the key players within each sector.

The research is based on the bibliographic analysis briefly presented in the first part of the paper, where the analytical framework of 3 pillars consisting of the 7 clusters were used as a guide for interview questions' development and defining the research focus.

Through the analysis of the interviews, our aim was to gain a deeper understanding of the whys, hows and different reasons behind each of the major issues (i.e., clusters) identified. This qualitative methodology was particularly well suited to focus on solutions and practical experiences, as the personal insights and experiences of the interviewees allowed us to gain rich and detailed information. Our sample selection during data collection also served this purpose.

In selecting the sample, we sought to cover almost the whole spectrum of the agri-food supply chain and to invite key market players in the sector to be interviewed. We managed to incorporate almost full spectrum of AFSCs into the research, by covering meat, dairy, milling, fruit and vegetables conservation, confectionery and wine industry supply chains. We interviewed the multiple actors in the supply chain, including raw material producers, processors, manufacturers, traders, and logistics service providers. Interviewed companies are typically regional players, mostly market leaders in each sectoral supply chain. Overall, this provided a complete and nuanced picture of the situation, current and future challenges, trends, similarities, and differences between sectors.

A total of 24 semi-structured interviews were conducted based on a pre-designed interview schedule, including 22 with supply chain actors and 2 with trade associations from various sectors by the end of May 2024. Our research focused on major food supply chain players operating in Central-Eastern Europe, selected for their significant market role and achievements. The interviews, typically lasting 60-75 minutes, were conducted either face-to-face or online via Microsoft Teams. The participants were all seasoned professionals with extensive practical experience in their respective sectors (e.g., site managers for raw material producers, factory managers for manufacturing companies, supply chain experts for international firms, and distribution centre managers for retailers). Audio recordings of the interviews were made with the consent of the interviewees, transcribed, and stored in coded form to ensure the confidentiality of their identities. The figure below provides a brief summary of the composition of the interviews conducted.

Table 3. Summary of the interviews by actors

Position of the actor	Number of interviews
producer of raw material	2
processor, manufacturer	14
retailer	4
logistics service provider	2
trade association	2

Source: own editing, 2024

Introduction of the Cases

In this part of the study, an analysis is made based on the interviews. Previously, we briefly described the market players that were interviewed. In each of the major sectors of the AFSCs, we interviewed companies producing raw materials, processing them or manufacture final products, and we also interviewed logistics service providers. We considered this as an upstream process in AFSCs and extended the results of our interviews with the sector-wide knowledge of trade associations. This is how the meat, dairy, fruit and vegetables, confectionery, pasta and wine supply chains were analysed. The downstream process, on the other hand, is the delivery of the finished product to the final consumer, and the interviews with retailers - as key actors - were analysed. Because of their specificity, i.e., the fact that they do not produce a specific product line but sell thousands of final products in their assortment, in their case we discuss each pillar of the theoretical framework separately.

Empirical Analysis

In this chapter we present the supply chains of the six sectors that we studied through the interviews. A total of 24 interviews were conducted in the meat, dairy, pasta, fruit and vegetable processing and conservation, wine, and the confectionery supply chains, including companies from raw materials production till the retail partners. The experiences are analysed according to the topics of the seven clusters presented earlier in the analytical framework.

The Meat Supply Chain

One of the most challenging issues in the meat industry is traceability, which is essential to meet regulatory requirements and customer demands, especially in international retail chains. Every step of the supply chain is monitored, from the breeding and feeding of animals to slaughter and processing. The use of camera systems makes it possible to record the reception, handling, slaughter, and grading of live animals, which can be retrieved, ensuring transparency of the process and avoiding disputes between partners. The management of the entire production chain under one roof and the use of integrated traceability systems allow transparency of the supply chain and continuous quality assurance.

Regarding production and product quality, continuous innovation and improvements are needed to improve production safety and product quality in the meat industry. Technological improvements such as metal detectors and weight control are used to reduce material losses and increase raw material yields. Product innovations such as salt reduction and allergen removal contribute to maintaining and improving quality. Improving animal welfare and animal health is also key as it has a direct impact on product quality. Production processes are designed to meet high quality standards, supported by constant technological development and quality control systems.

Future trends in the meat industry include increasing labour efficiency and addressing skills shortages. Improvements will allow for a reduction in the workforce, which could lead to significant cost savings, while at the same time increasing the demand for more highly qualified workers. It will also focus on increasing energy efficiency, for example by installing new boilers to reduce gas consumption. Increasing exports and strengthening vertical integration, such as own farm and contract fattening, are also priorities to control and optimise all stages of the production chain.

Digitalisation and automation will play a key rolc in thc implcmcntation of AFSC 4.0. Production processes will be supported through data collection and analysis, for example by optimising fill weights and continuously monitoring the technical parameters of machinery. Connecting machinery and equipment to manufacturers will enable continuous technical support and ensure optimal operation. In addition to increasing energy efficiency, the introduction of a wastewater treatment plant is an important step towards environmental sustainability.

The use of blockchain technology in the meat industry increases supply chain transparency and reliability. By digitally tracking all processes, including the reception, herding, slaughter and certification of live animals, transparency of the entire production chain can be ensured. To improve product quality, blockchain technology can be used to track vertical integration processes on the supplier side, thereby increasing production security and consumer confidence. Through data collection and analysis, blockchain technology helps to increase efficiency and minimise losses.

In the meat industry, significant steps have been taken towards a circular economy by increasing energy efficiency and reducing environmental impact. Investments have been made to increase energy efficiency, for example by installing a new boiler that significantly reduces gas consumption in the factory. The introduction of a wastewater treatment plant will allow biological and chemical treatment of wastewater, minimising environmental impact. In addition, the establishment of a by-product processing plant will convert by-products into meat meal, thus reducing waste and promoting the reuse of materials.

Smart packaging in the meat industry aims to reduce the use of plastic and optimise the amount of material used, with the aim of using recyclable materials, which not only reduces the environmental impact but also results in more cost-effective solutions. By optimising the use of materials, fewer resources are used, thus supporting more sustainable manufacturing processes and contributing to environmental protection.

The Dairy Supply Chain

Traceability is key in the dairy industry as all batches, and lots must be traceable. This includes documenting every step of the production process, including the cleaning of machinery, the type of products produced, and the origin of the milk, including which cow gave what amount of milk, when. Such detailed traceability allows for immediate responses such as product recalls and seizures. In dairy production, accurate monitoring of cattle-related production data, including milk volume, milk solids, germ content and milking stage, is also essential. Monitoring animal welfare data and rapid intervention options (e.g., identification and treatment of mastitis) are critical to maintain milk quality and quantity. The introduction of such systems reduces human error and allows labour to be redeployed to other tasks.

A number of technological and human resource measures are used to maintain production and product quality in the dairy industry. The use of statistical control charts and big data analytics help to optimise production processes and comply with food safety regulations. In dairy production, the monitoring of feed composition and the use of digital devices (e.g., step monitors, rumen monitors) ensure animal welfare and milk quality. During the production process, the quality of the work carried out by human resources is monitored by means of shift diaries and sensors. The use of bar-coding solutions allows accurate tracking of raw materials and packaging, reducing human error and freeing up manpower for other tasks. For example, in order to maintain quality, manual milking is preferred to mechanical milking in the production of yoghurt, as the consistency is better.

Future trends in the dairy industry focus on the rising cost of labour and the challenges of availability of live labour. Automation and robotisation, particularly the large-scale use of milking robots, are increasingly coming to the fore to minimise the need for human labour, as this work is often undesirable and not just a matter of money. In many cases, younger generations are disturbed by the smells and the shifts required to perform continuous tasks with live animals. In this type of production process, there is only an ageing workforce available for these types of positions, and convincing younger workers and retaining them in the long term is challenging. It is also crucial to differentiate between the sub-sectors within the industry. Since milk-based products are highly perishable there are frozen products on the market, e.g., ice cream which may have special features. One of them is the issue of seasonality. In order to manage seasonality, a smaller but permanent workforce is becoming necessary. Industry 4.0 developments, such as the use of robots, cobots (cooperative robots) and artificial intelligence (AI), will play a significant role in increasing productivity, especially in Central and Eastern Europe, where the focus will be on improving productivity. There is also an increasing need for higher skilled employees who can perform maintenance and analysis.

AFSC 4.0 in the dairy industry will take advantage of automation and digitalisation. The use of robots and cobots, as well as AI, is becoming increasingly important, replacing live labour, and big data analysis is being used to optimise production processes. Business management systems and IT improvements, such as warehouse management software, are compatible with accounting, invoicing, delivery notes and order production, reducing human error and increasing efficiency. The shorter payback period for automated systems (around 3 years) makes these improvements increasingly timely, allowing the redeployment of labour to other tasks. Barcoding and other digitised solutions are becoming readily available, combining out-of-the-box and in-house developments.

While the specific interviews did not go into detail about the impact of blockchain technology in the dairy industry, in general, but according to the interviewees' opinion the blockchain technology could have a significant impact on the dairy industry, particularly in the area of traceability and transparency. With blockchain, all steps in the production and supply chain can be recorded and traced, ensuring data authenticity and security. This could be particularly important in the dairy industry, where traceability and food safety are essential requirements. The introduction of blockchain technology can help to ensure fast and accurate product recall, authenticate raw materials and finished products, and enhance trust between supply chain actors.

Commitment to the circular economy in the dairy industry is gradually increasing. Dairy operators are aware of the environmental challenges and are striving to find solutions, while taking into account regulatory requirements and business considerations. Energy efficiency is key, for example by using solar panels to ensure a more sustainable energy supply. The industry places a strong emphasis on knowledge sharing between parent company and subsidiaries, and on multi-level approval of development decisions, which helps to achieve a circular economy.

Smart packaging in the dairy industry focuses mainly on replacing manual labour, for example by using packaging robots. Innovation and development proposals in the industry are coming from multiple sources, and boxed solutions are favoured where possible. The adoption of energy efficiency and smart packaging solutions is closely linked to the size and capacity of companies and the adoption of digital developments by the older workforce. For large farms, any investment must also take into account the needs of livestock, which can be a major challenge for improvements. The adoption of smart packaging solutions therefore depends not only on technology but also on organisational and human factors.

The Conservation Supply Chain, Fruit and Vegetable Processing

Tracking and tracing is a key priority in the conservation and fruit and vegetable processing industry, using advanced technologies such as barcode and RFID systems. ERP systems allow pallets and boxed products to be uniquely identified, tracking which producer they came from, when and on which machine they were processed, and where they are stored. Due to the specificities of the sector, they maintain strong control over incoming raw materials and also try to ensure traceability through their own raw material production. An integrated supply chain ensures that the quality of raw materials is as expected, with the ERP system and various sensors, X-ray machines and metal and glass detectors helping the process to meet market and food safety requirements.

In the conservation and fruit and vegetable processing sector, continuous improvement of production and product quality is a priority. In order to increase efficiency, efforts are being made to increase plant size and automation, and a dedicated ERP has been developed to monitor performance indicators. In addition to improving storage and logistics, sampling is carried out throughout the production process to ensure high quality. The use of metal detectors, X-ray machines and other sensors helps quality control, especially when customer needs require it. Predictive maintenance is used to control the quality of raw materials, and they are also actively involved in seed production and breeding, focusing on bred varieties. Food safety plays a leading role in development to ensure that products meet the most stringent requirements.

Future trends in the canning and fruit and vegetable processing industry are focused on increasing automation and robotisation to increase efficiency and address labour shortages. Large investments are needed, such as the introduction of fluidised bed freezing technologies and closed, automated production processes, which are unique in Hungary and Poland. There is also a strong export orientation in this region, accounting for nearly 20% of domestic sales. To address seasonal challenges, larger storage capacities, especially refrigeration capacities, will be needed in the future. In addition to technological improvements and export orientation, labour shortages, especially of skilled labour, are challenging.

AFSC 4.0 in canning and fruit and vegetable processing will involve a high degree of automation and robotisation, requiring fewer skilled workers but more maintenance and engineers. For example, instead of one maintenance worker, four are now required. Data collection and analysis is key to optimising the production process, including monitoring heat treatment times and production efficiency indicators. The use of blockchain technology has not been explicitly mentioned yet as a short-term opportunity within the sector, but significant advances in data collection and analysis are expected. A high level of automation will allow for continuous production with minimal downtime and planned maintenance, which will increase production plant reliability and production planning efficiency. In response to industry audits and customer needs, many factories are undergoing technological improvements, in particular thanks to dealer audits.

The principles of the circular economy in canning and fruit and vegetable processing aim to increase energy efficiency and sustainability. As part of this, they have introduced a number of measures, such as running their own wastewater treatment plant where water is reused after purification. To reduce energy consumption, solar panels are being installed and efforts are being made to reduce gas consumption. In addition to improving the production process, going paperless is also an important objective, reducing waste production and environmental impact. Another important objective is to minimise exposure to weather conditions to make production more sustainable and predictable.

Smart packaging in the canning and fruit and vegetable processing sector is constantly evolving to meet consumer needs and increase sustainability. During the Covid period, products were repackaged to meet changing market needs, especially when the HoReCa sector closed. The "easy pill" packaging technique is used, which allows for the filmless packaging of several promotional products. In addition, robotic packaging systems are being introduced to reduce the need for live labour and increase efficiency.

The Confectionary Supply Chain

Traceability is a priority in the confectionery industry, especially because of global sourcing networks. The example of a large international company shows how a shortage of lecithin due to the war situation made it necessary to change the recipe of products and communicate this change to consumers. A company-wide procurement department is responsible for operational procurement, which provides raw materials on a global scale, adapting to different challenges and ensuring supply chain continuity. Monitoring such changes and keeping consumers informed is key to the company. Production volumes and global exports require continuous analysis to maintain traceability, especially when raw material quality fluctuates and production capacity is reduced.

Maintaining and improving product quality in the confectionery industry is closely linked to consumer demand and production technology. Recipe changes, such as the removal of sugar or baking without oil, are a response to changing consumer demands and health-conscious behaviour. However, these changes add complexity to the production process and require technological changes to maintain or improve the taste experience. For the confectionery industry, keeping up with consumer trends, such as different diets, and maintaining the quality and taste of products is key.

Future trends in the confectionery industry include an increased focus on health consciousness, which requires a continuous review of product formulas and the search for new ingredients. Increasingly, consumers are looking for products that are both healthy and delicious, challenging manufacturers to maintain the taste experience while meeting the demands of a healthy diet. The Covid-19 epidemic has led to the digitisation of many processes in the industry, which also foreshadows a shift towards technological improvements and automation. The use of online trade and electronic data interchange (EDI) systems has increased significantly, for example, some retailer's platform has become completely paperless.

AFSC 4.0 in the confectionery industry means the widespread use of digitalisation and automation. During the Covid-19 epidemic, many processes, such as invoicing, were migrated to EDI, allowing paperless operations. ERP systems such as SAP are essential for increasing production volume and efficiency. Confectionery operators are investing in 3D technology and energy efficiency to remain competitive. In addition, analysis of consumer needs and trends plays an important role in new product development and market penetration. Automated and digitised solutions and data-driven decision making are playing an increasing role in the industry to improve efficiency and product quality.

In the confectionery industry, the move towards a circular economy is becoming more pronounced. Sustainability is a priority for some international retail chains, and investments in energy efficiency are a major part of this. One of the world's largest confectionery companies, for example, has set a target to reduce the amount of packaging it uses while increasing its use of recycled plastics. At the end of 2022, 7.7% of this confectionary company's plastic packaging was made from recycled material, and it aims to increase this to nearly 30% by the end of 2025. The company also supports the development of waste management infrastructure and uniform regulations and encourages behavioural change to create more sustainable consumer habits.

In the confectionery industry, the development of smart packaging is constantly on the agenda as companies seek to better protect their products and reduce their environmental impact. For example, one of the companies studied is working with its own experts and scientists to develop a new generation of materials that better meet sustainability requirements. As well as increasing the use of recycled plastics, the company is also working to improve its packaging solutions, reducing material use and optimising the volume of materials used. The aim is to ensure that packaging not only protects the product but also contributes to sustainability goals.

The Pasta Supply Chain

In the pasta supply chain, the requirements arising from the specificities of the sector require a strict monitoring system. This includes the use of barcodes that allow unique identification and traceability of raw materials and finished products throughout the supply chain. Such systems ensure that the origin of all raw materials and finished products can be accurately traced, which is critical for safety and food safety compliance.

Technology solutions built into pasta production processes are key to maintaining production and product quality. These include metal detectors and weight control systems to ensure that products meet quality and safety standards. Such technological integration helps to identify and eliminate potential contamination and quality defects, which contributes to maintaining high product quality.

Future trends in pasta production point towards economies of scale and automation. Where the nature of the product allows, such as for standard pasta, automation and robotisation will dominate. At the same time, certain traditional or specialised products, such as hand-crushed noodles or small batches of lasagne pasta, will remain in manual production. Automation and robotisation of repetitive steps, such as packaging, will also come to the fore. In addition, organisational changes are needed to adapt to new jobs that require greater numbers and higher digital and technical skills.

AFSC 4.0 solutions in the pasta industry focus on further automation and robotisation of high-volume production. These technologies support both manufacturing and back-end processes, including big data analytics. Although the data is already available, companies still need analysts to process and use it. Integrating I40 solutions will increase efficiency and productivity, while minimising human error and optimising resource utilisation.

In pasta production, the circular economy aspects include the preference for energy-saving solutions where possible. Although this is not the main focus, such solutions are used for cost reduction purposes, such as heat recovery and heat rotation during the drying process. This helps to optimise energy use and reduce production costs, which is a very important competitive factor.

Smart packaging in pasta production was not particularly emphasised in the interviews. Only the automatic palletiser was mentioned as a substitute for live work. There is also a strong emphasis on energy efficiency in packaging processes to reduce the environmental impact of packaging. Such approaches help minimise waste and improve the environmental footprint of packaging.

The Wine Supply Chain

Traceability in the wine sector is a priority, especially due to the specific challenges of each production region. Digital tools and technologies are being introduced to support traceability, ensuring that all products can be fully traced, helping to maintain and improve product quality.

Although the use of I40 solutions in viticulture is limited (e.g. spraying with drones over wines is not ideal because it does not deliver the sprayed material to the right place and cannot cover large areas with a heavy load), digital tools are increasingly used during winemaking process to improve and maintain product quality. These tools allow accurate monitoring of production processes, quality control and increased efficiency. In order to maintain and improve the quality of the region's wines, production data is constantly monitored to ensure the high quality of every bottle of wine.

Future trends in the wine sector focus on capacity expansion and maintaining or improving quality. This challenge will be met by supporting digitalisation tools. Digital solutions for winemaking are becoming increasingly important, as these technologies allow manual work to be facilitated or completely replaced, such as the use of separately controllable cooling controls for tanks and variable speed pumps.

AFSC 4.0 integrates digitalisation and automation tools into processes to increase efficiency and quality. Digitalisation provides tools that facilitate or completely replace manual tasks as a convenience. Examples include cooling controls that can be adjusted separately for each tank and variable speed pumps, which contribute to the efficiency of production processes and improve product quality.

Energy efficiency is a key priority, as reducing environmental impact is a fundamental objective. To this end, solutions are being implemented that contribute to optimising energy use and achieving sustainability goals.

Smart packaging in the wine sector is constantly evolving. Smart packaging solutions aim to preserve the freshness and quality of products while reducing the environmental impact. Innovative packaging techniques include using recyclable materials and reducing packaging waste. In addition, smart packaging technologies allow products to be tracked and provide important information to consumers.

The Key Actors in the Downstream Process of Food Supply Chains: The Retailers

Tracking and tracing plays a key role in the retail sector, which is ensured by the use of Serial Shipping Container Code (SSCC) codes and RFID technology. SSCC codes allow the tracking of store delivery trolleys from the warehouse to the store. RFID tags enable tracking of shipments and rolls, making the supply chain more efficient. However, tracking is currently only present at the warehouse to store level and relies on supplier performance to meet agreements.

To ensure product availability and quality in the retail sector, there are some good practices and results at retailers who own their own fleet of ships, chocolate and paper factories etc., and a number of own-brand products, practically, vertically integrate the chain. From one hand, for many retailers, fresh fruit and vegetables play a key role in maintaining the competitive advantage. On the other hand, the own-brand products mean a real value too. Customers are committed to own-brand products, and stores strive to offer them a transparent choice and good quality. The temperature of chilled goods is monitored by a digital thermometer system during storage, picking and transport, which helps to maintain product quality. A well-designed and managed sequence of steps is in place to control blockages and recalls, ensuring fast response times due to the high volume of daily transactions.

The increase in demand for frozen products in the retail sector is a significant trend, particularly since the Covid outbreak. Customers are cost-sensitive and there is little strategic cooperation between retailers and suppliers. Labour shortages are also a challenge. In addition, producers often build their own warehouses with subsidies, but are not always able to use them properly due to a lack of appropriate skills. 24-hour logistics are becoming increasingly important in managing complex tasks and online trade also played a significant role during Covid, although it declined afterwards. Sustainability is also an increasingly important consideration.

Digitalisation and automation are key to retailers, too. Modern large floor space warehouses, automated material handling systems and cold storage contribute significantly to the success of the company. Warehouse systems can plan picking routes, track forklift uptime and scheduled maintenance, and use electronic labels on shelves. Analysis of online checkout and point-of-sale loyalty cards enables the creation of customised offers for customers. Automated systems and artificial intelligence can help forecast and automate orders (regular replenishments), but the lack of a unified enterprise system makes reporting difficult.

In the retail sector, there is a strong emphasis on designing and operating energy-efficient and environmentally friendly warehouses. For example, warehouses recover the heat generated by refrigerators and use it to heat the facility, buildings are sunken, trees are planted around them and ponds are created. Waste is turned into biogas, bake-off products are turned into compost or potting soil, and paper is collected and recycled, baling 3-4 truckloads a day. Companies are striving for sustainability, but there is no single set of actions to achieve this.

Smart packaging in the retail sector focuses on to optimise packaging and avoid unnecessary waste. This involves removing packaging where possible, reducing it where possible, reusing and recycling where possible. Their aim is to ensure that all own-brand packaging is fully recyclable. Where packaging removal is not possible, they will work with suppliers to minimise the materials used and use paper and board from sustainable sources.

Table 4. summarizes the comparison of AFSC supply chains.

Supply chains	Meat	Dairy	Conservation	Confectionary	Pasta	Wine	Retail
Traceability challenges	Increases transparency, tracking all processes	RFID in animal tracking, increasing the traceability and transparency	Barcode and RFID systems for pallets and boxes	Global sourcing networks, procurement adaptation, continuous analysis of production volumes	Barcodes all around the processes for unique identification and traceability	Wine region identification and traceability	SSCC codes and RFID tags track shipments between warehouses and stores
Quality management	Continuous improvement to increase production safety and product quality	Statistical control charts, big data analytics, bar-coding	ERP, sensors, X-ray, metal and glass detectors	Recipe changes, technological adjustments for quality	Metal detectors, weight control systems, data collection and analysis for production optimisation, predictive maintenance	Quality assurance	Digital thermometer system for tempered goods, sequence control to manage blockages and recalls
Agrifood 4.0	Production automation and safety (metal detector), increase animal welfare. Data collection and predictive maintenance	Automation, digitalisation, robots, cobots, AI, warehouse management software	Automation and robotisation to handle labour shortage. Data collection to optimize production planning	Digitalisation, automation, ERP systems, 3D technology, data-driven decision making	Automation, robotisation, big data analytics	Increasing importance in production control	Digitalisation, automation, automated material handling, AI for forecasting and order automation
Future trends	Product improvement to meet new customer expectations (salt reduction)	Automation, robotisation, addressing labour shortage	Increasing plant size, automation	Health consciousness, recipe adjustments, technological improvements	Automation for standard products, manual production for traditional items	Increasing digital support in production and quality control, automation of repetitive work	Increase in demand for frozen products, 24-hour logistics, addressing labour shortages

continued on following page

Table 4. Continued

Supply chains	Meat	Dairy	Conservation	Confectionary	Pasta	Wine	Retail
Impact of blockchain	Could be a good tool to increase transparency and reliability, tracks cold chain	Could enhance traceability and transparency, ensures data authenticity	No example in the cases, but for global traceability it is a good opportunity	No example in the cases, but firms admit the importance in traceability	Barcodes are good basis for a future blockchain adaptation, but it is not on agenda yet	Could enhance traceability and transparency, ensures wine authenticity	It is not applicable throughout the whole food chain, players are on different level. But barcodes and RFID are good starting points
Smart packaging	Reducing the use of plastic, increasing the share of recyclable materials	Packaging robots, innovation, energy efficiency	Easy pill packaging, robotic packaging systems	Protects products, reduces environmental impact, new materials development	Packaging size optimisation, packaging robots, automation in palletising	Reduces environmental impact, uses recyclable materials	Optimises packaging, avoids waste, fully recyclable own-brand packaging
Sustainability	Energy efficiency, wastewater treatment, by-product processing	Circular economy, energy efficiency (solar panels)	Energy efficiency by solar panels and gas consumption reduction, and wastewater reuse	Energy efficiency, reduced packaging, recycled materials, waste management	Energy-saving solutions, heat recovery from drying process	Energy saving technologies in production	Energy-efficient warehouses, waste converted to biogas, bake-off products to compost, recyclable paper

Source: own editing, 2024

DISCUSSION

Current Challenges of Agri-Food Supply Chains

The AFSCs face several pressing challenges, with traceability and transparency being paramount across various sectors, such as meat, dairy, and fruit and vegetable processing. Ensuring detailed documentation and monitoring every step of the production process is essential to meet regulatory standards and consumer demands. Technologies like barcode systems, RFID, and blockchain (in the future) play a crucial role in enhancing traceability and transparency (Rahman et al., 2021). In line with work of Ben-Daya et al. (2021), quality control and safety are also significant challenges, necessitating the use of advanced technologies such as metal detectors, weight control systems, big data analytics, and digital devices to monitor feed composition and animal health. Labour shortages and the need for skilled workers are another major issue, particularly in the context of increasing automation and technological advancements. This challenge is evident across the meat, dairy, and fruit and vegetable sectors, where automation efforts aim to reduce reliance on manual labour. Additionally, seasonal variations pose significant challenges, especially in the fruit and vegetable processing sector, necessitating efficient supply chain management and adequate storage capacities during peak seasons.

State-of-the-Art of Digitalisation in Agri-Food Supply Chains

Digitalisation is revolutionizing the AFSCs, with automation and robotization taking centre stage, this is in line with the bibliometric analysis performed previously by Freund (2023). In the dairy industry, milking robots are becoming increasingly common, while the fruit and vegetable sector invests in advanced technologies such as fluidized bed freezing and closed automated production processes. The confectionery and pasta sectors are also embracing automation for repetitive tasks like packaging. The use of big data analytics is integral to optimizing production processes and ensuring quality control. For instance, dairy production meticulously tracks data on milk volume, solids, and germ content, while the fruit and vegetable sector employs data analysis to monitor production efficiency and quality indicators. Despite the emphasis in the literature on the potential of blockchain technology (Astill et al., 2019; Gad et al., 2022), our empirical research does not support this. According to our results, blockchain technology is gaining attraction, particularly in the meat industry, to ensure supply chain transparency and enhance consumer trust by digitally tracking all processes. Although not extensively detailed in all sectors, blockchain holds significant potential for improving traceability and authenticity across various AFSCs. The technologies on which the blockchain will later be built - barcodes, RFID - are there, but the next level has not yet been reached. Integrated management systems, such as ERP systems, are crucial for tracking and managing the supply chain, reducing human error, optimizing resource utilization, and ensuring compliance with food safety regulations.

Sustainability in Agri-Food Supply Chains

Sustainability is a core focus in the AFSCs, with energy efficiency and the adoption of renewable energy being key priorities. It can be concluded that sustainability is more strongly represented in the supply chains analysed than in the cluster identified by Freund (2023), despite the fact that the results are in line with those results reported in the literature, e.g. efforts in terms of waste management (Ada et al., 2021; Esmaeilian et al., 2018) in all areas. The meat and dairy sectors are investing in new boilers and solar panels to reduce gas consumption and ensure a sustainable energy supply, while the fruit and vegetable sector emphasizes energy-saving measures like heat recovery during drying processes. Waste reduction and the promotion of a circular economy are also critical goals. The meat sector, for instance, is developing by-product processing plants to convert waste into useful products like meat meal, while the dairy and fruit and vegetable sectors invest in wastewater treatment plants to minimize environmental impact. Smart packaging solutions aim to reduce the environmental footprint by using recyclable materials and minimizing resource usage. The confectionery industry is making significant strides in this area by increasing the use of recycled plastics and developing new packaging materials to meet sustainability goals. Some retailers are adopting strategies to optimise packaging and ensure that it becomes fully recyclable in the forthcoming years. Furthermore, efforts to reduce the environmental footprint include various measures implemented across sectors, such as energy efficiency initiatives and waste-to-energy conversions. These collective efforts highlight the ongoing commitment to enhancing the efficiency, transparency, and sustainability of AFSCs through technological advancements and innovative practices.

CONCLUSION

Many sectors in the AFSCs, such as meat, dairy, pasta, fruit and vegetable processing, conservation and confectionery, are facing different challenges. In order to achieve transparency and efficiency in the supply chains, it is necessary to improve traceability, which can be achieved through the use of barcodes, SSCC and RFID technologies. In addition, the cost sensitivity of customers and the complexity of supply chains further complicate the situation.

Digitalisation is playing an increasingly important role in AFSCs. Digitised traceability systems help maintain product quality and safety, precision farming and processing technologies increases efficiency, robotisation and automation aim to achieve cost-efficiency and reduce labour shortages. Through the use of data analytics and big data, companies are able to better forecast demand and optimise supply chains.

Sustainability is becoming a growing focus in AFSCs. Following the principles of the circular economy, companies are striving to reduce and recycle waste. Policymakers should prioritize enhancing traceability and food safety standards across all sectors of the AFSC. This involves mandating the adoption of advanced technologies like blockchain, RFID, and barcode systems to ensure comprehensive transparency and safety from farm to table. Providing subsidies or incentives to smaller businesses for implementing these technologies can further support this initiative. Additionally, promoting sustainable practices through regulations and incentives is crucial. This includes supporting renewable energy projects, waste reduction initiatives, and the adoption of circular economy principles to minimize environmental impact.

Investment in digital infrastructure and training programs is also essential. Policymakers should encourage the integration of I40 technologies, such as big data analytics, AI, and IoT, into AFSCs. This can be facilitated through grants, tax incentives, and public-private partnerships to foster innovation and enhance productivity. Addressing labour shortages and skill gaps is another critical area. Developing educational programs and vocational training focused on the skills needed for modern AFSCs will ensure the workforce is equipped with necessary expertise in automation, digital technologies, and advanced manufacturing processes.

Researchers should focus on developing and refining technologies that enhance traceability and transparency within the supply chain. This includes exploring the potential of blockchain, IoT sensors, and AI to create integrated, secure, and easily accessible traceability systems. Sustainable production technologies should also be a primary research focus, with an emphasis on innovative methods to increase energy efficiency and reduce waste in food production and processing. This includes studying renewable energy sources, waste-to-energy technologies, and sustainable packaging materials.

Additionally, research in automation and robotics should aim to improve the efficiency and reliability of production processes, developing scalable and affordable robotic solutions for various supply chain stages. The use of big data and AI to optimize supply chain operations is another critical research area, involving the creation of algorithms for predictive maintenance, yield optimization, and real-time decision-making to enhance productivity and reduce losses. Finally, understanding consumer preferences and market trends is vital for developing products that meet demand while adhering to sustainability goals. Researchers should study consumer behaviour related to health consciousness, sustainability, and the adoption of smart packaging, as well as innovations that improve the nutritional value and safety of food products.

By focusing on these recommendations, policymakers and researchers can collaboratively enhance the efficiency, sustainability, and resilience of AFSCs, ensuring a robust and secure food system for the future.

ACKNOWLEDGEMENT

Supported by the ÚNKP-23-4-II-CORVINUS-18 New National Excellence Program of the Ministry for Culture and Innovation from the source of the National Research, Development and Innovation Fund.

REFERENCES

Ada, N., Kazancoglu, Y., Sezer, M. D., Ede-Senturk, C., Ozer, I., & Ram, M. (2021). Analyzing barriers of circular food supply chains and proposing industry 4.0 solutions. *Sustainability (Basel)*, 13(12), 1–29. DOI: 10.3390/su13126812

Anderson, K., & Gaston, K. J. (2013). Lightweight unmanned aerial vehicles will revolutionize spatial ecology. In *Frontiers in Ecology and the Environment* (Vol. 11, Issue 3, pp. 138–146). DOI: 10.1890/120150

Apriliyanti, I. D., & Alon, I. (2017). Bibliometric analysis of absorptive capacity. *International Business Review*, 26(5), 896–907. DOI: 10.1016/j.ibusrev.2017.02.007

Aria, M., & Cuccurullo, C. (2017). bibliometrix: An R-tool for comprehensive science mapping analysis. *Journal of Informetrics*, 11(4), 959–975. DOI: 10.1016/j.joi.2017.08.007

Astill, J., Dara, R. A., Campbell, M., Farber, J. M., Fraser, E. D. G., Sharif, S., & Yada, R. Y. (2019). Transparency in food supply chains: A review of enabling technology solutions. *Trends in Food Science & Technology*, 91(July), 240–247. DOI: 10.1016/j.tifs.2019.07.024

Bayramova, A., Edwards, D. J., & Roberts, C. (2021). The role of blockchain technology in augmenting supply chain resilience to cybercrime. *Buildings*, 11(7), 283. Advance online publication. DOI: 10.3390/buildings11070283

Ben Ayed, R., & Hanana, M. (2021). Artificial Intelligence to Improve the Food and Agriculture Sector. *Journal of Food Quality, 2021*(Ml). DOI: 10.1155/2021/5584754

Ben-Daya, M., Hassini, E., Bahroun, Z., & Banimfreg, B. H. (2021). The role of internet of things in food supply chain quality management: A review. *The Quality Management Journal*, 28(1), 17–40. https://doi.org/https://doi.org/10.1080/10686967.2020.1838978. DOI: 10.1080/10686967.2020.1838978

Béné, C. (2020). Resilience of Local Food Systems and Links to Food Security – A Review of Some Important Concepts in The Context of COVID-19 and Other Shocks. *Food Security*, 12(4), 805–822. https://link.springer.com/article/10.1007/s12571-020-01076-1. DOI: 10.1007/s12571-020-01076-1 PMID: 32837646

Bibi, F., Guillaume, C., Gontard, N., & Sorli, B. (2017). A review: RFID technology having sensing aptitudes for food industry and their contribution to tracking and monitoring of food products. *Trends in Food Science & Technology*, 62, 91–103. DOI: 10.1016/j.tifs.2017.01.013

Chen, R. Y. (2017). An intelligent value stream-based approach to collaboration of food traceability cyber physical system by fog computing. *Food Control*, 71, 124–136. DOI: 10.1016/j.foodcont.2016.06.042

Chen, S., Brahma, S., Mackay, J., Cao, C., & Aliakbarian, B. (2020). The role of smart packaging system in food supply chain. *Journal of Food Science*, 85(3), 517–525. DOI: 10.1111/1750-3841.15046 PMID: 32056210

Drago, E., Campardelli, R., Pettinato, M., & Perego, P. (2020). Innovations in smart packaging concepts for food: An extensive review. *Foods*, 9(11), 1628. Advance online publication. DOI: 10.3390/foods9111628 PMID: 33171881

Duong, L. N. K., Al-Fadhli, M., Jagtap, S., Bader, F., Martindale, W., Swainson, M., & Paoli, A. (2020). A review of robotics and autonomous systems in the food industry: From the supply chains perspective. *Trends in Food Science & Technology*, 106(May), 355–364. DOI: 10.1016/j.tifs.2020.10.028

Esmaeilian, B., Wang, B., Lewis, K., Duarte, F., Ratti, C., & Behdad, S. (2018). The future of waste management in smart and sustainable cities: A review and concept paper. *Waste Management (New York, N.Y.)*, 81, 177–195. DOI: 10.1016/j.wasman.2018.09.047 PMID: 30527034

EU Green deal. (2020). Farm to Fork Strategy. In *European Comission*. https://ec.europa.eu/food/sites/food/files/safety/docs/f2f_action-plan_2020_strategy-info_en.pdf Accessed: 28. 06. 2024

European Comission CAP. (2021). *A Greener and Fairer CAP (Common Agricultural Policy)*. https://agriculture.ec.europa.eu/system/files/2022-02/factsheet-newcap-environment-fairness_en_0.pdf Accessed: 28. 06. 2024

FAO. (2018). *TRANSFORMING FOOD AND AGRICULTURE TO ACHIEVE THE SDGs 20 interconnected actions to guide decision-makers*. https://www.fao.org/3/I9900EN/i9900en.pdf Accessed: 28. 06. 2024

FAOSTAT. (2022). *Commodities Production by country*. https://www.fao.org/faostat/en/#rankings/commodities_by_country Accessed: 28. 06. 2024

Freund, A. (2023). Effects of digitalisation on food safety. *Studies in Agricultural Economics (Budapest)*, 125(3), 154–168. DOI: 10.7896/j.2498

Gad, A. G., Mosa, D. T., Abualigah, L., & Abohany, A. A. (2022). Emerging Trends in Blockchain Technology and Applications: A Review and Outlook. *Journal of King Saud University. Computer and Information Sciences*, 34(9), 6719–6742. DOI: 10.1016/j.jksuci.2022.03.007

Hashem, I. A. T., Yaqoob, I., Anuar, N. B., Mokhtar, S., Gani, A., & Ullah Khan, S. (2015). The rise of "big data" on cloud computing: Review and open research issues. In *Information Systems* (Vol. 47, pp. 98–115). Elsevier Ltd., DOI: 10.1016/j.is.2014.07.006

Hassoun, A., Prieto, M. A., Carpena, M., Bouzembrak, Y., Marvin, H. J. P., Pallarés, N., Barba, F. J., Punia Bangar, S., Chaudhary, V., Ibrahim, S., & Bono, G. (2022). Exploring the role of green and Industry 4.0 technologies in achieving sustainable development goals in food sectors. *Food Research International*, 162(August), 112068. Advance online publication. DOI: 10.1016/j.foodres.2022.112068 PMID: 36461323

Jámbor, Z., & Nagy, J. (2022). Resilience of Food Supply Chains – a Dynamic Capability Approach. *Ekonomska Misao i Praksa*, 31(2), 473–486. DOI: 10.17818/EMIP/2022/2.6

Liaghat, S., & Balasundram, S. K. (2010). A Review: The Role of Remote Sensing in Precision Agriculture. *American Journal of Agricultural and Biological Sciences*, 5(1), 50–55. DOI: 10.3844/ajabssp.2010.50.55

Maditati, D. R., Munim, Z. H., Schramm, H. J., & Kummer, S. (2018). A review of green supply chain management: From bibliometric analysis to a conceptual framework and future research directions. *Resources, Conservation and Recycling*, 139(August), 150–162. DOI: 10.1016/j.resconrec.2018.08.004

Magalhães, A. E. V., Rossi, A. H. G., Zattar, I. C., Marques, M. A. M., & Seleme, R. (2019). Food traceability technologies and foodborne outbreak occurrences. *British Food Journal*, 121(12), 3362–3379. DOI: 10.1108/BFJ-02-2019-0143

Marchiori, D., & Franco, M. (2020). Knowledge transfer in the context of inter-organizational networks: Foundations and intellectual structures. *Journal of Innovation and Knowledge*, 5(2), 130–139. DOI: 10.1016/j.jik.2019.02.001

Miranda, J., Ponce, P., Molina, A., & Wright, P. (2019). Sensing, smart and sustainable technologies for Agri-Food 4.0. *Computers in Industry*, 108, 21–36. DOI: 10.1016/j.compind.2019.02.002

Musa, S. F. P. D., & Basir, K. H. (2021). Smart farming: Towards a sustainable agri-food system. *British Food Journal*, 123(9), 3085–3099. DOI: 10.1108/BFJ-03-2021-0325

Mustapha, U. F., Alhassan, A. W., Jiang, D. N., & Li, G. L. (2021). Sustainable aquaculture development: A review on the roles of cloud computing, internet of things and artificial intelligence (CIA). *Reviews in Aquaculture*, 13(4), 2076–2091. DOI: 10.1111/raq.12559

Nasirahmadi, A., & Hensel, O. (2022). Toward the Next Generation of Digitalization in Agriculture Based on Digital Twin Paradigm. *Sensors (Basel)*, 22(2), 498. https://doi.org/https://doi.org/10.3390/s22020498. DOI: 10.3390/s22020498 PMID: 35062459

Nemes, S. A., Szabo, K., & Vodnar, D. C. (2020). Applicability of agro-industrial by-products in intelligent food packaging. *Coatings*, 10(6), 1–26. DOI: 10.3390/coatings10060550

Óskarsdóttir, K., & Oddsson, G. V. (2019). Towards a decision support framework for technologies used in cold supply chain traceability. *Journal of Food Engineering*, 240, 153–159. DOI: 10.1016/j.jfoodeng.2018.07.013

Qu, C. C., Sun, X. Y., Sun, W. X., Cao, L. X., Wang, X. Q., & He, Z. Z. (2021). Flexible Wearables for Plants. *Small*, 17(50), 2104482. Advance online publication. DOI: 10.1002/smll.202104482 PMID: 34796649

Rahman, L. F., Alam, L., Marufuzzaman, M., & Sumaila, U. R. (2021). Traceability of sustainability and safety in fishery supply chain management systems using radio frequency identification technology. *Foods*, 10(10), 2265. Advance online publication. DOI: 10.3390/foods10102265 PMID: 34681313

Raj, M., Gupta, S., Chamola, V., Elhence, A., Garg, T., Atiquzzaman, M., & Niyato, D. (2021). A survey on the role of Internet of Things for adopting and promoting Agriculture 4.0. *Journal of Network and Computer Applications*, 187(April), 103107. DOI: 10.1016/j.jnca.2021.103107

Rana, N. P., Chatterjee, S., Dwivedi, Y. K., & Akter, S. (2022). Understanding dark side of artificial intelligence (AI) integrated business analytics: Assessing firm's operational inefficiency and competitiveness. *European Journal of Information Systems*, 31(3), 364–387. DOI: 10.1080/0960085X.2021.1955628

Saunders, M., Lewis, P., & Thornhill, A. (2009). *Research methods for business students*. Pearson education.

Tsolakis, N., Aivazidou, E., & Srai, J. S. (2019). Sensor applications in agrifood systems: Current trends and opportunities for water stewardship. *Climate (Basel)*, 7(3), 44. Advance online publication. DOI: 10.3390/cli7030044

Weber, D. M., & Kauffman, R. J. (2011). What drives global ICT adoption? Analysis and research directions. *Electronic Commerce Research and Applications*, 10(6), 683–701. DOI: 10.1016/j.elerap.2011.01.001

Wolfert, S., Ge, L., Verdouw, C., & Bogaardt, M. J. (2017). Big Data in Smart Farming – A review. *Agricultural Systems, 153*, 69–80. rrrrrhttps://DOI: 10.1016/j.agsy.2017.01.023

Yousefi, H., Su, H.-M., Imani, S. M., Alkhaldi, K., Filipe, C. D. M., & Didar, T. F. (2019). Intelligent Food Packaging: A Review of Smart Sensing Technologies for Monitoring Food Quality. *ACS Sensors*, 4(4), 808–821. https://doi.org/https://doi.org/10.1021/acssensors.9b00440. DOI: 10.1021/acssensors.9b00440 PMID: 30864438

APPENDIX

Table 5. Lexicon of digital technologies

Name of technological solution	Definition	Reference
Artificial Intelligence (AI)	AI is "a field in computer science that aims to equip machines with learning mechanisms that can enable them to make decisions based on past experiences by mimicking the cognitive functions of living things."	Mustapha et al., 2021, p2078
Big Data	Big data "is a term utilized to refer to the increase in the volume of data that are difficult to store, process, and analyse through traditional database technologies. The nature of big data is indistinct and involves considerable processes to identify and translate the data into new insights."	Hashem et al., 2015, p100
Blockchain	Blockchain "is a distributed ledger feature, every record in this ledger is secured by rules of cryptography which makes it more secure and tamper-free."	Gad et al., 2022, p2
Cloud Computing	Cloud computing "refers to the practice of employing a network of remote servers hosted on the internet for storage, management, and data processing rather than on a local server or personal computer."	Mustapha et al., 2021, p2077
Cyber Physical Systems (CPS)	CPS is "a computational operation with the surrounding physical object across downstream/upstream industry in collaboration environment."	Chen, 2017. p125
Digital Twin	Digital twin approaches "make it possible to model, design, simulate, and develop agricultural machinery that would yield more productive machines in terms of energy and power efficiencies."	Nasirahmadi & Hensel, 2022, p6
Drones/UAVs	Unmanned aerial vehicles (UAVs) "can carry various imaging or non-imaging payloads to provide spatial datasets for a variety of end users."	Anderson & Gaston, 2013, p138
Flexible Wearables for plants	"In agriculture, flexible electronics are expected to not only solve the incompatibility between the sensors and the fragile parts of a plant but also conformably adhere to the plant surface to achieve continuous and accurate detection of growth state, physiological or biochemical parameters of plants."	Qu et al., 2021, p2
Information and Communication Technologies (ICT)	ICT can be defined as "technologies that support data and information processing, storage and analysis, as well as data and information transmission and communication, via the Internet and other means"	Weber & Kaufman, 2011 p684
Internet of Things (IoT)	IoT "refers to the concept of creating a network of physical computing devices capable of information gathering and sharing."	Mustapha et al., 2021, p2077
Machine Learning	Machine-based learning systems "are capable of managing a wide variety of inputs and nonlinear operations."	Raj et al., 2021, p2

continued on following page

Table 5. Continued

Name of technological solution	Definition	Reference
Precision Farming	PF "is an integrated, information- and production-based farming system that is designed to increase long term, site-specific and whole farm production efficiency, productivity and profitability while avoiding the undesirable effects of excess chemical loading to the environment or productivity loss due to insufficient input application."	Liaghat & Balasundram, 2010, p50
Robotics and Autonomous Systems (RAS)	RAS "allow the use of specialised machinery that operates in the field carrying out complete work cycles in tasks such as sowing and harvesting, the operation of semi-automatic and automated greenhouses, as well as great progress in the crop techniques."	Miranda et al., 2019, p21
Radio Frequency Identification (RFID)	RFID "is used for product identification and for gathering information on items automatically, without a physical connection or line of sight as required for other technologies such as barcodes."	Bibi et al., 2017, p92
Sensor	Sensors "can form a system that collects and transmits sensing data from physical or environmental conditions."	Óskarsdóttir & Oddsson, 2019, p155

Source: own collection based on literature, 2024

Chapter 8
Agricultural Warehousing: Functional Insights, Challenges, and Best Practices

Priyanka Pandey
https://orcid.org/0000-0002-1144-8957
DBS Global University, Dehradun, India

Satish Chandra Pant
https://orcid.org/0000-0002-1503-804X
DBS Global University, Dehradun, India

Parminder Singh
https://orcid.org/0009-0001-0742-1073
CCS HAU, Hisar, India

Hema Yadav
https://orcid.org/0009-0006-6862-4866
Vaikunth Mehta National Institute of Cooperative Management, India

ABSTRACT

The chapter reviews the current state of agriculture warehousing worldwide; this is particularly important in reducing postharvest losses and increasing food safety. It points to the economic and social consequences of faulty storage solutions in developing countries. It stresses the necessity for upgrading agricultural storage, sharing technologies, incorporating online storage, and applying modern storage facilities, especially among smallholder farmers. The issue of agri-warehousing, however, runs into numerous challenges in infrastructure, technologies, and regulatory frameworks in both developed economies as well as in emerging economies. Given the emphasis on theory, policy, and best practices, it is hoped that this chapter will provide a necessary foundation for better understanding of the agricultural

DOI: 10.4018/979-8-3693-4330-2.ch008

warehouse system from a more practical perspective.

1. INTRODUCTION

Globally, one-third of the edible part of food meant for human consumption, which amounts to 1.3 billion tonnes, is lost throughout the supply chain (FAO, 2017). A significant portion of this loss is due to improper storage, which can account for 10-20% of overall wastage (Jha et al., 2015). This type of loss directly reduces farmers' income and standard of living. Additionally, it squanders a substantial portion of the nation's food supply, creating unnecessary waste (Olorunfemi and Kayode, 2021). Recognizing this, the agriculture sector has placed increasing importance on warehouses as critical to upholding food quality and ensuring food security (Lydia et al., 2022). High-quality agricultural warehousing directly combats post-harvest losses, leading to the consistent availability of seasonal crops (Naik et al., 2022). This, in turn, stabilizes prices and strengthens food security across the globe (Mobolade et al., 2019).

Due to inferior agricultural infrastructure, post-harvest losses in developing countries are significantly higher than in developed countries. The loss of post-harvest loss in Sub-Saharan Africa varies from 30% to 50% (Makule et al., 2022). This includes inefficient farming practices, poor transportation networks, outdated storage, lack of processing technology, and under-managed market systems (Abass et al., 2014). Table 1 shows that the developing world experiencing substantial food losses in comparison to developed world in post-harvest operations of the supply chain (FAO 2017). These losses have far-reaching consequences, including direct weight loss from spoilage, degraded product quality, reduced nutritional value, compromised seed viability, and ultimately, setbacks to the commercial viability of farming (Kumar and Kalita, 2017). Insufficient access to modern warehousing particularly impacts small and marginal farmers within the supply chain (Naik et al., 2022).

Table 1. Percentage distribution of post harvest losses region wise

Region	Post Harvest Loss (in %)	Total food loss and waste (in %)	Percentage of Post-harvest loss in total food loss
Europe including russian federation	3.5	32.4	10.8
North America/Oceania	3.4	31.4	10.83
Japan/Republic of Korea/China	6.6	33.4	19.76
Latin America	7.5	36.2	20.72
North Africa/West and central Asia	7.8	36	21.67
Sub Saharan Africa	12.7	35.6	35.67
South Southeast Asia	9.6	27.2	35.29

Source: FAO 2017

Nevertheless, grain spoilage during storage contributes significantly to undernourishment and food insecurity, tragically impacting millions worldwide (Mobolade et al., 2019). In case of African countries, outdated conventional practices including poor storage facilities was identified as a prime reason for food spoilage. These methods of storing food and grains often required extensive physical labor, resulting in inefficiency, time constraints (Lydia et al., 2022), and limited parameter monitoring (Rai, 2022). Most of the focus has been on boosting crop yield within the water user associations (WUAs), with minimal attention given to enhancing postharvest management. (Díaz-Valderrama 2020). As food grains are essential for humans and livestock consumption, particularly in developing nations, significance of post-harvest management is paramount and thereby agri-warehousing plays pivotal role in minimising post-harvest losses. Effective grain storage techniques can greatly reduce food loss for smallholder farmers, leading to significant improvements in their livelihoods (Manandhar et al 2018).

For grain storage, warehousing is treated as a regulated agricultural service across the globe. United States warehousing Act (USWA) is one of the oldest regulations that was basically enacted to provide licencing to warehouseman to help producers (Schultz, 1981). Although, licencing is voluntary in US. Applicants agree to be licensed under the USWA. The purpose of this act was not limited to regulate the warehouse for supervising and control of storage practices but has several other purposes. Firstly, it encouraged farmers to store food products to minimise the losses due to lack of post-harvest practices. Secondly, to create unified system of warehousing for uses of warehouse receipt system (Yohe, 1924). Similar legislations were percolated in the developing countries with the aim to promote scientific warehousing practices and more specifically to establish a system of negotiable warehouse receipts in the country.

In emerging markets like India dedicated authority to regulate warehouse operations are in place Ex: Warehouse Development Regulatory Authority (WDRA) aims to make warehouse receipts a key trading tool and facilitate financing against storage commodity with prior assaying facility. This process, not only improve supply chain operations also enhance the grading and quality of goods, ultimately leading to higher returns for depositors. These receipts have been made mandatory by the WDRA, wherein all warehouseman has to ensure issuance of electronic negotiable warehouse receipt (eNWR) to prevent any fraudulent activity in the pledge financing.

In consideration of the context provided above, the purpose of this chapter is to provide a comprehensive discussion centered on the theory, policy, and practices for beginners aiming to comprehend the agricultural warehouse ecosystem worldwide. This chapter serves as a crucial guide for understanding contemporary operations in the agricultural warehousing sector. The chapter begins with a section defining the concept of warehouse and warehousing, followed by a section outlining the functions of agri-warehouses. The fourth section addresses the challenges in agri-warehousing, followed by a section detailing the best practices in agri-warehousing, and ends with a concluding section

2. DEFINITION AND CONCEPT OF WAREHOUSE IN AGRICULTURE

In the agricultural supply chain and agricultural marketing process, large quantum of agricultural commodities is placed in the custody of warehouses for several purposes such as food security and agricultural growth. Warehouses are defined as a structure of scientific storage build with the specific goal of preserving the quality and quantity of stored items (Acharya and Agarwal 2016). The area of warehouse in agriculture is under the control of law of land or regulations either partially or fully depend on the State legislation. Hence, largely warehouses are regulated structures. Further, regulations for the United States warehouse Act define warehouse as *"a structure or other authorized storage facility, as determined by Agricultural Marketing Service (AMS), in which any agricultural product may be stored or handled for the purpose of interstate or foreign commerce."* This Act is one of the earliest federal laws protecting food, ensures safeguarding the property of its owners or intended recipients by requiring warehouse operators to be licensed and financially secured.

In India, warehouses are regulated under the Warehousing (Development and Regulation) Act 2007 (WDRA Act). The act defines the warehouse as *"any premises (including any protected place) conforming to all the requirements including manpower specified by the Authority by regulations wherein the warehouseman takes custody of the goods deposited by the depositor and includes a place of storage of goods under controlled conditions of temperature and humidity"* under the WDRA Act 2007. Similarly, every nation has its own definition of warehouse which includes all the requirements of scientific storage facilities under certain regulation due to its importance in the food security as an important area of national interest.

The function of a warehouse for storage is crucial in preserving agricultural produce and maintaining its quality and quantity. Storing harvested crops for extended periods is vital for the advancement of agriculture (El-Ramady et al, 2015). In ancient times, hermetic grain storage was practiced in dry regions like the Middle East, Africa, and India, using methods such as underground pits and local cribs (Olarufemi and Kayode, 2021). Even today, many smallholder farmers utilize traditional methods like woven granaries and underground pits to protect their grain from spoilage and pests (Manandhar et al, 2018). Proper storage infrastructure and systems are essential for successful agricultural marketing and managing efficient supply chains. This is aligned with the role of futures markets in helping agents decide when to store (El- Ramady et al, 2015).

3. FUNCTIONS OF AGRI WAREHOUSES IN THE SUPPLY CHAIN

Agricultural produce is highly perishable in nature, it must reach consumers quickly to ensure optimal quality. Warehouses provide essential storage solutions, extending the freshness of produce while awaiting distribution, stabilize price in the market, provide market intelligence, and support in credit facilities. The important functions of agri warehouses in the supply chain are as under-

i. Scientific Storage: Food security is constantly threatened by food loss and waste; scientific storage practices are a vital tool to combat this issue. Warehouses offer secure spaces to store agricultural produce, preventing losses and wastages due to pests, weather, fire, or theft. The primary goal is to keep the crop fresh for as long as possible (Kumar et al 2021). Scientifically designed warehouses utilize temperature control, humidity management, and pest control measures to preserve the quality of perishable goods like fruits, vegetables, and grains.
ii. Agri-Financing: Agricultural production and trade are inherently risky due to low margins and high uncertainties. This makes it challenging and expensive for farmers to secure financing through traditional channels (WDRA, 2015). Warehouse receipt financing offers a powerful solution by enabling farmers to leverage stored commodities as collateral. By issuing warehouse receipts against their produce, farmers can obtain loans from banks and other financial

institutions, smoothing out income fluctuations and accessing much-needed capital at critical junctures. This liquidity injection empowers them to invest in better inputs, improve yields, and strengthen their overall financial well-being.

Warehousing systems offer farmers an alternative to selling their crops at low harvest-time prices that means it minimizes distress sale. By storing their produce in public, private or co-operative led warehouses and using warehouse receipts as collateral, farmers can obtain short-term loans to cover immediate financial needs (See figure 1). Later, when market prices improve, they can sell their products and repay the loans (Kovačević et al., 2023). This system reduces risk for banks approving loans against warehouse receipts, as the stored commodity guarantees delivery. As a result, the warehouse receipt financing system lowers costs and speeds up loan processing. Although, the use of warehouse receipt is limited in many developing countries because of institutional and structural shortcomings (Lacroix & Varangis, 1996).

Figure 1. Agri-financing through negotiable warehouse receipt (NWR)

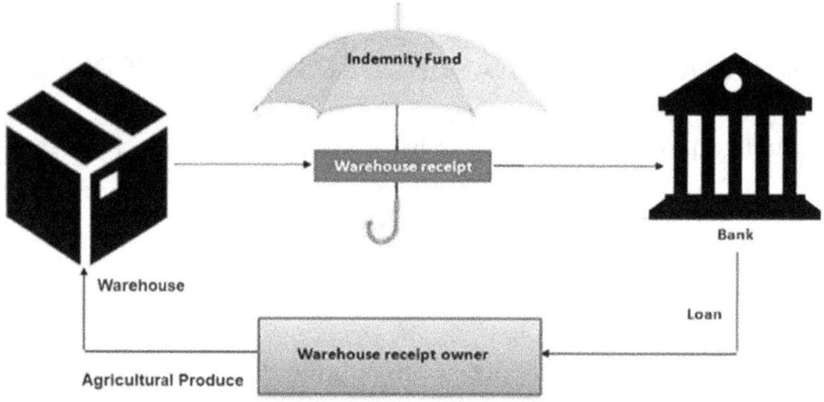

Source: Authors compilation based on industry practice

International institutions such as International Finance corporation (IFC) has joined hands with its unique program i.e. global warehouse finance program (GWFP), to increase working capital financing by using stored commodities as collateral. This initiative is a part of IFC's efforts to improve financial access for farmers and promote agricultural development to combat poverty (IFC, n.d.). In Latin America, Mexico has reformed its warehouse receipt system as centralized registry (Pledge bond) known as RUCAM. In Asia, Pakistan Mercantile Exchange Ltd. (PMEX) is in progress to provide end to end trading facility to provide eNWR to its farmers through Collateral Management Company (CMC). Emerging economies such as

India through digital India campaign introduced electronic negotiable warehouse receipts (e-NWRs) under the Warehousing (Development and Regulation) Act, 2007, e-NWRs are digital documents issued by WDRA-registered warehouses. These digital documents act as proof of ownership for stored commodities and allow farmers to use them as collateral for securing affordable credit. e-NWRs also streamline trade with electronic transfers, promoting price discovery and reducing post-harvest losses. While wider adoption of e-NWRs hinges on increased warehousing infrastructure and farmer awareness, they hold great potential to transform India's agricultural financing landscape, offering benefits for both farmers and the broader economy.

iii. Price Stabilization: To achieve price stability, governments can accumulate stock. Then, excess or shortage in demand and supply would be modulated by the above stock. While this does not ensure that prices will never move to a level outside the target, it improves the chances of the fluctuation level being a band within the chosen limits (Johnson, 1975). There is a lack of capacity to adequately address agri-warehousing in terms of adequately stabilizing prices. There is an increase in uncertainty when there is a shortage of ideal warehouse space or when basic information about storage is unavailable to buyers and sellers. Such uncertainty interferes with the price discovery mechanism, resulting in high volatility and broader gaps between wholesale and retail prices. On the other hand, with adequate warehousing facilities and efficient information flows, farmers can keep the crops in storage to wait for periods of low supply. NCAP- staff training and extension component, IIM Bangalore (India) report related to the Indian agrarian market suggested that more warehouses could reduce the disparity between wholesale and retail prices. In addition, it could address the long-term instability of agricultural commodity prices (Naik et al., 2020). Controlled distribution of goods from storage facilities helps avert too much supply and too little supply to the market, thus achieving price control. Gradual release of goods from warehouses prevents sudden market gluts or shortages, ultimately helping to stabilize prices and protect both farmers and consumers (Acharya and Agarwal 2016).
iv. Market Intelligence: Warehouses provide an invaluable service by offering market information to those who store their goods with them. They guide and inform them about prevailing prices and recommend opportune moments for selling their products, empowering them to make informed and profitable decisions (Acharya and Agarwal 2016). Further, commodity stored in warehouses linked with future market are providing market intelligence for many farmers, traders, and buyers. For example- National Commodity Derivative Exchange (NCDEX) of India has a reason to serve as an efficient platform not only for Price Discovery and Price Risk Management but has been instrumental in providing market intelligence

to many stakeholders on spot and future prices of listed commodities over the past two decades. The prices on NCDEX are widely considered a standard in both the domestic and international commodities market (Naik et al., 2022). Similarly, it has been reported that the presence of delivery warehouses has a favourable effect on the price discovery performance of futures in the Chinese futures market (Yang et al., 2020).

v. Value addition: Warehouses improve the shelf life of agricultural produce, thereby contributing to the value addition process and assisting in price realization at the producer's level. In many cases, they help them link with institutional buyers. For example, the NCDEX platform of accredited warehouses helps farmers eliminate intermediaries in the chain and link producers to the market (Chengappa, 2018).

4. CHALLENGES IN AGRI WAREHOUSING

4.1 Infrastructural Challenges

The storage of perishable produce is the biggest challenge as it requires strong infrastructure. The absence of enough infrastructure causes poor supply chain performance, which leads to significant post-harvest loss. There was a study conducted conducted by . (Olomu et al., 2020). That indicates the strong impact of poor infrastructure in Nigeria; the similar study conducted by (Singh et al.,2023) highlights that 35 to 40% of post-harvest losses are due to poor infrastructural only which puts a substantial burden on farmers and lack of storage is one of the reasons for the lack of overall benefits (Singh et al, 2023). However, this area exhibits great potential for growth despite requiring a small initial investment and offering the possibility of high income.

Modern warehousing provides an excellent advantage in terms of environmental benefits by reducing waste and improving the efficiency of the food value chain. It also promotes the effective utilization of natural resources, boosts economic growth by saving energy and costs, and raises the earnings of farmers in rural areas (Makule et al, 2022). Further, in countries like India, micro warehousing practices such as the setup of hermitage warehouses at the farm gate level facilitate the connection between traders/processors and farmers for selling their produce at a later stage when prices of the produce may increase soon. This approach not only provides broader accessibility to farmers' produce by connecting it to the alternate market channels but also minimizes distress sales among farmers. Investors took notice of micro-warehousing, resulting in funding for expanding operations in emerging markets.

Also, the warehouse requirements corresponding to the agricultural distribution of the produce also change as commodities, say grains require different levels of control; but perishables like fruits and vegetables technological preservation control parameters are different. Different commodities have different requirements such as temperature control, humidity control, oxygen control etc. which makes the warehousing of such commodities very complicated and so is the handling of those crops. Such warehousing operations have seasonal characteristics as regards the crops grown, requiring sufficiently large amounts of capital for almost every form of construction and marketing of agriculture in the region. In addition, other factors such as warehouse purely receipt system related costs: transportation, handling and storage fees, interest rate, etc. may limit farmers' participation as a challenge (Adjognon et al, 2019).

Moreover, there are changing marketing models, such as direct-to-consumer models, E-commerce, and farmer groups that produce agricultural goods and market them, which require warehouses to satisfy consumers' continuous demand. Warehouses feel the pressure to optimize their workflows and adapt technologies because of the requirements of e-commerce and direct-to-customers.

4.2 Technological Challenges

Storing food stock in the warehouse for a more extended period is not helpful anymore because of the wastage that comes with such long storage periods. Such losses tend to grow with time incrementally, and a study done recently by Huss et al. (2021) indicates that up to 25.6% of maize production is lost during the postharvest stage when suitable storage technologies are not utilized (Huss et al., 2021). Reducing such storage losses will enable small-scale farmers to secure their produce longer, increasing the food proportion in households and communities. High stock levels help farmers prepare for inevitable shortfalls, like during lean periods in agriculture, and also help deal with unexpected shocks in supply. In Tanzania, improved on-farm storage practices have alleviated food insecurity for smallholder farmers throughout the lean season.

4.3 Regulatory Challenges

Agricultural warehousing faces numerous regulatory challenges globally, encompassing safety standards, certification requirements, and other compliance issues. Here are some key regulatory challenges-

Safety Standards

The storage of perishable items has its peculiarity as it includes products necessary for everyday lives and actual characteristics like time of availability and regulation by specific food safety and quality standards. As it is with food security, safety, and good quality in most countries, it has been established that during the procedures of storage of commodities, some basic problems like moisture from the respiration of the grain, chambers, and pores in concrete walls for the passage of insects or rats, dampness caused by rainfall, and the heat of the summer sun beating directly on that section of the house where the storage of the commodities is done are still rampant across the nations. (Hamel et al., 2020). Additionally, research indicates that pests significantly damage stored commodities, contributing to approximately 10-20% of total storage losses (Guru et al., 2022). However, a well-constructed warehouse with special impermeability treatments can address these challenges (Hamel et al., 2020). Therefore, for governing the physical infrastructure of the warehouse, the majority of the legislation has a mandatory provision for using warehouse structure standards such as design and construction standards, flooring standards, roofing standards, maintenance standards, etc., to achieve efficiency and effectiveness in the warehouse safety and standardization (WAI, 2022).

Quality Standards

As of now, however, the buyer appreciates low prices along with high quality food. More than ever, they want to know more about nutrition and especially, how it is grown, how it is processed and stored. Over the last few decades, a number of both public and private quality assurance (QA) systems have become operational initially in the agri-business and subsequently within the food industry. Most notable in this regard are Hazard Analysis and Critical Control Point i.e. HACCP, Good Agricultural Practices (GAPs) and ISO (International organization for standards). These approaches are very important since they guarantee hazard prevention and the quality of food hence considered as standards that must be in the industry. They protect, guarantee, and ensure foods' safety, quality, and reliability which are core for food businesses (Lehtinen, 2017).

Also, it is noted that, it does not matter how these parameters are defined, such aspects as climate differences, and available technology can affect food quality and, at the same time, harsh quality standards tend to become a barrier for developing and emerging economies who wish to diversify their trades and penetrate into the export markets (Aksoy and Beghin, 2005). Besides, biological factors are also responsible for storage produce spoilage. Axata paper Insect pests, molds and other fungal contamination are other factors which contribute approximately thirty percent spoilage of harvested crops because of poor drying as well as post-harvest storage management. Although farmers make efforts to control these problems, inadequate storage facilities may lead significant losses in both quantity and quality of stored food crops (Srivastava & Mishra, 2021).

Regulatory Compliances

Regulatory compliances in the agri-warehouse industry are wide, such as food safety and quality, environment, licensing, trade and export, and health and safety, which creates obstacles in interstate or international trade. For developed countries, these compliances are stringent but somewhat flexible for developing countries. For example, In the United States, the Food Safety Modernization Act (FSMA) mandates that food storage facilities, including agricultural warehouses, comply with strict standards to prevent contamination and ensure food safety (Gordon et al., 2020). Similarly, the General Food Law Regulation (EC) 178/2002 of the European Union sets comprehensive food safety standards that agricultural warehouses must adhere to, ensuring traceability and accountability throughout the supply chain (Latino et al., 2022). In the case of India, it is the Warehousing (Development and Regulation) Act of 2007 which ensures the scientific storage of agricultural produce and mandates warehouses to register with the WDRA (Naik et al., 2022).

Further, moving to the sustainability and good governance mandate, such as environment related compliances, few countries are highly concern and have stringent compliances such as Australian government follows the Environment Protection and Biodiversity Conservation Act 1999 (EPBC Act) which requires agricultural warehouses to adhere to strict environmental guidelines to prevent land and water contamination. Further, Canadian government is highly conscious about trade and export related compliances and the regulation under the Safe Food for Canadians Act (2012) aims to provide better consumer safety and protection, to strengthen and streamline the Canadian agri-food sector and enhance international market (Gordon et al, 2020). In the area of health and safety compliances, warehouses in Japan strictly comply with the Industrial Safety and Health Act, ensuring the safety of workers in agricultural storage facilities through rigorous standards and regular inspections. These varied regulations poses a threat to the entire supply chain including agri-warehousing industry to meet the expectation of different countries in terms of several compliances.

5. BEST PRACTICES IN AGRI WAREHOUSING

5.1 Rural Warehouse Condominiums (Condos) of Brazil

The warehouse condominiums are a new business model in Brazilian agribusiness. Condos are formed by neighbouring rural producers who become founders and owners, working together to achieve common goals. This model has revolutionized agricultural practices by empowering smallholders and addressing various challenges. These condominiums introduce innovative mechanisms to reduce transaction costs, enhance participation, and support rural development while positively impacting cooperative membership. They help reduce warehousing deficits, storage facilities, logistical support, and infrastructure development, thereby enabling smallholders to generate income and participate actively in the market. These condos offer a cost-effective and efficient solution while integrating smallholders into the economy (Fillipi et al., 2019).

5.2 Modern Storage Infrastructure Development

Silos, controlled atmospheric storage, refrigerated warehouses (cold storage) are new developments in modern storage technique. Globally, silos are a very popular storage solution, often combined with systems for bulk handling, aeration, and pest control (Vishwakarma et al., 2022). They can be constructed from masonry, concrete, or metal, and come in either flat-bottomed or hopper-bottomed designs. For maximum strength, cylindrical silos are preferred. This technologically advanced storage method is standard in countries like the USA and Japan and is now being adopted in developing countries like India (Naik et al., 2022). Further, silo bags, similar to hermetic storage, is a new development, based on the principle of creating an airtight environment in the recent time. In these small silo bags small storage can be efficiently and effectively managed at farm gate for small and marginal farmers. When grain is sealed in this way, the natural respiration of both the grains and any organisms present depletes the oxygen and increases carbon dioxide. This creates an inhospitable environment for pests. Thus, hermetic storage can be achieved with underground structures, above-ground silos, or specialized silo bags (Vishwakarma et al., 2022). This type of storage, especially Silo bags has been in use globally and now penetrating in the developing countries too. For ex- introduced in India in 2011, such hermetic storage is especially good for preserving aroma-sensitive commodities like coffee, cocoa, and spices (Naik et al., 2022).

5.3 Warehouse Receipt and Financing

Smallholder farmers in developing countries, including Ghana and Tanzania, have found that warehousing receipts offer several benefits. These include safe storage for surplus produce, the ability to sell stored commodities at a later date for higher prices, using stored goods as collateral for loans, and the potential for quality improvement through the storage process (Miranda et al., 2019). Warehouse receipt financing through NWR, in many countries, has been adopted as policy initiative to protect small farmers who cultivate parcels of land and earn meagre profit, sometime get suffered with losses. The policy of adapting warehouse receipt system helping farmers to improve their access to credit, avoid distress sales, and reduce risks for financial institutions. In India, the regulations for NWR are defined by the WDRA (Warehousing Development and Regulatory Authority) Regulations of 2011. These regulations outline the standards, procedures for issuance, surrender, maintenance, and duplication of NWRs. As of March 2021, a total of 1973 warehouses with a combined capacity of 11 million tons are registered with WDRA (Naik et al., 2022). Further, in India, these warehouses are also declared as a sub-yard market in some States (Naik et al., 2022).

5.4 Interoperability in Warehouses in India

Warehouses play a crucial role in the storage of food commodities for sustained and prolonged use. In India, interoperability in warehouses operations can be seen as connecting link between farmers and market. As these warehouses are acting as storage cum market with the better connectivity, the flagship online cum spot market of the country popularly known as Electronic National Agriculture Market (eNAM) along with India's commodity future market i.e. the National Commodity and Derivative Exchange (NCDEX) of India are getting benefitted with the warehouse regulations and interoperability. This interoperability allow farmers to store produce, preventing immediate sales at lower prices and stabilizing prices by reducing supply fluctuations (Garg et al., 2023).

5.5 Adoption of Technology to Improve Efficiency

Agricultural storage facilities have transitioned from traditional paper-based processes to digital platforms, including web and mobile apps, to gather and organize data for the storage and transport of agricultural goods (Arishenbagam, 2024). Many individual storage facilities also embrace new methods to streamline and improve warehousing. In developing countries, agricultural storage has traditionally required much physical labor, which was time-consuming and inefficient. This task involves two important critical factors: crop protection and risk management. The integration of Industry 4.0 technologies, such as the Internet of Things (IoT), blockchain, autonomous robots, drones, and big data analytics, is instrumental in revolutionizing warehouse operations and driving increased profitability through informed decision-making.

5.6 Use of Artificial Intelligence (AI) in the area of Agri-Warehouse Operations

AI can enhance logistics and reduce transportation costs in agri-warehouse operations by examining various factors such as storage conditions, demand patterns, and transit routes (Saxena et al., 2023a). This enhancement boosts overall efficiency and reduces spoilage and food waste throughout the supply chain (Iyer, 2021). AI's ability to analyze market trends and historical pricing data enables the prediction of future agricultural commodity prices, aiding farmers and stakeholders in making informed decisions (Paul & Garai, 2021; Paul et al., 2023).

5.7 Use of Blockchain in Agri-Warehouse Operations

Food safety is not just a concern but a critical challenge for global supply chains. Cross-contamination and outbreaks of foodborne illnesses, often triggered by improper storage conditions during international shipping, are a stark reality. However, by leveraging blockchain technology to track and monitor all aspects of food items meticulously, we can effectively mitigate these risks, thereby fortifying the safety of agricultural products (Xu et al., 2019). The use of blockchain in the area of agri-warehousing is crucial because combining blockchain technology with agri-storage enables traceability from farm to fork, allowing customers to learn more about the origin and quality of the goods they buy, including best warehousing practices across the supply chain (Pant & Kansra, 2023).

5.8 Use of Robotics in the Area of Agri-Warehouse Operations

Robotics is the technological advancement in Agri 4.0 wherein one can observe, control, and monitor the warehouses that allow the users to manage data in real-time. Robotics promise a wealth of opportunities in food supply chain logistics. They offer increased productivity, heightened efficiency, optimized resource utilization, end-to-end tracking and traceability, improved warehouse logistics, and minimized health and safety risks (Chauhan et al., 2022). It is crucial to consider several factors when managing a warehouse. These include product perishability, quality-driven demand forecasting, pricing and stock management, food loss prevention, and sustainability demands in real-time (Meneghetti and Monti, 2014). These factors create a significant need for efficient material flow, which can be addressed by implementing a robotic mobile fulfillment system (RMFS). An RMFS can automatically sort inventory and quickly adapt the warehouse layout (Chauhan et al., 2022). Such automated systems not only help farmers assess stock and alert them to any mismanagement, but they also provide automatic control through sensors and computer vision to prevent accidents and issue alerts (Kumaresan, 2022).

5.9 Sustainable Practices in Warehouse Operations

Sustainable practices in warehouse operations has become increasingly important, given its significant impact on supply chain performance and cost. According to Richards (2017), warehouse operations constitute 24% of logistics costs, making it imperative for businesses to focus on sustainability in this area to reduce overall expenses. Ali et al. (2021) further emphasizes the importance of sustainable practices in warehouse operations, highlighting the need to address design, facilities, operations, equipment, recycling, waste management, and human resource management to achieve sustainability. This approach not only helps in minimizing food waste but also leads to energy and resource efficiencies. While challenges such as waste minimization and energy conservation exist, identifying and implementing essential green strategies can ultimately lead to cost savings and environmental benefits in warehouse operations.

6. CONCLUSION AND FUTURE DIRECTIONS

In conclusion, agricultural warehousing plays a critical role in addressing post-harvest losses and food security challenges. The implementation of technological innovations such as electronic negotiable warehouse receipts (e-NWRs) and hermetic storage bags has shown promising results in improving agricultural financing and reducing storage losses, particularly in smallholder farming regions. Hermetic storage bags are particularly beneficial in regions vulnerable to climate change-related shocks to staple food production. These bags help reduce waste and carbon emissions, provide efficient use of natural resources, and accelerate economic growth through energy and cost savings. They are a low-cost technology that has been shown to minimize storage losses, particularly in smallholder farming regions. Similarly, standardized warehouse structures are essential for maintaining the quality and safety of agricultural goods, and they significantly enhance supply chain operations and the quality of goods.

Additionally, uses of blockchain, IoT, RFID, AI, and Robotics are instrumental in the efficient technology led practices on the ground. Artificial Intelligence optimizes logistics and reduces transportation costs by analyzing storage conditions, demand patterns, and transit routes. It also aids in market trend analysis and price prediction, enabling informed decision-making for farmers and stakeholders. Blockchain technology ensures traceability from farm to fork, enhancing food safety by meticulously tracking and monitoring all aspects of food items. It enables customers to learn about the origin and quality of the goods they buy, including best warehousing practices across the supply chain. Robotics offers increased productivity, efficiency, and optimized resource utilization in Agri warehousing. It provides end-to-end tracking and traceability, improves warehouse logistics, and minimizes health and safety risks. Automated systems help assess stock, prevent accidents, and issue alerts.

Future areas of study should focus on the impact of interventions on small and marginal farmers as they contribute significantly to the process of storage of agri-food and their effect on reducing food losses and wastages. Research in this area can provide valuable insights into developing efficient and sustainable storage practices that address the challenges posed by climate change and ensure food security. Furthermore, exploring the potential integration of emerging technologies, such as blockchain for supply chain transparency and IoT devices for real-time monitoring of stored goods, can offer new avenues for enhancing the efficiency and safety of agricultural warehousing. The study suggests that the Impact on food losses and wastage and the impact on intervention on post-harvest handling practices, including storage, are some potential areas (Saxena et al., 2023b) for future researchers. Additionally, the interoperability of Warehouse operations with spot market alternatives, commodity future markets can be examined.

By delving into these future research areas, scholars can contribute to the development of comprehensive strategies that not only mitigate post-harvest losses but also promote sustainability, resilience, and equitable access to agricultural financing throughout the supply chain and thereby advancing global food security and the well-being of farming communities.

REFERENCES

Abass, A. B., Ndunguru, G., Mamiro, P., Alenkhe, B., Mlingi, N., & Bekunda, M. (2014). Post-harvest food losses in a maize-based farming system of semi-arid savannah area of Tanzania. *Journal of Stored Products Research*, 57, 49–57. DOI: 10.1016/j.jspr.2013.12.004

Acharya, S. S., & Agarwal, N. L. (2021). Agricultural Marketing. In *India* (7th Revised edition). CBS Publishers & Distributors.

Adjognon, G. S., Gassama, A., Guthoff, J., & Pouliquen, V. (2019). Implementing Effective Warehouse Receipt Financing Systems: Lessons from a pilot WRS project in the Senegal River Valley. The World Bank, 139445, 1-28. retrieved from https://documents1.worldbank.org/curated/en/173921563521093845/pdf/Implementing-Effective-Warehouse-Receipt-Financing-Systems-Lessons-from-a-Pilot-WRS-Project-in-the-Senegal-River-Valley.pdf

Aksoy, M. A., & Beghin, J. C. (2005). *Global agricultural trade and developing countries*. The World Bank., Retrieved from https://cisp.cachefly.net/assets/articles/attachments/01518_agriculturaltrade.pdf#page=98

Arishenbagam, P., Tamil Selvi, S., & Ganapathy Ram, B. (2024). Wireless Geofenced Inventory Management for Agricultural Warehouse. *Journal of the Institution of Engineers (India): Series B, 105*, 375–383. https://doi.org/DOI: 10.1007/s40031-023-00967-z

Chauhan, A., Brouwer, B., & Westra, E. (2022). Robotics for a Quality-Driven Post-harvest Supply Chain. *Current Robotics Reports*, 3(2), 39–48. DOI: 10.1007/s43154-022-00075-8

Chengappa, P. G. (2018). Development of agriculture value chains as a strategy for enhancing farmers' income. *Agricultural Economics Research Review*, 31(1), 1–12. DOI: 10.5958/0974-0279.2018.00001.0

Díaz-Valderrama, J. R., Njoroge, A. W., Macedo-Valdivia, D., Orihuela-Ordóñez, N., Smith, B. W., Casa-Coila, V., Ramírez-Calderón, N., Zanabria-Gálvez, J., Woloshuk, C., & Baributsa, D. (2020). Postharvest practices, challenges and opportunities for grain producers in Arequipa, Peru. *PLoS One*, 15(11), e0240857. DOI: 10.1371/journal.pone.0240857 PMID: 33147234

El-Ramady, H. R., Domokos-Szabolcsy, É., Abdalla, N. A., Taha, H. S., & Fári, M. (2015). Postharvest Management of Fruits and Vegetables Storage. In: Lichtfouse, E. (eds) *Sustainable Agriculture Reviews*. Sustainable Agriculture Reviews, vol 15. Springer, Cham. https://doi.org/DOI: 10.1007/978-3-319-09132-7_2

FAO. (2017). *Report on The Future of Food and Agriculture: Trends and Challenges*. Food and Agriculture Organization, Rome. Retrieved from https://www.fao.org/3/a-i6583e.pdf

Filippi, A. C. G., Guarnieri, P., Carvalho, J. M., Reis, S. A., & da Cunha, C. A. (2020). New configurations in Brazilian agribusiness: Rural warehouse condominiums. *Journal of Agribusiness in Developing and Emerging Economies*, 10(1), 41–63. DOI: 10.1108/JADEE-12-2018-0178

Garg, M., Singhal, S., Sood, K., Rupeika-Apoga, R., & Grima, S. (2023). Price discovery mechanism and volatility spillover between National Agriculture Market and National Commodity and Derivatives Exchange: The study of the Indian agricultural commodity market. *Journal of Risk and Financial Management*, 16(2), 62. DOI: 10.3390/jrfm16020062

Gordon, A., DeVlieger, D., Vasan, A., & Bedard, B. (2020). Technical considerations for the implementation of food safety and quality systems in developing countries. *Food Safety and Quality Systems in Developing Countries*, 1-40. https://doi.org/ DOI: 10.1016/B978-0-12-814272-1.00001-2

Guru, P. N., Mridula, D., Dukare, A. S., Ghodki, B. M., Paschapur, A. U., Samal, I., Nikhil Raj, M., Padala, V. K., Rajashekhar, M., & Subbanna, A. R. (2022). A comprehensive review on advances in storage pest management: Current scenario and future prospects. *Frontiers in Sustainable Food Systems*, 6, 993341. DOI: 10.3389/fsufs.2022.993341

Hamel, D., Rozman, V., & Liška, A. (2020). Storage of Cereals in Warehouses with or without Pesticides. *Insects*, 11(12), 846. DOI: 10.3390/insects11120846 PMID: 33260599

Höllinger, F., Rutten, L., & Kiriakov, K. (2009). The use of warehouse receipt finance in agriculture in transition countries. Presented at the World Grain Forum 2009, FAO. https://www.fao.org/4/i3339e/i3339e.pdf

Huss, M., Brander, M., Kassie, M., Ehlert, U., & Bernauer, T. (2021). Improved storage mitigates vulnerability to food-supply shocks in smallholder agriculture during the COVID-19 pandemic. *Global Food Security*, 28, 100468. DOI: 10.1016/j.gfs.2020.100468 PMID: 36568028

IFC. (n.d.). Global Warehouse Finance Program (GWFP). Retrieved from https://www.ifc.org/en/what-we-do/sector-expertise/financial-institutions/global-trade/global-warehouse-finance-program

Iyer, L. S. (2021). AI enabled applications towards intelligent transportation. *Transportation Engineering*, 5, 100083. DOI: 10.1016/j.treng.2021.100083

Jha, S. N., Vishwakarma, R. K., Ahmad, T., Rai, A., & Dixit, A. K. (2015). Report on assessment of quantitative harvest and post-harvest losses of major crops/commodities in India. ICAR-All India Coordinated Research Project on Post-Harvest Technology. ICAR-CIPHET, PO: PAU, Ludhiana, India.

Johnson, D. G. (1975). World Agriculture, Commodity Policy, and Price Variability. *American Journal of Agricultural Economics*, 57(5), 823–828. DOI: 10.2307/1239087

Katunze, M., Kuteesa, A., Mijumbi, T., & Mahebe, D. (2017). Uganda Warehousing Receipt System: Improving Market Performance and Productivity. *African Development Review*, 29(S2), 135–146. DOI: 10.1111/1467-8268.12268

Kovacevic, V., Rajnovic, L., & Vukovic, P. (2023). Warehouse receipt finance in the agricultural sectors: Lessons learned in Serbia. *Ekonomika (Nis)*, 69(1), 59–68. DOI: 10.5937/ekonomika2301059K

Kumar, C., Ram, C. L., Jha, S. N., & Vishwakarma, R. K. (2021). Warehouse Storage Management of Wheat and Their Role in Food Security. *Frontiers in Sustainable Food Systems*, 5, 675626. DOI: 10.3389/fsufs.2021.675626

Kumar, D., & Kalita, P. (2017). Reducing Postharvest Losses during Storage of Grain Crops to Strengthen Food Security in Developing Countries. *Foods*, 6(1), 8. DOI: 10.3390/foods6010008 PMID: 28231087

Kumaresan, S. J., Vikneswaran, M., Surai, P., & Vignesh, S. (2022). Agri Warehouse Management System. In *2022 International Interdisciplinary Humanitarian Conference for Sustainability (IIHC)* (pp. 705-709). IEEE. DOI: 10.1109/IIHC55949.2022.10059944

Lacroix, R., & Varangis, P. (1996). Using warehouse receipts in developing and transition economies. *Finance & Development*, 33(3), 36–39.

Latino, M. E., Menegoli, M., Lazoi, M., & Corallo, A. (2022). Voluntary traceability in food supply chain: A framework leading its implementation in Agriculture 4.0. *Technological Forecasting and Social Change*, 178, 121564. DOI: 10.1016/j.techfore.2022.121564

Lehtinen, U. (2017). Sustainable Supply Chain Management in Agri-food Chains. In Bhat, R. (Ed.), *Sustainability Challenges in the Agrofood Sector.*, DOI: 10.1002/9781119072737.ch7

Lydia, J., Sherwin Vimalraj, L., Monisha, R., & Murugan, R. (2022). Automated food grain monitoring system for warehouse using IOT. *Measurement. Sensors*, 24, 100472. DOI: 10.1016/j.measen.2022.100472

Makule, E., Dimoso, N., & Tassou, S. A. (2022). Precooling and Cold Storage Methods for Fruits and Vegetables in Sub-Saharan Africa—A Review. *Horticulturae*, 8(9), 776. DOI: 10.3390/horticulturae8090776

Manandhar, A., Milindi, P., & Shah, A. (2018). An Overview of the Post-Harvest Grain Storage Practices of Smallholder Farmers in Developing Countries. *Agriculture*, 8(4), 57. DOI: 10.3390/agriculture8040057

Meemken, E., Barrett, C. B., Michelson, H. C., Qaim, M., Reardon, T., & Sellare, J. (2021). Sustainability standards in global agrifood supply chains. *Nature Food*, 2(10), 758–765. DOI: 10.1038/s43016-021-00360-3 PMID: 37117971

Meneghetti, A., & Monti, L. (2014). Greening the food supply chain: An optimisation model for sustainable design of refrigerated automated warehouses. *International Journal of Production Research*, 53(21), 6567–6587. DOI: 10.1080/00207543.2014.985449

Miranda, M. J., Mulangu, F. M., & Kemeze, F. H. (2019). Warehouse receipt financing for smallholders in developing countries: Challenges and limitations. *Agricultural Economics*, 50(5), 629–641. DOI: 10.1111/agec.12514

Mobolade, A. J., Bunindro, N., Sahoo, D., & Rajashekar, Y. (2019). Traditional methods of food grains preservation and storage in Nigeria and India. *Annals of Agricultural Science*, 64(2), 196–205. DOI: 10.1016/j.aoas.2019.12.003

Naik, G., Raghuram, G., Rajan, J., Bansal, M., Gopikuttan, G. S., Tawri, P., & Singh, R. (2022). Agricultural Warehousing in India: Trends, Constraints, and Policies. IIM Bangalore. Retrieved from https://ncdex.com/downloads/Research/casestudies-conceptpaper/Agricultural%20Warehousing%20in%20India,%202022-%20A%20report%20by%20IIMB.pdf

Olomu, M. O., Ekperiware, M. C., & Akinlo, T. (2020). Agricultural sector value chain and government policy in Nigeria: Issues, challenges and prospects. *African Journal of Economic and Management Studies*, 11(3), 525–538. DOI: 10.1108/AJEMS-03-2019-0103

Olorunfemi, B. J., & Kayode, S. E. (2021). Post-Harvest Loss and Grain Storage Technology- A Review. *Turkish Journal of Agriculture -. Food Science and Technology (Campinas)*, 9(1), 75–83. DOI: 10.24925/turjaf.v9i1.75-83.3714

Pant, S. C., & Kansra, P. (2023). Indian Agriculture Sector and Blockchain Technology: A Systematic Review Envisaging Business Opportunities. *Empirical Economics Letters, 22* (Special Issue 1), 21-35. https://doi.org/DOI: 10.5281/zenodo.8266156

Paul, R. K., & Garai, S. (2021). Performance comparison of wavelets-based machine learning technique for forecasting agricultural commodity prices. *Soft Computing*, 25(20), 12857–12873. DOI: 10.1007/s00500-021-06087-4

Paul, R. K., Yeasin, M., Kumar, P., Paul, A. K., & Roy, H. S. (2023). Deep learning technique for forecasting the price of cauliflower. *Current Science*, 124(9), 1065. DOI: 10.18520/cs/v124/i9/1065-1073

Rai, B. K. (2023). IoT based humidity and temperature control system for smart warehouse. *Gazi University Journal of Science*, 36(1), 173–188. DOI: 10.35378/gujs.993959

Richards, G. (2017). *Warehouse management: A complete guide to improving efficiency and minimizing costs in the modern warehouse*. Kogan Page Publishers.

Sagar, N. A., & Pareek, S. (2020). Safe Storage and Preservation Techniques in Commercialized Agriculture. *Natural Remedies for Pest, Disease and Weed Control*, 221-234. https://doi.org/DOI: 10.1016/B978-0-12-819304-4.00019-1

Saxena, R., Pant, D. K., Pant, S. C., Joshi, L., Paul, R. K., & Singh, R. (2023a). Artificial intelligence-led innovations for agricultural transformation: A scoping study. *Agricultural Economics Research Review, 36* (Conference Number), 1-18.

Saxena, R., Pant, D. K., Pant, S. C., & Singh, R. (2023b). Mapping the Global Research Landscape: Bibliometric Analysis of Agri-food Systems and Nutritional Security. *Indian Journal of Agricultural Economics*, 78(3), 328–342. https://isaeindia.org/wp-content/uploads/2024/02/03-RAKA-SAXENA.pdf. DOI: 10.63040/25827510.2023.03.003

Schultz, H. W. (1981). United States Warehouse Act. In *Food Law Handbook*. Springer., DOI: 10.1007/978-94-011-7373-5_17

Singh, A. S., Bharti, N., & Kumar, P. (2023). Is Micro-warehousing a Solution to Agricultural Commodity Storage? *Asian Journal of Management Cases*. https://doi.org/DOI: 10.1177/09728201231195441

Srivastava, S., & Mishra, H. N. (2021). Ecofriendly nonchemical/nonthermal methods for disinfestation and control of pest/fungal infestation during storage of major important cereal grains: A review. *Food Frontiers*, 2(1), 93–105. DOI: 10.1002/fft2.69

USDA. (n.d.). USWA Licensed Warehouse. Agricultural Marketing Service 2022. Available online: https://publicdashboards.dl.usda.gov/t/ MRP_PUB/views/WCMDDashboard/WCMDDashboard?:isGuestRedirectFromVizportal=y&:embed=y (accessed on 15 August 2024).

Vishwakarma, R. K., Kumar, N., Sharma, K., Kumar, Y., & Kumar, C. (2022). Storage. In Sharma, H. K., & Kumar, N. (Eds.), *Agro-Processing and Food Engineering: Operational and Application Aspects* (pp. 353–413). Springer Singapore., DOI: 10.1007/978-981-16-7289-7_9

WAI. (2022). e-Handbook on Warehousing Standards. Retrieved from https://warehousingindia.org/wp-content/uploads/2022/09/Handbook-on-Warehousing-Standards.pdf

WDRA. (2015). Report on warehousing in India Study commissioned by the Warehousing Development and Regulatory Authority. Retrieved from https://wdra.gov.in/documents/32110/38476/Report12.pdf/fcfa2f56-a673-a5f6-a870-f73292eab04a

Xu, J., Guo, S., Xie, D., & Yan, Y. (2019). Blockchain: A new safeguard for agri-foods. *Artificial Intelligence in Agriculture*, 4, 153–161. DOI: 10.1016/j.aiia.2020.08.002

Yang, J., Li, Z., & Wang, T. (2020). Price discovery in Chinese agricultural futures markets: A comprehensive look. *Journal of Futures Markets*, 41(4), 536–555. DOI: 10.1002/fut.22179

Yohe, H. S. (1924). The Banker and the United States Warehouse Act. Retrieved from https://archive.org/details/CAT31297878/page/4/mode/2up

Chapter 9
Overcoming Supply Chain Barriers in Onion Through Technology

Sangeeta Shroff
Vaikunth Mehta National Institute of Cooperative Management, India

Rohini Ashok Vilhekar
https://orcid.org/0009-0005-8981-7707
Vaikunth Mehta National Institute of Cooperative Management, India

ABSTRACT

The onion economy is characterized by major price volatility due to seasonal production, while there is daily consumption by almost every Indian household. The crop is highly perishable and depends upon stored produce for about 6-7 months as there is no harvest between May to November. The crop besides suffering from considerable storage losses, is also susceptible to unfavorable monsoons which can lead to huge crop loss and high prices. Price rise is addressed through government agencies releasing buffer stocks, fixation of Minimum Export Price and even regressive measures such as export ban. This paper, examines the price fluctuations and the status of demand, supply and exports of onion crop. The paper highlights that the supply shocks must be addressed through technology, which can increase yields and also be leveraged for creating advanced storage structures in place of traditional ones, often used by farmers. This serves as the real solution rather than export bans, considering that India has export potential and several countries are dependent upon India for supplies.

DOI: 10.4018/979-8-3693-4330-2.ch009

1. STATUS OF ONION: GLOBAL PERSPECTIVE

Onion is one of the most consumed vegetables and a staple ingredient in the cuisine of many cultures worldwide. The world demand for fresh onions has shown a rising trend driven by increase in population, rising per capita income, urbanization, *etc.* especially in several developing countries. The large majority of world consumption takes place in Asia, notably China and India due to demographic factors and hence these two countries dominate the world consumption as well as production of onion. The importance of onions in the cuisines of Asian countries across all levels of income groups is likely to perpetuate and strengthen the demand for this vegetable crop.

Onions are a versatile crop and differ in color and taste with some varieties more pungent than others. Yellow onions make up 75% of all cultivated onions in the world, while other colors are red and white. India is the top producer in the world known for pungency in the varieties cultivated. The country also has the advantage of cultivating the crop throughout the year in three seasons. In Bangla Desh onion is considered as a major spice crop while in Indonesia onions are also used for medicinal purposes. The country wise share of onions in world area and production is indicated in Table 1.

Table 1. Major Onion Producing Countries in the World (2020)

Country	Area (Million Ha)	% share to Total Area	Production (Million Tonnes)	% share in Production	Productivity (Tonnes / ha)
India	1.43	26.09	26.74	25.58	18.7
China	1.08	19.71	23.66	22.63	21.91
USA	0.05	0.91	3.82	3.65	65.6
Egypt	0.09	1.64	3.16	3.02	35.11
Türkey	0.07	1.28	2.28	2.18	32.57
Pakistan	0.15	2.74	2.12	2.03	14.13
Iran	0.05	0.91	2.06	1.97	41.2
Bangladesh	0.19	3.47	1.95	1.87	10.26
Sudan	0.11	2.01	1.95	1.87	17.73
Indonesia	0.19	3.47	1.82	1.74	9.58
Others	2.07	37.8	34.99	33.47	16.90
World	**5.48**	100	**104.55**	100	**19.08**

Source: ICAR-National Institute of Agricultural Economics and Policy Research, New Delhi (2020)

From Table 1, it can be observed that India and China together constitute 45.8% of world area and 48.21% world production of onions. However, with respect to productivity, the picture strongly differs as the productivity of USA is about 3 times more than that of India and China while share in world area is less than 1%. Countries such as Iran, Egypt and Turkey which have negligible area under onion, have yields that are little less than double of India and China.

Despite low productivity levels, India is the largest producer of onions in the world and the crop is often termed as 'Queen of Kitchen" as it is consumed daily and mostly in fresh form across all income groups. Infact traditionally, onion is considered to an important part of the diet of the vulnerable section and hence is a politically sensitive crop. The price of onion is closely monitored by Reserve Bank of India due to its contribution to inflation and public sentiment. In years when prices escalate, there is considerable media coverage while the issue is also discussed in Parliament.

2. OBJECTIVES

While onion is a major vegetable crop in the consumption basket of majority of Indian households, source of foreign exchange and contribution to stability in the economy, the crop is characterized by major fluctuations in prices and storage losses due to fluctuations in yield, low shelf life and easy perishability. The crop is characterized by crisis every few years brought about by supply shocks and price rise.

In view of the above, this paper attempts to study the role that technology can play in increasing the supply of onion. The increase in supply will not only meet the requirements of the domestic economy, but the country will also be able to capitalize on export potential and higher trade share in global markets. The paper firstly attempts to explain the fluctuations in price of onion and the knee jerk reactions by the government through interventions in the domestic and export markets. The use of technology to address this issue is further discussed as a measure to overcome supply shocks. The paper observes the supply constraint that can be mitigated by increasing yield of onion and also through the use of scientific storage structures or early detection of deterioration of the crop. This will help to reduce post harvest losses. Certain policy measures to be addressed are also suggested.

3. RESEARCH METHODOLOGY

The paper is based on secondary data and qualitative description on certain onion storage structures that are devised to reduce losses. Secondary data was collected on various variables of onions such as Area, Production, Yield, Wholesale Price Index and Exports. The source of data was mainly from APEDA website and www .indiastat.com. Simple statistical methods were used to analyse the secondary data which throw light on the onion economy relating to demand, supply and exports. In order to understand recent technological intervention in storage of onions, a case study of Godaam Innovations was observed which prevents the onions from spoilage through early detection using IOT sensing. An interview was also conducted with the innovator of the product to understand the advantages, cost and reduction in losses due to technological intervention. The study thus aims to find solutions to overcome supply constraints through technology so that stability can be imparted to the Indian onion economy with the help of secondary data as well as qualitative analysis.

4. ONION ECONOMY IN INDIA: PRICE FLUCTUATIONS AND GOVERNMENT INTERVENTION

Onion is consumed raw, cooked or processed in the daily Indian diet. This horticultural crop is market sensitive leading to interventions. A detailed analysis with respect to these issues is discussed in this section.

4.1 Government Purchase and Sale Operations to Reduce Price Fluctuations

The onion economy is characterized by volatility in prices, both inter seasonal and intra seasonal. Inter seasonal prices occur because while bumper crop is experienced in certain years, the following years face shortage. These situations are triggered by climatic factors as well as changes in area causing cobweb phenomenon. Intra seasonal price fluctuations depend upon the quantity stored. However, the quantity stored also depends upon the quality of storage structure and technology used, as these factors help in stabilizing prices and boosting exports.

Since onion is a sensitive crop, consumed daily by households and any price rise contributes to inflation, there is constant intervention by the government to ensure that prices always remain affordable. Interventions are mainly two types, namely government purchase and release operations and also export policy.

In order to regulate the price volatility of sensitive and important agri-horticultural commodities, a Price Stabilization Fund was set up in 2014-15 by state agencies such as National Cooperative Consumers Federation (NCCF) and National Agricultural Cooperative Marketing Federation (NAFED). A strategic buffer stock is maintained by NCCF and NAFED which directly purchase from farmers/farmers' associations at farm gate prices for subsequent calibrated release for moderating price volatility and discourage hoarding and speculation (Ministry of Consumer Affairs, Food & Public Distribution, PIB, 2024).

The Wholesale Price Index(WPI) for onion is indicated in Graph 1 which also clearly reveals the inter seasonal price variations from a low of 135.05 in 2013-14 to a peak of 312.27 in 2019-20 and year on year fluctuations. In early 2023, the prices of onion had crashed to unduly low levels and the WPI (Base 2011-12) which was 200.5 in January 2023 began declining and crashed to 138.1 in March 2023, i.e decline by 31%. In order to arrest this decline, NAFED and NCCF purchased about 6.4 lakh tonnes of onion during 2023-24 for buffer stock which enabled the farmers to get remunerative prices (Ministry of Consumer Affairs, Food & Public Distribution, PIB, 2024). The stocks were released during the lean months of October and November 2023, when prices began to move upwards. The WPI increased by 80% from September to November 2023 but gradual release of stocks from government agencies and export policy helped to bring about price reduction by 58% in February 2024 from the November 2023 peak. The coefficient of variation in WPI during the twelve year period 2012-13 to 2023-24 is 29.52% indicating considerable inter seasonal fluctuations in prices.

Figure 1. Wholesale Price Index of Onion

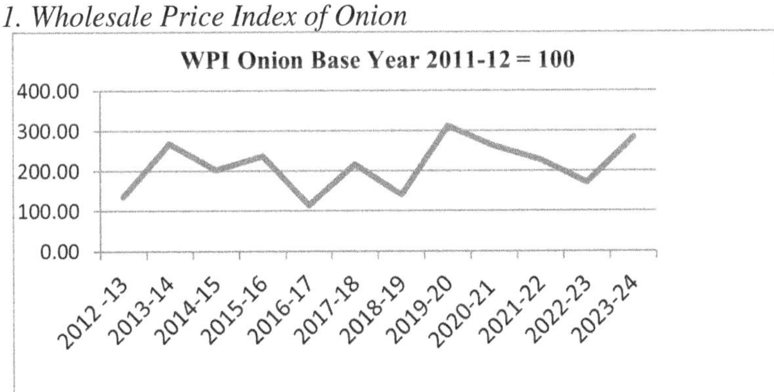

Besides, inter seasonal prices of onions, the crop also suffers from intra seasonal fluctuations in prices. The intra seasonal fluctuation in WPI for onion can be observed from Graph 2. The coefficient of variation in WPI in 2022-23 season which was 19.37% increased to 34.36% in 2023-24. In 2019-20 season, the coefficient of variation was as high as 55.92%.

From the above, it is clear that onion crop is subject to considerable volatility in prices. High prices bring a strain on the consumer's budget while low prices can cause losses for farmers. In order to keep the prices under check and prevent huge fluctuations in prices, the government often intervenes in onion markets.

Figure 2. Intra Seasonal Fluctuations in WPI-Onion

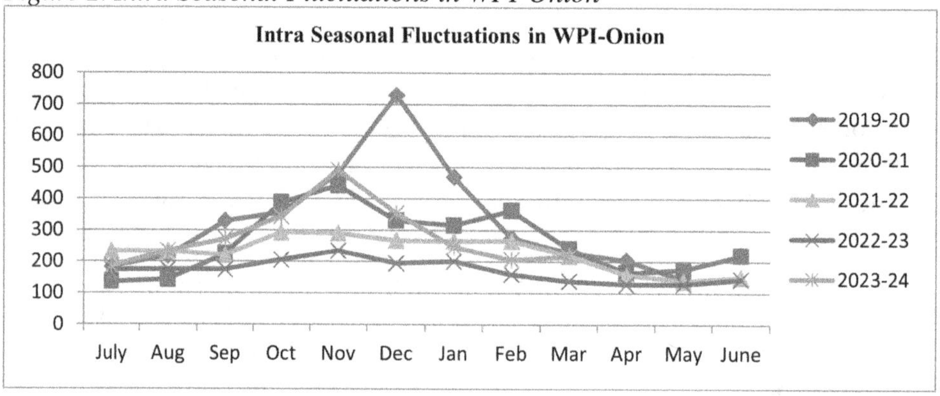

Source: Data on WPI obtained from indiastat.com

4.2 Export Policy and Status of Onion Exports in India

India normally produces more onions than it consumes and has always been an exporter of onions. The quantity exported in the recent past has ranged from 1.23 million tonnes in 2014-15 to 2.53 million tonnes in 2022-23. On an average since last decade, India has exported 1.7 million tonnes of onions. This indicates that the country still has potential to export more, but the same is restricted by the government in order to contain domestic prices.

There are frequent changes in export policy of onion, normally in the form of export ban when prices begin to rise. However, in certain years the government fixes a Minimum Export Price (MEP) and exporters are thus not permitted to export at prices below MEP. Table 2 explains the Government intervention in onion export policy.

Table 2. Government intervention in Onion Exports

Year	MEP fixed	Remarks
29th June 2012	MEP not imposed for exports	Exports encouraged
14th August 2013	$650 / metric tonne	Sharp rise in price from 2012-13 season to 2013-14 season.
19th September 2013	$ 900/metric tonne	MEP raised due to further price rise
1st November 2013	$ 1100/metric tonne	MEP further raised to curtail sharp rise in price.
16th December 2013	$800 /metric tonne.	MEP reduced as prices showed downward trend
19th December 2013 tonne	$ 350/metric	MEP further reduced with decline in domestic prices
4th March 2014	MEP was removed.	Prices showed downfall
17th June 2014	$300 /metric tonne	MEP again imposed due to increase in prices
2nd July 2014.	500/metric tonne.	MEP increased
21st August 2014	$300/metric tonne	The MEP was lowered due to price fall
7th April 2015	$250/metric tonne	The MEP lowered to promote exports
21st June 2015	$425/metric tonne	MEP raised to ensure adequate domestic availability
24th August 2015	$700/metric tonne	MEP increased as domestic prices had firmed up.
11th December 2015	$400/metric tonne.	MEP reduced as domestic prices began declining
24th December 2015	MEP removed	The domestic prices declined sharply
23rd November 2017	$ 850 / metric tonne	till 31/12/2017 in order to arrest the rise in prices
February, 2018,	MEP abolished	Prices moved downwards
September 2019	Ban on exports	Domestic prices began to sharply increase
March 2020	Ban lifted and MEP abolished	Fall in prices
August 2023	40% Export duty	To stabilize domestic prices
November 2023	$800/metric tonne	Prices began to rise
8th December 2023	Export Ban	To arrest price rise
4th May 2024	MEP $550/metric tonne	Also Export ban lifted and 40% duty on exports to ease domestic prices
March 2024	Limited Exports to selected countries	Limited quantities to be exported to Bhuta, Sri Lanka, UAE, Mauritius, Bahrain through National Cooperative Export Limited
13th September 2024	MEP withdrawn	Export duty lowered to 20%

Source: MEP obtained from Directorate General of Foreign Trade notifications (Ministry of Commerce & Industry)

The fixation of MEP also prohibits the export of onion to distant western markets where export price is high but quantity exported is low. The same is indicated in Table 3 where it is observed that recently in 2023-24, while India exported 42.2% of total exports to Bangla Desh at a unit price of INR 21.47/kg, the volume of exports to distant European countries such as France was negligible but price fetched was INR 59/kg.

Table 3. India's Exports of Onions- Country wise Share in Volume and Export Price (INR Per kg)

Country	2020-21		2021-22		2022-23		2023-24	
	Volume %	Export Price INR Kg	Volume %	Export Price INR/Kg	Volume %	Export Price INR./Kg	Volume %	Export Price INR/Kg
Bangladesh	35.01	13.68	42.84	19.75	26.58	13.37	42.20	21.47
Malaysia	12.55	23.29	11.09	28.78	15.58	21.57	11.72	27.17
UAE	10.81	19.37	7.99	22.72	15.97	19.46	10.71	19.67
Sri Lanka	9.17	21.64	10.60	25.26	10.71	16.71	10.12	23.82
Nepal	7.20	14.51	10.89	16.70	6.92	15.32	4.89	17.95
France	0.01	38.17	0.01	51.30	0.01	41.96	0.01	59.05
Others	25.25	20.09	16.58	26.25	24.23	20.77	20.35	25.55

Source:https://agriexchange.apeda.gov.in/indexp/Product_description_32head.aspx?gcode=0201

The volume of world trade in onion is to the tune of 9.74 million tonnes. Although India is the largest producer of onion in the world, the average price at which exports take place are not very competitive and other major exporters are able to realize higher prices. The same can be observed from Table 4.

Table 4. Country-wise Share in World Trade of Onions and Export Price (INR / kg) Year: 2021

Country	% Share in World trade	Export Price (INR /kg)
Netherlands	17.45	34.52
India	14.68	23.63
China	7.2	49.40
Mexico	5.1	66.70
USA	3.6	57.56

Source: World Integrated trade Solution (https://witsworldbank.org)

It can be observed from Table 4 that Netherlands has highest share in exports of onions which is 17.45% followed by India which has a share of 14.68%. While other countries such as China, Mexico and USA have lower share in volume of exports, they are able to realize much higher prices in the international markets. Hence, it is clear that India is not able to capitalize on lucrative world prices of onion. It was also earlier observed that rather than taking advantage of international prices, the country often fixes MEP or bans exports. Instead of addressing such issues through interventions, the real answer lies on the supply side of the produce which will satisfy the domestic market and promote exports.

5. ONION CONSUMPTION IN GLOBAL PERSPECTIVE

Onion consumption is reported to be on the rise globally with increasing trend in demand as onions are an essential vegetable in all continents with a variety of uses in food preparation. While India is a major producer and consumer of onions, the per capita consumption and price vary widely across countries. There is considerable world trade in onions as observed earlier with Netherlands being the major exporter, followed by India. The Netherlands has strategically nurtured an extensive network of international trade partnerships which encompass both the traditional and emerging markets. Several countries such as Bangla Desh import a major part of their onion consumption.

5.1 Global Consumption of Onion

Onion as a vegetable crop has a prominent place in several cuisines worldwide and is also a staple partner of many culinary preparations. This crop has several health benefits and medicinal uses and therefore cultivated in most countries for consumption in different forms including fresh, dry, shallots, powder, etc. The annual per capita consumption of onion varies widely across countries, besides variation in prices. The same is indicated in Table 5.

Table 5. Onion Consumption (Per Kg, Per Capita) in Selected Countries and Price Per Kg (2021)

Country	Per Capita/Per Kg Consumption	Price ($ Per Kg)
Tajikstan	60.4	0.43
USA	44	3.07
Eyypt	26	0.49

continued on following page

Table 5. Continued

Country	Per Capita/Per Kg Consumption	Price ($ Per Kg)
South Korea	25.74	3.11
UAE	25.3	1.24
India	16.3	0.43
Bangla Desh	15.6	0.63
Netherlands	15.4	2.03
China	14.6	0.94
Japan	11.04	2.45
Germany	9.94	1.64
Mexico	7.25	1.69
World	12.1	Between 3.1 and 0.40 across all countries

Source: FAO Statistical Database & https//numbeo.com

It can be observed from Table 5 that average onion consumption in the world is 12.1 kg and the same varies across countries. The highest consumption is 60.4 kg per capita in Tajikstan while there are some countries such as Cambodia, Laos, Togo, *etc.* where the consumption is negligible. The price of onion also varies across countries and the highest price was observed in South Korea which is 623% higher than the lowest price in Tajikstan and India.

India is a major onion producing country, but the exports are mostly to Bangla Desh where the produce is not sold at attractive prices. The prices prevailing in other countries clearly reveal that India must capitalize on export potential by increasing productivity and also accessing major markets in USA, Europe, etc through innovation and technology which expand the shelf life of onions. India must also diversify the production of onions to yellow varieties which are favored in many developed economies. The price of onion in Japan is $2.45 per kg which is 470% higher than that in India. Hence if India produces varieties that are less pungent, the crop can easily penetrate into the Japanese markets.

5.2 Onion Consumption in India

Onion is a major vegetable crop in India, largely consumed in fresh form as it is available throughout the year and cultivated in three seasons.

The onion consumption of the country can be observed in Table 6.

Table 6. Total Consumption of Onion in India (Million Tonnes)

Onion	2017-18	2018-19	2019-20	2020-21	2021-22	2022-23	2023-24
Production	23.26	22.81	26.09	26.64	31.69	30.21	25.47
Household Consumption	13.73	13.88	14.03	14.19	15.00	15.00	15.00
Processing (6%)	1.39	1.37	1.57	1.60	1.9	1.8	1.53
Bulk Consumption (15%)	3.49	3.42	3.91	3.99	4.8	4.5	3.82
Wastage/Loss (8.48% minimum loss)	1.97	1.93	2.21	2.26	2.69	2.56	2.16
Total Domestic Consumption	20.58	20.6	21.72	22.04	24.39	23.86	22.51
Exports	1.67	2.18	1.15	1.58	1.54	2.53	1.7
Domestic Consumption +Exports	22.25	22.78	22.87	23.62	25.93	26.39	24.21

Source: compiled from www.indiastat.com, APEDA.

From Table 6, it can be observed that total consumption of onion has ranged from 22.25 million tonnes in 2017-18 to 26.39 million tonnes in 2022-23 and by and large the production of onion is able to satisfy the domestic consumption. However, two major issues that emerge from Table 6 are that there is considerable loss due to wastage and spoilage each year and India is unable to capitalize on its export potential. The important point to note is that in Table 6, wastage is assumed at 8.48% which causes on an average a loss of 2.3 million tonnes of onion. This is the minimum loss that farmers incur, assuming that they use well designed storage structures. In reality however, several studies, (YASHADA 2019 & PoCRA 2022) have noted that use of traditional structures is still prevalent and losses range between 30 to 45% and in some cases even up to 60% for several farmers. This is a crying problem for the onion economy which reduces supply and consequently leads to price rise. Further, the crop also undergoes physiological weight loss due to shrinkage, etc. The Ministry of Consumer Affairs noted that due to improper storage systems, annual losses of onion amount to Rs 11,000 crores (Ministry of Consumer Affairs, Food and Public Distribution, PIB, 2024).

It may also be noted that India is experiencing increase in population, urbanization, rising per capita income which are fuelling the demand for onions. These features of the economy are coupled by increasing demand for processed foods for which onion is used as a major ingredient such as onion flakes or onion powder. In order to capitalize on this increased demand, it is important to increase the yield of onions.

6. BRIDGING THE YIELD GAP IN ONION CULTIVATION IN INDIA

While India is the top producer and exporter of onion in the world, the yield of the crop is much below that of several countries. In fact, in India the area and yield varies across states as can be observed from Graph 3 & 4.

Figure 3. State Wise % Share in Area under Onion

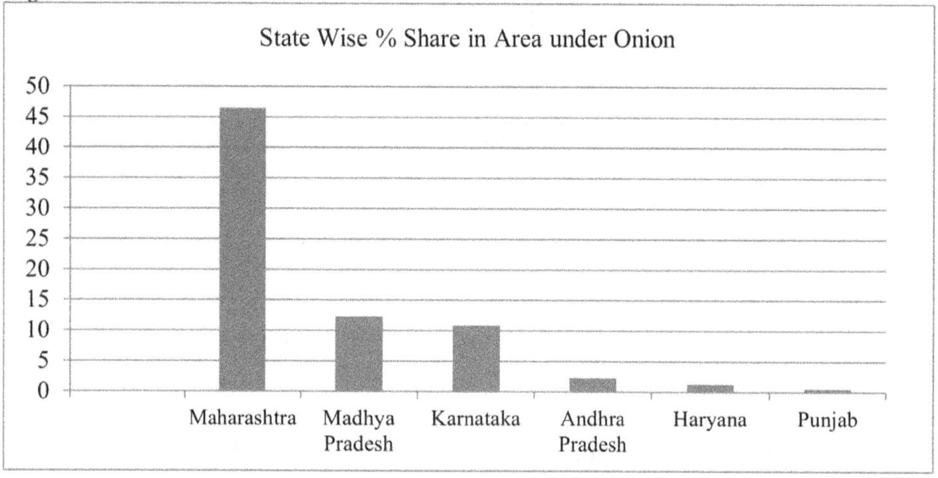

Source: www.indiastat.com

Figure 4. State Wise Yield (tonnes/hectare) of Onion (2022-23)

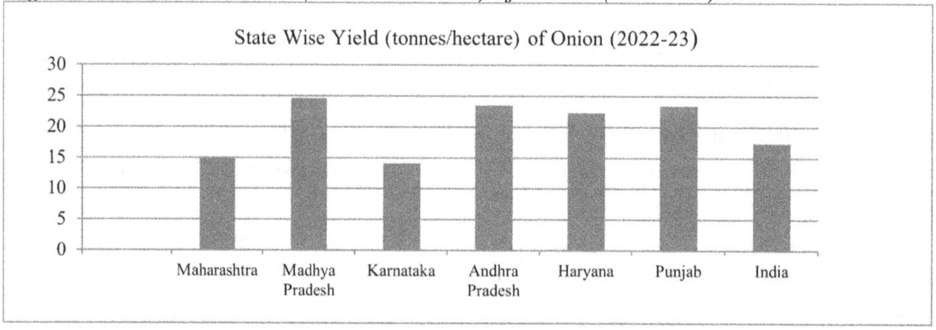

Source: www.indiastat.com

Maharashtra and Karnataka together cultivate 57.85% of the area under onion. However, the yield in both these states is much lower than in other states such as Madhya Pradesh, Haryana, etc. and even below All India average. Madhya Pradesh which ranks second in area under onion, has a yield which is 65% higher than that of

Maharashtra which ranks first in area. There is scope to increase the yield through adoption of suitable technology. A study (Rajnath & Sharma, 2021) on yield gap analysis in Maharashtra revealed a yield gap of 46.47% on an average across four districts studied. The study noted that farmers continue to use traditional farming methods, and there is poor transfer of technology from laboratory to farm. The farmers are unable to access irrigation systems and the crop is dependent upon monsoons. In certain years there is drought while in some years there is excessive rainfall. This leads to poor quality and quantity of produce. The quality of the kharif crop especially, is very poor due to high moisture content. Since production is seasonal, but consumption is throughout the year, storage of the crop assumes importance. However, the study noted that the storage facilities were poor, leading to huge storage losses. If farmers in Maharashtra produce the potential yield there is huge scope to increase the production of onions. Increase in yield, coupled with suitable marketing strategies can increase the incomes of onion farmers.

Other studies (PoCRA 2022) identified several causes such as poor quality of onion seed, mixing of varieties, heavy and uncertain rain, defective onion bulbs, etc. Hence several factors are responsible for low yield and need to be addressed.

The country-wise production of onions reveals that the yield of onions in India is lower than that in several countries. While developed countries like USA have a yield as high as 76.40 tonnes per hectare, less developed countries such as Egypt, Turkiye and Iran have yields that are 88%, 74% and 120% respectively higher than India. Hence India has the potential to increase yield with adoption of scientific technology and better cultivars, mechanization, drought and flood management, *etc*.

The National Horticulture Research &Development Foundation (NHRDF) aims at disseminating advance technology for increasing productivity of horticultural crops through research and extension services. The NHRDF has centres located in different parts of India and through its research has observed the yield of several varieties of onion. Varieties such as Agrifound Light Red can yield 30-35 tonnes per hectare while NHRDF-Red can yield 35 to 40 tonnes per hectare (NHRDF, 2022).

From the above, it can be concluded that if yield of onion increases through scientific practices and extension services, the supply shocks in onion can be substantially reduced. It was also observed earlier (Table 1) that several countries, not only developed, but also less developed like Egypt and Iran, experience yields that are much higher than India. Besides lower yields, India also suffers from storage losses as the storage practices are often unscientific. Hence devising improved and scientific storage practices is important.

7. DEVISING IMPROVED AND SCIENTIFIC STORAGE STRUCTURES FOR ONION

Onion is a seasonal crop, but consumed throughout the year. Hence scientific storage of onions assumes importance so as to minimize on losses and ultimately maintain price level. In several countries onions are stored in open ventilated warehouses with forced air ventilation. Curing and irradiation of onion bulbs to reduce microbial losses is used in most onion producing countries to preserve onions.

7.1 Global Onion Storage Practices

A storage technology used in Africa and devised by Sasakawa Africa Association maintains the shelf life of onions by keeping the onions dry through a solar-powered fan which keeps the onion neck tissue and outer scales dry. The system is programmed to work twelve hours per day and the fans are operated every alternate hour to avoid too much air flow across the bulbs. In Netherlands, onion is harvested only in one season and hence in order to store the produce the process of condensation drying is undertaken through which moist air is drawn through the evaporator and cooled, causing the moisture in the air to condense and be removed. Reheating the cooled air allows the moisture to be absorbed.

Hence normally technologies to store onions are based on curing and reducing the moisture content in the crop so that the shelf life is increased. In India, storage of onions is very important because in the lean season, the price tends to rise, making the crop often unaffordable to the common man.

Due to unscientific design of storage structures, onion is exposed to extreme weather that causes huge damage of the crop which begins to rot. Often, the onions are stored in ventilated storage structures but the temperature and humidity is not controlled. A low cost bamboo storage structure with roof made of sugarcane leaves is commonly used for storage, but leads to 42% losses of onions during four months of storage.

In view of the above, Indian Council of Scientific Research (ICAR) in collaboration with Directorate of Onion and Garlic research have devised cold storage structures where onions are stored in controlled conditions having proper ventilation that enhances their storage life significantly by utilizing low energy while minimizing the losses. The temperature is maintained at 27 degrees C and relative humidity of 60 to 65% with air circulation system. This structure significantly reduces the physiological weight loss, unwanted sprouting and rotting of stored bulbs which reduces the wastage losses to 7%. This structure is also useful in storing *kharif* onions which have high moisture content (ICAR, 2022).

The project report by PoCRA (2022) presented details of Climate Agnostic Smart Integrated Onion curing and storage structure, developed by Indian Institute of Technology, Bombay, for climate resilient districts of Maharashtra. The structures aim to provide facilities to farmers to store their onion for a long period without undergoing major losses. Before storage, the freshly harvested onions are cured and the process involves uniform and steady heating of onion to maintain curing temperatures. The air with consistent temperature, relative humidity and air flow rate is allowed to circulate with onion bulbs in a controlled system. The storage structure is designed for stacking system with optimum strength and easy handling by workers. The motorized conveyor system allows fast and precise loading and unloading of onions. There is provision to attach a facility for sorting and grading of onion. The results of these storage structures reveal the losses to be in the range of 10-12% during field experiments. The study also noted that from field surveys, interviews, etc that in traditional storage structures the losses were to the tune of 35% to over 60%.

In order to address the problems of sprouting and rotting of onions and the need to minimize on storage losses, technologies are now being introduced to stabilize the warehouse monitoring and crop storage ecosystem so that the shelf life of the crop is increased and the farmers are able to earn higher returns. Godaam Innovations is one such Agri tech startup which leverages IoT technology to avoid onion waste through spoilage by 20 to 25%. This startup was founded by a Nashik based entrepreneur in 2018 and partnered with National Cooperative Marketing Federation (NAFED) and NABARD to understand issues relating to warehousing.

It was observed by the Startup that most farmers rely on identifying onion spoilage through smell when about 30% of the crop is already wasted and farmers suffer losses. Godaam, an IoT based sensing hardware device is installed by the Startup that does real time micro climate monitoring and identifies early spoilage of the crop.

The features of the technology are:

(a) The Godaam device integrates sensors and IoT technology to monitor key environmental parameters essential for onion storage, including temperature, humidity and three gas parameters. Field trials have proved the ability of the device, to provide instant alerts and notifications when spoilage begins to take place.
(b) Before installation of the device, they also retrofit the warehouses and improve their ventilation.
(c) The device consists of sensors that continuously track the gases released by the stored onions. When the crop begins to rot, there is a spike in certain gases and sulfides which are picked up by the sensors. This real time data is notified to the farmers and they become aware of spoilage of the crop at a very early stage.

(d) The sensors send the data to a server where the data analysis team runs an algorithm analysis on the data. Based on the insights from the analysis, the team sends an alert to the farmer/warehouse manager vis text, whats app, Godam's dashboard and the mobile application in their preferred language.
(e) The device regularly feeds the dashboard with sensor data so that the farmer can remotely monitor the stored crop.
(f) The device has wireless connectivity to send data to the cloud based server.
(g) Since the farmer becomes aware of spoilage at a very early stage, he can take corrective steps and dispose the crop.
(h) The device has low power consumption, therefore economizing on the use of energy and is also compatible with various power sources.
(i) The device is compact making it easy to install and use in warehouses.

The major benefits due to the use of this device are:

(a) Early detection of rotting through continuous monitoring of gases and micro climatic conditions which allow for suitable action to be taken to prevent further spoilage.
(b) Reduction in Post-Harvest Losses which arise due to rotting of onions when stored. By identifying and segregating affected batches promptly, farmers can minimize revenue losses and thus maximize their returns from sale of onions.
(c) Improved Storage Efficiency is possible as the onions are stored in a conducive environment with consistent temperatures and humidity levels made possible because of automated monitoring and control mechanisms. This mitigates the risk of spoilage and thus increases the shelf life of the onions.
(d) The use of this innovation has enhanced the marketability of onions by ensuring superior quality and freshness and therefore sales at competitive prices.

Discussion with the Startup also revealed that Godaam devices detected rotting as early as 1% and the shelf life of onions has increased by approximately 2 months of storage. Further, the recovery of A Grade onions has increased by 10%. The team of Godaam connects with farmers, mainly through Farmer Producer Organizations (FPOs) and other collectives so that they can access huge stocks of stored onions and fit their device. The client locations are presently concentrated in the major onion growing districts of Nashik and Pune.

This technology using IoT and sensors is newly introduced and the first prototype had necessary corrections made based on feedback from stakeholders. The cost of installing the device is INR 15,000 plus GST for 10 metric tonnes besides a maintenance charge of 30% of the cost of the device (INR 4500) annually. The device requires maintenance from the second year onwards. Alternatively, the product is also available in purely service based approach by charging the customer INR 170 per metric tonne per month, which includes all facilities. It appears that the cost of utilizing the device is quite cost effective. It was revealed by the Startup that rather than selling the device to individual farmers, they prefer to deal with FPOs or other collectives. This will perhaps ensure economies of scale and it is easier for the startup to deal with big customers than with several small and marginal farmers. It was also indicated that even if the device is used in traditional storage structures, it helps to reduce storage losses by at least 50% (Godaam Innovations Pvt. Ltd., 2024).

Figure 5. Working pattern of Godaam Sense (IoT based Hardware) on the field

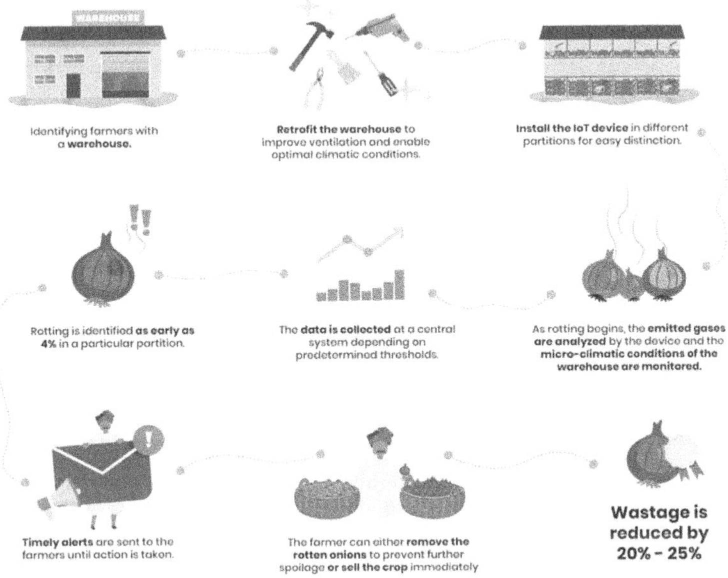

Source: https://www.godaaminnovations.com/product/

Overall discussions with the Startup revealed that while the technology is still in the infant stage, the results are encouraging even in traditional storage structures. The technology is cost effective, and has the potential to detect very early deterioration of the crop, so that the farmers can take corrective measures such as removing the spoilt onions or make decision for timely sale so that further losses are prevented.

8. CONCLUSION AND POLICY

The onion crop is often characterized by imbalances in demand and supply due to seasonal nature of production, poor storage facilities coupled with daily requirement of the vegetable in almost every household globally. When the crop fails in any country due to climate or other reasons, imports are resorted to, in order to meet the domestic requirements. India is the largest producer of onion in the world and the country is not only self sufficient in production but also a major exporter. This crop has an important place in the diet of most Indian households, especially the low income groups. Onion is also an economically and politically sensitive crop in India. However, the onion economy is characterized by huge fluctuations in prices. In times of price downtrend, the farmers realize low revenue, while during lean periods when prices firm, the consumers undergo hardship. In order to balance the interest of various stakeholders, viz. farmers, consumers and exporters, the government intervenes not only by releasing buffer stocks, but also imposing MEP and export ban. Such measures are not only detrimental to the interests of farmers and exporters, but the country loses on gains from trade. The real answer therefore lies in increasing production through technology and shifting from traditional storage units to advanced technology. In years of excess production, in order to restrict price fall, processing of onions should be encouraged. Efforts should also be made to ensure quality onion seed and suitable post harvest technology at the farm level, valorization and other value addition methods to prevent wastage and loss for farmers. Since storage of onion is paramount, in view of seasonal nature of production but regular consumption, storage structures must be technologically advanced. The government is also making all round efforts to maintain supplies of the crop during lean season by providing subsidies on onion storage structures and encouraging farmers to avail the same. As yields increase with transfer of technology from research stations to land and scientifically designed storage structures are created for storage of produce, the supply shocks in onion value chain will be overcome. India will then face less price fluctuations and also capitalize on its export potential in international markets. In order to further increase global market share, India must also tune the production pattern to meet the varieties required by other countries. For example yellow onions are preferred in many countries and hence these varieties must be cultivated for export. Onions are highly priced in several countries such as USA, South Korea and Japan which provides an opportunity for India to increase market share in world onion trade by exporting less pungent varieties to these countries. Such policy measures aimed at promoting technology in production of onions and storage will benefit both farmers as well as consumers in domestic and international markets.

REFERENCES

Centre starts releasing onions from buffer stock to rein in prices (2023). The Economic Times. Retrieved from https://economictimes.indiatimes.com/news/economy/foreign-trade/centre-starts-releasing-onions-from-buffer-stock-to-rein-in-prices/articleshow/102641853.cms?from=mdr

CRISIL. the Onion Paradox (2022). Retrieved from https://www.crisil.com>documents>2022/11.

Deepika, M. (2024). *Nurturing India's Agriculture: A Closer Look at Onion Price Dynamics*. BigHaat.

Essential Commodities Amendment. Bill 2020 (5th June 2020). Retrieved from: https://loksabhadocs.nic.in/Refinput/New_Reference_Notes/English/10092020_163243_102120474.pdf

Gadge, S. S., & Lawande, K. E. (2012). Crop Damage Due to Climatic Change: A Major Constraint in Onion Farming. *Indian Research Journal of Extension Education*, 2(Special Issue), 38–41.

Horticultural Statistics at a Glance, GOI. (28th May 2024). Retrieved from https://agriwelfare.gov.in/Documents/CWWGDATA/Agricultural_Statistics_at_a_Glance_2022_0.pdf

India Export of Agro Food Products. (2021). APEDA Agri Exchange. (2nd June 2024). Retrieved from https://agriexchange.apeda.gov.in/IndExp/PortNew.aspx

Market Intelligence and Analytics. (2023). *The Onion Paradox: Output falls, but so do prices*. CRISIL An S & P Global Company.

Ministry of Consumer Affairs. Food & Public Distribution (2024). Government has directed NCCF and NAFED to initiate procurement of 5 lakh tonnes of onion for the buffer requirement directly from the farmers. Retrieved from https://pib.gov.in/PressReleasePage.aspx?PRID=2016405#:~:text=In%20the%20current%20year%2C%20the,started%20arriving%20in%20the%20market

Ministry of Consumer Affairs, Food and Public Distribution. (2024). Grand Challenge for development of "Technologies for Primary Processing, Storage and Valorization of Onions". Retrieved from https://doca.gov.in>Grandchallenge-Document.

Monthly Report of Onion. (2017). Horticultural Statistics Division, Department of Agriculture, Cooperation and Farmers Welfare, Ministry of Agriculture & Farmers Welfare, Government of India.

Namiban Agronomic Board. (2021). Market Intelligence Report Onion. 3: pp 1-21. Retrieved from https://www.nab.com.na/wp-content/uploads/2022/04/Market-Intelligence-Report-Onion.pdf

Nanaji Deshmukh Krishi sanjeevani Prakalp (PoCRA) and Indian Institute of Technology, Bombay (2022), 300 MT Climate Agnostic Onion Storage Structure, (Project of Government of Maharashtra in Partnership with World Bank). Retrieved from https://mahapocra.gov.in>DPR.

Plant Genetic Resources and Crop Improvement, Onion (2023). National Horticultural Research & Development Foundation. Retrieved from: www.nhrdf.ora/Onion.php

Retail onion price up 57 pc; Centre steps up buffer onion sale to provide relief to consumers (2023).The Indian Express. Retrieved from https://www.newindianexpress.com/business/2023/Oct/27/retail-onion-price-up-57-pc-centre-steps-up-buffer-onion-sale-to-provide-relief-to-consumers-2627557.html#:~:text=NEW%20DELHI%3A%20As%20the%20all,to%20provide%20relief%20to%20consumers

Rise in prices of Domestic Products (2021). Ministry of Consumer Affairs and Public Distribution. Retrieved from www.pib.press.release.

Sasakawa Africa Association. (2024), retrieved from https://saa-safe.org/newsletter/saa-new-technology-helps-farmers-improve-onion-shelf-life/

Saxena, R., & Kumar, R. (2020). *Commodity Profile on Onion*. ICAR-National Institute of Agricultural Economics and Policy Research.

Sharma, D., Banyal, S. K. and Chauhan, A. (2020). Kharif Onion Cultivation – a boon for the farmers of Northern India. *Indian Horticulture*, ICAR. PP 20-23.

Sharma, D., & Chauhan, A. (2022). Kharif Onion Production in India-Present Status and Future Potential: A Review. *Agricultural Reviews*. Retrieved from https://arccjournals.com/journal/agricultural-reviews/R-2455

Socio-Economic Statistical Information about India. (15th May 2024). Retrieved from www.indiastat.com

Srivastava, R., Meena, K., Tiwari, A., Singh, N., & Behra, T. K. (2022). Yield and Economics of Kharif Onion (Allium cepa L.) under Front Line Demonstration in Eastern Plain Zone of Uttar Pradesh, India. *International Journal of Plant and Soil Science*, 34(23), 1034–1040. DOI: 10.9734/ijpss/2022/v34i232513

The Economic Times News. (2023). How the onion rolls: The politics and economy of India's kitchen staple. Retrieved from https://economictimes.indiatimes.com/news/economy/agriculture/how-the-onion-rolls-the-politics-and-economy-of-indias-kitchen-staple/articleshow/104498132.cms?from=mdr

Tolsma Grisnich-Onion Storage. (2024). retrieved from https:/www.tolsmagrisnich.com/pages/onion-storge

Vivek, T. R., & Mohammed, F. (2024). Why India Faces a Constant Onion Crisis. The Plate. Retrieved from https://theplate.in/why-india-faces-a-constant-onion-crisis/

Yashada, (2019), International Symposium on Edible Alliums: Challenges and Opportunities, Feb 9-12. Retrieved from https://krishi.icar.gov.in>jspui>bitstream.

Chapter 10
Sweet Sorghum Supply Chain for Bio-Ethanol Production

Karnatakapu Manoj Kumar
Vaikunth Mehta National Institute of Cooperative Management, India

Sneha Kumari
https://orcid.org/0000-0002-7228-1986
Symbiosis School of Economics, Symbiosis International University, Pune, India

ABSTRACT

20% ethanol blending is targeted for petroleum fuel (EBP-20) by countries. However, traditional feedstocks like sugarcane, rice, and maize are facing constraints due to climate change and food security concerns. To bridge this gap, this research explores that Sweet Sorghum as an alternative potential feedstock for bio-ethanol production. While having lower conversion efficiency and ethanol price compared to traditional crops, Sweet Sorghum offers advantages such as multiple cropping cycles per year, high temperature tolerance, and low water requirements. This makes it a sustainable choice for bio-ethanol production and augmenting the ethanol supply.

INTRODUCTION

Global Sweet Sorghum Production

Sweet Sorghum production is high in the United States, Nigeria, India, Mexico, and Brazil. The top-producing countries are shown in table 1.

DOI: 10.4018/979-8-3693-4330-2.ch010

Table 1. Top Producing Countries for Sweet Sorghum

Country	% of Global Production	Total Production (2023/2024, Metric Tons)
United States	14%	8.07 Million
Nigeria	11%	6.4 Million
India	8%	4.74 Million
Mexico	8%	4.49 Million
Brazil	8%	4.43 Million
Ethiopia	7%	4.01 Million
Sudan	5%	3.06 Million
China	5%	3 Million
Argentina	4%	2.5 Million
Australia	4%	2.2 Million

Source: USDA, Foreign Agricultural Service, U.S. Department of Agriculture, https://fas.usda.gov/data/production/commodity/0459200

Ethanol Production

India is the third largest energy-consuming country in the world (Dhinakaran & Kesavan, 2020). India also imports crude oil to meet the requirements. India's share in global energy consumption is going to double by 2050 (Anandarajah & Gambhir, 2014). Increasing trends of energy consumption and import may lead to issues like security and health welfare. Over dependence on oil imports can be avoided by mixing domestic produced ethanol with conventional fossil fuels. India launched the Ethanol Blending Programme (EBP) in 2003 with the idea of saving the overuse of fossil fuels. The ethanol blending program was developed for the sugarcane by product like molasses. This has set an example of circular economy for producing from byproducts. The sugar industry has been able to generate revenue from its byproducts. By 2015, 5% blending with ethanol was achieved. In 2018, the National Biofuel Policy was revised, introducing feedstocks other than C molasses for ethanol production. By 2019, 10% blending was achieved, and the target to achieve 20% blending has been set for 2025. By March 2024, India achieved an 11.6% ethanol blending with petrol in the first four months of the 2023-24 period (Mukherjee, 2024). The commercial sources of Bio-Ethanol production are sugarcane syrup, broken rice, maize, and molasses.

When the government of India announced the Biofuel policy, the availability of sugarcane and rice was surplus, but due to erratic changes in the climate and the El Nino effect in 2022-23, the production of Sugarcane and rice has drastically reduced. Therefore, the availability of these sugarcane and rice for the conversion of

Bio-Ethanol as feedstock is at risk. In 2023, due to the El Nino effect, the monsoon rains were limited to 6% only, as half of the agriculture cropping in India is rainfed. This led to a reduction in the yield of rice by 4 million tonnes (MT) and a reduction of 55 million tonnes (MT) of sugarcane in India. As a result, the Government of India banned the export of rice and made restrictions on the export of sugar up to a maximum of 6 million MT only as per the government's first advance estimates.

The competition between the industries of sugarcane, rice, and maize has increased. Rice and maize are required for feed manufacturing industries for the food security of the people. In the case of sugar mills sugarcane is required for the production of sugar and it's export. The sugar sector is in high demand due to high remunerative prices in international markets. On the other hand, oil marketing companies require sugarcane for the production of the Bio-Ethanol to blend in petrol. The competition for maize is fierce as different industries need more of the crop. The trade-off in the case of rice is between meeting its needs and exporting it.

The strategy of using water-guzzling crops to produce green fuel seems misguided. The processing of cane in mills requires a lot of water. Before deciding on the use of such crops to produce green fuel, it is important to conduct a comprehensive impact study on emissions and resource use across the crops life cycle.

Crop yields are being affected by climate change. Projections of future supplies are difficult because of the unpredictable yields. Thus, Bio-Ethanol from these high water-demanded feedstock crops has deviated from their targets.

Among these challenges, Sweet Sorghum is an alternate source of Bio-Ethanol. The chapter aims to explore the supply chain of sorghum for ethanol production.

This study is based on secondary research through scholarly reports, research articles, government reports, and government websites.

Sweet Sorghum Supply Chain

1. **Cultivation:** Sweet Sorghum can be cultivated on diversified lands with minimal inputs like water and fertilizers. Good agricultural practices like timely irrigation, intercultural operations, and harvesting enhance the overall yield. Famers play a crucial role in this supply chain by providing the key raw material for the production of Bio-Ethanol.
2. **Transportation:** Robust and timely logistics services are required for the supply of sweet sorghum from producer to distilleries. In this supply, there will be different stakeholders involved between producer and processor. The government ensures a structured supply chain to improve the efficiency of the output.

3. **Processing:** The sweet sorghum is processed into Bio-Ethanol in distilleries by fermentation of sweet sorghum plant vegetative stalks, which contain sweet juice, as explained in Figure 1; during the processing, various by-products are produced, which have various industrial uses and generate additional revenue.
4. **Distribution:** After the production of Bio-Ethanol, the government of India procures from the distilleries at a specified price, and it will be distributed to the Oil Marketing Companies (OMCs) for the blending of Bio-Ethanol with petrol.

Circular Economy Perspective in Sorghum Supply Chain

The circular economy is the concept that aims to reduce waste and increase revenue in sustainable ways throughout the supply chain. In the supply chain of sweet sorghum, the circular economy is involved in the reduction of waste and using its byproducts for extra revenue generation.

Reduction in Carbon Emissions Usually in all the crops, after harvesting the economical part of the plant the remaining plant parts are considered as the residue which will decompose and releases carbon dioxide, but in the case of the sweet sorghum it is fast growing crop and it is C4 plant which utilizes the carbon dioxide in the efficient manner also the plant body (vegetative stalks) acts as feedstock for the bio-ethanol industry which helps in carbon sequestration and carbon credits are generated.

Usage of Residual Matter: During the processing of the Sweet Sorghum to Bio-Ethanol, various by-products like bagasse and Distilleries Dried Grain Solid (DDGS) will be released, which has industrial value. Bagasse is used in the paper industry, power generation, and also in animal feed production. DDGS is a Distilleries Dried Grain Solid waste, which is a residual matter left from crushing and fermentation of vegetative stalks of sweet sorghum during the ethanol extraction, which will be used in the animal feed industry.

OVERVIEW OF SWEET SORGHUM AS A BIOFUEL CROP

Figure 1. Sweet Sorghum Flowchart

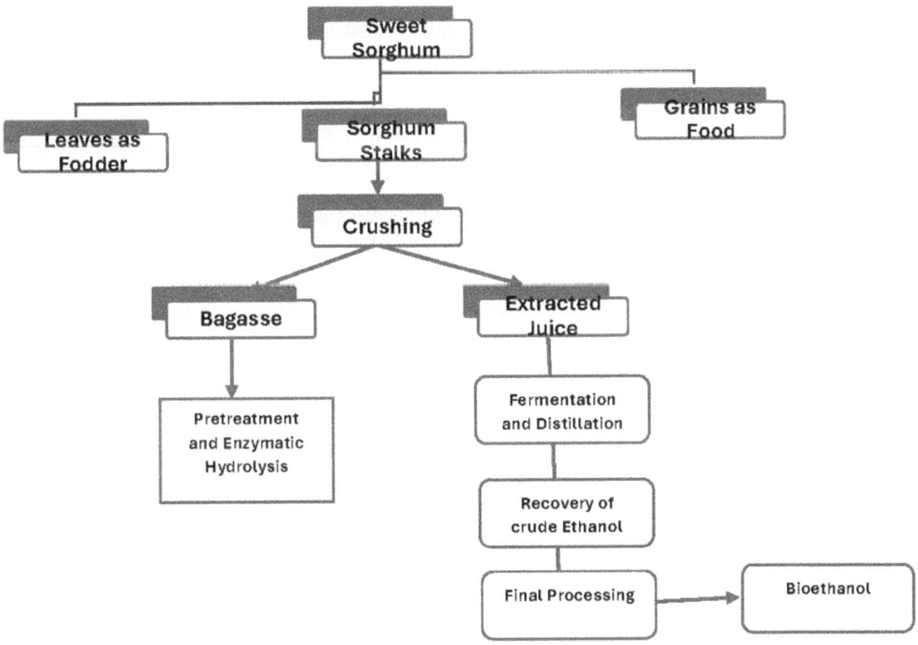

Sweet Sorghum grains are similar to maize kernels (Quintero et al. 2008). After harvesting the crop, the grains are used as food or animal feed, but the stalks are crushed and milled to extract the plant juice with sugars. This extracted juice is further saccharified using enzymes like amylase and glucoamylase and then fermented. After that, it will distillate to extract the Bio-Ethanol. The by-products are Dried Distillers Grain Solids (DDGS). The DDGS is a by-product of the ethanol production process and a high-value feed used by the livestock industry.

Sweet Sorghum is an annual plant that can be grown in diversified climates and in various soil conditions. It thrives in tropical, subtropical, and semi-arid regions. Sweet Sorghum is efficient in converting sunlight into energy because it is a C4 plant, and has a superior photosynthetic rate compared to C3 grasses. It has a robust root system, which makes the plant more resilient and helps with deep penetration of 1.5 to 2.5 meters. It also spreads laterally to approximately one meter. It has a tap root system which spreads into the subsurface of the soil and absorbs moisture reserves (Damasceno et al., 2014). The plant can endure prolonged periods of drought without wilting because of its extensive root network. The plant can go into the dormant stage whenever the water is not available and can resume its growth with the presence of adequate moisture again. Besides drought tolerance, sweet sorghum also gives a good yield. It has good capability to produce a substantial number of juicy stems per hectare, even on barren lands that are not suitable for cultivating conventional food crops; the stalk is the main part of the plant that can be used to extract the Bio-Ethanol which is an excellent source as sustainable bioenergy production.

Sweet Sorghum is underrated as feedstock for the Bio-Ethanol industry as many research results clearly show a significant positive correlation between the quantity of fresh biomass produced and the quantity of Bio-Ethanol that can be extracted. Sweet Sorghum attains maturity lately and it is sensitive to long day length but gives good yield of biomass even when the amount of rainfall is very limited. Smith and Buxton, (1993) research have shown that, on average, temperate climates crop produces fresh biomass of 89.2 tons and 65.5 tons per hectare in irrigated and dryland conditions, respectively. The stalk of the plant contributes to the major biomass of the sweet sorghum, 70% of the biomass of sweet sorghum is contributed by the stalks and stalk weight is correlated with height, thickness, and juiciness (Audilakshmi et al., 2010). The combination of high biomass productivity, drought tolerance, and suitability for marginal lands positions sweet sorghum as a valuable crop for addressing global challenges related to food security, energy production, and environmental sustainability.

Economic Viability

The production of ethanol, a biofuel that can reduce the release of greenhouse gas emissions and reduce dependence on fossil fuels, has garnered significant attention. Among the various feedstock options, sweet sorghum has emerged as a compelling candidate due to its exceptional sugar content and adaptability to diverse agroclimatic conditions. Unlike the production of ethanol from grain starch, which necessitates energy-intensive processes for the depolymerization of carbohydrates, the conversion of sugar into ethanol is notably more efficient. This fundamental difference translates into a lower overall energy input for ethanol derived from sugar crops, such as sweet sorghum, when compared to corn-based ethanol.

Theoretical ethanol yields from sweet sorghum have demonstrated competitiveness with corn while requiring a comparatively lower energy investment. Under optimal growing conditions, sweet sorghum can produce up to 13.2 tons of total sugars per hectare, which equates to an impressive 7682 litters of ethanol (Jackson et al., 1980; Murray et al., 2009). While these figures represent maxima, field trials, and commercial-scale operations have yielded substantial ethanol quantities.

In the United States, the total sugar yields from sweet sorghum have ranged from 4 - 10.7 megagrams per hectare, with some regions, such as Hawaii, achieving even higher levels of up to 12 megagrams per hectare. These figures translate to theoretical ethanol yields estimated between 3000 and 4000 litres per hectare (Lueschen et al., 1991). Practical ethanol production from sweet sorghum stalks has yielded comparable results, with studies reporting ethanol yields ranging from 3450 to 4132 litres per hectare (Serna-Saldivar et al., 2012). India has been at the forefront of sweet sorghum research and ethanol production. The Indian Institute of Millets Research (IIMR) has spearheaded numerous pilot-scale studies in collaboration with sugarcane distilleries. These research reports show that Bio-Ethanol production from sweet sorghum has huge potential under Indian climatic circumstances. While initial results were promising, with Bio-Ethanol yields ranging between 35 to 40 litres per ton (Rao et al., 2014) of crushed stalks, challenges related to feedstock supply and ethanol pricing hindered the sustained commercialization of sweet sorghum-based ethanol production. The International Crops Research Institute for the Semi-Arid Tropics (ICRISAT), India has also played a pivotal role in producing Bio-Ethanol from sweet sorghum. Through its partnership with Rusni Distilleries, which has established the world's first Sweet Sorghum-based Bio-Ethanol distillery plant. This plant successfully extracted the Bio-Ethanol of 45 litres per ton by crushing the Sweet Sorghum stalks (Basavaraj et al., 2012). However, economic factors and feedstock availability challenges ultimately led to the closure of this distillery.

COMPARATIVE ANALYSIS

Sweet Sorghum has the potential to be one of the sustainable feedstocks for the production of Bio-Ethanol (Table 2). Further research for development of improved varieties with lengthy stalks and more sugar content coupled with good government policies are required for unlocking the complete potential of Sweet Sorghum as feedstock for Bio-Ethanol.

Table 2. Comparative analysis of different aspects of sugarcane, sugar beet, corn and sweet sorghum

Indicators	Sugarcane	Sugarbeet	Maize (Corn)	Sweet Sorghum
Crop duration (in months)	12-14	5-6	3-4	4
Growing Season	One season	One season	All season	All seasons depending on water availability
Propagation	Setts (40000 ha^{-1})	Seed (3.6 kg/ha)	Seed (8kg/ha)	Seed (8kg / ha)
Soil Requirement	Grows well in drained soil	Grows well in sandy loan	Grows well in sandy loam	All types of drained soil

continued on following page

Table 2. Continued

Indicators	Sugarcane	Sugarbeet	Maize (Corn)	Sweet Sorghum
Water Management	Water throughout the year 36000 m3/ha	18500 m^3/ha	12000 m^3/ha	800 m^3/ha
Crop Management NPK in kg/ha	250-400:125:125	120:60:60	120:60:60	80-90:40:0
Yield (t/ha)	65-80 stalk	85-100 beet	5-10 grain	40-55 for one cycle/ year (stalk) 80-110 for two cycles/ year stalk
Sugar content (weight basis) %	10-12	15-18	na	7-12
Sugar Yield (t/ha)	5-12	11-18	na	4-6 for one cycle per year 7-12 for two cycles/ year
Ethanol yield from juice or grain (L/ha)	4350-7000	7100-10500	2150-4300	2000-3500 for one cycle per year 4000-7000 for two cycles per year
Harvesting	Mechanically	mechanically	mechanically	Manual and mechanically

Source: Kumar et al., 2012

Sweet Sorghum is a versatile crop and is one of the solutions for the contemporary issues of limited fossil fuel reserves and increasing greenhouse gas emissions. Utilizing the potential of the Sweet Sorghum for extraction of Bio-Ethanol and green electricity production leads to a significant rate of reduction in the consumption of fossil fuels. Let us assume that cultivating Sweet Sorghum for both purposes of producing ethanol and food grains can result in energy savings, equivalent to 2300 liters of crude oil per hectare. Furtherly, by cultivation of Sweet Sorghum, carbon dioxide sequestration can be done at 1.4 to 22 kg per hectare depending upon various factors like crop management practices, land usage, and yield of the crop. When compared to Bio-Ethanol extracted from the Maize, Sweet Sorghum Bio-Ethanol gives a better return on investments and a good margin. It gives an exceptional 12 to 16 British Thermal Units(btu) of energy for every btu that has invested in its production process using fossil fuel in contrast to extracting the ethanol from Maize which offers 1.3 to 1.8, which is very low as compared to to the Sweet Sorghum ethanol (https://www.uky.edu/Ag/CCD/introsheets/sorghumbiofuel.pdf).

Beyond biofuel production, sweet sorghum proves to be a multi-faceted crop with diverse applications. Its juice, rich in glucose, fructose, and sucrose, serves as an ideal substrate for direct ethanol fermentation or the synthesis of other bio-based chemicals. The remaining parts of the plant, including leaves, bagasse, and grain, find utility in various sectors. Bagasse, in particular, holds immense value as it can be converted into biofuels, animal feed, or utilized for power generation and the production of valuable byproducts like pulp and particle (Somani and Taylor, 2003) board. Notably, sweet sorghum bagasse surpasses sugarcane bagasse in nutritional content, making it a superior feed for livestock. Furthermore, it also has an opportunity for the extraction of second-generation biofuels from cellulose and hemicellulose, as shown in figure 1. The accompanying table provides a comprehensive overview of the diverse processing options available for this remarkable crop.

Besides ethanol, sweet sorghum is gaining attention because of its capability to produce a wide range of valuable products beyond bio-ethanol. Through fermentation processes, sweet sorghum can be transformed into hydrogen, butanol, acetone, butyric acid, lactic acid, and methane. Additionally, this remarkable plant offers several native products, including cellulose for proteins, allelopathic compounds like sorgoleone, paper production, and waxes (Whitfield et al., 2012). The wide range of applications highlights the prospects of sweet sorghum contributing to a sustainable and bio-based economy, not only for the production of bio-ethanol.

Policy Support and Regulatory Framework

Governments should recognize and consider sweet sorghum as one of the major sources for the extraction of Bio-Ethanol, and the government should provide incentives to farmers and business entities in the supply chain to promote sweet sorghum as a viable source for Bio-Ethanol production. These incentives will help to frame subsidies or various schemes for farmers to encourage the development of sweet sorghum. Incentives to establish distilleries to extract ethanol from sweet sorghum should be facilitated. In addition to that, competitive prices should be given for the bioethanol extracted from sweet sorghum vis a vis from other crops (table 3).

Table 3. Comparison of the Price per litre

Type	Price per liter at exit of distilleries
Maize	Rs. 72/-
Sugarcane Syrup	Rs. 66/-
B Molasses	Rs.61 /-
C Molasses	Rs. 56/-
Rice	Rs. 64/-
Sweet Sorghum	Rs.48.5/- to 49.5/-

Source: PIB

A robust supply chain should be established from farmers to distilleries is required, like NCCF and NAFED procure the maize from the farmers and supply to the distilleries in the same way They should facilitate the sweet sorghum supply to distilleries to produce Bio- Ethanol.

REFERENCES

Ashok Kumar, A., Reddy, B.V.S., Blummel, M., Anandan, S., Reddy, Y.R., & Reddy, Ch.R., Srinivasa Rao, P., Reddy, P.S. (2010). On-farm evaluation of elite sweet sorghum genotypes for grain and stover yields and fodder quality. *Animal Feed Science and Technology*, 10S, 69–78.

Audilakshmi, S., Mall, A. K., Swarnalatha, M., & Seetharama, N. (2010). Inheritance of sugar concentration in stalk (brix), sucrose content, stalk and juice yield in sorghum. *Biomass and Bioenergy*, 34(6), 813–820. DOI: 10.1016/j.biombioe.2010.01.025

Basavaraj, G., Rao, P. P., Basu, K., Reddy, C. R., Kumar, A. A., Rao, P. S., & Reddy, B. V. S. (2012). Assessing viability of bio-ethanol production from sweet sorghum (No. 30, p. 20). Working paper series.

Bonin, C. L., Heaton, E. A., Cogdill, T. J., & Moore, K. J. (2016). Management of sweet sorghum for biomass production. *Sugar Tech*, 18(2), 150–159. DOI: 10.1007/s12355-015-0377-y

Damasceno, B. M. C., Schaffert, E. R., & Dweikat, I. (2014). Mining genetic diversity of sorghum as a bioenergy feedstock, plants and bioenergy. *Adv. Plant Biol.*, 4, 81–106.

Hoffmann-Thoma, G., Hinkel, K., Nicolay, P., & Willenbrink, J. (1996). Sucrose accumulation in sweet sorghum stem internodes in relation to growth. *Physiologia Plantarum*, 97(2), 277–284. DOI: 10.1034/j.1399-3054.1996.970210.x

Hunsigi, G., Yekkeli, N. R., & Kongawad, B. Y. (2010). Sweet stalk sorghum: An alternative sugar crop for ethanol production. *Sugar Tech*, 12(1), 79–80. DOI: 10.1007/s12355-010-0017-5

Hunter, E. L., & Anderson, I. C. (1997). Sweet sorghum. In Janick, J. (Ed.), *Horticultural Reviews* (Vol. 21, pp. 73–104). John Wiley and Sons. DOI: 10.1002/9780470650660.ch3

Jackson, D. R., Arthur, M. F., Davis, M., Kresovich, S., Lawhon, E. S., Lipinsky, W. T., & Rudolph, A. 1980. Research Report on Development of Sweet Sorghum as an Energy Crop, vol. 1. Agricultural Task. US DOE, US Department of Commerce, Battele, Columbus.

Kim, M., & Day, D. (2011). Composition of sugar cane, energy cane, and sweet sorghum suitable for ethanol production at Louisiana sugar mills. *Journal of Industrial Microbiology & Biotechnology*, 38(7), 803–807. DOI: 10.1007/s10295-010-0812-8 PMID: 20803247

Lueschen, W. E., Putnam, D. H., Kanne, B. K., & Hoversted, T. R. (1991). Agronomicpractices for production of ethanol from sweet sorghum. *Journal of Production Agriculture*, 4(4), 619–625. DOI: 10.2134/jpa1991.0619

Murray, S. C., Rooney, W. L., Hamblin, M. T., Mitchell, S. E., & Kresovich, S. (2009). Sweet sorghum genetic diversity and association mapping for brix and height. *The Plant Genome*, 2(1), 48–62. DOI: 10.3835/plantgenome2008.10.0011

Quintero, J. A., Montoya, M. I., Sanchez, O. J., Giraldo, O. H., & Cardona, C. A. (2008). Fuel ethanol production from sugarcane and corn: Comparative analysis for a Colombian case. *Energy*, 33(3), 385–399. DOI: 10.1016/j.energy.2007.10.001

Rao, S. S., Patil, J. V., Rao, B. D., Ratnavathi, C. V., Umakanth, A. V., Mishra, J. S., & Chapke, R. R. (2014). 14. Current status of sweet sorghum biofuel industrial experiences for pilot–production of bioethnol. Proc. of "Model Training Course on "Improved sorghum cultivation and value-addition perspectives, 22, 29.

Sanjana Reddy, P., Reddy, B. V. S., & Srinivasa Rao, P. (2011). Genetic analysis of traits contributing to stalk sugar yield in sorghum. *Cereal Research Communications*, 39(3), 453–464. DOI: 10.1556/CRC.39.2011.3.15

Serna-Saldívar, S. O., Chuck-Hernández, C., Pérez-Carrillo, E., & Heredia-Olea, E. (2012). Sorghum as a multifunctional crop for the production of fuel ethanol: Current status and future trends. *Bioethanol*, 2012, 51–74.

Smith, G. A., Bagby, M. O., Lewellan, R. T., Doney, D. L., Moore, P. H., Hills, F. J., Campbell, H. J., Hogaboam, G. J., Coe, G. E., & Freeman, K. (1987). Evaluation of sweet sorghum for fermentable sugar production potential. *Crop Science*, 27(4), 788–793. DOI: 10.2135/cropsci1987.0011183X002700040037x

Smith, G. A., & Buxton, D. R. (1993). Temperate zone sweet sorghum ethanol production potential. *Bioresource Technology*, 43(1), 71–75. DOI: 10.1016/0960-8524(93)90086-Q

USDA. Foreign Agricultural Service, U.S. Department of Agriculture, https://fas.usda.gov/data/production/commodity/0459200

Vinutha, K. S., Rayaprolu, L., Yadagiri, K., Umakanth, A. V., & Srinivasa Rao, P. (2014). Sweet sorghum research and development in India: Status and Prospects. *Sugar Tech*, 16(2), 133–143. DOI: 10.1007/s12355-014-0302-9

Whitfield, M. B., Chinn, M. S., & Veal, M. W. (2012). Processing of materials derived from sweet sorghum for biobased products. *Industrial Crops and Products*, 37(1), 362–375. DOI: 10.1016/j.indcrop.2011.12.011

Wu, X., Staggenborg, S., Propheter, J. L., Rooney, W. L., Yu, J., & Wang, D. (2010). Features of sweet sorghum juice and their performance in ethanol fermentation. *Industrial Crops and Products*, 31(1), 164–170. DOI: 10.1016/j.indcrop.2009.10.006

Chapter 11

Overcoming Fertilizer Supply Chain Disruptions in Agriculture With Special Reference to India

M. P. Sukumaran Nair
https://orcid.org/0000-0002-7368-025X
Centre for Green Technology and Management, India

ABSTRACT

Agriculture gradually depletes the soil of the essential nutrients for the growth of plants resulting in low crop productivity and yields. Soil is to be continuously fortified by the application of fertilizers to maintain food production. Therefore, an unbreakable supply chain is essential for continuous production of food from the fields. The supply chain gets disrupted following global economic downturn, pandemic or war situations. Due to this disruption supplies are affected, farmers are overburdened by rise in fertilizer prices beyond their reach and it could possibly lead to global food insecurity. India is a major foodgrain producing country and nearly 55% of its 1.44 billion population depends on agriculture for survival. This chapter discuss importance of al fertilizers in Indian agriculture and how the consumption is managed through domestic production and imports. The disruption in global supplies due to Covid 19 and the Ukraine war and problems encountered in making adequate supplies and government efforts to streamline the fertilizer supply chain are discussed.

DOI: 10.4018/979-8-3693-4330-2.ch011

INTRODUCTION

Food is the prime requirement of all living beings. It is produced by plants during its lifetime through photosynthesis by absorbing the required minerals from the soil in which it is growing.

Nitrogen, phosphorus and potash are the primary nutrients, sulphur (S), calcium (Ca) are magnesium (Mg) are secondary nutrients and boron (B), zinc (Zn), chloride (Cl), copper (Cu), iron (Fe), manganese (Mn), selenium (Se) molybdenum (Mo) and nickel (Ni) are micronutrients required for plant growth and productivity. Agriculture gradually depletes the soil of the essential nutrients for the growth of plants resulting in low crop productivity and yields. The soil is fortified by the application of mineral fertilizers. Therefore, an unbreakable supply chain is essential for the continuous production of food from the fields. The raw materials to produce mineral fertilizers are not evenly distributed across the globe. Some countries are blessed with natural gas, rock phosphate and potash. Others must manage a regular supply of these materials for their agricultural development. It is at this point that the management of supply chain and logistics become important. The fertilizer supply chain management ensures the availability of an uninterrupted supply of raw material to processing plants, its conversion to specific products for consumption by the farmers, as well as the associated flow of information all along the supply chain. It includes sourcing raw materials, its procurement and movement, processing, storage, and distribution to the end user. Disruptions occur due to economic downturn, pandemic or geo-political conflicts and limit availability and free movement of supplies.

After the Covid 19 pandemic, the fertilizer demand in major agrarian economies including US, EU, Brazil, China, and India have gone up considerably threatening global availability and prices reaching sky high. There was an unprecedented hike in natural gas demand and prices, world over. The conflict between Russia and Ukraine, coupled with China's prohibition on fertilizer exports, caused disruptions to the global fertilizer industry and the flow of raw materials and finished goods. The war in Ukraine will have a catastrophic effect on the already fragile global food production system because of climate change, population growth, and the pandemic. According to Svein Tore Holsether, President and CEO of Yara International, one of the largest fertilizer companies in the world, - This could lead to a significant rise in food insecurity in less developed nations. Farmers in several countries are overburdened by the increases in fertilizer prices beyond their reach, and consequently, many of them are struggling to pay for key components necessary for their production, adding to supply chain disruptions. There are instances of farmers abandoning cultivation which could possibly lead to a global situation of food insecurity. Economically backward countries in Africa, Southern America, and Asia are worst impacted, as observed by UN Secretary-General António Guterres in “Response to the Multiple Challenges to Global Food Security" in 2022. No country will be immune to the social and economic repercussions of such a catastrophe, he noted.

Indian Scenario

Indian population stands at 1.44 billion in January 2024 and is growing at 0.9% annually. Around 65% of the population is in the working age limit of 14 to 65 years of age and thus the demand for food is constantly rising. Furthermore, 67% of the population must get a free ration entitlement of 5 kg of food grain per person per month under the National Food Security Act (NFSA), 2013. In 2023–2024, India produced an astounding 332 million tonnes of food grains. The use of mineral fertilizers is directly responsible for between 50 and 55 percent of the increased grain production. These fertilizers were crucial to the success of India's green revolution, which began in the 1980s and enabled the nation to become food self-sufficient by the year 2000. Since then, the need for fertilizers has increased significantly year over year over the previous few years, which has allowed the domestic sector to thrive and resulted in massive imports of goods and inputs.

India is currently the world's largest importer, producer, and consumer of fertilizer materials, ranking second globally. Even still, India continues to apply far less mineral fertilizer per acre than the majority of developed and developing nations.

Table 1. India Fertilizer Scene: 2023-'24 (million metric tons-MMT)

Consumption	N+P2O5+K2O	30.64	Global 2nd largest
Production	N+P2O5	21.99	Global 3rd largest
Import	N+P2O5+K2O	9.75	Global largest

N- Nitrogen, P2O5- Phosphate, K2O- Potash
Source: Fertilizer Association of India

In the past three decades, fertilizer consumption in India grew at a compounded average growth rate (CAGR) of 3.3%. During the period production of both nitrogenous and phosphatic fertilizers grew only by 2.18% necessitating huge imports. Consequently, imports of N and P fertilizers also grew at 5.81%. As we do not have any domestic resources, the entire requirement of Potash (P) is met by imports alone. In 2016, the Union Government decided to revive five brownfield urea production units which were closed down earlier to increase domestic production by an additional 6.35 MMT per annum and reduce import to that extent.

A. DEVELOPMENT OF FERTILIZER INDUSTRY IN INDIA

Urea, diammonium phosphate (DAP), various grades of NP and NPK complex fertilizers, ammonium sulphate, potash, and single super phosphate (SSP) are the major fertilizer products used in Indian agriculture. The history of fertilizer production and use in India dates to 1906 when the first single super phosphate (SSP) manufacturing unit was established in Ranipet, Tamil Nadu. In the behest of the widespread famine in the country in the 1940s the erstwhile Government of the princely state of Travancore, established a fertilizer company, the Fertilizers And Chemicals Travancore (FACT) Ltd in 1944 at Kochi to achieve self-sufficiency in the production of food grains. It started producing ammonium sulphate in 1947. Another factory to produce ammonium sulphate nitrate and urea was also set up by the Fertilizers Corporation of India (FCI) at Sindri in Bihar in the 1950s. During the 1960s and 1970s several small and medium fertilizer plants which used indigenous raw materials came to being in the public and private sector which catalyzed the Green Revolution in the country.

A major boost to the production of fertilizers was initiated with the adoption of the Retention Pricing Policy (RPS) adopted by the Government of India in 1978 upon the recommendation of the S.S. Marathe committee. The committee investigated the basis of existing pricing of fertilizers and recommended a policy which assured a post-tax 12% return on net worth on investments in a sustained manner. In those days the economy was not liberalized and there were only few investment options open to private entrepreneurs in the national economy. The retention pricing scheme attracted several new investors, most of them from the private and cooperative sector, and incentivized the setting up of a number of large scale fertilizer plants in different parts of the country. This was given a further fillip with discovery of natural gas in the Bombay High region and establishment of the Hazira- Bijaipur-Jagadishpur (HBJ) pipeline carrying natural gas from Gujarat to UP. Most of these units became productive by the 1990s and the country achieved self-sufficiency in the production of urea in 2000. Indian agriculture was modernized incorporating scientific techniques during the period and with the increased application of mineral fertilizers on its fields, foodgrain production increased heaps and bounds and the country achieved food self-sufficiency by 2000 and subsequently became an exporter of rice.

Worldwide, natural gas is the feedstock of choice for production of ammonia. In India domestic natural gas resources were limited and therefore, had to essentially depend on liquid feedstock like naphtha and fuel oil and solid raw materials like coal and lignite. The capital and operating costs of these plants compared to the natural gas-based ones, were high and the reliability of operation was low, on account of which the cost of ammonia production was high. The installed capacities of the plants were also lower when compared to the demand from the agriculture sector.

Fertilizer Subsidy System

During the early decade of the century, the Government of India adopted a massive program for development of indigenous sources of natural gas and decided to import liquefied natural gas by setting up LNG terminals at coastal locations. Most of the ammonia plants operating or naphtha and fuel oil got changed over to natural gas depending upon availability. Some plants were closed on account of loss of financial viability due to high energy consumption and increased cost of production.

The cost of production of fertilizer materials being high, as such it is not affordable to the farmers. Therefore, the Government of India instituted a fertilizer subsidy scheme in 1976 to make fertilizer materials available to the farmers at affordable costs. The above subsidy scheme envisaged to compensate the producer companies for each tonne of product they make available to the farmers at the price fixed by the government. Under this administered pricing, the quantum of this compensation is estimated by the Fertilizer Industry Coordination Committee (FICC) in the Ministry of Fertilizers, Govt of India depended on the cost of production plus approved return on investment minus the price realized from the market.

Besides finished fertilizer materials, the country also needed to import large quantities of fertilizer raw materials and intermediates for downstream processing for which domestic facilities are already put up. During 2023 the country had to import 21% of its urea and 36% of all phosphate fertilizer requirements. It also imported 12% of its ammonia and 60% of its phosphoric acid requirement. The import dependance for sulphur and rock phosphate was over 95%. As we do not have domestic resources, India imports its entire requirement of potash from overseas sources.

On an average 3 to 5% increase in consumption of fertilizers is registered every year. Thus, the import dependance on fertilizers increased heavily resulting in huge outgo of foreign exchange. Increase in commodity prices, foreign exchange variations, keeping the consumer end price very low in farmer's interest and the increase in yearly consumption contributed to increase the subsidy burden of the Government over the years.

Table 2. Installed capacity and production of fertilizers

Year	Nitrogen (N) (million tonnes)				Phosphate (P2O5) (million tonnes)			
	No of plants	Installed Capacity	Production	Capacity Utilization %	No of plants	Installed Capacity	Production	Capacity Utilization %
2021-'22	56	15.35	13.87	91.9	119	7.1	4.71	68
2022-'23	60	17.28	15.74	95.2	122	7.28	5.01	70.6

Source: Fertilizer Association of India

The additional urea capacity of 6.5 million tonnes created through brownfield projects since 2021 has contributed to substantially reduce urea imports. Commissioning of the Talcher fertilizer plant based on coal gasification is scheduled for this year.

At the industry level, farming and the manufacture of food grains are not highly lucrative enterprises in India. The majority of Indian farmers are little, marginal farmers who grow for their personal needs and sell any surplus they have to the market, despite the fact that certain houses control huge farm holdings. As a result, there is a strong reliance on the government's support system, which includes minimum support prices for grain purchased by government agencies and fertilizer subsidies. A significant portion of government funds are allocated to fuel subsidies and food fertilizer in the annual budgets that are submitted to Parliament. The fertilizer industry's supply chain has been completely disrupted as a result of the war in Ukraine and the deterioration of the Middle East's geopolitical situation.

Consequently, prices of fertilizer raw materials and finished product when the sky high.

Table 3. Production, import & consumption of fertilizers

Year	2021-'22 (million tonnes)				2022-'23 (million tonnes)			
Nutrient	N	P2O5	K2O	Total	N	P2O5	K2O	Total
Production	13.87	4.71	-	18.58	15.74	5.01	-	20.75
Import	5.38	2.78	1.66	9.82	5.22	3.76	1.46	10.44
Consumption	19.44	7.83	2.53	29.80	20.21	7.92	1.72	29.84

Source: Fertilizer Association of India

Fertilizer Movement

Fertilizer being a controlled commodity, its movement across the country is regulated by the Government. Making adequate and timely supply of fertilizers to the farmers in all parts of the country is the responsibility of Department of Agriculture & Farmers Welfare (DA&FW), Govt of India. The requirement of chemical fertilizers i.e. Urea, DAP, Muriate of Potash (MOP) & NPK for each season in the region is assessed by the Department. Thereafter, Fertilizer materials are allocated to various agricultural regions based on a government directive. Distribution and interstate movement of fertilizer products are regulated by the Government. Retail shops, widely distributed and owned by producer companies or farmers' cooperatives stock the material and distribute the farmers at subsidized rates.

The movement and distribution of subsidized fertilizers, from various fertilizer plants and ports is coordinated by the Department based on agreed supply plans from manufacturers/Importers on a monthly basis. The Department also monitors the movement of all major subsidized fertilizers through a web-based monitoring system i.e. integrated Fertilizers Monitoring System (iFMS).

Decontrol of Phosphatic and Potassic Fertilizers

In the past two decades, the Government modified and revised the fertilizer policy several times with the objective to attract new investments, increase domestic production and reduce the subsidy burden. The results were not at all encouraging. Instead, certain major private players in the fertilizer business sold off their assets and exited the business due to the inordinate delay in the disbursement of subsidy amount which hit them hard on working capital management.

Phosphatic and potassic fertilizers were decontrolled in 2008 and a Nutrient Based Subsidy (NBS) scheme specifying a rate of subsidy for each of the nutrients contained in the product was instituted. Under this scheme, producers are allowed to determine market prices for the products, but only under government oversight. With the introduction of nutrient-based pricing and partial decontrol of phosphatic and potassic fertilizers in 2008, product prices went up sharply imposing considerable financial strain on the farmers.

B. IMPORTANCE OF SUPPLY CHAIN IN THE FERTILIZER INDUSTRY

Mineral fertilizer products are essential to maintain soil fertility and thus increase crop yields. In the nature, nutrients that support plant growth and increase food production are not distributed evenly across continents. Such materials are made available in major agricultural locations by ensuring a steady and reliable traded flow of products. However, in recent times, this crucial supply chain has experienced significant disruptions due to a variety of reasons – economic crisis, pandemic, and war- sending repercussions throughout the agricultural industry. Consequently, fertilizer prices increased over three folds in several countries. The disruption in the fertilizer supply chain raises concerns about global food security, as it could potentially lead to decreased agricultural productivity and increased vulnerability to food crises, particularly in regions heavily dependent on imported fertilizers.

Whenever inputs cost goes up, naturally fertilizer prices in the market will go up and it will adversely impact the farm economics and erode the profits. When farming turns out to be unremunerative, producers in distress may reduce extent of farming or even opt to abandon agriculture during such times and seek other means of livelihood. Even reduced application of fertilizers or use of low-quality products may lead to lower crop productivity and in some cases lower quality produce. Both situations impact food security and the availability of affordable agricultural products.

C. FACTORS CONTRIBUTING TO SUPPLY CHAIN DISRUPTION

Widespread disruption in the fertilizer supply chain arise out of geopolitical tensions, natural disasters, transportation issues, and disruptions in raw material production. It's political impact on national governments will also be serious. Ensuring a more resilient and diverse supply chain for critical agricultural inputs becomes a challenge to governments to keep off civic disturbances citing vulnerability of food supply in the markets.

Crisis in the National Economy 2019

According to the Reserve Bank of India's (RBI) annual report 2018–19, which verified that the Indian economy had really entered a difficult phase, a crisis is apparently developing in the Indian economy as of late 2019. As evidenced by the collapse of the auto industry, an increase in non-performing assets (NPAs) in banks, weak consumer demand, and a collapsing manufacturing sector, the GDP growth rate fell to 5% in the first quarter of the 2019–20 fiscal year, the lowest level in over six years. For the first time in seven years, India's manufacturing growth turned negative in the June 2017 quarter. In the June 2019 quarter, manufacturing growth is virtually zero (0.6%). Meanwhile, due to the lack of impact from the 2014 introduction of the "Make in India" program, imports from China have doubled between 2014 and 2019. Over time, the US-China trade war grew more intense, which decreased global trade and, consequently, Indian exports. Above all, high Goods and Services Tax (GST) rates, a liquidity crisis in Non-Banking Financial Companies (NBFC), and a change in labor behavior brought about by the influx of youth have made saving less common.

Thousands of jobs were lost across sectors and under-performance is evident in almost every sector.

Impact of Covid 19

The global pandemic Covid 19 struck the world by the end of 2019 resulting in cessation of economic activity in almost all sectors. Energy and fuel demand was impacted heavily and oil price dropped to negative levels for the first time in history. Energy consumption in general, and oil demand in particular, have decreased as offices shut and industrial activity slowed sharply with government travel and work restrictions to curb the spread of the coronavirus. Furthermore, the Russia–Saudi Arabia oil price war and failed negotiations between Russia and the Organization of the Petroleum Exporting Countries (OPEC) to reduce daily barrel production have also played critical part in the collapse of the oil markets. On April 21, 2020, the price of U.S. crude oil turned negative for the first time in history, forcing producers to pay buyers to take the barrels that they could not store due to the oversupply of oil.

Following effective vaccination by June 2020, the spread of the coronavirus came under control and economic activity slowly started regaining by the summer of 2020. With the planned rollout of multiple Covid-19 vaccines, more and more countries emerged out lockdowns and hence demand for oil and gas started picking up. By then, the Organization of the Petroleum Exporting Countries agreed to major cuts in crude oil production which resulted in the increase of oil spot prices. Globally gas prices were on an increasing trend since June 2021 due to the demand surge post-Covid 19 and consequently, international fertilizer prices were also hardening. The restricted supply of gas disrupted conventional supply chains and forced the fertilizer producers and other consumers to go for high-cost imported LNG.

China's Ban on Fertilizer Exports

As in all commodity markets, fertilizer prices are also susceptible to the interplay of supply and demand. While high and volatile energy prices, export restrictions, disruptions in movement and trade impact on the supply side, subsidies, and grain prices impact on the demand side. According to the International Fertilizer Association, China, Russia, the United States, India, and Canada accounted for more than 60% of the world's fertilizer production. In terms of fertilizer trade, Canada, Russia, Belarus, Morocco, and the United States accounted for around 90% of the world's exports of potassic fertilizers, while China, Russia, the European Union, Morocco, and the United States accounted for more than 75% of phosphatic fertilizer exports.

China, a major exporter of phosphates and urea-imposed measures including export quotas and lengthy inspection requirements on the fertilizer ingredients to stabilize domestic prices, and secure supplies for own use in 2021. Export restrictions reduce global fertilizer supply and consequently prices are up beyond the affordability of import country farmers. Furthermore, the International Monetary Fund (IMF) found that the reduced current and anticipated supply of food and fertilizers triggered a sharp rise in global food prices, creating food shortages in countries that could not shift supply quickly and absorb pressure on prices. Such trade policies on fertilizer affordability and food security, that are contrary to WTO stipulations harmfully impact poorer countries, the most.

China remains a major source of imported fertilizers for India. Both availability and prices were affected adversely following the suspension of fertilizer exports by the Chinese government in August 2021. Already supplies were restricted due to the pandemic and the demand for nutrients went strong from the US, Brazil, China, and India resulting in international prices run up significantly since last year. However, in India alternative supplies organized from Russia, Oman, and the United Arab Emirates have helped to offset the lower shipments from China.

The Russia-Ukraine War

"The conflict between Russia and Ukraine, which erupted in February 2022 and is still going strong in 2024, severely impacted the previously mentioned circumstance. As mentioned above, India is heavily dependent on the import of fertilizer materials from Russia, Belarus, and Ukraine; the Baltic trade in these goods is also noteworthy. The war has skyrocketed prices and jeopardized material availability. Common folks and small-scale farmers are the most negatively impacted."

According to the International Food Policy Research Institute (IFPRI), global share of potassium fertilizers exported from Belarus halved between 2021 and 2023. The gainers were Russia and other smaller exporting countries like Germany, Israel, Jordan, and Lithuania. Belarus, which had a share of 31% in 2021 in the Indian market, fully lost it by 2023. It was captured by Russia and Canada. Russian share increased from 4.4% to 25% where the share of Canada was up from 23% to 35%. Canada lost 11% of its potash export share to China. Russia and Belarus are the gainers.

The 2023 market shares of top phosphate producers (Morocco, China, Russia, the U.S and other smaller producers did not significantly change from 2021.

In the case of nitrogenous fertilizer exports, Russian share shrunk by 4% which was captured by Saudi Arabia. Bulk of the dramatic reduction was contributed by export of anhydrous ammonia due to the closure of the Togliatti-Yuzhny ammonia pipeline running through Ukraine to the port of Odessa which served to carry ammonia before the war. Russia is building a new ammonia export terminal on Black Sea at Taman Peninsula to tide over the situation and also constructing a bridge between Nizhneleninskoye, in far eastern Russia, and Tongjiang in China's Heilongjiang province on Amur river to increase fertilizer trade from Russia and Belarus to China.

The Hamas-Israel War

The latest fertilizer material supply disruption occurred due to the intensification of armed conflict and fighting between Hamas and Israel, which started on the 7th of October, 2023. The hostilities posed potential risks to major phosphate and potash production facilities operated by ICL and Haifa Chemicals in southern Israel, both produce and export a wide array of raw materials and finished fertilizers. Israel is the largest muriate of potash (MOP) exporter after Canada, Russia, and Belarus. "The largest disruption to shipping since the COVID-19 outbreak has resulted from the war in the Red Sea. On December 21, Yemen's Houthis targeted ships traveling through the Red Sea. Major container shipping companies have either drastically scaled back or stopped operating in the Red Sea region as a result of the circumstances. Ships had to be rerouted around the Cape of Good Hope due to persistent threats to the safety of the crew and cargo, which caused significant delays and increased shipping expenses. A little over 12% of world trade passes via the Suez Canal, and the attacks have caused interruptions in this vital route. Delivery periods have extended by two weeks, while containerized costs have tripled. Numerous industries were impacted by this disturbance, including those that supply fertilizer materials. Surcharges and insurance premiums are examples of ancillary costs that have increased; the cost of war risk insurance has doubled in a single week."

Impact on Fertilizer Prices and Subsidies

Following the above disruption, "For example, over the 12-month period ending in March 2022, the price of phosphoric acid and rock phosphate climbed by 92% and 99%, respectively, and the price of potassium tripled in the previous year due to concerns about impending shortages caused by sanctions and logistical bottlenecks. This went against the general assumption in the Indian industry, which was for the second half of this fiscal year to see a 3% increase in the demand for fertilizers and a reasonable decrease in raw material costs. Given the state of the world today, reliable supplies are essential for the upcoming farming season."

Governments in agriculture subsidized economies will have to bear the increased financial burden due to the rise in the prices of fertilizers. The entire burden of increased fertilizer prices could not be passed on to the fertilizer producers and the farmers instantly. To maintain production, Governments and policymakers will have to address the rising fertilizer prices considering its implications on agricultural sustainability, food security, and economic stability. This situation will lead to absorption of input price rise beyond the normal within the state's economy for which effective policies to balance these interests have to be worked out. The price fluctuations fertilizer products in the market is highly volatile and susceptible to speculations exacerbating the situation. Therefore, maintaining stability of fertilizer prices is a complex task, as it involves multiple market players and global dynamics.

"In India, fertilizers are substantially subsidized. Fertilizer subsidies are payments given to manufacturers that make up the difference between the cost of production, assuming a satisfactory return on investment for the producer, and market realization upon selling it to customers at government-mandated prices." As a result of subsidized supplies of electricity, water, fertilizers and other crucial inputs for the farm sector and the introduction of scientific methods of agricultural practices, India, once dependent on foodgrain imports, could step up its foodgrain production steeply and attain self-reliance over the decades. This has resulted in a corresponding spurt in demand for fertilizers, which prompted the then governments to realize the strategic importance of fertilizers as a lifeline for foodgrain production in the public sector.

In the wake of the liberalized economic policies, it was decided to make no more state investments in the fertilizer sector and thus production stagnated. It also resulted increased imports to satiate the farmers' demand. The situation worsened in 2002 when seven fertilizer plants which operated on coal, naphtha, fuel oil, and lignite were closed on account of high specific energy consumption and increased cost of production. This decision also resulted in reducing domestic availability leading to higher quantum of imports.

Today, India is the second-largest fertilizer consumer, and producer and the world's largest importer of fertilizer materials. The per hectare application of mineral fertilizer in India remains much lower than that of most of the developed and emerging economies, despite increase in consumption on an average 4 to 5 percent every year. As imports getting difficult on account of the war and the pandemic, a renewed strategy may be needed to ensure availability of plant nutrients to the farmers. Putting up more plants based on conventional technologies may hamper the decarbonization agenda of the country. Therefore, optimal production and realization of full capacity utilization, diligent use based on the 4R nutrient stewardship with a focus on better nutrient efficiency, and zero wastage become inevitable. The rise in energy and material costs these years have put tremendous pressure on the finances of the government due to the mounting subsidy burden. The payment towards fertilizer subsidy which was Rs 81124 crores during 2019-20 increased to Rs 127922 crores during 2020-21, to Rs 153758 crores during 2021-22 and further to Rs 251340 crores during 2022-23. Since last year, certain decline is evident in the recent budgets. The revised budget for 2023-'24 provided Rs 188947 crores towards fertilizer subsidy and the budgeted provision for 2024-'25 was Rs 164150 crores. The government is making an all-out effort to ensure steady supplies by talking to major producers and overseas governments to see that fertilizer materials are always available to the farmers at affordable costs to them.

D. OVERCOMING THE DISRUPTION

Increasing Domestic Production

To counter the supply chain disruption, the first step is to increase domestic production. Accordingly, the Government of India, in 2016 decided to rebuild five brownfield urea plants having a total output of 6.35 million MT of urea per year with natural gas as feedstock. The public sector units at these locations were closed in the early 2000 due to the loss of technical and financial viability of the plants. Of the above, two plants - Ramagundam, and Gorakhpur - are already commissioned in 2021. FCI's Sindri fertilizer plant was renovated and commissioned in 2022 and Hindustan Urvarak and Rasayan Ltd (HURL) Barauni unit in March 2024. The coal-based urea plant at Talcher expected to be commissioned in 2025. The Matix plant (1.27 million MT) though commissioned in 2017 became fully operational in 2021. Once all these plants become productive at the rated capacities by 2025, the country's urea import will reduce substantially. Besides, the government has also approved a 0.86 million tonnes per year urea plant for the public sector Brahmaputra Valley Fertilizer Corporation Ltd (BVFCL) as a replacement of two lower capacity old plants

Revamp Existing Plants for Higher Output

Most fertilizer manufacturers have adopted and implemented projects and schemes intended to improve capacity utilization, operational reliability, and energy efficiency so as to achieve the best onstream factor. Retrofit and revamp options, whenever available and economically feasible, such as improved designs of equipment, new generation material of construction, and better-performing catalysts, are easily incorporated into existing plants. Some have adopted advanced operational practices such as rigorous plant monitoring, digitized process control and online corrosion monitoring.

Joint Ventures Abroad

As has been stated earlier, India's import dependency is to the extent of 25% of our requirement of urea, 90% in case of phosphates, either as raw material or finished fertilizers and 100% in case of potassic fertilizers. The Government has been encouraging Indian producers to establish joint venture manufacturing facilities abroad in countries that are rich in fertilizer resources with buy back arrangements and to enter into long term agreement for supply of fertilizers inputs or finished products to India. Producing urea in countries where raw material natural gas is cheaply available was a successful option considered by the government to ensure the availability of urea. Subsequently, the Oman India Fertilizer Company SAOC (OMIFCO), a joint venture initiative by the Government of Oman and major Indian producers IFFCO and KRIBHCO was established in 2005. The plant located at the Sur Industrial Estate in the Sultanate of Oman has the capacity to produce 1.7 million MT of urea per year of which now, one million tonnes will be exported to India. In the phosphatic sector, four phosphoric acid plants (a total of 1.82 million MT) were put up through JVs involving Jordan, Morocco, Tunisia, and Senegal. There were proposals also for acquisition of the fertilizer assets – production units, mines of raw materials etc—especially for the acquisition of potash and phosphate mines in Russia, Canada, and Belarus. As the Department of Fertilizers could not formulate viable incentive schemes, such acquisition proposals have not yet taken off.

Import of Fertilizers

India is heavily dependent of imports in the fertilizer sector and therefore maintaining a steady supply chain is important for the country's agricultural productivity. The deficit in production which is nearly 30 percent of the consumption is met by import of raw materials intermediates and finished products. Imports from surplus region is undertaken through canalized agencies by the Government to maintain the supply chain. Private entities are not allowed to import. Public sector companies Indian Potash Ltd (IPL), National Fertilizers Ltd (NFL) and Rastriya Chemicals and Fertilizers (RCF) Ltd are the authorized agencies for canalizing import of urea on government account. Phosphate and potash fertilizers (DAP, MOP and NPK) are under open general licenses (OGL) as per Foreign Trade Policy. During 2022-'23 India imported 19.52 MMT of fertilizer materials besides raw materials sulphur and rock phosphate and intermediates ammonia, phosphoric and sulphuric acids.

During the financial year 2022-'23, India imported 7.58 million MT of urea, 6.58 million MT of DAP and 1.87 million MT of potash. Urea import was from Oman, China,, Finland, Qatar, Saudi Arabia and UAE, phosphatic fertilizers from Saudi Arabia China, Morocco, Russia and Jordan and potash from Canada, Israel Jordan, and Russia. During the year urea import was reduced by 17 percent whereas the import of DAP increased by 21 percent.

Regarding fertilizer intermediates, during 2022-'23 India imported 2.34 MMT of ammonia from Saudi Arabia, Qatar, Ukraine, Egypt, Turkey, Indonesia, Iran and Oman. India also imported 2.69 MMT of phosphoric acid during the same year from Morocco, Jordan, Senegal, and Tunisia. Apart from the above, during the same year India imported 9.09 MMT of rock phosphate and 1.34 MMT of sulphur for use in the fertilizer plants. Most of the rock phosphate was imported from Jordan, Morocco and Egypt and sulphur from Qatar, UAE, and Oman.

Achieving Better Nutrient Efficiency

Efficient use of fertilizers is important to optimal crop productivity, and economic gain to farmers and for avoiding pollution from leaching and farmland runoff due to excessive application. Serious efforts are required to achieve better nutrient use efficiency of applied fertilizers in the field based on nationwide soil testing and balanced administration of fertilizers. The government nowadays insists on a scientific administration of plant nutrients and promotes site and crop-specific nutrient management, the use of coated and delayed-release materials.

An optimal supply of soil nutrients over time and space to match the requirements of crops through the 4R principles can be achieved through crop and site-specific nutrient management. These broad principles of such kind of nutrient administration to the soil were developed by the International Plant Nutrition Institute in 1988. They are:

1. Right product: Apply the required nutrient depending on the crop's requirement and type of soil to ensure a balanced supply of nutrients.
2. Right rate: Administer the exact quantum of fertilizer to crop needs, based on soil analysis to compensate the nutrient deficit in the soil. Excessive application of fertilizer leads to environmental pollution, including runoff, leaching and gaseous emissions, as well as wastage of money. An inadequate supply to replenish nutrient loss reduces soil fertility and result in gradual degradation of the soil.
3. Right time: Ensure that nutrients are available when crops need them by assessing crop nutrient dynamics. Split applications of mineral fertilizers or combining organic and mineral nutrient sources to provide slow-releasing sources of nutrients may become necessary
4. Right place: Placing and keeping nutrients at the optimal distance from the crop and soil depth so that crops can fully absorb them with minimal loss of nutrients. The ideal method – soil surface or foliar application- depends on characteristics of the soil, crop, tillage regime and type of fertilizer.

Coating of urea with ingredients like sulphur, neem oil (India) or polymers and addition of micronutrients allow controlled release of nutrients to the soil over an extended period. Nutrient release rate and duration are guided by coating thickness and soil temperature. Besides, farmers are educated on the benefits of advanced techniques such as fertigation and integrated nutrient management.

Large-scale use of compost and farmland manure, biofertilizers, fortified fertilizers, and micronutrients must be promoted. Over the years, the fertilizer sector has undergone all the strains associated with the government's neo-liberal policies. It needs to undergo a transformation by absorbing the new trends, scientific advancements, and strategies that are associated with the production, crop nutrition, and use of mineral fertilizers around the world.

Modified/Fortified/Specialty Fertilizers

Normally, only a part of the nitrogen administered to the soil as fertilizers is absorbed by the plant, and the rest leaches to the environment as nitrate or lost to the atmosphere as ammonia or nitrous oxide, a potential climate-distorting greenhouse gas. Various techniques are available to improve the nitrogen efficiency of applied fertilizers by product modifications. Controlling the release of nutrients to the soil by coating the product, stabilizing the product with additives, delivering of product as water-soluble or in liquid forms, and chelation with crop-specific micronutrients are all efforts to add more value to common fertilizers to significantly improve nitrogen use efficiency. The International Fertilizer Association (IFA) listed controlled release fertilizers (CRFs), slow release fertilizers (SRFs), sulphur coated urea (SCU), stabilized nitrogen fertilizers (SNFs), water soluble fertilizers (WSFs), liquid NPKs, and chelated micronutrients and boron as specialty fertilizers.

Nano Urea

India's fertilizer major IFFCO has developed nano fertilizer grade urea which is expected to revolutionize the application of nitrogen fertilizers. Nano urea in liquid form having particles of 20 to 50 nanometers (1 nano meter is 10^{-6} mm) when sprayed on the plant leaf gets readily absorbed by the plant through the leaf stomata and releases nitrogen inside the plant. They also stimulate the enzymes involved in nitrogen metabolism inside the plant cells. It is also expected to minimize the environmental footprint by reducing the loss of nutrients from agriculture fields in the form of leaching and gaseous emissions which used to cause environmental pollution leading to climate change. If widely adopted, it will also significantly reduce the necessity for massive future imports or granular urea.

E. CONCLUSION

We must address the above challenges and ensure a steady supply of fertilizer materials to farmers without any disruption in the supply chain to sustain agricultural productivity, food security, and environmental sustainability. Therefore, a strong collaboration involving multiple stakeholders – government departments, agricultural producer organizations, and the private players in the distribution channel has to be developed. The recent geo-political developments have resulted in massive supply disruptions resulting in large-scale increases in the cost of mineral fertilizers. Several informed agencies have forecasted that the prevalence of a situation of this order may lead to loss of food security and famine in import dependent countries. The following observations and suggestions become relevant in ensuring the availability of fertilizer products to farmers at affordable costs.

1. The war between Ukraine and Russia disrupted the fertilizer supply chain as both the countries largely have huge deposits of fertilizer raw materials and are major producers of intermediates and finished products. In India, luckily, the prevailing administered pricing system for urea prevented supply disruptions and saved the farmers from any massive price rise during the period. An increased

allocation of nutrient based subsidy from the government also prevented huge price rise for imported phosphate and potassic fertilizers from the international markets.

2. The need of the hour is to increase domestic production of fertilizers. This becomes more relevant in the context of the ongoing energy transition as ammonia, the major fertilizer input material is slated to become a fuel of the future for long-distance hauling of ships etc.
3. A massive program to decarbonize the fertilizer sector may be taken up in line with the country's energy transition and proposed achievement of net zero emissions by 2070. As of now, domestically produced natural gas is being preferentially allocated to household consumption and for automobiles. The fertilizer sector needs a fairer treatment in the allocation of natural gas in line with the government commitments to reduce emissions from major manufacturing operations.
4. The public sector fertilizer industry, being critical to agriculture and food security, may be taken out of the government's privatization agenda, and its systematic expansion and diversification may be ensured.
5. A balanced and scientific application of fertilizer nutrients in the agricultural fields is the key to sustainable farming. Our experience is that the bulk of the nutrients applied to the farmlands get leached away without being absorbed by the crops. Therefore, a more scientific administration of fertilizers based on an assessment of the crops' requirements as revealed through soil testing, etc., is warranted. Besides reducing wastage, it will also ward off soil degradation and environmental pollution.
6. Promoting low nutrient content fertilizers such as ammonium sulphate, mono ammonium phosphate vis a vis high nutrient urea and DAP assumes paramount significance in the sense that such products supply only the required quantum of nutrients during the cropping season, the whole of which will be absorbed by the crops leaving little residue.
7. Several fertilizer plants operating in the country have the capability to produce beyond their nameplate capacities due to overdesign. Under the prevailing subsidy regime, producers are not encouraged to produce beyond their licensed capacities and claim more capital subsidies. The present situation calls for increased domestic production to tide over the prevailing exorbitant import prices.
8. Commissioning of the urea plants under construction may be expedited and existing plants shall undergo debottlenecking of to increase production and improve efficiencies.
9. Massive urbanization taking place across the country has led to increased availability of city composts in our urban locations. Together with agricultural waste, it will be able to partly meet the demand for manures in the farmlands.

Therefore, more and more municipalities and corporations may be encouraged to set up compost plants. Besides ensuring cleanliness around cities and neighborhoods it will also yield green organic manure for horticulture etc.

10. High nutrient content fertilizers such as urea and DAP may be blended with city composts so that the product makes the soil coarse so that assimilation of nutrients by the plant is rendered easy.
11. The Government shall institute a countrywide program to promote the use of agricultural wastes, green leaves, and farmland manure as readily available sources of manure for crops.
12. We do not have high-grade phosphate rock mineral deposits used for the manufacture of phosphoric acid and phosphatic fertilizers. Available low-grade rock phosphate deposits (Rajasthan, etc.) may be effectively used through advanced beneficiation, etc.
13. The salt pans are a source of gypsum which is a good soil conditioner and contains several plant nutrients that are essential for plant growth. It is also used as a product modifier in cement plants. Natural gypsum from salt pans may be reserved for fertilizer use. Cement plants can also use phospho-gypsum or mined gypsum.
14. Being the second largest producer of fertilizers globally, our ammonia industry has, over the years, acquired vast expertise in the design, execution, handling, storage, and safety of handling hydrogen. The National Green Hydrogen Mission intended to decarbonize the energy and industrial operations shall build on the above strength residing in our fertilizer plants.
15. Nano liquid fertilizers developed through independent research by Indian Farmers Fertilizer Cooperative (IFFCO) and Coromandel International are a breakthrough achievement in the sense that they align with the national goal of enhancing agricultural productivity along with low nutrient loss and ensuring farm sustainability. It needs to be promoted for reasons of optimal nutrient delivery and increasing crop yield.

Modern agriculture is criticized for being energy and carbon-intensive and contributing to massive emissions and, therefore, not at all environmentally sustainable. A paradigm shift is needed to innovate current agricultural practices, with an emphasis on long-term sustainability. The government, together with the scientific community, shall explore alternatives to traditional fertilizers, such as organic and bio and nano-based options or precision agriculture techniques. It will also free up conventional fertilizer disruption products and raw materials supply chains arising every now and then on account of some reason or the other.

REFERENCES

Agriculture Dive. (n.d.). Israel-Hamas war: Fertilizer exports and shipping through Gaza. https://www.agriculturedive.com/news/israel-hamas-war-fertilizer-exports-shipping-gaza/696567/

Chandrasekaran, N., & Raghuram, G. (2014). Agribusiness supply chain management. CRC Press.

Chandrasekaran, N., & Raghuram, G. (2014). *Agribusiness supply chain management.* CRC Press.

Department of Agriculture, Government of Andhra Pradesh. (2017). *Agriculture action plan 2017-2018.* https://apagrisnet.gov.in/2017/Agri%20Action%20Plan%20 2017-18%20(English)/5%20Fertilizers.pdf

Department of Fertilizers, Government of India. (2023). *Annual report 2022-2023.* https://www.fert.nic.in/sites/default/files/2020-082024-01/Annual%20Report%20 fertilizer%20new%20new17%20%20%2029-07-2023.pdf

Department of Fertilizers, Government of India. (2022-2023). *Annual report 2022-2023.* https://www.fert.nic.in/sites/default/files/2020-082024-01/Annual%20Report %20fertilizer%20new%20new17%20%20%2029-07-2023.pdf

Department of Fertilizers, Government of India. (n.d.). *Phosphatic and potassic (P&K) policy.* https://www.fert.nic.in/phosphatic-and-potassic-pk-policy

Department of Fertilizers, Government of India. (n.d.). *Official website.* https://www.fert.nic.in/

Department of Food and Public Distribution, Government of India. (n.d.). *National food security act.* https://dfpd.gov.in/WriteReadData/Other/nfsa_1.pdf

. Dept of Fertilizers, Govt of India, Annual Report 2022-'23,

European Central Bank. (2022). *Economic bulletin article: The Russia-Ukraine war and its impact on fertilizers.* https://www.ecb.europa.eu/press/economic-bulletin/articles/2022/html/ecb.ebart202204_01~7b32d31b29.en.html

Fertilizer Association of India. (2023). *Fertilizer Statistics* (68th ed.).

Fertilizer Association of India. (2023). *Fertilizer statistics* (68th ed.). Fertilizer Association of India.

Food and Agriculture Organization. (n.d.). *Title unknown.* https://openknowledge.fao.org/server/api/core/bitstreams/9ddf4aab-bc18-4d48-aeb6-ab14efaed1fd/content

India Forum. (n.d.). Looming crisis: Fertilizers. https://www.theindiaforum.in/article/looming-crisis-fertilisers

International Food Policy Research Institute (IFPRI). (2023). Global fertilizer trade 2021-2023: What happened after war-related price spikes? https://www.ifpri.org/blog/global-fertilizer-trade-2021-2023-what-happened-after-war-related-price-spikes

Press Information Bureau. (2020). [Press release]. https://pib.gov.in/PressReleasePage.aspx?PRID=1601273

Press Information Bureau. (2024). [Press release]. https://pib.gov.in/PressReleasePage.aspx?PRID=1968931

Press Information Bureau. (n.d.). [Press release]. https://pib.gov.in/newsite/PrintRelease.aspx?relid=191212

Press Information Bureau. (n.d.). [Press release]. https://pib.gov.in/PressReleasePage.aspx?PRID=1886054

PRS Legislative Research. (n.d.). *System of fertilizer subsidy*. https://prsindia.org/policy/report-summaries/system-of-fertilizer-subsidy

S&P Global Ratings. (2022). *The Russia-Ukraine war is reshaping the fertilizer industry*. https://www.spglobal.com/ratings/en/research/articles/220912-the-russia-ukraine-war-is-reshaping-the-fertilizer-industry-12484018

The Hindu Business Line. (2024). Fertilizer subsidy crosses 70% of budget allocation. https://www.thehindubusinessline.com

United Nations. (2022). [Press release]. https://press.un.org/en/2022/sgsm21350.doc.htm

Yara. (2023). One year on: The impact of the war in Ukraine. [Press release]. https://www.yara.com/news-and-media/news/archive/2023/one-year-on-the-impact-of-the-war-in-ukraine/

Chapter 12
Sustaining Livestock Supply Chain in Namibia:
Global Demand and Supply Scenario, Prospect of Lucrative Livestock Farming in Namibia

Saryamani Roul
FEED, India

ABSTRACT

Namibia has a tremendous potential in the livestock sector which has come a long way in developing its competitive advantage and USP with a good animal identification and traceability system. This potential is likely to be harnessed with Namibia's progressive policies, plans and actions to integrate the existing three livestock marketing systems -small communal farmers in north, emerging small scale farmers and existing commercial farming in the south, gain access to high-end markets particularly by supporting small farmers in north part of the country on integrated crop-livestock systems, range land management plans, and strengthen knowledge and capacity of the farmers. Integration of the three systems is expected to provide ample opportunities on Out grower models, central processing facilities like Agro Processing Hubs and Agri Transformation Centers nearer to the small holder farmers as one-stop shops to do business and bring prosperity.

DOI: 10.4018/979-8-3693-4330-2.ch012

INTRODUCTION: PROSPECT OF LUCRATIVE LIVESTOCK FARMING IN NAMIBIA

The agriculture sector in Namibia plays a crucial role in the country's Gross Domestic Product (GDP). On an average, during last ten years, the agriculture sector has contributed to 4.8% of GDP very consistently and uniformly, with livestock farming, primarily cattle and small stock contributing to 2.9%. As per Namibia Statistics Agency (NSA), the agriculture, forestry and fishing sectors contributed 7.7% of the 2023 GDP amounting to NAD17.85 billion.[2]

In Namibia, around 80% livelihoods and 64 million ha land are dedicated to livestock production. The Northern Communal Areas (NCA) constitute around 55% smallholders which is located north to the veterinary cordon fence.. The NCA is well acclaimed for its extensive livestock farming, particularly with cattle, sheep, and goats.[3] Coincidently Namibia's vast and arid landscapes offer significant grazing potential for livestock.

The livestock especially the cattle sector accounted for NAD 6.3 billion which accounts for 54.7% of the GDP with maximum contribution from the beef sector. The Namibian beef industry contributes about to 70% of the Agricultural GDP[4]. In 2020 GDP, Meat-processing itself has contributed NAD 985 million accounting for 5.1% of manufacturing activities.[5]

Namibia exports 85% of beef to South Africa, the European Union, Norway, China and Angola, with the rest exported to other African and emerging markets valued at NAD 2.6 billion.

[6]As per the report by HEI, Namibia has marketed a total of 4,432,695 cattle between 2010-2022, of which 1,718,289 were slaughtered and 2,712,380 were exported live, mainly to South Africa. Around 46% of the export market is accounted by the European Union with other important export markets in South Africa, Norway and China.[7] MeatCo, as an apex organisation plays a big role in beef export from Namibia. Thus, Namibian beef caters to a variety of markets-both global and domestic segments. This extensive portfolio allows Namibia to select the lucrative markets including South Africa being its immediate neighbouring country. In pursuit of exploring market opportunities, Namibian authorities continue to remain engaged in periodic scoping assessments. The findings of a study undertaken by the Meat Board of Namibia in 2020 recently renamed as Livestock and Livestock Products Board (LLPB) recommended on the viability of exporting bone-in fresh or chilled as well as frozen beef to Middle Eastern countries such as Kuwait, Saudi Arabia, Bahrain, the United Arab Emirates and Qatar, as well as China. It is reported in a study that abattoirs can attain reasonable profit margins from exploiting these markets. The LLP Board and the Directorate of Veterinary Services (DVS) are also engaged with

efforts to open up these markets for the benefit of farmers particularly the north which has big potential of bringing equity in the beef value chain in both the regions.

However, in international scenario, Namibia still has comparatively small share in the global beef market as compared to big players such as South America and Australia. The Veterinary Cordon Fence in Namibia creates two distinct livestock value chains-in South and North of the country. The fence stretching over 1500 km separates Namibia's northern regions from the rest of the country covering parts of Oshana, Kavango East, Omusati, Zambezi, Omaheke, Kunene, and parts of Khomas and Oshikoto regions. As a result, there are two distinctive beef value chains: the commercial value chain predominantly in south of the VCF and the communal lands value chain in northern regions. The northern areas have limited access to market because of occurrence of FMD in the region.

Critical Factors for Sustainable Management of Livestock[8]: Beef Supply Chain in Namibia

While seeing from the view of economic importance of beef value chain, it is crucial to look at it in sustainability lens covering environmental, social, and economic aspects of supply chain. To ensure a sustainable beef supply chain, the following five major considerations are of great significance which are further discussed in subsequent sections.

- Production Systems
- Market Systems
- Animal Health Care management system
- Water Resource Management
- Efficient use Land Resources

Livestock Farming System in Namibia

The different land tenure prevalent in Namibia divides the livestock sub-sector into two distinct segments-Commercial and Communal. The producers in Namibia can be broadly categorized into following three types:

1. Commercial Producers in the south of VCF: are those producers who have large farms and rear animals for commercial sale and employ advanced systems and methods to rear animals for profit maximization.
2. Small-scale commercial producers in the south of VCF: are like commercial producers but on a relatively small scale.

3. Small-Scale Communal & Subsistence producers in the north of VCF are those farmers who keep just a few animals mostly for own consumption or sale in the local market.

Production Linked Marketing Systems for Beef Animals in Namibia

Broadly, there are four major marketing system for beef animals in Namibia, which are briefly described below:

Weaner System-This is known as the 'engine room' of beef production. Under this system, a cow herd is run on the farm and their calves are sold once they reach the age of six to eight months for further processing. Under this system, the number of weaners produced depends on the size of the cow herd, which is determined by the grazing available on the farm to keep the cows in good condition throughout their production cycle.
Long-yearling system (Tollies)- In this system, before the bull calves are marketed, they are castrated and put into the veld. They stay in the farm for 12-18 months. The herd composition is normally determined by the age at which the tollies leave the farm. In most cases, the gender ratios of the animals are slightly shifted towards steers rather than cows.
Two-year-old system- This system is slightly different from the Long-Yearling system under which the castrated bull calves are kept for a longer duration on the farm. This system is also popularly known as 'oxen-off veld' system. The farmers need to ensure that the they have to consider both oxen and cows for grazing. In this system, the castrated bull calves (oxen) are marketed at a much later stage when an age greater than two years is reached.
Buying-in systems- Considering the frequent occurrence of droughts in Namibia, farmers have started adopting this system wherein they buy animals from an auction, irrespective of the age, and keep them for a certain period during which they are fed to gain body condition or mass and then marketed. This system is gaining popularity amongst the farming community helping them in mitigating the risks particularly addressing drought situations.[9]

Marketing Systems for Beef Animals

The prevalent marketing system for beef animals in Namibia is broadly clubbed under four broad heads as below.

• *Auction Sales*-Auctions are formally organized for farmers to sell their livestock to the highest bidders. The auction sales are comprised of more than one buyer, thus, creating competition for livestock which then raises the prices. The buyers are middlemen or speculators buying animals and raise or fatten for slaughter. These are predominant in areas SVCF.
• *Permit Day Sales*-This type of system is formally organized but with only one buyer involved. The price per kilogram per animal category is already determined or it does not increase beyond the set price. The buyers are middlemen who buy animals to raise or fatten for slaughter. In recent years, MeatCo has also embarked on buying animals via permit days. This is predominant in SVCF.
• *Slaughter Abattoir Sale*- In this channel, farmers who raise or fatten their own animals deliver directly at abattoirs for slaughter. This type of marketing does not include a middleman. This is predominant in SVCF areas, and recently practiced in the NVCF areas, the farmers are delivering cattle directly at abattoirs going through a process via quarantine camps.
• *Meatco Buying System*- Meatco's marketing system involves purchasing weaners from producers, fattening them in the feedlot, and then slaughtering and processing the animals. This system ensures a consistent supply of high-quality meat products by controlling the entire process from purchase to processing.

However, the market systems for Beef Animals in north and south of the VCF are distinct and different following formal and informal market channels. Main actors in the value chain in the formal market are Producers (both commercial and communal), Auctioneers, Abattoirs, Butcheries, Feedlots, International market actors, Local market players, Livestock traders/Speculators, Transporting companies/individuals, Livestock and Livestock Products Board of Namibia and MAWF (Directorate of Veterinary Services). The informal market constitutes open and street markets where fresh and cooked meat is sold in open and street stall markets. Informal meat market margins are generally narrow, which partly reflects the competitiveness of the market.

Export and Import Status

The beef sub-sector contributes significantly to the country's export earnings and overall economic growth. Namibian beef is valued for its natural grazing, hormone-free production, and strict quality control measures. Namibia is one of the leading exporters of beef to European Union and America because of its USP of high quality being its major strength. Next to South Africa, Namibia is the only country in the African region to meet the stringent import conditions for bone-in beef to China. Thereby, Namibia is now the first African country to export red meat to the United

States. The maximum earnings i.e. up to 70% earning is from meat exports to the UK and Norway markets.

Cattle Export: The major export from Namibia is that of live animals which are exported to South Africa and Botswana where they are kept in feed lots for fattening and further processing and beef export. This is primarily attributed to the insufficiency of infrastructure such as feedlots, cattle feed processing units, and slaughterhouses to process meat etc. in Namibia. This situation underscores the potential for investing on modern infrastructure to enhance the livestock value chain, including upgrading cattle feed processing units, constructing feedlots, establishing cattle slaughterhouses, and implementing other necessary support structures.

Beef Export: In year 2023, a total of 16,916.50 MT of beef has been exported to all Namibian destinations. The destinations include African countries such as South Africa, Angola, Zambia, Botswana, Democratic Republic of Congo (DRC), Ghana, Tanzania. Other destinations for export mainly include the European Union (EU), Norway, and the United Kingdom. The EU has the biggest share of 9,051.25 MT (54%) of the total export share. Despite its dry climate, Namibia exports meat to the UK and Norway market, where it earns 70% of its income showcasing robust animal health system and high operational standards. Namibia has also the highest level of export specialization in frozen beef carcasses.[10]

Beef Import Status: In contrast, Namibia also imports beef products. The imported beef consists of different processed products of beef that are Beef Carcasses, Beef Cuts, Beef Livers, Beef Offal, Beef Pastries, Beef Polonies, Beef Pies, Canned Beef, Cooked & Uncooked Meat and Smoked Beef. Among which maximum import by product is canned beef followed by beef offal and beef livers. A total of 1233.46 MT of beef has been imported to Namibia from different nations of the world in year 2023 indicating opportunities for the Namibian meat processing sectors.

With regard to *Animal Health Care management system,* Namibia has a well-established system of veterinary services for ethical treatment of animals which is fundamental towards better product quality and consumer trust particularly for export purpose.

Land and Water Management

Namibia's vast and arid landscapes are well-suited for extensive cattle farming in both commercial and communal lands.[11] It offer significant grazing potential for livestock. It is estimated that the Stocking rate is 67 cattle/1,000 ha as compared to 83 cattle /1,000 ha in Botswana and 200 cattle/ 1,000 ha in South Africa. However, as per the information from a report by NAU-NNFU on Rangeland Management, it suggests that even though the stocking rate is low, the animal performance is high.[12] In context of SVCF, the production system comprises of organised large and medium farms as well as small commercial sector. In case of NVCF regions, it is largely dominated by smallholders and state-owned land.

VALUE CHAIN OF BEEF IN NAMIBIA

Based on a combined approach of desk research and field insights from Namibia, an understanding and description of the three types of Value Chains for Beef are explained below.

A. Commercial Producers in South of VCF

Commercial producers breed, raise, and manage cattle herds, focusing on maximizing production efficiency. They further send the cattle to the feedlots where the cattle are fed with high-quality diet to enhance growth and fattening of the cattle before sending it to the export abattoirs. The second node in the Value Chain is Pre-abattoir quarantine. The producers send cattle to pre-abattoir quarantine where applicable and available only. The quarantine stations have been established by large abattoirs and processors which are under the supervision of Veterinary personnel including collaborative efforts of farmers' associations and police.

Pre-slaughter quarantine is one of the stringent regulations that are in place to ensure that FMD-infected cattle are not slaughtered and contaminated meat is not moved from the infected /protected zones to the FMD-free zone south of the VCF and international destinations. Animals are quarantined for the period of twenty-one (21) days pre-slaughter.

MeatCo, the biggest beef trader in the country, has its own feedlots and its own quarantines. These feedlots are also used as quarantine facilities to keep the health in check and add value to the cattle. There are several other operators like BeefCo and Savana Beef (operating in the south part only) who also have their own feedlots. In addition, some individual business people also run their feedlots, buy weaners at auctions and keep them until a certain weight before they slaughter them or sell them to others like MeatCo, BeefCo and Savana Beef.

In case of Export Abattoirs, the animals are put in quarantine facility and then healthy animals are taken for slaughter following standard procedures set by the DVS. The abattoirs sell meat to the exporters and domestic traders. From there onwards, the processed meat is further exported to international markets such as EU, Norway, China, US etc. A part of meat also goes into Local Market and regional exporters and domestic markets. (RSA refers to Republic of South Africa).

Figure 1. Flowchart

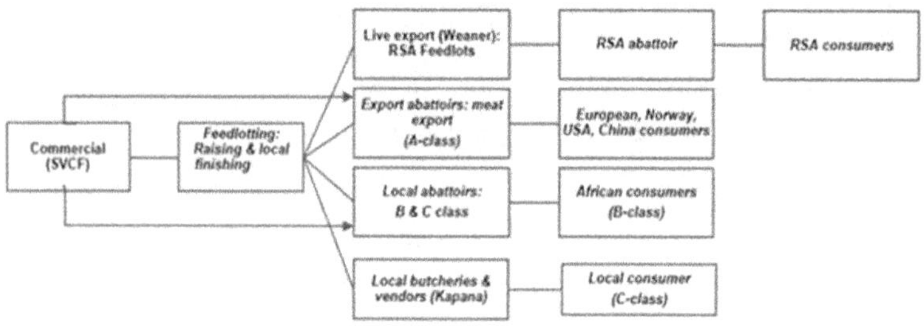

B Small-Scale Commercial Producers in South of VCF

Small scale commercial farmers represent a significant segment of beef production in Namibia, particularly in rural areas where communal farming practices prevail. The farmers mainly rear the cattle on communal land. And once the cattle reach maturity, they are either sold at auctions or directly to abattoirs for processing. These abattoirs are strategically located to serve different regions and ensure efficient processing and distribution of beef products. These facilities ensure that the beef meets international standards for hygiene and quality. The Livestock and Livestock Products Board is responsible for regulating and monitoring the meat industry in the country.

Figure 2. Flowchart

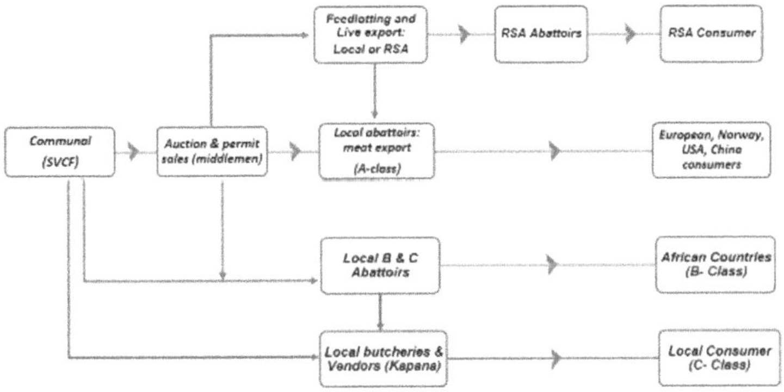

C. Subsistence and communal producers in North of VCF

Subsistence farming is mainly practiced in the 'communal lands' of Namibia's north. To ensure the marketability of livestock and meat products from the NVCF, the DVS has developed a manual on Standard Operating Procedures (SOPs) that aligns with World Organization for Animal Health (WOAH) standards. However, farmers from the NVCF face the challenge of having limited participation in livestock marketing due to diseases outbreaks, limited local formal markets, and poor road infrastructures amongst others. There is an emergence of small-scale commercial farmers in the north region as well. Both subsistence and small-scale in NVCF, send the cattle mostly to the local butchers who are into slaughtering cattle and selling beef in the informal markets, (open market, pension day market, local exhibition).

Figure 3. Flowchart

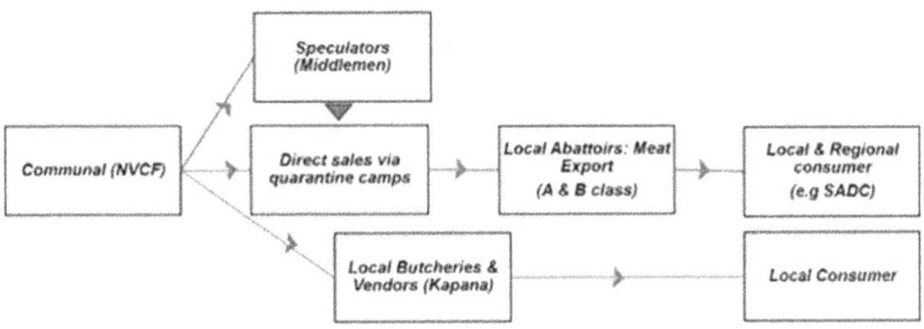

COST OF BEEF PRODUCTION AND BENEFIT TO FARMERS

Two types of analysis are presented in order to get an insights on primary and secondary production and costing.

As per the data from secondary source,[13] the producer unit cost amounts to NAD 39.59 per kg of beef. With a weighted average producer price is NAD 44.91 per kg, producers are able to receive a NAD 5.32 per kg contribution margin. However, there is an estimated financing cost of NAD 9.27 per kg. Thus it implies that producers are not able to attain a positive contribution margin during the first five years of their operations due to loan repayments.[14]

Secondary production-For benchmarking purpose, and in consideration of the export markets share of Botswana and Australia for benchmarking purpose, LLPB has compared operational efficiency of the state-based meat-processing companies. It has been found that among the state-based processors in the three countries, operational efficiency is reported highest in Namibia along with lowest operational cost at 8% of its weighted export price. The operational cost in Australia is 26% and 33% in Botswana which are significantly higher than Namibia.

An alternative to the cost based method to equity determination to farmers and other stakeholders as explained above, a benefit pass-through approach where export earnings are compared with producer prices so that a producer's income may be compared with the processor's income (export earnings). The transfer of benefits to producers in Namibia is only 55% while state-based processors. But in other countries transfer between 66% and 75% of their earnings from exports.[15] This approach suggests that Namibian producers ought to receive between NAD 54.32 and NAD 66.12 per kg.

Competitiveness of the Sector

Because of Namibia's strong quality compliance credibility, it is comparable to countries like Australia, New Zealand and Uruguay. South Africa is a crucial market for Namibia, particularly for live cattle and beef products like frozen beef carcasses while the UK serves as the primary destination for fresh or chilled beef. This creates an opportunity to slaughter and process the weaners in Namibia for export to Europe, USA, and China and in turn benefit from high prices paid for processed meat.[16] The policy environment is conducive to import essential feed ingredients and veterinary medicines (though expensive). NVCF is less competitive due to prevalence of FMD, domination of communal growers who have less resources & skill set for commercial production. However, this can be converted into high potential by taking adequate steps at Government and producer's level.

Namibia also has its uniqueness on natural production environment and animal health, welfare & Traceability with a world class veterinary systems based on incident specific data collect by the veterinarians. Namibian Livestock Identification and Traceability System (NamLITS) is used to track the origin of cattle, monitor their movements, and respond effectively to any disease outbreaks or other emergencies. Namibia's animal welfare strategy, 'from farm to fork' is implemented through a set of farming standards. The country is widely-known for its world-class commercial meat industry. Namibia's beef industry has gained global recognition for its good standards. The Farm Assured Namibian Meat (FAN Meat) standards provide the institutional framework for this implementation.

However, there are a few limiting factors too. The average carcass mass of cattle in NVCF (168.07 kg) is less than the average carcass mass SVCF (258.07 kg), which leads to lower carcass prices in the market. It requires high investment for setting up of secondary processing units for value-added products like canned beef, beef pies, offal etc. The cost of transport within Namibia is quite high that becomes challenging for the small holder farmers to bear the cost.

MAJOR CHALLENGES AND CONSTRAINTS IN THE BEEF SECTOR

Even though the Namibian beef subsector is a vital part of country's economy, the sector faces certain challenges across the production, post-production, processing and marketing stages, that impacts the overall productivity and profitability of the industry. A range of challenges including climatic, institutional, socio- economic and technical factors greatly limit production and productivity. Smallholder farmers particularly faces issues such as market access, financing, land ownership, farm machinery, lack of agro-inputs and irrigation Some of the key constraints that affects the beef sector in Namibia.

Production Constraints

- *High cost of imported animal feed* (e.g., yellow maize) and feed additives from South Africa and Zambia poses a significant challenge to the establishment of the large-scale feedlots. The reliance on live export has become a dominant marketing segment for producers in Namibia due to financial constraints imposed by costly animal feed.

- *Bush encroachment and the degradation of pasture land* continue to pose risks to profitable and sustainable beef production in the country. Climate change and rise in seasonal droughts is leading to reduction of natural grazing that is a concern for the long-term sustainability of the industry.[17]
- *Disease and pest outbreak*- some regions (e.g., Zambezi) are constantly constraint to market livestock optimally due to outbreak of FMD outbreaks.
- *Climate Change:* When there is flooding in the northern regions it contributes to poor grazing conditions, pests and diseases, poor livestock conditions and deaths. Conditions of water scarcity or drought, dry boreholes and low levels of groundwater including salinity are also major challenges faced by the farmers in many farming areas.
- *Lack of Skill or Knowledge* to cope with increasing demand, technology and the changing environment. Besides being the largest employer in the country, the agricultural sector is still low owing to a low skill set.
- Poor extension services hamper development of production efficiencies and adoption of technology and know-how for better management of cattle by small holders.
- Access to financial services to access loans to improve their production or acquire breeding materials, and production inputs.

Processing Constraints

- Processors in Namibia have not been able to fully capitalize the existing processing units. Meeting the production volumes is still one of the critical to complete throughput and supply volumes. Due to this, the beef sub-sector struggles to supply the cattle to the domestic processing units. Distance and related logistics is a factor affecting required supply to the existing abattoirs or having no abattoirs operating in the region, thus, resulting in lack of accessibility of the units.
- Inadequate infrastructure and facilities like feed lots, feed units, storage/cold chain, processing plants, scientific warehousing etc.

Marketing Constraints

- One of the important factors that has been limiting the performance of the market is the inadequate logistical connectivity within the regions. Markets are not accessible due to remoteness and poor road infrastructure within Namibia. The supporting network and facilities for distribution including cold chain storage and transportation of animals and meat products, remains limited.

- Unavailability of livestock marketing facilities (auction / kraals) near to production clusters.
- There is no proper systematic market intelligence that provides information and analysis on current market conditions, consumer preferences etc.
- Namibia's export specialization is for its frozen beef carcasses. However, the current limited range of Namibia's export of only frozen and chilled beef carcass to South Africa, Angola and Europe, indicates that there is an opportunity for product diversification both for export and local markets. Investments on infrastructure and processing of products like processed canned beef, beef offal, livers, beef pastries etc. can be good import substitutes.

CONCLUSIONS AND SUGGESTED ACTIONS

Drawing references from the above with regard to the current scenario and potential growth in livestock and beef sector, there is an ample scope and space on bigger opportunities in global market for Namibia. However, it relies on transformative measures that can enhance its competitiveness and sustainability. The focus should be on promoting a sound business environment and improving value chain productivity while maintaining the product quality and safety especially in context of the northern region. Collective actions particularly on strengthening of the Cooperatives, Producers Organizations and Collection Centers with integrated service provisioning can act as strong pillars of success to ensure access to better livestock farming practices, extension support, veterinary services, feedlot facilities, auctioning system, processing facilities in central locations. There are varieties of good experiences and practices on functioning of the community structures but limited in scale which need close examination for up-scaling in context of Namibia. In specific, ten suggestive actions are made hereunder for all-round development of the sector in sustaining the existing supply chain.

1. *Promoting and integrating the cattle and beef value chains into One*- As explained above, Namibia currently operates three over-lapping livestock marketing systems. While marketing of livestock and infrastructure linked with the supply chain in south of the VCF is well developed having access international markets, conditions in the NVCF still restrict market access due to FMD in both the protected and infected zones. For tapping better market opportunities, it is worth considering explorations of avenues in strengthening the current supply chain, identifying new markets and diversifying to new products to improve market orientation in export operations. Strategies and programs aimed at addressing the production constraints should align with broader export market objectives

while maintain a balance by adopting to climate change conditions, incorporating integrated feed crops and fodder, and utilising irrigated agriculture systems for ensuring water availability.

2. *Balance approach on Rangeland Management-* For Namibia, it is essential for timely and rapid off-take of livestock from the rangeland. An effective marketing system is required that supports marketing of livestock with realisation of good prices but with adequate forage with no pressure on land resource which is already under pressure, due to occurrence of frequent drought. Therefore, promotion of geo-clusters for feedlots to address fodder needs particularly in adjacent areas of the Green Scheme of the Government possibly with private and community participation, can bring solutions in addressing the issues.
3. *Introducing and promoting Climate Smart Cattle Production Systems-*Measures such as improved feed management, animal health care, balanced rangeland management, and use of drought-tolerant breeds to reduce the industry carbon footprint are to be looked at in a manner that aligns with Namibia's clear statement of commitments under updated the Nationally Determined Contributions (NDCs) for climate-smart livestock practices.
4. *Promotion of Small-Scale Feedlots-* To improve animal health and cattle quality for export and other markets, Namibia should transition from operating large-scale feedlots which are becoming increasingly capital intensive and unsustainable in long term. Encouraging the promotion and development of small-scale feedlots would also provide business opportunities to emerging producers and private investors. In addition, an effective strategy to optimize throughput for abattoirs, pilot community-based quarantine feedlots can be established to supply NVCF abattoirs and local markets. This can potentially provide the triple wins of smallholder-led enterprises, rural job creation, and direct private investments in production operations.
5. *Differentiated processing and branding–* Meatco is responsible for meat export operations in Namibia. While centralized processing in Namibia has a good reputation for quality and authenticity of meat products, it has limited diversification of other grades of meat products that can potentially serve domestic and African markets. Thus, a differentiated processing of high-grade meat is needed that establishes multiple graded scales for other meant products not destined for EU markets. There is a need to establish a branding strategy for its products through coordinated and globally compatible branding and labelling system, a consistent upgrading of branding strategy will allow emerging consumer demands such as low plastic use, safety and nutrition labels.
6. *Animal health, welfare, and traceability-*Namibia is well known for adhering to all health standards and traceability systems. However, it still needs a lot of work in context of small-scale farmers in NVCF areas. Towards promoting youth and

agri entrepreneurship, a community-based animal health delivery system could be one of the ways forward to incentivise them as rural service providers.

7. *Access to finance-* To increase the participation of private sector in the beef industry, the two main government sponsored initiatives are the Market Share Promotion (MSP) Scheme and the Green Schemes. The MSP is being implemented by the AMTA, and the Green Schemes operate under oversight for the AgriBusDev, which is a parastatal organization. In this regard, AgriBank, the apex bank for agri-financing has a role to play in motivating youth and women with customised financial products, introducing value chain-based instruments for each of the actors (such as logistics, storage, feed, seed, fodder etc.) and also, for promoting public- private -community partnerships. It can be extended by other commercial banks that can offer blended financing products, guarantees and index-based insurance framework for risk management.
8. *Policy and institutional reforms-*There has been a number of initiatives taken by the Namibian Government to undertake policy and institutional reforms. It has been observed that in order to support beef industry's transition to inclusive and green, few steps are required especially there is a need to reforms the Communal Land Act, promotion of community-based group rights for rangeland management through land lease tenure arrangements. There is a need to privatise Green Schemes for irrigated agriculture projects and also establish pilot programs on integrated crop-livestock systems that constitute biogas, and conservation agriculture practices in the NVCF.
9. *Awareness, Training, Extension on sustainable livestock management-* There is a need to raise awareness among communal farmers on good rangeland management practices. This will be through proper customised extension and advisory services Tools like Namibia Rangeland Early Warning System (N-REWS) needs to be introduced to the farmers for decision support system. These tools need to be customised and made user friendly as per different farmer groups and unions. To increase the production and productivity, it becomes imperative to adopt new practices and technologies that are climate resilient such as improved breeding, feed management and herd management.
10. *Allocation of higher public investments and coordinated action-* Above all, the sector needs coordinated planning and action for increasing the size of the cake. The beef industry stakeholders including the farming community and their associations, Meatco, the Ministry of Agriculture, Water and Land reform, the Ministry of Industrialisation, Trade and SME development, the Ministry of Foreign affairs and the LLPB of Namibia can strive to tap access to premium markets in the European Union, Norway, United States, China and South Africa, and golf countries in a big way. Namibia has access to diverse range of beef markets. To get better prices, high value cuts can be exported other than offal.

To ensure value chain equity, it is important to boost the existing economic output needs to guarantee sufficient throughput for export facilities, thereby, ensuring economies of scale for meat processors. The retaining of live cattle can best be attained through a market-based price system that competitively rewards livestock producers encouraging a shift from weaner exports to export abattoirs. In the beef processing segment, it also requires bringing a system to achieve both operational efficiency and profitability together in most of state-owned beef processing units.

Thus, Namibia has a tremendous potential in the livestock sector which has come a long way in developing its competitive advantage and USP with a good animal identification and traceability system. This potential is likely to be harnessed with Namibia's progressive policies, plans and actions to integrate the existing three livestock marketing systems-small communal farmers in north, emerging small-scale farmers and existing commercial farming in the south, gain access to high-end markets particularly by supporting small farmers in north part of the country on integrated crop-livestock systems, range land management plans, and strengthen knowledge and capacity of the farmers. Integration of the three systems is expected to provide ample opportunities on Out grower models, central processing facilities like Agro Processing Hubs and Agri Transformation Centers nearer to the small holder farmers as one-stop shops to do business and bring prosperity.

Therefore, in a quest for value addition and processing, Namibia needs to look at more agro-beef businesses through processing and value creation facilities such as beef products including slaughtering, canning, tanning and many other value addition activities in order to create employment rather than to export of live animals. As per the African Continental Free Trade Area (AfCFTA) agreement, Namibia will require more beef to export to the rest of the continent withlarge volumes are needed. [18] Collective actions particularly strengthening of the Cooperatives, Producers Organizations and Collection Centers with integrated service provisioning can act as strong pillars of success to ensure access to better livestock farming practices, extension support, veterinary services, feedlot facilities, auctioning system, processing facilities in central locations. There are varieties of good experiences and practices on functioning of the community structure which can be further studied and examined for making the supply chain sustainable and higher impact at farmers' level.

The sole purpose of this chapter is to put centrality of efforts advocating for inclusive value chains system in place that emphasises more and more engagement of communal smallholder livestock producers particularly women and youth. It intends to elicit interest of cattle value chain actors in Namibia and solicits interest of policy makers, think tanks and development organisation for designing both short- and long-term roadmap which remains dynamic as well as system driven with pro-producer centric measures to boost production, productivity, and export remaining market savvy and maintaining quality standards blended with climate resilient pragmatism in view planning towards a sustainable supply chain of cattle and beef.

REFERENCES

https://growthlab.hks.harvard.edu/sites/projects.iq.harvard.edu/files/growthlab/files/2023-10-cid-fellows-wp-154-namibia-agriculture.pdf, as retrieved on 24th July 2024

By Staff Reporter; New Era Live; Namibia: Two Beef Value Chains and the Shifting of the 'Red Line'; https://neweralive.na/namibia-two-beef-value-chains-and-the-shifting-of-the-red-line/; as retrieved on 19th August 2024

Fortunato, A. & ENCISO VALDIVIA, S. H. E. Y. L. A. (2023). Food for Growth: A Diagnostics of Namibia's Agriculture Sector. CID Research Fellows and Graduate Student Working Paper Series.

Size, F. A. M. (2020). Share & Industry Analysis. By Type (Class F and Class C), by Application (Cement & concretes, Fills & embankments, Waste Stabilization, Mining, Oil field Service and Road Stabilization and Others), and Regional Forecast, 2027.

Helmuth Stehn; Rangeland Management, Farmers' Support Project (FSP); Joint Presidency Committee (NAU and the NNFU) Private Bag 13255, Windhoek, Namibia; https://cms.my.na/assets/documents/p1bk4fuusf1eud1qcifql1e5s1rer2.pdf; as retrieved on 19th August 2024

Institute for Public Policy Research to the Namibia National Farmers Union. Final report: Value Chain Analysis of Livestock and Livestock Products in the Northern Communal Areas of Namibia. Windhoek, Namibia, May 31 2019

Intelligence, H. E. Beef Industry in Namibia June 2023; https://hei.com.na/wp-content/uploads/2023/08/Updated-beef-report.pdf, as retrieved on 24th July 2024

Monasa Advisory & Associates; Study to investigate the impact of the cattle marketing scheme in Namibia, Windhoek, Namibia, February 2019

Olibile, O. Okapuka Feedlot mentorship student, Beef Production Systems, MEATCO Namibia, https://www.meatco.com.na/news/714/Beef-Production-Systems/, as retrieved on 19th August 2024

Paul, S., & Ace, M. Comparisons of the beef value chain in Namibia, Botswana and Australia; Namibia Economist, https://economist.com.na/63512/agriculture/towards-value-chain-equity-in-the-beef-sector/, as retrieved on 24th July 2024

Paul, S., & Ace, M. Namibia: Towards Value Chain Equity in the beef Sector; https://allafrica.com/stories/202108090805.html, as retrieved on 24th July 2024

Syed, T., Goodman, P., Mataruka, Z., Evans, D. E., & Pachena, H. T. Roadmap for Sustainable Livestock Value Chains in Southern Africa (English). Washington, D.C.: World Bank Group. https://documents.worldbank.org/curated/en/099430006082232802/P174621055c7850b80808d09c38c18d5ed9, as retrieved on 24th July 2024

ENDNOTES

1 Managing Trustee of Forum of Enterprises for Equitable Development (FEED) based in New Delhi. Views expressed are personal and not the associated organisation.

2 https://allafrica.com/stories/202108090805.html

3 https://documents.worldbank.org/curated/en/099430006082232802/P174621055c7850b80808d09c38c18d5ed9

4 HEI (High Economic Intelligence) Report -Updated-beef-report.pdf (hei.com.na)

5 Namibian Economist, August 10, 2021, and Data from the Namibia Statistics Agency (NSA)

6 https://documents.worldbank.org/curated/en/099430006082232802/P174621055c7850b80808d09c38c18d5ed9

7 https://hei.com.na/wp-content/uploads/2023/08/Updated-beef-report.pdf

8 The livestock sector in Namibia comprises of four sub sectors- namely Beef, Poultry, Small Stock (Sheep and Goat) and Piggery. However, the chapter limits its scope only to cattle and beef part of it.

9 https://www.meatco.com.na/news/714/Beef-Production-Systems/

10 https://documents.worldbank.org/curated/en/099430006082232802/P174621055c7850b80808d09c38c18d5ed9

11 https://www.farmersweekly.co.za/animals/cattle/the-challenges-facing-namibias-cattle-producers/#

12 https://cms.my.na/assets/documents/p1bk4fuusf1eud1qcifql1e5s1rer2.pdf

13 Namibian Economist, Aug 10, 2021, Comparison of the Beef Value Chain in Namibia, Botswana and Australia written by Paul Strydom and Ace Mutelo.

14 https://economist.com.na/63512/agriculture/towards-value-chain-equity-in-the-beef-sector/

15 https://economist.com.na/63512/agriculture/towards-value-chain-equity-in-the-beef-sector/

16 https://documents.worldbank.org/curated/en/099430006082232802/P174621055c7850b80808d09c38c18d5ed9

17 https://cms.my.na/assets/documents/p1bk4fuusf1eud1qcifql1e5s1rer2.pdf

18 https://neweralive.na/namibia-two-beef-value-chains-and-the-shifting-of-the-red-line/

Chapter 13
Microfinance Serving Agriculture Supply Chain in Dapaong, Togo in Western Africa

Amandine Laré
Métis Lab, E.M. Normandie Business School, France

Malick Kana
University of Le Havre Normandy, France

Roland Condor
https://orcid.org/0000-0003-0353-3788
Métis Lab, E.M. Normandie Business School, France

ABSTRACT

This chapter examines the crucial importance of microfinance in the agricultural supply chains of developing countries. The study aims to determine the impact of microfinance on agricultural activities in Dapaong city in northern Togo. We used a case study approach by conducting semi-structured interviews with a representative group of 15 local farmers. We found that microfinance plays a key role in funding new projects and improving farmers' living and working conditions. However, agricultural financing is still limited due to the risks and specific challenges of rural areas. Additionally, microfinance hasn't yet helped farmers shift to more eco-friendly practices, as many still use chemical inputs. Consequently, microfinance does not fully contribute to the sustainability of global supply chains. Therefore, additional funds are necessary to enable this transition.

DOI: 10.4018/979-8-3693-4330-2.ch013

INTRODUCTION

Microfinance is a core element of the agricultural supply chain in developing countries. Indeed, as mentioned by Mentzer et al., (2001), financial organizations are part of the "ultimate supply chains", ie. a supply chains integrating not only suppliers, customers and intermediaries but also companies involved in marketing, transportation and financial activities. By financing agricultural activities (purchase of seeds, investment in advanced mechanical equipment and digital tools for example), microfinance contributes to "provide the right products (quantity and quality), in the right amounts, to the right place, at the right time, and at a competitive cost" (Jaffe and al., 2010). By doing so, microfinance acts for food security but also for social and environmental issue whereas food supply chains have to integrate these topics in their day-to-day operations (Sjah and Zainuri, 2020).

Despite the development of microfinance activities in developing countries, financing agricultural activities (ie. one the first link in the agrifood supply chains) in rural areas remains unusual (Hollinger, 2004). The caution of microfinance institutions in financing agricultural activities can be explained by the characteristics of rural areas: dispersion, low population density, remoteness from urban areas, heterogeneity, lack of transport and communication infrastructure, climatic hazards, invasions of crop-destroying insects, lack of market outlets, low profitability, low level of education, lack of appeal for graduates to rural areas, and finally the absence of land titles that could serve as collateral (Diagne and Zeller, 2001; Lapenu, 2008; Nakano and Magezi, 2020; Villalba et al., 2023). Because of the risks inherent in the agricultural sector, financial institutions like the Word Bank are often more involved in financing agricultural activities (Jaffe and al., 2010).

However, as Kendro (2012) shows, microfinance has a positive influence on the process of improving agricultural productivity and reducing poverty. More specifically, microfinance can contribute to the financing of three major categories of innovations commonly observed in the African agricultural sector that work towards increasing productivity and reducing poverty: 1) the use of improved inputs adapted to climatic and local constraints (seeds, organic composts, biopesticides) (Nakano and Magezi, 2020 ; Serpantié, 2009); 2) organizational innovations such as pooling factor costs (purchasing and sharing equipment, inputs, storage spaces), combining agriculture and livestock, and the knowledge and skill dissemination system from relay farmers (Zoundi et al., 2005); 3) the use of digital technology (thanks to mobile phones, computers connected to the internet, African farmers can receive information on agricultural product prices and weather forecasts, communicate with potential buyers, and get advice on farming techniques) (Baumüller, 2015; Subervie and Galtier, 2018).

Although microfinance is recognized as useful for economic and social development and innovation is seen as necessary to increase productivity and improve living conditions, few authors have focused on the financing of agricultural innovation through microfinance. Addressing this issue is necessary in a book aiming to sustain the Global Agricultural Supply Chain. Indeed, microfinance can sustain agricultural activities and contribute to the social and environmental issues of agrifood supply chains.

In this work, we ask how microfinance can enable the adoption of new agricultural technologies and which ones. We contribute to the literature on microfinance by focusing on agriculture and, more specifically, on the financing of innovative technologies. We also contribute to the literature on the adoption of agricultural innovations in Africa by addressing the issue of their financing through microfinance. Finally, we expose how microfinance can sustain agricultural supply chain and provide a social value added to this one.

To move in this direction, we analyze the impact of microfinance on agriculture in the commune of Dapaong, in the extreme north of Togo. Through interviews with 15 farmers, we seek to determine whether microfinance promotes the adoption of new agricultural technologies and which ones. We show that microfinance is necessary and can be used to finance both current and future investments in innovation and to improve the living and working conditions of farmers.

To reach this conclusion, we have structured the chapter into two parts. Section 1 presents the challenges of financing the agricultural sector through microfinance. The case Study in Dapaong is discussed in Section 2.

SECTION 1. THE CHALLENGES OF FINANCING AGRICULTURE THROUGH MICROFINANCE

Whereas microfinance aims to reduce poverty (Laré et al., 2021), farmers are neglected by organizations managing the microcredit offer. This situation is damaging because farming is one of the main economic activities in developing countries.

1.1. Microfinance as a Development Tool

Microfinance, initiated by Muhammad Yunus (Nobel Peace Prize laureate), originated in Bangladesh in the late 1970s. Initially, microfinance aimed to enable populations excluded from the conventional banking system to access small loans, commonly known as "microcredits" (Lelart, 2006). It then evolved to include other financial services such as savings, insurance, and money transfers. This service is traditionally oriented towards a poor clientele and significantly contributes to reducing income inequalities within populations (Bourguignon, 2004; Beck et al., 2007; McKenzie and Woodruff, 2008). According to Littlefield et al., (2003), it fosters the development of economic activities and the social promotion of individuals. Thus, the initial objective of microfinance institutions was to meet social needs by providing financial services to disadvantaged populations.

However, with the development of this industry, Mersland et al. (2010) observed a mission drift. This drift led to an opposition between two schools of thought: the welfarists and the institutionalists. Welfarists like Yunus and Hatch, founder of FINCA International[1], advocate for maintaining social objectives, while institutionalists such as Rhyne (1998), Robinson (2001), Littlefield and Rosenberg (2004) opt for financial profitability to ensure the sustainability of microfinance institutions. This debate has led to the emergence of differentiated microfinance offerings based on the targeting populations and the objectives set by the funding organizations. Three major categories of microfinance have thus emerged: those with financial objectives like a conventional bank, those with social objectives, and those with both financial and social objectives. This growth in microfinance, along with a certain degree of differentiation, should have addressed the needs of different populations, including farmers. Unfortunately, the financing of agricultural activities in rural areas remains weak (Hollinger, 2004)

1.2. Financing African Agriculture Through Microfinance

"Despite the hopes raised by the emergence and growth of MFIs, the microfinance offer aimed at financing agricultural activities has proven to be marginal and/or unsuitable in many contexts" (Gorse et al., 2008). Microfinance seems poorly suited to investments in farms. Indeed, very few farmers take out microfinance loans for investments in their farms and several factors explain this. The first factor is related to the characteristics of rural areas that hinder MFIs' efforts in financing agricultural activities. These are low-density population areas lacking infrastructure, which complicates the work of MFIs (Gorse et al., 2008; Nakano and Magezi, 2020; Villalba et al., 2023).

The second factor concerns the very low return on investment combined with the risks and/or costs of borrowing, which compel MFIs to turn to other categories of clients. The interest rate must be below the expected return on investment to cover the risk of loss and leave a reasonable surplus for the farmer (Gorse et al., 2008; Villalba et al., 2023).

Finally, dependence on climatic conditions, the timing of production cycles, the seasonality of incomes, low monetary incomes, the volatility of agricultural product prices, unreliable guarantees both legally and economically, climate change, and natural disasters such as locust invasions are bottlenecks in the process of financing

African farmers (Diagne and Zeller, 2001; Gorse et al., 2008; Jaffee et al., 2010 ; Lapenu, 2008; Nakano and Magezi, 2020).

However, Diagne and Zeller (2001), while mentioning the reluctance of MFIs in financing agricultural activities, show that MFIs can finance agricultural activities when they adopt a risk minimization strategy. Additionally, Wampfler and Lapenu (2002) identify three categories of farmers' needs that MFIs can finance:

i) Short-term needs, which include inputs, fungicides, vegetable crops, labor, sharecropping, storage, and processing of production;
ii) Medium-term needs, which consist of purchases of animals such as oxen, land, marketing, and equipment for intensification;
iii) Long-term needs, which concern the creation of larger plantations of perennial crops.

Due to the size of microcredit and the subsistence incomes of farmers, only short- and medium-term needs are likely to be met (Hayami and Rutton, 1998). Therefore, it is possible to finance agricultural activities to a lesser extent.

After reviewing the limited literature on the financing of the agricultural sector through MFIs, we now turn to the presentation of the study area and data collection.

SECTION 2. CASE STUDY IN DAPAONG (TOGO)

2.1. The Agricultural Sector in Togo: Perspectives and Financial Challenges

Located in the sub-Saharan region of Africa, Togo is a small coastal country situated between Ghana to the west, Benin to the east, and Burkina Faso to the north. It covers an area of 56.785 square kilometers, stretching approximately 650 km from north to south and nearly 70 km from west to east along the Atlantic coast (Gulf of Guinea). According to the World Bank, its population was estimated at 8.5 million inhabitants in 2024, making it one of the smallest countries in Africa in terms of population (Nigeria: around 220 million inhabitants approximately). According to the United Nations Development Programme (UNDP), in terms of human development, Togo ranked 162nd out of 188 countries in 2024. Its economy is primarily based on the services sector, with port and airport traffic serving as the main drivers of economic growth, particularly in Lomé, the capital situated on the Gulf of Guinea coast. These "economic hubs" are used for the exportation of agricultural products like cotton, coffee, cocoa and palm oil. The country's public debt represented 59.9% of GDP in 2022 (AfDB, 2023). The Togolese economy showed remarkable resilience in 2023, recording a growth rate of 5.4%. However, this growth is expected to slightly slow to 5.1% in 2024 due to budgetary consolidation efforts, weak external demand, and regional uncertainty. Despite this, poverty is expected to decrease from 42.4% in 2023 to 36.1% in 2026 (World Bank, 2024). However, poverty levels remain problematic in rural areas (58.8% compared to 26.5% in urban areas). Agriculture plays a significant role in the country by providing income and subsistence for local populations.

Togo faces challenges such as unemployment, poverty, social inequalities, and risks of regional instability, financial turbulence, and frequent climate shocks. Agriculture remains a crucial pillar of the Togolese economy, engaging over 60% of the workforce and contributing 40% to GDP. Food crops, representing 70% of agricultural production, are dominated by maize (51% of cereals) and cassava (61.8% of tubers). However, only 10% of farms have access to bank credit, limiting their capacity for modernization. Initiatives such as the Agricultural Financing Incentive Mechanism (MIFA), which granted over 3.8 billion CFA francs in credit in 2019, aim to improve this situation (Togo First, 2024).

2.2. Microfinance Organizations in Dapaong

The commune of Dapaong, in the extreme north of Togo, was chosen as the study area. This choice is due to the extreme poverty of its inhabitants, who have long faced enormous difficulties in accessing basic services. This exclusion is largely explained by the geographical location of this commune. It is part of the prefecture of Tône in the savannah region (the poorest region of the country), located 650 km north of Lomé. Additionally, it has a tropical-Saharan climate with a long dry season that severely constrains agriculture, despite its predominantly agricultural population. This vulnerability has motivated NGOs, civil society, and microfinance institutions to come to the aid of the populations of this small town. Nowadays, the city of Dapaong has 5 microfinance institutions – UMECTO (Union of Mutual Savings and Credit of Togo), Federation of Cooperative Savings and Credit Unions of Togo (FUCEC-TOGO), Cooperative Savings and Credit of Assemblies of God (COOPEC-AD), Cooperative Savings and Credit - Support for Women's Self-Promotion Initiatives (COOPEC-SIFA), and Union of Mutual Savings and Credit Unions of the Savannas (USMECS). A geographical map of the area is provided in Figure 1.

Figure 1. Geographical map of the Dapaong area

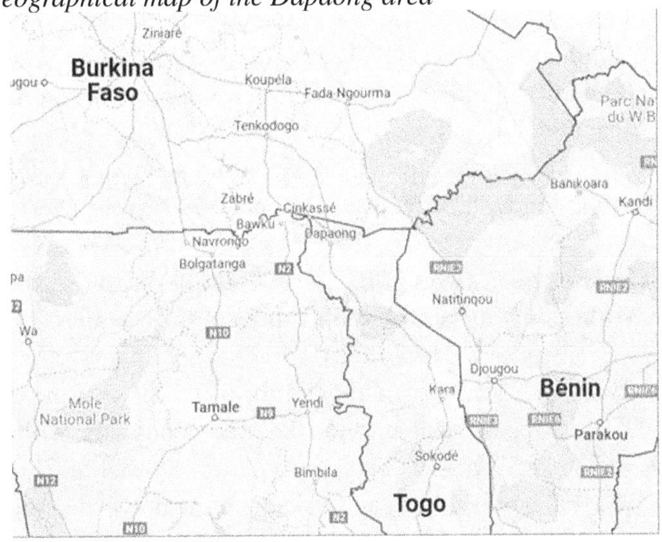

Source: Google Maps

The illustration of this chapter is based on a qualitative study conducted in 2017 and 2018 in Dapaong. This study unfolded in two stages: a series of interviews with the directors of MFIs in the study area, followed by a second series of interviews with farmers who had benefited from microcredit.

Box 1: Overview of the Main MFIs in Dapaong

1. **UMECTO (*Union des Mutuelles d'Epargne et de Crédit du Togo*) Union of Savings and Credit Mutuals of Togo**

UMECTO, established in 2000 by SOCODEVI in partnership with AFD, is a mutual initially intended for women and primarily managed by women. It comprises 15 mutuals and 21 service points across Togo, including one in Dapaong-City. UMECTO has 56,000 active members, with a total savings of 3.7 billion FCFA and a loan portfolio of 4 billion FCFA. The proportion of women among the members has decreased from 95% to 54% today. UMECTO offers regular, seasonal, and traditional holiday loans, with interest rates ranging from 5% to 20% depending on the collateral.

2. **COOPEC-AD (*Coopérative d'Epargne et de Crédit de l'Eglise des Assemblées de Dieu du Togo*) Cooperative of Savings and Credit of the Church of the Assemblies of God of Togo**

Established in 1994, COOPEC-AD is a financial institution stemming from the evangelical church of the Assemblies of God. It has 28 branches across Togo, with a total loan portfolio of 14 billion FCFA and deposits of 15 billion FCFA. Loans range from 30,000 to 50 million FCFA, and interest rates vary according to the type of loan, with a maximum of 24% set by the BCEAO. COOPEC-AD has diversified its activities to reach populations in remote areas and adopts an approach of environmental responsibility with 27 branches equipped with green energy.

3. **UCSMEC (*Union des Caisses Mutuelles d'Épargnes et de Crédit des Savanes*) Union of Mutual Savings and Credit Banks of the Savanes**

Founded in 2004, UCSMEC has roots dating back to 1986 with the Belgian NGO Ile de Paix. UCSMEC focuses primarily on the rural population, with 15 branches in the Savanes region and 90 agents. Its main products are trade credit and agricultural credit. UCSMEC works to preserve the environment by encouraging women to diversify their activities to reduce deforestation. Clients are required to plant a tree when they obtain a loan.

4. **COOPEC-SIFA (Coopérative d'Epargne et de Crédit pour le Soutien aux Initiatives des Femmes pour l'Autopromotion) Savings and Credit Cooperative for the Support of Women's Initiatives for Self-Promotion**

Established in 2010 from an initiative of the Catholic Church, COOPEC-SIFA evolved from the JARC/SIFA project. Primarily aimed at women (88% of clients), it is open to all without distinction of religion. COOPEC-SIFA aims to combat poverty by providing viable and sustainable financial services, with a portfolio of 3 billion FCFA, reaching over 47.000 beneficiaries, 92% of whom are in rural areas. The average loan is 60.000 FCFA, and group lending is the most popular. Although not formally offering environmental products, the cooperative incorporates environmental considerations into its activities.

These MFIs play a crucial role in providing financial support to the Togolese population, particularly in rural areas and among women, thereby contributing to poverty reduction and local economic development.

Based on the 15 interviews conducted with farmers, we summarize the contribution of microfinance to sustainable agricultural supply chains.

2.3. The Contribution of Microfinance to Sustainable Agricultural Supply Chain

The results show that microfinance help farmers to invest in new technologies but also to improve their working conditions. As such, microfinance contributes to an agricultural supply chain more sustainable, mixing economic and social issues.

The sources of financing for agricultural activities are self-financing and micro-credit. Farmers in Dapaong sell surplus crops and/or raised animals to finance their activities. All farmers are aware of microfinance, but not all of them benefit from this service. Some do not benefit from it at all but may have an account. They rely on self-financing for their agricultural activities. Others, who have an account, have benefited or continue to benefit from microfinance services to supplement financing for their activities. This type of farmers represents two-thirds of our sample. Going into detail, we observe that microcredit is used to: 1/ improve the living and working conditions of farmers, 2/ acquire digital tools, and 3/ prepare for the adoption of future digital and/or agricultural techniques (see Appendix 2 for the coding results of the interviews)

The first category in the Appendix 2 shows that microcredit serves to improve both the living conditions of farmers and their working conditions. As the four subcategories indicate, these two conditions are interrelated: microfinance cannot be considered solely as a tool for agriculture. Microfinance also serves to improve comfort and to purchase equipment used both for daily life and work.

The second category mainly refers to the purchase and use of mobile phones, also for professional and personal reasons. Digital agriculture is thus limited to this aspect in the context of the study. Finally, the last category refers to modern agricultural techniques. While microfinance can hardly allow the purchase of large equipment, it allows farmers to envision a modernized agricultural world and prepare for future investments. As such, microfinance appears as a brick in the building of a sustainable agricultural supply chain.

2.3.1. Improvement of Living and Working Conditions

The improvement of living and working conditions appears clearly in the results. Here are some excerpts from farmers' statements: "*The loan brought a lot of change. I was able to acquire a means of transportation»; "The loan allowed me to build a six-room house»; "This loan is very beneficial; I was able to buy fertilizer and also bulls for farming*". The farmers' verbatim show that microcredits are not solely used to operate the farm. They allow for engaging in other professions in parallel and, more generally, to meet family needs.

In terms of agricultural activity, by benefiting from microcredit, farmers have expanded their farms, hired external labor, used more fertilizer, and bought oxen. These new contributions have increased their production, part of which is sold to generate more income, used to build new homes and/or buy motorcycles. In terms of other activities, the study shows that farmers engage in multiple activities: "*I am a farmer and I trade*"; "*I am a driver but during the rainy season, I devote myself to farming*"; "*I am a farmer. I learned masonry and do gardening in the dry season*". The interviewed individuals need microcredit for these activities.

The interviews highlight that farmers benefiting from microcredits have increasingly turned to digital agriculture. They report using mobile phones to contact their fertilizer suppliers, easily hire labor, and negotiate prices with their customers. One farmer interviewed stated, "*My phone is very useful in my farming activities; I call my friends to help me, and when we have community work, we communicate over the phone to organize our work*". Another interviewee said, "*I often use my mobile phone to call my workers and fertilizer vendors*". A third declared, "*With my mobile phone, I handle my farming activities; I use it to call the storekeepers before going to get fertilizer, and I also contact friends to find sharecroppers, sell my maize, and conduct my business*". Thus, the use of mobile phones allows farmers to receive information on agricultural product prices and weather forecasts, communicate with potential buyers, and receive advice on farming techniques (Baumüller, 2015; Kitole et al., 2024; Subervie and Galtier, 2017). Moreover, today, the development of several digital tools can serve inclusive finance and development.

The interviews reveal that farmers have increasingly turned to digital agriculture. However, this mainly involves communication tools, especially mobile phones. Connected devices, as used by major agricultural nations, were not mentioned at all. Farmers report using mobile phones to contact their fertilizer suppliers, easily hire labor, and negotiate prices with their customers. One farmer interviewed stated, "*My phone is very useful in my farming activities; I call my friends to help me, and when we have community work, we communicate over the phone to organize our work*". Another interviewee said, "*I often use my mobile phone to call my workers and fertilizer vendors*". A third declared, "*With my mobile phone, I handle my farming activities; I use it to call the storekeepers before going to get fertilizer, and I also contact friends to find sharecroppers, sell my maize, and conduct my business*". Thus, the use of mobile phones allows farmers to receive information on agricultural product prices and weather forecasts, communicate with potential buyers, and receive advice on farming techniques (Baumüller, 2015; Subervie and Galtier, 2017). Moreover, today, the development of several digital tools can serve inclusive finance and development.

2.3.2. Projection in the Adoption of New Agricultural Technologies and Cultivation Techniques

In general, farmers in Dapaong practice mixed farming and animal husbandry, mainly engaging in subsistence farming. They use ancestral methods and rudimentary tools such as hoes, machetes, digging sticks, plows, and oxen to furrow the land. Additionally, there is a notable absence of irrigation systems; farmers conduct their activities according to the natural water cycle.

The verbatims show that microcredit has sparked farmers' desire to modernize their agricultural operations. One farmer said, "*If there are new methods or tools for the fields, I will acquire them to progress like others*". Another added, "*I would like, if possible, with the support of a project, to buy a tractor to cultivate more land*". The verbatims reveal intentions rather than actions. Microcredit cannot facilitate the acquisition of large equipment or significantly contribute to agricultural modernization. However, perhaps due to new communication means, farmers are more open or aware of new techniques: they can inform themselves about these techniques and possibly start learning how they work.

The improvement of living and working conditions is one of the main contributions of microfinance. Microfinance help farmers to invest in day-to-day life, materials, inputs and new mobile technologies. However, the main limit is that it doesn't prevent the use of chemical intrants. In that sense, the contribution to sustainability is more social than environmental. Another limit is that microfinance cannot provide the means to invest in advanced agricultural equipment. It just allows to imagine how the farm might evolve, without having the capacities to radically transform the farm.

CONCLUSION

This chapter demonstrates that microfinance contributes to more sustainable agricultural supply chains. From an economic perspective, microfinance helps farmers to invest in new technologies which are essential to sustain global food supply chains. For example, digital technologies are essential today to collect market data and raw material prices. It is also useful to communicate with suppliers and buyers. The traceability is another issue. If our study doesn't focus on, it will be probably an important challenge for farmers in the future because of the need to inform customers about the origin of the products they buy. Microfinance can contribute to face all these challenges as other financial solutions.

At a social and environmental level, the study shows that microfinance can contribute to the corporate and social responsibility of agricultural supply chains. Indeed, microfinance solutions facilitates the working and living conditions of the farmers. The actors of the supply chains can then valorize their sustainability by showing the improvement of working conditions. These ones should probably complete the financing of this transition because of the limitations of microfinance.

However, microfinance has not proven its capacity to finance the agroecological transition (Bateman, 2010). Farmers still buy chemical inputs, so microfinance isn't fully boosting the sustainability of global supply chains. Environmental challenges are significant for African farmers and everyone in food supply chains, including microfinance institutions (FAO, 2020; Kpenekuu et al., 2024). Unpredictable weather, such as droughts and floods, and soil issues affect farming productivity (Kpenekuu et al., 2024). These problems increase risks for microfinance and make it harder to fund sustainable agricultural projects (Bateman, 2010). Traditional farming methods that use chemicals also worsen environmental issues like soil and water pollution (Rajan Selin and Shanmugasundram, 2024). To tackle these problems, we need to invest in sustainable farming practices like agroecology and improve farmers' ability to use eco-friendly methods. This will need more funding and close cooperation between microfinance institutions, governments, and local groups to create financial solutions suited to the realities of rural Africa (Li Sun et al., 2021).

REFERENCES

AfDB. (2023). *Perspectives économiques au Togo*. Retrieved from https://www.afdb.org/fr/countries/west-africa/togo/togo-economic-outlook

Allard-Poesi, F. (2003). Coder les données. In Y. Giordano (Ed.), *Conduire un projet de recherche une perspective qualitative* (pp. 320). Edition EMS Management & société.

Bateman, M. (2010). *Why Doesn't Microfinance Work? The Destructive Rise of Local Neoliberalism*. Bloomsbury Publishing. DOI: 10.5040/9781350223974

Baumüller, H. (2015). Assessing the role of mobile phones in offering price information and market linkages: The case of m-farm in Kenya. *The Electronic Journal on Information Systems in Developing Countries*, 6(68), 1–16. DOI: 10.1002/j.1681-4835.2015.tb00492.x

Beck, T., Demirgüç-Kunt, A., & Levine, R. (2007). Finance, inequality, and the poor. *Journal of Economic Growth*, 12(1), 27–49. DOI: 10.1007/s10887-007-9010-6

Bourguignon, F. (2004). Trade exposure and income volatility in cash-crop exporting developing countries. *European Review of Agriculture Economics*, 31(3), 369–387. DOI: 10.1093/erae/31.3.369

Diagne, A., & Zeller, M. (2001). Access to credit and its impact on welfare in Malawi (Research Report No. 116, 168 p.). International Food Policy Research Institute.

FAO. (2020). Food and Agriculture Organization of the United Nations. *The State of Food and Agriculture*, 2020, •••. https://www.fao.org/interactive/state-of-food-agriculture/2020/en/

Gorse, L., Chambas, G., & Raynaud, E. (2008). *Microfinance et agriculture en développement: Études de cas*. FinDev Gateway. Retrieved from https://www.findevgateway.org/sites/default/files/publications/files/mfg-fr-etudes-de-cas-microfinance-et-agriculture-en-developpement-09-2008.pdf

Hayami, Y., & Ruttan, V. W. (1998). *Agriculture et Développement, une approche international*. QUAE Editions.

Hollinger, F. (2004). *Financing agriculture term investments*. Agricultural Finance.

Jaffee, S., Siegel, P., & Andrews, C. (2010). Rapid agricultural supply chain risk assessment: A conceptual framework. *Agriculture and rural development discussion paper, 47*(1), 1-64.

Kendro, S. (2012). Développement du secteur financier et financement de l'activité agricole dans un contexte de crise alimentaire: Quelle place pour la micro-finance? *Développement Durable & Territoires*, 3(3), 1–19.

Kitole, F. A., Mkuna, E., & Sesabo, J. K. (2024). Digitalization and agricultural transformation in developing countries: Empirical evidence from Tanzania agriculture sector. *Smart Agricultural Technology*, 7, 100379. DOI: 10.1016/j.atech.2023.100379

Kpenekuu, F., Antwi-Agyei, P., Nimoh, F., Amoakowaah Ose, M., & Guodaar, L. (2024). Understanding drought risk management in vulnerable communities in dryland farming systems: Evidence from northwest Ghana. *International Journal of Disaster Risk Reduction*, 100, 104186. DOI: 10.1016/j.ijdrr.2023.104186

Lapenu, C. (2008). Evolutions récentes dans l'offre et les stratégies de financement du secteur. Presented at the *conference on How to Regulate and Supervise Microfinance*, Kampala, Uganda, 22-26 November.

Laré, A., Lessoua, A., Diaw, D., Boutabba, A., & Gomado, K. (2021). Microfinance et pauvreté: Le cas des quartiers périphériques de Lomé au Togo. *Revue d'Economie du Developpement*, 29(3), 79–116. DOI: 10.3917/edd.353.0079

Lelart, M. (2006). *De la finance informelle à la microfinance*. Agence Universitaire de la Francophonie.

Li Sun, S., & Liang, H. (2021). Globalization and affordability of microfinance. *Journal of Business Venturing*, 36(1), 106065. DOI: 10.1016/j.jbusvent.2020.106065

Littlefield, E., Murdoch, J., & Hashemi, S. (2003). *Is microfinance an effective strategy to reach the Millennium Development Goals?* CGAP Focus Note.

Littlefield, E., & Rosenberg, R. (2004). Microfinance and the poor: Breaking down walls between microfinance and formal finance. *Finance & Development*, •••, 38–40.

McKenzie, D., & Woodruff, C. (2008). Experimental evidence on returns to capital and access to finance in Mexico. *The World Bank Economic Review*, 22(3), 457–482. DOI: 10.1093/wber/lhn017

Mentzer, J. T., DeWitt, W., Keebler, J. S., Min, S., Nix, N. W., Smith, C. D., & Zacharia, Z. G. (2001). Defining supply chain management. *Journal of Business Logistics*, 22(2), 1–25. DOI: 10.1002/j.2158-1592.2001.tb00001.x

Mersland, R., & Strom, R. O. (2010). Microfinance mission drift. *World Development*, 38(1), 28–36. DOI: 10.1016/j.worlddev.2009.05.006

Morduch, J. (2000). The microfinance schism. *World Development*, 28(4), 617–629. DOI: 10.1016/S0305-750X(99)00151-5

Nakano, Y., & Magezi, E. F. (2020). The impact of microcredit on agricultural technology adoption and productivity: Evidence from randomized control trial in Tanzania. *World Development*, 133, 104997. DOI: 10.1016/j.worlddev.2020.104997

Rajan Selin, A., & Shanmugasundram, S. (2024). *Traditional approaches of agriculture soil remediation: boon or bane for global food security. In Bioremediation of Emerging Contaminants from Soils*. Elsevier.

Rhyne, E. (1998). The yin and yang of microfinance: Reaching the poor and sustainability. *Microbanking Bulletin, 2*, July.

Robinson, M. (2001). *The microfinance revolution: Sustainable banking for the poor*. The World Bank. DOI: 10.1596/0-8213-4524-9

Serpantié, G. (2009). L'agriculture de conservation à la croisée des chemins en Afrique et à Madagascar. *Vertigo: La revue en sciences de l'environnement, 9*(3), 21.

Sjah, T., & Zainuri, Z. (2020). Agricultural supply chain and food security. In *Zero Hunger* (pp. 79–88). Springer International Publishing. DOI: 10.1007/978-3-319-95675-6_82

Subervie, J., & Galtier, F. (2017). L'information sur les prix agricoles par la téléphonie mobile: Le cas du Ghana. Retrieved from https://hal.archives-ouvertes.fr/hal-01594670/

Togo First. (2024). Panorama de l'agriculture au Togo: Aujourd'hui et demain.... Retrieved from https://www.togofirst.com/fr/panorama-agriculture/1301-4703-panorama-de-l-agriculture-au-togo-aujourd-hui-et-demain

Villalba, R., Venus, T. E., & Sauer, J. (2023). The ecosystem approach to agricultural value chain finance: A framework for rural credit. *World Development*, 164, 106177. DOI: 10.1016/j.worlddev.2022.106177

Wampfler, B., & Lapenu, C. (2002). *Le financement de l'agriculture familiale dans un contexte de libéralisation. Quelle contribution de la microfinance? Actes du Séminaire de Dakar, 21-24 janvier*. CIRAD-CERISE.

World Bank. (2024). *Togo - Vue d'ensemble*. Retrieved from https://www.banquemondiale.org/fr/country/togo

Zoundi, J., Hitimana, L., & Hussein, K. (2005). Économie familiale et innovation agricole en Afrique de l'ouest: Vers de nouveaux partenariats. *SAH / D, 550*, 106.

APPENDIX 1: KEY CHARACTERISTICS OF THE INTERVIEWED FARMERS

Table 1.

Farmer	Gender	Age	Marital Status	Number of Children	Microfinance Institution
Farmer No. 1	M	37	Married monogamous	4	UCMECS
Farmer No. 2	M	30	Married polygamous	3	UCMECS
Farmer No. 3	M	27	Married monogamous	0	UCMECS
Farmer No. 4	M	47	Single	2	UMECTO
Farmer No. 5	M	35	Married monogamous	4	COOPEC-AD
Farmer No. 6	F	45	Widow	4	COOPEC-SIFA
Farmer No. 7	M	34	Married monogamous	5	COOPEC-SIFA
Farmer No. 8	M	61	Married polygamous	13	COOPEC-SIFA
Farmer No. 9	M	21	Married monogamous	4	COOPEC-SIFA
Farmer No. 10	M	48	Married polygamous	8	UMECTO
Farmer No. 11	M	36	Married monogamous	3	UMECTO
Farmer No. 12	M	42	Married polygamous	7	UMECTO
Farmer No. 13	F	44	Widow	6	UCMECS
Farmer No. 14	M	32	Married monogamous	3	FUCEC
Farmer No. 15	M	58	Married monogamous	5	COOPEC AD

APPENDIX 2: MAIN USES OF MICROCREDIT ACCORDING TO INTERVIEWED FARMERS

Table 2.

Categories	Subcategories	Examples of Verbatim
Improvement of Living and Working Conditions	Purchase of Transportation Means	*The loan brought a lot of changes. I was able to acquire a means of transportation.*
		I made a profit with the loan.
		Using microcredit led to a change in my living conditions, it allowed me to get a motorcycle.
	House Construction	*Despite these difficulties, the loan allowed me to build a six-room house.*
	Improvement of Working Conditions and Purchase of Inputs	*This microcredit allowed me to equip myself, this money allowed me to buy fertilizer.*
		This loan is very beneficial. I was able to buy fertilizer and also bulls for farming.
		I took a loan from UCMECS to buy fertilizer.
	Productivity	*This microcredit allowed me to increase my agricultural production.*
Agriculture numérique	Communication with Workers and Input Sellers	*I often use my mobile phone to call my workers and fertilizer sellers.*
		With my mobile phone, I contact people for my farming activities, I use it to call warehouse keepers before going to get fertilizer, and I contact friends to find sharecroppers, to sell my maize, and conduct my business.
	Managing Agricultural Activities Remotely	*I can continue to manage my agricultural activity remotely with my phone.*
		I use my phone to reach out to my friends who have oxen to plow my field.
		I greatly use my phone to call my workers and request labor.
		My phone is very useful in my farming activities, I call my friends for help, and when we have community work, we communicate via phone to arrange work.
Projection in Adopting New Technologies or Farming Techniques	Desire to Learn New Methods	*If there are new methods or tools for the fields, I will acquire them to progress like others.*
	Desire to Mechanize Agricultural Activities	*I would like, if possible, with the support of a project, to buy a tractor to be able to cultivate a larger area.*
		If I find this equipment, I can finance up to 500.000.

[1] FINCA is a non-profit microcredit association based in the USA and very active globally.

Chapter 14
A Case of Climate Financing in Agriculture in the Global South:
Imperative to Develop Innovative Financing Instruments

Anup Samal
Climate Policy Initiatives, India

Labanya Prakash Jena
Xavier School of Management, India

ABSTRACT

Climate change is a threat to the agricultural sector; rising temperatures, altered precipitation patterns, and increased frequency of extreme events lead to reduced crop yields, and food security. Further, the agriculture sector contributes to greenhouse gas emissions that lead to climate change. The sector requires substantial investments in climate-smart practices and technologies to address these challenges. However, it has traditionally been underserved by the financial system, so farmers and other stakeholders don't have access to the financing to follow climate-smart agriculture practices. This chapter identifies financing challenges in following CSA practices and explores critical role innovative financial mechanisms can play in alleviating these challenges. This includes development of tailored financial products, use of risk-sharing instruments, and promotion of public-private partnerships. It highlights importance of capacity building of FIs and farmers to design and implement climate-smart interventions and develop robust metrics to measure impacts of these investments.

DOI: 10.4018/979-8-3693-4330-2.ch014

1. INTRODUCTION

The agricultural sector constitutes a significant source of greenhouse gas (GHG) emissions, contributing approximately 15 billion tons annually, equivalent to roughly 20% of global GHG emissions[1]. Forecasts suggest a potential increase in emissions by 15 to 20 percent by 2050, attributed to population growth and rising food consumption per capita if mitigation measures are not implemented[2]. The sector's emissions are primarily attributed to the production, processing, and transportation of agricultural products and inputs such as fertilizers and pesticides[3]. The farm sector in India has experienced significant growth in recent decades, driven by the need to meet the food demands of a rapidly growing population. India, being home to over 1.4 billion people, and the country's agricultural production has had to keep pace with this expanding population[4].

Figure 1. Sources of emissions in the agriculture sector

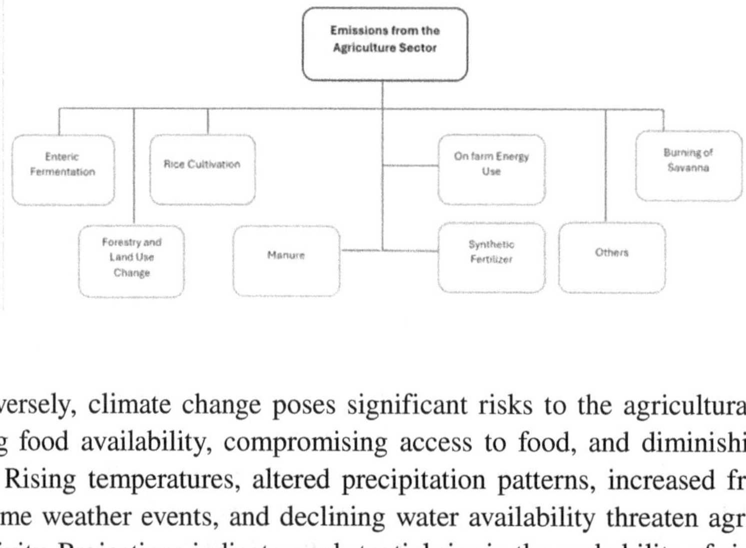

Conversely, climate change poses significant risks to the agricultural sector, reducing food availability, compromising access to food, and diminishing food quality. Rising temperatures, altered precipitation patterns, increased frequency of extreme weather events, and declining water availability threaten agricultural productivity. Projections indicate a substantial rise in the probability of significant yield decreases for major crops in countries like India, with potential implications for food security[5]. The International Food Policy Research Institute (IFPRI) has highlighted the possibility of increased hunger among millions due to reduced agricultural production and disruptions in food supply chains by 2030[6].

This is particularly concerning for staple crops such as rice, wheat, and maize, which are critical for food security in many regions (NMSA 23). The National Mission for Sustainable Agriculture (NMSA) in India has projected a significant decline in rice yields by 2050 and 2080, with irrigated rice yields expected to reduce by 3.5% and 5%, respectively, and rainfed rice yields expected to decline by 20% and

47%[7]. Similarly, wheat yields will decrease by 19.3% and 40% in 2050 and 2080, respectively [7]. Climate change exacerbated challenges related to food insecurity and disruptions related to the food distribution systems, which has a ripple effect on vulnerable populations, particularly in the developing and least developed countries where food security still poses a significant threat (World Bank 2023)[8].

These challenges can be tackled through the accelerated adoption of climate-smart agricultural (CSA) practices, as it would aid in the reduction of not only GHG emissions but also in the various adaptation measures (FAO, 2023)[9]. The adverse impacts of climate change are significant on agriculture; however, some strategies could be adopted to tackle these impacts. These strategies can broadly be divided into adaptation and mitigation measures. The adaptation measures primarily include deploying various agricultural techniques such as precision agriculture and agroforestry, developing climate-resilient cropping systems, etc. Similarly, the mitigation measures would comprise ramping up the renewable energy capacity in the countries' total energy mix, promoting sustainable land usage practices, using DRE in agriculture, etc.

2. BARRIERS TO CLIMATE-SMART INVESTMENTS IN AGRICULTURE

Several barriers often hinder the deployment of climate finance to support various initiatives on CSA. This section sheds light on some key challenges and discusses their implications for the allocation and effectiveness of climate finance in promoting CSA. The key barriers to climate-smart investments are as follows:

I. **High perceived risk:** The climate-smart investments' high potential risk factor often demotivates the investors from supplying the capital for these initiatives. This risk primarily stems from the uncertainty related to the financial impact of every CSA practice (ORF, 2024)[10].

II. **Limited availability of concessional financing:** The availability of concessional finance from various public sources in emerging markets and developing economies like India in the climate domain has been very crunched. This leads to limited availability of such finances for CSA initiatives essential to attract commercial capital (ORF, 2024)[10, 11, 12].

III. **Lack of enabling policies and governance structure:** The adoption of climate-smart practices follows a top-down approach, where climate policies and governance structures such as robust monitoring and verification techniques, effective risk management strategies, the establishment of proper carbon pricing mechanisms, etc., could catalyze the on-ground adoption of such practices.

However, the absence of such enabling practices hinders the adoption and the flow of finance to such activities [10, 12, 14].

IV. **Prioritizing climate mitigation over adaptation:** Historically, most climate finances have been flowing to combat climate mitigation through various sectors, such as energy efficiency, clean transportation, and clean energy, as these sectors are commercially viable. As a result, there needs to be more public investments in the vulnerability and adaptation sectors [13].

V. **Limited availability of climate data:** Climate-related data is the prerequisite to conducting any climate-related risk assessment, which further would aid in determining the investment necessity in that geography. However, the limited availability of reliable localized climate data creates a barrier in the risk assessments arising from climate change, which hinders the flow of capital. Thus, robust data on climate-related risks, impacts, and effectiveness of various climate-smart practices are necessary to unlock commercial capital [12].

VI. **Heterogenous land-holding pattern:** Indian farmers have very heterogeneous land-holding capacities. The majority of the farmers are either small or marginal farmers. This composition of land holding in the country's farming sector poses a significant barrier to investments in achieving climate-positive outcomes at scale. The fragmentation of these farmers' lands makes implementing CSA practices on a scale prerequisite to attaining climate-related benefits challenging. The lending institutions often show reluctance to lend capital to smallholder farmers due to the high-risk profile and even higher transaction costs associated with agriculture loans[11, 12].

VII. **Capacity constrained:** The limited capacity within the agricultural sector to gauge the exact financial need for climate adaptation and mitigation poses a significant challenge in unlocking capital from the market. This is particularly a problem in the developing economy, where institutional and technical capacity constraints exist. Most of the stakeholders in the farming sector need more technical and financial expertise and tools to gauge the financial requirements for implementing climate-smart practices, which in turn leads to the failure to attract capital. Furthermore, the climate finance unlocking mechanisms are complex and often demand stringent requirements for access to various climate funds, which exacerbate the challenges. For instance, many such instruments require robust cost-benefit analysis, detailed project proposals, and risk assessments, which lie beyond the capacity and purview of a farmer and local food producer organization to produce [13, 14].

VIII. **Incomplete linkage between key stakeholders:** Unlocking capital for farmers in the agricultural sectors requires a perfect linkage between some key stakeholders, such as farmers, SMEs, financiers, or lenders. This leads to a fragmented finance delivery channel where capital fails to reach the end users, especially in remote areas [13].

IX. **Absence of pipeline of the projects:** The absence of an attractive pipeline in climate-smart agricultural projects poses another significant challenge, especially for investors. Many potential CSA projects fail to secure funding because they need to meet climate finance institutions' rigorous standards and criteria. This issue is compounded by the fact that CSA investments are often context-specific, requiring tailored solutions that address different regions' unique environmental, social, and economic conditions [12,13,14].

X. **Inadequacy of MRV tools:** No standardized Monitoring, Reporting, and Verification (MRV) metrics and tools are available to measure CSA practices' impact on agricultural production and yield gap. Thus, an accurate demonstration of the outcomes and benefits of CSA practices could not be gauged, which is essential for securing climate finance and demonstrating the effectiveness of these interventions. Furthermore, the existing tools often fail to capture the full range of environmental, economic, and social impacts [11,12,13,14].

3. ROLE OF CLIMATE FINANCE IN THE CLIMATE-SMART AGRICULTURE

Climate finance is the key to tackling the impacts of climate change in the agricultural sector by mobilizing capital through public and private channels to support various climate-smart farming practices. Some of the key solutions of climate finance include providing appropriate insurance for various crop failures in case of climate emergencies, providing loan guarantee schemes, managing climate-related risks via various risk management tools, etc. These solutions always aim to strengthen the linkages between financiers and farmers through financial support to their later counterparts. This includes direct financing to farmers and support for agricultural value chains and market infrastructure[15].

Climate-smart agricultural technologies contribute to GHG mitigation efforts and aid in adaptation efforts in the sector while increasing the overall resilience of the agriculture sector, particularly in pursuing sustainable livelihoods. However, the full benefits of the CSA have yet to be realized as its investments have remained limited. Most of these potential investments face barriers, as discussed in the above section 2. These barriers deter investors from putting their capital in the sector and stifle the growth and penetration of innovative climate-smart technologies. Thus, along with technical assistance and a capacity-building program, proper financing would be required to design CSA projects to help farmers, lenders, and other stakeholders implement the projects and benefit from them successfully. This section describes the role of various financial instruments in facilitating greater adoption of climate-smart practices.

a) Aligning Climate Smart Practices With Global Climate Financial Institutions' Agenda

Any agricultural investments that aim toward delivering adaptation and mitigation goals are eligible for public finance. Public finance can be defined as the sources of finance that are either government-owned or administered institutions, such as national, bilateral, and multilateral funding (World Bank 2016). These financial institutions are designed to support projects that aid in GHG emissions and increase the resilience of the overall agricultural systems. These projects also draw contributions from Development Finance Institutions (DFIs) and dedicated international climate funds such as the Global Environment Facility (GEF), the Climate Investment Funds (CIFs), particularly the Pilot Program for Climate Resilience (PPCR-CIF), and the Green Climate Fund (GCF)[16].

For instance, the GEF primarily funds projects integrating sustainable land management practices; similarly, the CIFs, through the PPCR-CIF, focus on enhancing the climate resilience of vulnerable communities by investing in adaptive agricultural practices[16]. The GCF, the largest global fund dedicated to helping developing countries respond to climate change, provides substantial financial resources to support CSA initiatives that align with national climate strategies and contribute to the broader goals of the Paris Agreement (World Bank 2016).

b) Importance of Public Finance in Climate-Smart Agriculture Practices

Finance from public sources is imperative. It would enable the capital to initiate, scale up, and implement CSA projects and facilitate private investment, which might only be possible with higher perceived risks and longer payback periods.

Public finance is also important for other key reasons. Firstly, it provides the much-required initial capital to implement climate-smart innovative technologies, which otherwise would be very expensive for farmers and FPOs in the developing and least developed countries. Public finance would also allow farmers to access new technologies such as precision farming, drought resistance crop varieties, and efficient irrigation systems[18,19]. Further, it would also aid in derisking the investments in climate-smart agricultural practices. Public funds would provide grants, concessional loans, guarantees, etc., thus reducing the financial risks associated with CSA projects and making these projects more attractive to private investors.

For instance, public finance would aid the development of insurance products to safeguard farmers from the climatic extremes related to crop loss, thus reducing the perceived risk of investing in CSA. Any solutions related to climate or any other sector require heavy research and development (R&D) before their potential can be fully realised. Thus, public finance can fund these R&D activities for innovative solutions vis-à-vis CSA. This would lead to further investments in the sector. Finally, capacity building and knowledge dissemination to the end users about innovative solutions are crucial for its accelerated adoption. Thus, public finance can facilitate these services by providing essential funding for capacity-building programs, demonstration of projects, and extension services, thus aiding farmers with the skills and technical know-how required to adopt the innovative CSA solutions[18,19].

(c) Smart Agriculture Practices Need Long-Term, Concessional, and Risky Climate Finance

Large financial obstacles challenge the agriculture industry, especially in developing nations, because of the CSA projects' long payback periods, limited profitability, and perceived high risks. By supplying mechanisms for risk mitigation and risk sharing, as well as financing on more benevolent terms or in subordinate positions that make private investment more feasible, climate finance can help remove these obstacles.

Instruments for sharing risks, like insurance, can also be quite important. The creation of novel insurance products that shield farmers from climate-related hazards, including droughts, floods, and pests, can be aided by climate finance. Farmers may feel more comfortable investing in CSA practices with the security of these insurance policies since they won't have to worry about suffering irreversible financial loss. Climate finance has the potential to generate large private investments by minimizing risks and addressing financing gaps. Furthermore, it can facilitate the support required to build the necessary capacity for the increased adoption of CSA practices[16,17]. Such capacity-building efforts can also include the development of customized financial products, training of lending officers, establishing partnerships between other public and private sector employees, etc. Climate finance could always pave the way for increased private sector engagement in innovative CSA solutions through building successful track records and derisking the sector[18].

Additionally, publicly funded climate finance can support the development of robust monitoring and evaluation systems that track the performance and impact of CSA projects, providing data and insights that can further reassure private investors[16]. In addition, this public climate finance can support building capacity in the agriculture sector through channeling capital to conduct training programs, provide extension services, and equipping farmers with the knowledge and skills needed to adopt and sustain CSA practices

4. INNOVATIVE WAYS TO ATTRACT ADDITIONAL CAPITAL TO CLIMATE-SMART AGRICULTURE

Identifying entry points for directing climate finance into agriculture ensures the financial resources reach the intended end-users and support the transition to CSA. One such key entry point is enabling climate finance through national and subnational climate finance strategies. The government would develop these strategies, which should prioritize investments in the agricultural sector and outline specific measures to mobilize and allocate funds. Further, these strategies should be aligned with the national development goals and climate targets mentioned in the climate action plans while ensuring that the agriculture sector receives enough attention and resources[12].

This section delves into seven other key entry areas for enhancing CSA's climate finance. By focusing on these areas, it is possible to foster a sustainable and resilient agricultural sector that can withstand the impacts of climate change.

a) Creating an Agricultural Financing Enabling Environment

Mobilization of finance via proper channels to end users can only happen when there is a conducive policy environment. This involves the creation of appropriate policy and regulatory framework that would support the flow of investments in the climate-smart agriculture domain. Thus, policymakers should create solutions that not only encourage investment in climate-smart agricultural practices but also incentivize the adoption of climate-resilient technologies[19]. For example, governments could implement policies that offer subsidies for farmers who deploy renewable energy solutions for their farming activities, such as solar-based irrigation systems and other decentralized renewable energy solutions.

Furthermore, developing the necessary structures to facilitate and accelerate CSA investments is crucial, such as creating an agricultural finance platform that brings various stakeholders together, such as banks, microfinance institutions, agribusinesses, and farmer-producer organizations [16]. These kinds of platforms can not only serve as hubs for knowledge exchange, capacity building, and promoting best practices in CSA but also could do a match-making job while bringing investors and investees together to unlock capital from the market. Additionally, creating mechanisms for monitoring and evaluating the effectiveness of agricultural finance policies is essential to ensure that they achieve the desired outcomes and make necessary adjustments[18].

b) Establishing Public-Private Partnerships

Establishing public and private partnerships (PPP) would be another crucial aspect of enabling the environment to flow finances into the sector. PPP not only leverages the strengths and weaknesses of both sectors to enhance the reach and impact of CSA initiatives. For example, a collaboration between government agencies and private financial institutions can lead to the development of innovative financial products tailored to the needs of smallholder farmers. These products can include microloans, climate risk insurance, and green bonds specifically designed to fund CSA projects[16].

c) Deploying Risk-Mitigating Instruments

The agricultural sector is considered inherently risky due to multiple influencing factors such as weather variability, market fluctuations, pest infestations, etc. Thus, strengthening the risk management systems to tackle such risks within financial institutions is crucial to unlock capital for the sector. So, financial institutions must develop robust risk management strategies to combat these risks. One such effective tool would be the creation of rural credit rating agencies [16,17]. This would aid in assessing a farmer's or FPO's creditworthiness, thus reducing the risk for the lenders and facilitating access to credit.

Similarly, the guarantees and insurance schemes would also be very useful in enhancing the resilience of the agricultural sector. For instance, guarantee funds could provide coverage (partial or full) for the loans provided to the farmers, thus reducing the perceived risks for the lenders and enabling them to lend more freely. Further, insurance could protect the farmers against the loss of crops due to climate extremes, thus ensuring their recovery of input costs and continuity of farming even in adverse conditions. Insurance products tailored to specific agricultural risks, such as crop and livestock insurance, can provide a safety net for farmers and enhance their financial stability[19].

d) Establishing a Value Chain Finance

Value chain finance is another important mechanism that supports agricultural investments. Through the linkage of financial services to the various stages of the farming value chain, financial institutions can provide targeted support to farmers, processors, and marketers. For example, warehouse receipts can be used as collateral for loans, allowing farmers to access credit based on the value of their stored produce. This improves farmers' access to finance, encourages better post-harvest management, and reduces post-harvest losses[16,17].

e) Integration of Climate-Smart Advisory Services

Climate advisory services use climate and weather data to provide early warning signals for future extreme events. Thus, integrating climate advisory services in lending-based decision-making can significantly reduce the perceived risk in agriculture. Lending agencies and other financial institutions could collaborate with such service providers as the Indian Meteorological Department and technology, which provides real-time information on weather patterns, pest outbreaks, and market trends[18]. This not only helps the financiers to make informed decisions but also aids farmers in optimizing their agricultural practices and reducing the risks associated with climate variability. Further, these data analytics could also assist financial institutions in creating precise risk assessment models and tailoring their financial products to the specific needs of farmers[16,17,18]. To tap multilateral climate finance institutions, there is a need to develop high-quality project proposals that meet the eligibility criteria and demonstrate the potential for significant climate and development benefits.

f) Leveraging Technology to Make Finance Accessible to Farmers

High transaction costs hinder the financial viability of an agricultural project. Thus, reducing transaction costs is critical to making financial services more accessible and affordable for farmers and FPOs. One of the ways to achieve this is via mobile financial services and branchless banking. Mobile financial services have revolutionized financial transactions in the agricultural sector. With the widespread use of mobile phones, farmers can now access various financial services, including payments, savings, loans, and insurance, without visiting a physical bank branch. One such example of digital financial services is "M Pesa," a mobile money platform that has reiterated the role of such platforms in promoting financial inclusion for farmers.17.

g) Specialized Financing Instruments

Financial institutions could play a significant role in channeling climate finance into agriculture, thus supplementing the various national and international climate finance mechanisms. Such institutions can create customised financial products and services catering to the needs of the agricultural sector. For example, they can create specialized green bonds that can fund CSA projects, which further can pave a path for impact investors who are looking to support sustainable agriculture initiatives. Moreover, the FIs could also explore other mechanisms, such as blended finance mechanisms that can use public finance as a derisking mechanism to attract private investors. [16].

h) Network Development

The linkage between FIs, small and marginal farmers, and agricultural SMEs is key for the effective delivery of climate finance. Thus, achieving this would be one of the most crucial aspects. Thus, this could be established by creating a value chain partnership between three key stakeholders, namely farmer cooperatives, agribusiness networks, and financial institutions. This linkage can act as a conduit to enhance the bargaining power of the farmers and can facilitate financial resource aggregation by improving market access. It would also provide an opportunity for FIs to collaborate with these entities to provide customized financial products and services such as group loans, cooperative credit schemes, and value chain financing. Moreover, they can work closely with the farmer producer organizations (FPOs), which will ensure the financial solutions can reach their intended beneficiaries [16,17,18].

5. CASE STUDY: USING CARBON PRICING IN CSA PRACTICES IN RICE FARMING

The System of Rice Intensification (SRI) is an innovative farming practice that intends to enhance the amount of rice harvest while using the resources sustainably, thus limiting the environmental impact. It was first introduced in Madagascar by French Jesuit Father Henri de Laulanié. This method focuses on optimizing the management of soil nutrients, plants, and water usage for the crop, which results in a sustainable and productive rice cultivation system while increasing the overall yield [21].

In this SRI framework, the potential for generating revenue from carbon offset could be beneficial for Least Developed Countries (LDCs) and Small Island Developing States (SIDS), which are facing maximum impacts of climate change and, thus, are more vulnerable to extreme events and, economic shocks [21,22].

The revenues that would be generated from integrating the carbon pricing system could then be leveraged towards supporting policy initiatives, including capacity building of farmers to encourage the greater adoption of SRI methods 23. Further, the implementation of a carbon pricing mechanism can aid the government in creating a tangible financial incentive for farmers to transition away from traditional rice cultivation practices to the SRI method, which could further reduce GHG emissions and promote sustainable agricultural practices.[24,25].

The above model illustrates potential pathways for the government to generate carbon revenue, depicting the flow of revenue to the end source. In this model, a farmer transitioning to a System of Rice Intensification (SRI) collaborates with intermediary verification agencies operating in the field. These agencies provide technical assistance, monitor progress, and facilitate working with companies responsible for managing and processing carbon credits. By establishing this linkage between farmers and carbon markets, monetizing their carbon potential becomes more streamlined. These intermediary organizations act as conduits for governmental oversight and validate the process, enhancing its security and credibility.

While the assessment of carbon revenue potential is subjective, it is noteworthy that nearly 100 countries have explicitly referenced carbon pricing in their Nationally Determined Contributions (NDCs) (UNFCCC, 2023)[26]. Carbon revenues are significant for many middle-income countries as they present an opportunity to generate substantial revenue streams. By supporting the development of infrastructure for international carbon markets, these countries can tap into significant flows of carbon revenues, enabling cost-effective mitigation strategies for developing nations (World Bank, 2019)[27].

CONCLUSION

On the one hand, climate change can disrupt food availability and increase food prices; on the other hand, the farming sector is amplifying climate change. The agriculture sector needs to adapt to climate change to supply food at an affordable price and reduce GHG emissions to mitigate climate change risk – a global challenge to save the planet. Climate-smart agriculture (CSA) practices can support the farming sector in adapting to climate change while reducing GHG emissions from farming practices. However, there are several challenges to CSA practices; access to finance and the cost of finance are some of the key obstacles to CSA practices. High perceived risk, limited availability of concessional financing, lack of enabling policies and governance structure, limited availability of climate data, small size loan, capacity constraints, poor linkage between key stakeholders, and absence of pipeline of the projects are the key bottlenecks to access to finance to adapt CSA practices.

Climate Finance, particularly international public climate finance, can accelerate the adoption of CSA practices in the global south, including India. They can offer concession capital, risk mitigation financial instruments, grants, and risky equity capital that can be deployed smartly to attract private capital to adopt CSA practices. These kinds of public capital can demonstrate the commercial viability of CSA practices and consequently attract private capital.

In addition, public climate finance can be used cleverly to attract private capital for climate-smart farming. It can support private financial institutions in developing and strengthening their risk management mechanisms for the farming sector following CSA practices. Besides, public climate finance can create adequate institutional and agricultural finance capacities and customize agricultural financial products and services to support CSA practices. Public finance can also be used to develop the capacity and skills of farmers by conducting training programs on CSA practices and climate risk management and imparting skills to develop proposals for accessing climate finance.

REFERENCES

Agriculture Sector - India GHG Program. India GHG Program. Retrieved from https://indiaghgp.org/agriculture-sector

Climate finance for agriculture. Retrieved from https://ccafs.cgiar.org/sites/default/files/202010/Module%203%20Presenter%20Guide.pdf

Food and Agriculture Organization (FAO). (2023). Climate-Smart Agriculture. Retrieved from https://www.fao.org/climate-smart-agriculture

Greenhouse Gas Emission from Indian Agriculture. krishi.icar.gov.in. Retrieved from https://krishi.icar.gov.in/jspui/bitstream/123456789/32431/1/GreenhouseGasEmissionfromIndianAgricultureHPathaketal%20(1).pdf

International Food Policy Research Institute (IFPRI). (2023). Climate Change and Food Security. Retrieved from https://www.ifpri.org/publication/climate-change-and-food-security

Jena., ... (2023). financing climate-smart agriculture for sustainable agri-food systems. Retrieved from https://www.nitk.ac.in/document/attachments/5611/T20_PB_TF6_6162_Jena.pdf

Jena., ... (2023). *How to mobilise climate-smart agriculture finance-World Economic Forum-* Retrieve from https://www.weforum.org/agenda/2023/05/climate-smart-agriculture-finance/

Khatri-Chhetri, A., Sapkota, T. B., Sander, B. O., Arango, J., Nelson, K. M., & Wilkes, A. (2021). Financing climate change mitigation in agriculture: Assessment of investment cases. *Environmental Research Letters*, 16(12), 124044. DOI: 10.1088/1748-9326/ac3605

Making Climate Finance Work in Agriculture - World Bank. Retrieved from https://www.worldbank.org/en/topic/agriculture/publication/making-climate-finance-work-in-agriculture

Mohieldin, M., Kenewendo, B., & Wambui, R. (2023). Breaking Financing Barriers for Just Climate Transition in Africa.

National Mission for Sustainable Agriculture (NMSA). (2023). Climate Change and Agriculture. Retrieved from https://www.nmsa.gov.in/climate-change-and-agriculture

Thakur, A. K., Mohanty, R. K., Paul, D. U., & Kumar, A. (2014). Impact of water management on yield and water productivity with system of rice intensification (SRI) and conventional transplanting system in rice. *Paddy and Water Environment*, 12(4), 413–424. DOI: 10.1007/s10333-013-0397-8

Thakur, A. K., Singh, R., & Ashwani, K. (2014). "The science behind the system of rice intensification (SRI)." Research Bulletin No. 69, Directorate of Water Management (ICAR), Bhubaneswar, Odisha.

Thakur, A. K., Singh, R., & Ashwani, K. (2014). "The science behind the system of rice intensification (SRI)." Research Bulletin No. 69, Directorate of Water Management (ICAR), Bhubaneswar, Odisha.

UNFCCC. (2023). Retrieved from: https://unfccc.int/sites/default/files/resource/Session%201b_Introduction%20to%20carbon%20pricing_MENA.pdf

World Bank. (2023). Climate-Smart Agriculture. Retrieved from https://www.worldbank.org/en/topic/climate-smart-agriculture

Chapter 15

Application of Digital Technologies in Agri-Supply Chain:
The Story of India and Comparative Narrative

Pankaj Pathak
https://orcid.org/0000-0002-5875-0387
Symbiosis Institute of Digital and Telecom Management, Symbiosis International University, India

Madhavi Damle
https://orcid.org/0000-0003-2649-4291
Symbiosis Institute of Digital and Telecom Management, Symbiosis International University, India

Pravin Dange
https://orcid.org/0000-0002-5346-1172
Symbiosis Institute of Digital and Telecom Management, Symbiosis International University, India

Samaya Pillai
https://orcid.org/0000-0002-8451-8936
Symbiosis Institute of Digital and Telecom Management, Symbiosis International University, India

ABSTRACT

Agricultural supply chains are undergoing a huge change due to the advancements in digital technology. These technologies have triggered the unknown concepts of

DOI: 10.4018/979-8-3693-4330-2.ch015

sustainability, efficiency, and transparency. The integration of digital technologies like the big data analytics, blockchain, artificial intelligence (AI), and the Internet of Things (IoT) in agricultural supply chains is discussed in the chapter. IoT devices provide for precision farming and less resource waste. Blockchain technology improves transparency and traceability guaranteeing food safety. It also enables to confirm the calibre of agricultural products. Big data analytics is helpful in the better decision-making. All these technologies address the major issues facing the agriculture industry. The issues include resource management, climate change, and food security. But the technology adoption has its own challenges. It is necessary to address issues including data protection, farmers' digital literacy, and the large upfront costs associated with implementing new technologies.

1 INTRODUCTION

The versions that include their attributes illustrate the developments as a data philosophy for agriculture. It's referred to as "Agriculture 4.0," "Digital Farming," or "Smart Farming," and it emphasizes data management in agriculture more than other fields. Agriculture 4.0 is based on Precision Agriculture principles, with producers using systems that generate data in their farms, which will be processed in such a way as to make proper strategic and operational decisions.

It is often referred to as Agriculture 5.0, with hyperintelligence embeddings like robotics and artificial intelligent (AI), deeper insights, and technical problems that usually lead to significant breakthroughs in disruptive agriculture-related technology. In contrast to farms that traditionally required a large workforce, the advancements expressed in the concept of Agriculture 5.0 imply that farms are using equipment that involves unmanned operations and autonomous decision support systems with robots and some forms of AI. They also follow the principles of Precision Agriculture. mostly seasonal, to maintain farm productivity and collect crops. A solution to the issue of a labour shortage that farms are suffering as a result of society shifting from being an agricultural society with huge numbers of people living on farms to people living in cities. (Rubio-Saiz, 2020).

In our discussion, we look at precision agriculture as the story of India, which has major challenges. Here, we discuss the systems that will help precision agriculture become robust in its formative stages. The challenges are complex due to weather patterns, fragmentation, and non-unified processes and structures,

2 TECHNOLOGIES FOR APPLICATION

2.1 Internet of Things: Collecting Information

The usage of Internet of Things (IoT) in agriculture refers to the process of turning major aspects and actions of farming into data via the use of sensors and other devices. According to reports, IoT solutions are being used by 10% to 15% of US farmers. They span almost 1200 million hectares and 250,000 farms. Agriculture 4.0 is powered by IoT. Agriculture can provide the useful data because of IoT technologies. The agriculture industry is anticipated to be greatly impacted by these technological advancements. By 2050, it's predicted that the IoT might boost agricultural productivity by 70% (2020 Saiz-Rubio.)The main advantages of using IoT are that it achieves higher crop yields and lower costs (Mohanraj et al., 2016), see Figure 1: IoT in Farming (Hash Technologies Article, 2021).

Figure 1. IoT in farming

(Source: Hash Technologies Article, 2021)

2.2 Data Analytics and Cloud Computing Solutions

Farmers may use data analytics to make judgments about resource management based on established trends, generate predictive analytics about future yields, and keep an eye on the health of their crops in real time. They are increasing revenue and cutting waste. Farmers can maximize crop yields and reduce stress by starting and harvesting their crops at the ideal times thanks to the intelligence that data analytics provides. Farmers can take demand changes into consideration rather than filling a whole plot.

Among the various tasks necessary to lower the risk of loss and raise profitability are keeping an eye on market movements, boosting yields, and forecasting the weather. Farmers are now able to precisely manage their enterprises by visualizing their land, crops, and management techniques in never-before-seen ways thanks to the tools of geospatial analysis and predictive modelling refer Figure 2: Informatics for Bio Life of Plant (Joshi, 2019).

In order to provide farmers with useful information that will help them better understand their land and make more informed decisions, access to spatial data has become a crucial part of farming operations. Online access to this data allows for the creation of intelligent maps that improve agricultural business operations.

Figure 2. Informatics for bio life of plant

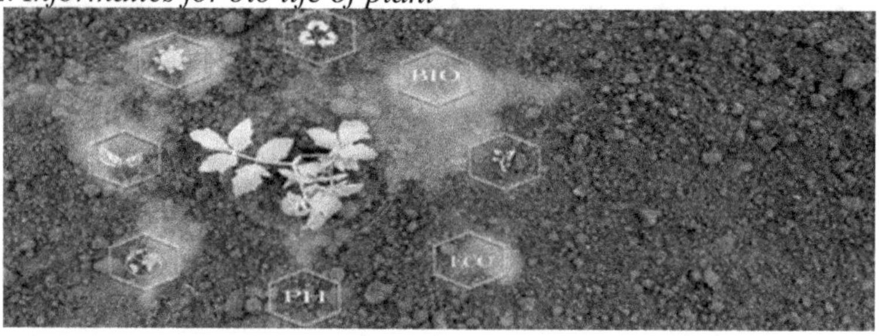

(Source: Joshi, 2019)

The advantages of this include improved analytical support for precision farming, a better comprehension of risk factors, increased revenue generation and cost recovery, increased efficiency through task automation, increased access to government services and data, improved decision-making accuracy, increased insight into policymaking, simpler reporting for government applications, and improved resource management due to regulatory compliance. The Web Soil Survey offers a large specifications database for 50 national standard interpretations for the terrain and the specific land environment. Some examples of interpretations are land applications, the productivity of cropland, rangeland, and forestland, and suitability for different recreational developments such as paths and trails, campsites, and picnic grounds, see Figure 3: Data and Information about the Harvest (Agro Informatics Article, 2020).

Figure 3. Data and information about the harvest

(Source: Agro Informatics Article, 2020)

2.3 LoRa Farm Network Used for Precision Agriculture

A cultivation management strategy based on monitoring, estimating, and responding to entomb and intra-field crop inconstancy is known as precision agriculture (PA), satellite cultivating, or site-specific crop management (SSCM).

Seed rearing has been an Israeli claim to fame since the start of the state. Researchers in the scholarly community for the Agricultural Research Organization have made numerous increasingly nutritious, high-return, tasty, and impervious seeds to dry spells and sickness. They have also had success in developing new fruit and vegetable varieties.

Few varieties of good-quality, high-protein seeds. The development of this iron- and calcium-rich seed will find new markets thanks to its new licensed sesame assortment. These latest discoveries offer enhanced dietary profiles and can be machine-reaped safely. Some seeds like the sesame seeds which need to be gathered by hand has prevented the yield to reach its maximum capacity. It causes a large portion of the harvest to be lost due to contamination when the units collide with the ground.

Soil conditions: It aims to use land resources efficiently. The Farmers are able to adjust their crops to the conditions of the soil. Thus, saving the costs and thereby boosting the production. This is also due to the development of the special sensors. These sensors can gather data about the land and analyse it. Israel has benefitted enormously by this technique, becoming highly productive, efficient, profit-yielding (Mohanraj et al., 2016).

Roots Sustainable Agricultural Technologies: In this technology, robotic water pipes are used. These pipes are set in the soil. They help to decide the comfort temperature for a particular piece of land. Here the soil temperature is managed -if it's warm, its cooled and if it's cool, its warmed. This Technology enables the yield of crops such as basil and strawberry.

Figure 4. IoT LoRa farm framework as a solution framework

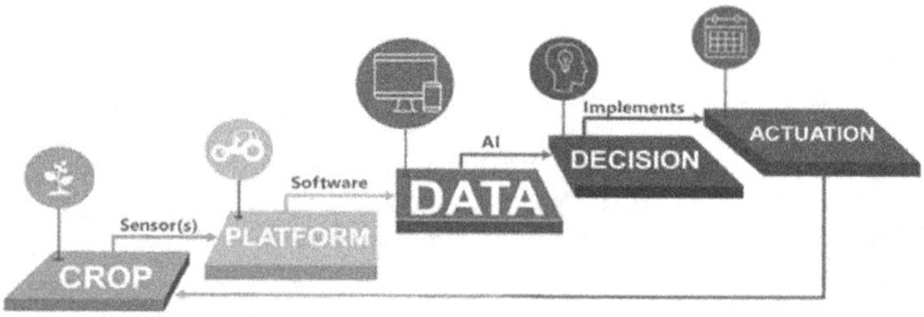

(Source: Adopted from Saiz-Rubio & Rovira-Más, 2020, "From Smart Farming towards Agriculture 5.0)

Drip irrigation provides small quantities of water and fertilisers to the crops at frequent and equal intervals. Drip irrigation has been hugely commercialised and is continuously developing. Israel adopted measures to create drippers that might be immune to clogging with this commercialisation of production and allow farmers to get better crop yields by providing very little water. Encouraging drip irrigation as a new process changes food production for the better under conditions such as water shortage and shortage in the space of plots. In this process, each plant is planted close-by to the dripper for technology components refer Figure 4: IoT LoRa Farm Framework as a Solution Framework.

Water is never a restricting element, and when the nutrients are provided with the water, the yields of drip irrigation rise while only a fraction of the soil surface gets wet. A large increase in crop manufacturing occurs when water and nutrients are available in the root regions at the maximum level. Specific, low-volume, high-frequency irrigation techniques reduce the harmful effects of water getting logged and salination of the water along with generating favourable moisture conditions even in dry kind of soils, e.g., coarse ones, sandy ones or gravelly ones and terrains or hilly areas that had previously been considered difficult for the cultivation of crops through other techniques.

Dripper is the most important element of this Technology, and its function is to give a constant and exact amount of water under pressure. This is a closed system where the pressure builds up along the lines and is not in open channel irrigation. Water is sent from rivers to level beds in the valleys via open channels using irrigation techniques. The advantages of a drip system are that it can be operated at very low pressures and can be customised and applied to any size of field or plot.

There are various water sources, such as marginal water, salt water and treated wastewater. These can be applied, provided the water is passed through filters to remove particles that are irregular in shape to prevent clogging of emitters (Mohanraj et al., 2016). This is achievable solely because the type of soil wetting underneath the dripper shifts the excess salts to the margins of the wetted plot. This helps maintain optimal conditions for root growth and uptake in the quantities of nutrients in the wetted zone.

The topography of the field needs to be considered in precision agriculture. Large landholdings exist in many countries; to achieve the greatest results, farmers need to be aware of the challenges that each area of the field faces. Satellite imagery can help you in precision agriculture by identifying areas of the field that require more watering, plants that have insect infestations, and other related information. This can help with resource or asset use more effectively. Therefore, by dousing the sick plant with insecticides, we can increase productivity if one plant is healthy and the one next to it is not.

2.4 Geographic Information Systems (GIS) and Global Positioning Systems (GPS)

A computer-based tool for mapping and analysing geographic data is called a geographic information system (GIS). Geographic and spatial data are stored in GIS for three distinct kinds of areas: Places with an area include farms, fields, wetlands, and neighbourhoods; places without an area include the site of a building, tractor, or grain bin; and locations with an end and a beginning include private roads, major highways, and streets, refer Figure 5: GIS – Aerial View (Indian Agriculture Archives I AgriBazaar, 2019).

Geographic information system (GIS) technology is used by GIS Solutions for the Agriculture industry to exchange data, boost yields, forecast results, and enhance operational procedures. Agricultural businesses that use GIS technology may better manage their resources and obligations, create data portals that provide a wealth of interactive maps and agricultural data, and assist rural communities. Through the creation of information-dense reports and maps that offer them a distinct view of their operations, producers utilize GIS to manage their farms more effectively.

Figure 5. GIS: Aerial View

(Source: Indian Agriculture Archives | AgriBazaar, 2019)

Numerous alternatives are available for measuring and tracking the consequences of farm management approaches, as well as for visualizing farming conditions, thanks to the robust analytical capabilities of GIS. GIS can accurately identify and manage inputs with the use of remote sensing technology, which lowers soil damage and prevents preventative costs. Moreover, farm managers utilize GIS to streamline lengthy, multi-step applications for government programs. GIS is a powerful communication medium that enables internal and external information sharing between government and consumers.

GIS is used in precision agriculture to handle geographical data and support decision-making for farmers. Soil type, elevation (topography), crop yield, crop quality, field borders, management zones, remotely sensed images, weed and pest locations, and historical land usage are a few examples of data layers that farmers could employ (Gharde et al., 2018).

2.5 Satellite Imagery

This aids in creating a bird's-eye perspective and images. With the transmission of soil survey data, they are integrating features like property borders to provide fast access to details about specific properties in their files. An assessment of the state of satellite technology accessible for agricultural applications may be found in Satellite Applications Catapult. This study will mostly address Earth observation (EO), with some coverage of satellite communication (Satcom) and global navigation satellite systems (GNSS) technologies as needed. Enhanced connection and observational capabilities are made possible by advancements in satellite constellations, payloads, and launches, refer Figure 6: Mapping in Aerial View (Agricultural Solutions, 2024)

Utilizing these advancements in conjunction with "smarter" data infrastructures, analytics, and computing expands the potential applications of satellite technology in the agriculture industry. This opens up new markets for goods, services, and decision support, but it also puts pressure on the agriculture industry to make sure that the newest technology is properly connected to production issues and, as a result, can be leveraged to provide the gains necessary to satisfy. Utilizing these advancements in conjunction with "smarter" data infrastructures, analytics, and computing expands the potential applications of satellite technology in the agriculture industry. The agricultural sector now faces a challenge in ensuring that the newest technology is appropriately linked to production challenges and, consequently, can be used to deliver the gains necessary to meet the societal, economic, political, and environmental needs. This presents new opportunities for products, services, and decision support.

Figure 6. Mapping in aerial view

(Source: Agricultural Solutions, 2024)

Satellite Earth observation (EO) is a type of remote sensing that gathers data from platforms up to 36,000 kilometres in orbit about the Earth's surface and atmosphere. The data comes from several satellites with distinct instruments and mission objectives rather than from a single satellite mission. The choice of EOS mission and equipment is determined by the intended application.

2.6 Robotic Technologies

In robotics, computer systems and robots are designed, built, operated, and used for information processing, control, and sensory feedback. Harvesting and picking are two of the most prevalent uses for robots in agriculture. Weed control. autonomous seeding, spraying, thinning, mowing, and trimming. Internet of Things Systems

The introduction of agricultural robots aimed to address these issues and meet the growing need for higher yields. The worldwide market for agriculture and crop production is growing thanks to robotic developments; the Verified Market Intelligence research states that agricultural robots will be able to do field activities more effectively than human interventions, refer Figure 6: Yield Measurements (Greentech, 2023), refer Figure 4: Robotic Technologies-Harvesting Bots (Patidar, 2024) to see the effective use of intelligent systems as robotic arms and process automation systems which have narrow intelligence to pick the harvest or look for areas of infestations refer Figure 7: Robotic Technologies-Harvesting Bots (Patidar, 2024) Figure 8: Intelligent Pesticides and Sprays (Agribots, 2009 Figure 9: Yield Measurements (Greentech, 2023).

Figure 7. Robotic technologies-harvesting bots

(Source: Patidar, 2024)

Figure 8. Intelligent pesticides and sprays

(Source: Agribots, 2009)

Figure 9. Yield measurements

(Source: Greentech, 2023)

2.7 Digital Twin Technology

Digital Twin Technology is a revolutionary concept that deals with creating intelligent and physical copies of tangible properties, frameworks or systems for reviewing and improving them in the real-world context. This technology combines information collected from various sensors and devices to recreate state, or operational and behavioural characteristics of the counterpart, in digital form. Digital twin is a replica of a physical object created with analytical, learning tools like IoT or machine learning that offers a look into the performance and the possible problems of the actual object (Mohanraj et al., 2016), refer Figure 10: Digital Twins in the agri-food supply chain (Adapted from Tzachor et al.,2022).. This makes it possible to plan the maintenance and usage of equipment to ensure effective resource utilisation and optimal production. Several industries of Engineering include manufacturing, healthcare, urban planning, and automobile industries etc., are moving towards digital twin technology for better and efficient output, cost controlling, new ideas, creativity etc. The feature of staging and exploring different changes without the reality consequences creates the appeal of the digital twin, making them an indispensable element of modern decision-making based on computer knowledge.

Figure 10. Digital Twins in the agri-food supply chain

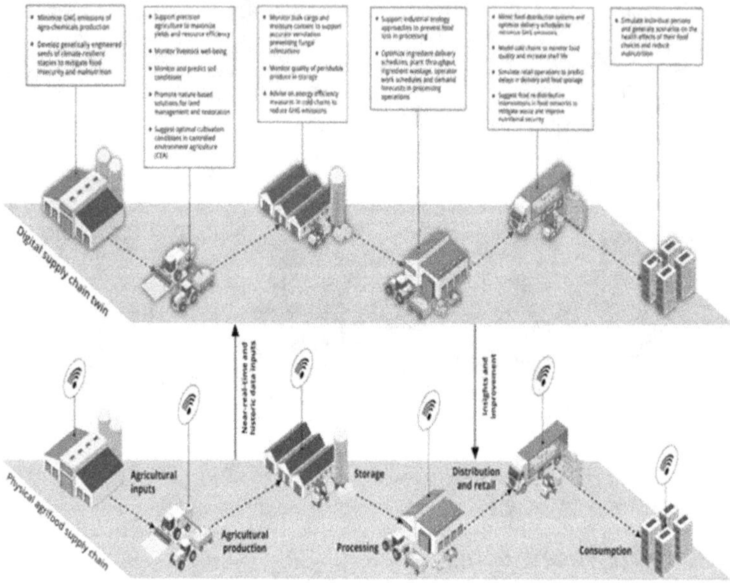

(Source: Adapted from Tzachor et al.,2022).

2.8 Blockchain

Blockchain is a synergistic and distributed database that enables transaction history to be registered across multiple computer systems in a manner that doesn't allow for subsequent alteration of the recorded activities. This guarantees privacy, openness, and invariability making it appropriate for numerous sectors as illustrated next. Blockchain is brought to savings in finance by cutting on procedural inefficiencies and fraudulent activities; ideal in supply chain because of its add-on value to traceability and easy flow; ideal in the medical field because of its capability in patient data safekeeping and information sharing. The technology is a basic peer-to-peer platform that eliminates communication with third parties and, as a result, the costs have been reduced, and the speed has been increased. Smart contracts, which can trigger certain transactions depending on the unfulfillment of certain conditions, are another application of blockchain that goes even deeper in creating possibilities for better automation and solutions. Thus, with the introduction of new improvements to the technology, a number of industries will start to experience a revolution in terms of trust, creation of risk minimization tools, and improvement of business procedures (Lazebnyk & Voitenko, 2020).

2.9 Drone and Other Aerial Imagery

Fixed-wing aircraft, helicopters, unmanned aerial vehicles (UAVs or "drones"), balloons, blimps, kites, parachutes, stand-alone telescoping poles, and vehicle-mounted poles are some of the platforms used for aerial photography. In order to enhance the aerial view of the regions utilized for agriculture, mounted cameras can be activated remotely or automatically. Hand-held images can be taken by the photographer, refer Figure 11: Drones in Farming Operations (Article – Farm Drone Adoption, 2023).

Figure 11. Drones in farming operations

(Source: Article – Farm Drone Adoption, 2023)

To classify an unmanned aerial vehicle, several criteria may be applied. Some of them are analysed for UAV flying, and if utilized for aerial spraying, factors including size, MTOW (maximum take-off weight), and range will be considered. The categorization is based on wing type; the level of autonomy has been determined by considering both wing type and autonomy level, which may be the most important factors in agricultural missions.

The following categories might be used to group the most often reported UAV missions: • Waypoint trajectory. capturing images and videos of the sky while in flight. • The mapping projects. A section of the 2D map designated for mosaic techniques.• An oblique mission is to represent a region in 3D from many perspectives. • Corridor mission: The flight path follows a target, such as a river, an aerial power line, or a railroad line. • An agricultural mission must span vast dimensional expanses.

Using a fleet of vehicles can be a way to shorten the overall time needed to accomplish the task and, consequently, produce more uniform photos owing to potential changes in light matter in agriculture, which requires monitoring over wide expanses. Currently, research centres are working to provide UAV fleets a high degree of autonomy through the deployment of sophisticated ground control systems that employ automatic monitoring procedures to decrease the number of operators needed to oversee the fleet.

Five firmly positioned cameras produce one vertical and four low-oblique images in picometres that may be combined.

Certain digital cameras used for aerial photogrammetry combine photos from many imaging elements—sometimes taken with different lenses—geometrically correcting them to create a single image.

Emerging spatial multimedia such as aerial films can be applied to object tracking and scene interpretation. The input video is captured by low-flying aerial platforms, which are usually composed of non-ground-plane constructions with significant parallax. Global positioning systems (GPS), automated image processing, and digital video integration will enhance data collecting while lowering accuracy and cost-effectiveness. A number of different airborne platforms are being looked into for the purpose of gathering data.

2.10 Application of Diverse Technologies in Farming Operations

Gaps in implementing Transforming Indian Agriculture Scenarios are significant. There is a significant effect on the yields and quality of manufacturing. By contrast, Israel's agriculture sector ranks among the world's most productive and developed based on nearly all metrics. This outcome has been made possible by the agro-industry's use of cutting-edge, progressive, inventive technologies as well as clever agricultural solutions. Less than 20% of Israeli land is suited for agriculture. The farmers meet the population's needs in food products by 95%. This is a form of agrarian phenomenon, which is worth the attention of farmers from other countries. The government of Israel strongly supports the country's agricultural sector, subsidising farmers by up to 40% of the purchase price and implementing new technologies. The smart farming approach has the following components: software, irrigation systems, and innovative harvesting equipment, which are cheaper due to the subsidy system. Agriculture is a unique business crop production dependent on many climate and economic factors. Some factors on which agriculture depends are soil, climate, cultivation, irrigation, fertilisers, temperature, rainfall, harvesting, pesticide weeds (Gharde et al., 2018) and other factors.

Typically, 70% of the water on Earth is used for water systems, primarily due to the fact that certain areas only employ outdated flood water systems. Israeli ag-tech companies like CropX, Saturas, Manna, and SupPlant assist customers all over the world in implementing efficient trickle water system projects that use less water and yield higher and more abundant crops. The trickling water system's development is never-ending. In order to establish best practices for using the drip water system as an accurate root-conveyance framework, Netafim announced in February 2019 that it would be collaborating for three years with the tech-move organization of Ben-Gurion University, Bayer, and regards to soil investigation, advanced forecast instruments, and Netafim innovation that is at the forefront of the industry.

There are several solutions for the technology adaptation. These are listed as a framework for the type of crop, the climatic conditions, and the process it will address.

Several smaller applications using these technologies have a significant impact on the accuracy and quality of the results. a method to cut back on or completely stop using pesticides, enabling farmers to sell their produce to developed nations while international regulations prevent the trading of chemically treated grain. Techniques for extracting wool and sorting it into several categories, as well as methods for working with leather that include liming, fleshing, splitting, deleting, and bating are all closely supervised. Harvesting, crushing and pressing, fermentation, clarifying, and finally aging and bottling are the five fundamental phases or steps in the winemaking process.

Birds are raised commercially or domestically in poultry and animal husbandry for their meat, eggs, and feathers. The main commercially important poultry are chickens, ducks, turkeys, and geese; guinea fowl and squabs are primarily of local relevance. Selection, inbreeding, and crossover are the three. The instruments have been employed to enhance the current breeds; they will be utilized to produce new types and breeds, improve current breeds even more, and increase the output of commercial animals for the four main industries of dairy, poultry, fish, and bee farming. With Bio Bee, technology contributes to the proliferation of different spider, bee, and fly breeds that prey on harmful insects.

These are a few steps in farming with the relevant technologies that can be used in the solutions shown in the Table 2: Precision Technologies Used in Farming

Table 1. Precision technologies used in farming

AREA OF DEVELOPMENT	PRECISION TECHNOLOGIES
Location Choice and area mapping	Using drones and GPS
Soil Preparation	sensor, Humidity Sensor, Volatile matter sensor, etc.
Addition of chemicals to the soil	Using sensors like temperature
	Soil Data and analysis
The Field preparation for future farming. (Ploughing and planking)	Robotic Arms for tractors and bullocks
	Agricultural robot for automatic
	Robotic arm for ploughing
The main tasks of Seeding and planting	Robotic arms depicting the hand tools

continued on following page

Table 1. Continued

AREA OF DEVELOPMENT	PRECISION TECHNOLOGIES
	Precision drills, Broadcast seeding,
	Seed drills, Air seeding
Watering the seeds and plants	Drip Irrigation systems with sensors/ IoT Devices/ Controllers
Using the Fertilisers and Pesticides	Robotic arm as a hand spray
	UAVs and UGVs, Sprayers, GPS tracking.
	Smartphone Applications and Remote sensing
Weed removal	Small Bots which are tracking and do it similar to hand tools
	Technologies Weeding Robots
Harvesting	Small Bots which are tracking and do it similar to Manual picking
	Robotic pick and place arm,
	Mechanical harvesting, limb shaker,
	canopy shaker, Abscission Chemical
Transit movement of Agri Goods	Digital Twin, Analytics and Blockchain

2.11 Adoption of These Technologies by India

The important foundation is the Government of India's Mission for Integrated Development of Horticulture (MIDH). The devised and possible solutions for a farmer in cases where he can probably move forward with his produce and enter into the market or where a farmer estimates the shelf life of the produce for about three months. This benefits a farmer because he can coordinate and collaborate with the private sector and state to bring a novel or feasible solution.

There are 28 centres in 12 states in India, through which an organised value chain has been designed. In this value chain, the product and an entire process that begins from the cultivation moves forward with irrigation, ends with post-harvest management, and is efficiently taken care of. The needs of both the farmers and markets are considered, and then solutions are designed and developed, keeping a concern for both sides.

The supply chain in India strengthened for goods from farmers' end to direct markets, which makes the supply chain robust for India to another country, say the UK. Hence, it must go through an organised packaging, transportation, and invoicing process. The global competition, if analysed by its exquisite style of packaging that can extend the life period, called the shelf life for three months for fruit as perishable produce, can maintain shelf life. This proves to be a turning point where the game changes, such that a farmer enjoys a window of three months where he can judiciously decide to sell when the prices of the product in the market touch the heights. In this manner, not only is wastage reduced, but it also increases the farmer's income.

Thus, when everyone works in unison, results are more fruitful. Thus, farmers should proactively work with the government, researchers and the private sector to increase effectiveness and efficiency.

Precision agriculture under the Agricultural Project of Indo-Israel in India has already been implemented. Weekly satellite images of the field are captured and analysed. This proves helpful for the project manager of a particular centre in analysing and determining where and to which extent irrigation has to be done.

Challenges in Adoption of the Technologies in India for Agriculture

Poor Adoption of Mechanization: Despite the rise in mechanization most agricultural tasks in India are still carried out by laborers. In India the highest percentages of mechanization—between 60 and 70 percent—are found in threshing irrigation harvesting and ploughing. Despite being developed for other agricultural tasks like weeding and sowing machinery is not widely used by farmers for crop production. Because they own a limited amount of land farmers find it difficult to use automation. The lack of knowledge and financial constraints faced by rural farmers are the root causes of this problem.

Credit Availability: Access to timely adequate and reasonably priced financing from financial institutions is essential for the agriculture sector especially for small and marginal farmers. In addition to all other inputs capital is one of the most crucial components for increasing agricultural output. One factor influencing agricultural productivity is the availability of credit. Farmers with capital limitations would use fewer inputs than farmers without such restrictions because they wouldn't spend money on better tools machinery and technology for their farming operations.

Poor irrigation facilities, Soil fertility depletion, Impact of climate change are few more factors which hamper the use of the available technology. Inadequate insurance plan coverage inadequate understanding of insurance policies estimating the amount of losses from agricultural losses and non-payment or postponed claim settlement are few more reasons of farmers coming under the brunt of financial debts and not able to invest in the newer technologies.

The solution would be for the private players to come ahead for newer innovative "Jugad" – "financial models" shaking hands with the government. Somme innovative – out of the box thinking will help in implementation of these technologies in the Indian scenario.

3 SOME SUCCESS STORIES

Rationale of the chapter: The chapter discusses the technological upgrades and its impact on agriculture. Even though the usage seems far-fetched at this point of time. As with all technologies, once it proves its viability, the usage is bound to grow- like with cell phones. Ideally the target group is the agriculture and technology fraternity, because it involves people from both areas

3.1 Case 1: John Deere's Precision Agriculture Technology and Climate Field View by Bayer

3.1.1 Overview

John Deer and Bayer are leveraging various advanced technologies for the agricultural sector, and the outcome is excellent. John Deer's precision agriculture technology offers various attractive and comprehensive solutions that enhance farming efficiency, productivity, and sustainability. These platforms utilize data analytics, real-time monitoring and automation (Goodman D., 2023). The advanced technologies in precision agriculture provide farmers with actionable insight and tools to optimize their operations.

3.1.2 Functionality

John Deere's Precision Agriculture Technology

JDLink is a telematics system that connects machinery to the internet and provides real-time data, location, and diagnostics. This system truly enables remote monitoring and management. Synchronisation should be used to facilitate communication between machines in the field. The synchronisation among the machines, such as combines and grain carts, allows synchronised unloading and improves harvest efficiency. To perform all these activities, a centralised operations centre is required, capable of data collection, visualising and analysing data from equipment and operations, and supporting informed decision-making.

Bayer performs the following functionalities.

- **Field Health Monitoring**: Crop health monitoring, identification of issues like pest infestations, and monitoring of nutrient deficiencies can be done through satellite imagery.
- **Weather Monitoring**: Weather monitoring is executed through analysis of hyper-local weather data, aiding decisions related to planting, irrigation and harvesting.

- **Yield Analysis**: Data analytics can analyse yield data to identify trends and assess the effectiveness of agronomic practices that guide adjustments for improved yields.
- **Variable Rate Technology (VRT)**: Provisions variable rate applications of seeds, fertilisers, and crop protection products, improving input use and minimising environmental impact.
- **Field View Drive**: A device plugs into the tractor's diagnostic port, collects data from various equipment, and sends it to the Field View platform for seamless data integration.

3.1.3 Impact

1. **Increased Efficiency**: One of the important impacts of automation and data-driven capabilities of these technologies is to reduce manual labour and to increase the precision of farming operations. Farming Utilises these technologies, leading to more efficient resource use.
2. **Cost Savings**: By using advanced technologies, optimising inputs, and reducing waste, farmers can lower their operational costs and enhance profitability.
3. **Improved Yields**: Activities like monitoring and managing field conditions lead to healthier crops and higher yields, contributing to better financial outcomes for farmers.
4. **Sustainability**: These technologies promote precision agriculture practices and reduce the environmental impact of farming by minimising the overuse of fertilisers and pesticides and conserving water, supporting long-term sustainability.

3.2 Case 2: eNAM (National Agriculture Market)

3.2.1 Overview

Implemented in April 2016 by the Government of India, eNAM is an electronic portal connecting the APMC mandis on Indian soils. The primary objective is to establish a single unified national cheapest mandis for effected agricultural commodities integration. eNAM is an online web based electronic trading portal for the entire nation of India. This means it connects the present APMC/mandis to form a unified national market for trading in agricultural commodities. The e-NAM (Kalamkar et al., 2019) is a digital national network connecting the various existing local markets such as APMC or Mandis for agricultural products at the national level. Incorporated and launched by the Prime Minister on the 14th of April, 2016, the e-NAM online portal is an initiative towards establishing the "One Nation One Market" wherein farmers can sell their produce. Managed by the Small Farmers

Agribusiness Consortium (SFAC) under the Ministry of Agriculture and Farmers' Welfare, e-NAM allows farmers to sell their yield online through nearby e-NAM markets, while traders can bid from anywhere. The portal offers several services, including commodity arrivals, quality checks using AI, online bidding, and direct payments to farmers. In Jammu and Kashmir, the Horticulture Planning & Marketing Department started e-NAM by integrating two markets in May 2020, and now 11 markets in the region are associated with e-NAM.

3.2.2 Functionality

eNAM provides an online trading portal for farmers, traders, and buyers. It authorises farmers to sell their produce nationwide without being restricted to their local mandis.

3.2.3 Impact

As of now, eNAM has integrated over 1,000 mandis across 18 states. The platform has facilitated better price discovery, improved auction process transparency, and reduced mediators' role. Farmers have described increased income and reduced transaction costs. Additionally, eNAM has permitted smoother interstate trade and helped distribute real-time market information.

3.3 Case 3: Agri Bazaar

3.3.1 Overview

StarAgri, an agri-services firm, introduced AgriBazaar, an online platform for agri-trading, in 2016. Offering end-to-end services, including warehousing, logistics, quality inspections, and payments, puts farmers and buyers directly in contact. The vision with which it was founded was to "Empower FARMERS by delivering efficiency, TRANSPARENCY and TRACEABILITY in the agri value chain through technology." The Values that they stand by are termed R-E-A-P where R stands for "Radical innovation Harness future-tech to innovate and disrupt the way buyers and growers engage in the agri--marketplace", E for " Execution -Act with speed, internal mobilisation and responsiveness to accelerate our time-to-market ", A for "Agility -Adapt decisively and efficiently to drive change in agriculture with next-gen technologies" and P for "Purpose- Empower stakeholders across the agri value chain to enhance small farm owners' incomes in a responsible, resourceful and reliable " From the agri-business experience, they understand small farm owners face challenges disproportionate to their size and resources. Technology providers mainly focus on the needs of large enterprises. At AgriBazaar, we aim to level the playing field. AgriBazaar started with the idea that a technology-driven platform could help the agri value chain operate with significantly fewer overheads and hidden inefficiencies. They digitised the agri value chain with proof of origin and authenticity, enabling traceability whilst bringing down agri-trade commissions from 2-3% to 0.5%. At AgriBazaar (Ahmad et al., 2008), the transformation of the agri value chain through technology for small farm owners by enabling digital access to a wider marketplace is happening. For buyers, quality assurance is managed while mitigating counterparty risk, reducing middle man and enabling secure payments. They started in 2016 by building an online platform that connected all parties with the farmers, such as traders, commission agents, banks, companies, and governments. In 2017, the award-winning wallet was created called the AgriPay, a crucial key that facilitates transactions of over $1 billion. Moreover, we now leverage AI-driven insights and analytics to offer customised credit-on-a-click, crop advisory and more. Furthermore, these are more efficient and get value chain; through direct linkages, we bring efficiency, transparency, and traceability like never before.

3.3.2 Functionality

Growers may display their goods for sale on the site, and customers may place direct orders. AgriBazaar guarantees quality and transparency in transactions by providing farmers with various services, including weather predictions, crop advice, and soil testing.

3.3.3 Impact

By guaranteeing farmers receive higher rates for their goods and minimising their reliance on neighborhood mandis, AgriBazaar has given farmers more authority. Through its logistics and warehousing capabilities, the platform has also assisted in reducing post-harvest losses. Farmers may make wise judgments using up-to-date information on crop pricing and market trends.

3.4 Case 4: DeHaat

3.4.1 Overview

DeHaat was founded in 2012 and is a B2C Agri-tech company operating in India with a vision to transform the agriculture industry (Sinha A. & Jha S., 2019) . It provides farmers with not only the complete solutions but also supports their farming process fully. This involves connecting farmers with Market linkages,

Technical support, and availability of Agriculture inputs including fertilizers, improved seeds and Crop protection products among others. Currently, DeHaat uses methods of technological progress to bring advanced farming practices, precise farming instruments, and excellent data analysis for its customers, which increase yield and production rates among farmers. The firm's digital platform facilitates a farmer-to-expert, and farmer-to-service provider interface that ensures that they receive advice, weather information, and a pest management, including pest-resistant seeds, that suits a particular farmer's particular preferences. Besides these services, DeHaat also assists farmers in obtaining the necessary financial services, like crop insurance and credit, thus providing farmers with the necessary funds to invest in their farms and protect against certain risks. The services provided by the company in the supply chain running from the point of procurement to distribution enable effective and efficient supply of agricultural inputs with lower costs. Regarding sustainability, it is obvious that DeHaat tries to demonstrate the improvement of the ecological agenda of farming and agriculture as a whole. With a vision of enabling farmer to practice sustainable agriculture and help them gain necessary knowledge and skills, DeHaat has been launched to help farmers and bring positive change in agriculture industry. Currently, DeHaat has been gradually growing its farmers base to more than 650000 across several states of India, thus making future achievements of its goals of changing agricultural practices. The firm has embraced versatility and has remained steadfast in its commitment to the welfare of farmers in the society to become a reference producer and manufacturer in the Agri-tech industry as it brings positive change that propels development of the rural settings.

3.4.2 Functionality

The crop management advice it offers to farmers entails recommendation of preferred crop and dosage based on the soil type, climate, among other factors using advanced technologies embraced by the DeHaat including AI and machine learning. While on the one hand, it connects the farmers to the input suppliers they need to buy the inputs; on the other hand, it connects the farmers to the markets or customers where they can sell their output.

3.4.3 Impact

Currently, DeHaat has over 6,50,000 farmers of India as active users of the network. By offering them better market access and specificity in agricultural consultation depending on their needs, these farmers can enhance their production and incomes. The untapped knowledge base of crop planning with the help of technology enables better usage of resources and production of higher yield and much better quality. Farmers benefit through being able to sell their produce and being assured of supply of the needed inputs.

COMPARATIVE NARRATIVES OF MAGRIC (UGANDA) VS. KISAN SUVIDHA (INDIA)

magRic

Overview

mAgric is an innovative solution, developed for small scale farmers in Uganda: a one-stop-shop that aims at improving small farmers' lives through giving them necessary information to improve their yields. These basics include weather updates for the upcoming days, post harvests market prices on products and crops, and information on the actual ways of farming. These services are available via text messages, which means that people even in the rural areas can access them because of text messaging does not require advanced internet connection. mAGrAg (Maime, 2015) also creates awareness on preferred rainfall time for planting, preferred time for irrigation, and correct time for harvesting thus minimizing effects of volatile weather conditions. Hence, through the provision of real time market prices, farmers are in a position to together set up their sales and supply patterns thereby pulling the best prices in the market and earning the maximum returns. Besides the weather and market information that directly affects the business, mAGrIC does offer advices on how to approach the new era of farming. This consists of recommendations on proper crop succession, ways of controlling pests infestations, and proper management of the soil as well as use of correct Organic farming techniques. The information provided is highly specific and trackable to the needs and circumstances of the farm-based communities that permit the platform, thus making it useful. In addition, the communities that mAGric creates for farmers encourage the expression of similar experiences, achievements and shortcomings. This peer group support provides for improved flow of knowledge and improved group problem solving which benefits the agricultural discipline as a whole and thus its dependability. To support small farmers, mAGrich`s program offers information on Microfinance title and insurance schemes available for them. This makes him or her certain of sources of funds that enable him or her to undertake improvements on her or his farm and make sure that livelihood is protected against any risk that may occur. By availing the son of comprehensive services, mAgric is already in a strategic position of revamping the farming industry in Uganda and therefore enhancing the livelihood of the many small-scale farmers.

Impact

While farmers often live in some of the more desolate regions of the country with less than perfect internet connectivity, thanks to the magic platform, farmers are now in a position where they could possibly receive relevant information on their fingertips, in their mailboxes to be precise at any given point in time. Such farmers have been able to boost crop production and income by becoming more informed in decision making regarding their farming activities due to the timely data that they have been receiving.

Kisan Suvidha

Overview

A similar application which delivers services different in interface but similar in functionality is Kisan Suvidha an Indian governmental project which was initiated for the sole purpose of empowering farmer community by providing them with necessary agriculture information and aid. With the features such as input costs, market prices and weather forecast among others accompanied with advisory services, the app is big toolbox for a farmer in taking his/her decisions. The information available in Kisan Suvidha's (Bisheko & Rejikumar, 2023) input cost database proves very useful for farmers to set up their operations in the most logical way possible, spending their money on seeds, fertilizer, pesticides, and other equipment. Through http input, the app will provide farmers with current market price hence enhance their ability to determine when to sell their produce and where they need to sell the product to make maximum profits. Information on the current weather is also important in Kisan Suvidha. This way farmers are able to acquire up to date information on the climate that acts as a guideline to planting, irrigation, and even the time to harvest the crops, to avoid the negative impacts of climate change. Customers will benefit greatly from this information because it helps to minimize crop losses and enhance the quality of yield. Besides the mentioned data services, Kisan Suvidha has expert advice from professionals on different issues related to farming. This covers matters relating to crop choice and cultivation, control of pests and diseases, concept of soil health, and other good agricultural practices. The app provides a link between farmers and others involved in the farming sector, enabling them to get assistance from qualified experts in the field, as well as the support of extension services that can help to solve most of the problems that the farmers encounter in the process of their activities due to their individual requirements and environmental conditions. This app also covers such functions as market linkage that facilitates the link farmers with buyers and suppliers to expand their market and business. In addition, by providing information regarding governmental plans such as subsidies and credit facilities for farmers, Kisan Suvidha also acts as a tool for financial literacy. These multiple services, if linkage together in a format that is easily accessible to the farmer through Kisan Suvidha are contributing to the process of aborigine agriculture in India. It increases the efficiency and productivity of the farmers by providing knowledge and related technologies required to improve productivity, income and sustainable farming to improve agriculture or the overall growth of agriculture in India.

Impact

Through availability of key information, the farmers are able to make improved decisions due to the assistance of the app in improving agricultural production. The current use of cell phone in India has made easy the access to Kisan Suvidha due to its availability to millions of farmers and enable them have access to materials which make their farming to be better and that can also help them to earn more.

Comparative Analysis

Both mAgric and Kisan Suvidha have made, since these applications provided life-critical information, the decision-making process has been reinforced and crop yields have risen. As for the delivery of information, there is a distinguish that separates two platforms the most. While Kisan Suvidha employs smartphone very much of India for its services, mAgric still depend on the simple means of SMS which is very efficient in areas of low internet connection.

COMPARATIVE NARRATIVE: TWIGA FOODS (KENYA) VS. WAYCOOL (INDIA)

Twiga Foods

Overview

An example is Twiga Foods (Kumar & Perepu., 2023) which is an Agri-tech start-up from Kenya to addresses the middle-bull(ion) connecting farmers and merchants through a digital platform thus breaking the long chain. It was established in 2014 and specializes in fashion, retail, food and beverage services, and technology industries. Twiga Foods is therefore able to improve the effectiveness of the agricultural market through the use of technology resulting into less of food wastage while on the other hand improving the standards of living of farmers because of better and proper market that is affiliated to them. Through the platform, merchants and suppliers are able to timely supply fresh produce and other products to the client thus enhancing the consistency of their supply. Twiga Foods has been able to incorporate mobile payments and data analysis into its operations in order to always embrace innovation and thus sustainable development in the agriculture sector. Through its link of producers and retailers, Twiga Foods has been instrumental in altering the face of Kenya's agriculture sector and offer its worthy input to the food security and economic growth cause.

Functionality

In the delivery process, Twiga Foods employs a logistics system to help in the collection of fruits from farmers and deliver to the sellers in urban areas. It will ensure fair prices to farmers and reliable supplies to vendors since it is organized on a platform.

Impact

Twiga has benefited the overall reduction of wastage and an enhanced supply chain for Twiga Foods. As urban retailers acquire fresh food articles at a favourable price, farmers are in a position to sell their products at a favourable higher price and consistently. Moreover, it has created jobs in production distribution and logistics job openings as well as in education.

WayCool

Overview

Navigate to WayCool An Indian start-up company in the agricultural technology sector it was established in the year 2015 and is involved in the automation of agricultural supply chain right from the farm to the consumers. Using integrated solutions in digital technologies, WayCool is focusing on creating a fully integrated digital food logistics system which will ensure the effective delivery of food products and minimize food waste. A set of the tools which the company uses include IoT devices, artificial intelligence, and analytical tools to manage the production cycle from farming to delivery. Another strategy with farmers is to foster mutually beneficial ways in which WayCool (Ahmad & Jamaludin, 2015) can help provide the necessary resources for better farming techniques that would generate improved yields and healthier produce. Through real-time data and statistics and using the principles of predictive analysis, the firm assists the farming community in making the right decisions and increasing his profit margin. Fair trade and farmers are prioritized also, which means that the farmers should be paid reasonable prices for their products. On the distribution side, WayCool has a particularly developed logistics system that helps it deliver its products to the retailers, and thus the consumers. This interlinked approach not only gives an added advantage of quality and freshness of the products but also helps in cutting down the overall costing and environmental expenses of the chain. Since that time, the company has evolved and increased its product line beyond fresh produce to cover the general needful of the consumers. Due to its visionaries of innovation coupled with sustainability, the firm has become a vital stakeholder in the Indian agri-tech market that has improved food security and progression among rural dwellers.

Functionality

The company also brought in technology in managing nearly all the supply chain processes from procurement, handling, transportation to sales. It offers protection services or Quality assurance, stock maintenance, and market access services.

Impact

WayCool has managed to lessen post-harvest loss and ensure timely delivery of fresh fruits by improving the chain supply in India. Farmers benefited through better prices and market while the consumers enjoyed the enhanced quality produce. Ackerman also supplies waste minimisation and reduced-impact agriculture practices.

Comparative Analysis

Due to the initiatives introduced by both WayCool and Twiga Foods, the amount of food waste has been minimized whilst the effectiveness of the supply chain has been greatly enhanced. Market issues such as market fragmentation in Kenya is well solved by Twiga's strategy while WayCool's comprehensive across the value chain services aims at providing solutions for the diverse and challenging Indian agricultural sector. Technology is used by both companies to fund and create the significant changes in their respective regions.

Twiga Foods targets to bring in efficiency in the agricultural value chain in Kenya by solving for inefficiencies of numerous smallholder farmers as well as multiple middlemen. Through connecting the farmers and the merchants, the platform limits the number of the middlemen hence reducing the costs of transactions and consequently, the losses occasioned by adverse preservation conditions. This increases accuracy and transparency with the help of mobile payments and real-time data tracking, thus making the platform more predictable in terms of supply chain.

On the other hand, WayCool's unique solutions for supply chain automation engulf all the segments starting with farm and ending with customers in a diverse setting of Indian agriculture. Incidentally, WayCool uses IoT devices, Artificial Intelligence, and big data to enhance productivity, supply chain, and delivery. This strategy not only optimizes the supply chain but is also environmentally friendly by empowering farmers to practice sustainable agriculture. Boasting of fair-trade practices and affordable prices, the farmers are also potential customers benefiting from WayCool's innovations and hence helping in rural development and averting economic fluctuations.

Both firms remain relevant in matters to with food security. Twiga Foods helps improve market information for farmers in Kenya while at the same time providing fresh produce to cities. By supplying a wide product portfolio of Halal, Vegetarian, FMCVG and Kosher food items and with successful supply chain management WayCool fulfils the dietary requirements of a huge population of Indian users and minimizes wastage of food items and delays in delivery.

4 CONCLUSION

A complicated network spans several stages, from production to marketing and sales. This forms the agricultural supply chain. For agricultural goods to be efficient, high-quality, and sustainable, each step is essential.

This supply chain might be completely changed by the introduction of digital technologies. The digital technologies will would increase efficiency, cut waste, and improve traceability. The difficulties in implementing these technologies, however, fluctuate depending on the situation. This is specifically evident in developing countries like India. Such issues are discussed in the supply chain application phase (Priyadarshini & Abhilash, 2020).

The summary-discussion gives a thorough rundown of the main roles played by the agricultural supply chain. It also gives the analysis of the difficulties for the unique Indian setting. The digital technologies improve the roles, their usefulness, and the challenges in deploying these technologies (Dayıoğlu & Turker, 2021). The potential effects are supported by comparable real-world examples.

The prospects emerge within the discussion of the specific function that can be taken forward with Digital Technologies, their Utility and Challenges in Adoption, and cases used for applied research areas.

REFERENCES

Agribots – How Robots Made Their Way onto Farms – Pigeonsfarm.co.uk. (2019, February 3).https://pigeonsfarm.co.uk/agribots-how-robots-made-their-way-onto-farms/

Ahmad, F., Zakaria, N. H., & Osman, W. R. S. (2008, April). Transforming Information-Based Agricultural Portal to Knowledge-Based Agricultural Hub. In 2008 3rd International Conference on Information and Communication Technologies: From Theory to Applications (pp. 1-4). IEEE. DOI: 10.1109/ICTTA.2008.4529914

Bisheko, M. J., & Rejikumar, G. (2023, March). A study on farmers' perceptions about the scope of the Kisan Suvidha App in improving agricultural sustainability. In *2023 Conference on Information Communications Technology and Society (ICTAS)* (pp. 1-5). IEEE. DOI: 10.1109/ICTAS56421.2023.10082741

Dayıoğlu, M. A., & Turker, U. (2021). Digital transformation for sustainable future-agriculture 4.0: A review. *Journal of Agricultural Sciences*, 27(4), 373–399.

Gharde, Y., Singh, P. K., Dubey, R. P., & Gupta, P. K. (2018). Assessment of yield and economic losses in agriculture due to weeds in India. *Crop Protection (Guildford, Surrey)*, 107, 12–18. DOI: 10.1016/j.cropro.2018.01.007

Greentech Americas boosts horticultural productivity. (2023). Www.greentech.nl. Retrieved June 20, 2024, from https://www.greentech.nl/press-releases/americas/greentech-americas-boosts-horticultural-productivity

Hash Technologies Article. (2021), The Agriculture Industry Is Being Transformed By The Internet Of Things, accessed on 23 March 2024, https://hashstudioz.com/blog/how-internet-of-things-is-transforming-the-agriculture-sector/https://primathon.in/blog/ai-agriculture-revolutionizing-the-future-of-sustainable-farming/

Indian agriculture Archives | AgriBazaar. (2019). AgriBazaar. https://blog.agribazaar.com/tag/indian-agriculture/

Informatics, A. (2020), Understanding Agro Informatics and the Way it is Transforming Agriculture, Posted On September 18 2020 / Agro Informatics, accessed 23rd March 2024, https://www.kisangates.com/agro-informatics.html

Izmailov, A. Y. (2019). Intelligent technologies and robotic means in agricultural production. *Herald of the Russian Academy of Sciences*, 89(2), 209–210. DOI: 10.1134/S1019331619020072

Kalamkar, S. S., Ahir, K., & Bhaiya, S. R. (2019). Status of implementation of electronic National Agriculture Market (eNAM) in selected APMCs of Gujarat. PROGRESS, 112.

Kumar, G., Engle, C., & Tucker, C. (2018). Factors driving aquaculture technology adoption. *Journal of the World Aquaculture Society*, 49(3), 447–476. DOI: 10.1111/jwas.12514

Kumar, K. B. S., & Perepu, I. (2023). Twiga-A B2B Agritech Startup's Innovative Solution to Improve Food Security in Kenya. IUP Journal of Supply Chain Management, 20(4).

Lazebnyk, L., & Voitenko, B. (2021). Digital technologies in agricultural enterprise management. *Financial and Credit Activity Problems of Theory and Practice*, 6(41), 203–210. DOI: 10.18371/fcaptp.v6i41.251440

Maime, M. J. (2015). *Evaluation of the fertility of communal bulls in Moretele district Northwest Province in South Africa*. University of Pretoria.

Mohanraj, I., Ashokumar, K., & Naren, J. (2016). Field monitoring and automation using IOT in agriculture domain. *Procedia Computer Science*, 93, 931–939. DOI: 10.1016/j.procs.2016.07.275

National Agriculture in Class Room NACR. (2024), Relevancy and Engagement, Text Book Tracing the Agriculture Supply Chain. Accessed May 2-24, https://agclassroom.org/matrix/lesson/print/838/

OEDC Report - Agricultural Policies in India, 2018. Accessed March 2024 https://www.oecd.org/regional/agricultural-policies-in-india-9789264302334-en.htm

Patidar, S. (2024, March 28). AI Agriculture: Revolutionizing the Future of Sustainable Farming.

Phillips, P. W. B., Relf-Eckstein, J.-A., Jobe, G., & Wixted, B. (2019). Configuring the new digital landscape in Western Canadian agriculture. *NJAS Wageningen Journal of Life Sciences*, 90(1), 100295. DOI: 10.1016/j.njas.2019.04.001

Priyadarshini, P., & Abhilash, P. C. (2020). Policy recommendations for enabling transition towards sustainable agriculture in India. *Land Use Policy*, 96, 104718. DOI: 10.1016/j.landusepol.2020.104718

Saurabh, S., & Dey, K. (2021). Blockchain technology adoption, architecture, and sustainable agri-food supply chains. Journal of Cleaner Production, p. 284, 124731.

Sinha, A., Jha, S., & Amritesh, A. (2019). Green Agrevolution Pvt Ltd: Delivering 360°"seed-to-market" solution. *Emerald Emerging Markets Case Studies*, 9(1), 1–23. DOI: 10.1108/EEMCS-06-2018-0148

Smidt, H. J., & Jokonya, O. (2022). Factors affecting digital technology adoption by small-scale farmers in agriculture value chains (AVCs) in South Africa. *Information Technology for Development*, 28(3), 558–584. DOI: 10.1080/02681102.2021.1975256

Solutions, A. (2024). Application technology and digital agriculture, article on" The best agribusiness solutions for your property!" Accessed on 23 March Agrocision.com.br. https://www.agrocision.com.br/

Somashekhar, I. C., Raju, J. K., & Patil, H. (2014). Agriculture supply chain management: A scenario in India. *Research Journal of Social Science and Management*, 4(07), 89–99.

Swain, P. K., Jamir, C., Dinerstien, M., & Narula, S. (2022). Impact of development of National Agriculture Market (e-NAM) on farmers. *Indian Journal of Agricultural Sciences*, 92(12), 1484–1489. DOI: 10.56093/ijas.v92i12.127701

Technology, F. M. (2021, November 24). What is Global Food Supply Chain - FMT Magazine. Food, Marketing Technology. https://fmtmagazine.in/what-is-global-food-supply-chain/

) Tzachor, A., Richards, C. E., & Jeen, S. (2022). Transforming agrifood production systems and supply chains with digital twins. npj Science of Food, 6(1), 47.

Yadav, V. S., Singh, A. R., Raut, R. D., & Govindarajan, U. H. (2020). Blockchain technology adoption barriers in the Indian agricultural supply chain: An integrated approach. *Resources, Conservation and Recycling*, 161, 104877. DOI: 10.1016/j.resconrec.2020.104877

Chapter 16
Supply Chain Through Machine Learning and Artificial Intelligence

Raj Krishna
NLIU Bhopal, India

Kumar Mukul Choudhary
School of Law, Galgotias University, India

ABSTRACT

The authors in this chapter have made a detailed study upon the usage of Artificial Intelligence in Supply Chain Mechanism System. Firstly, the authors have gone through the existing literature and practices to know about the current situation. Furthermore, the review of existing literature also helped the authors in identifying the loopholes in the present system. Apart from that the authors have made a comparative study of United States, Africa and India to know how various countries have used this technology to revolutionize agriculture in their respective countries. Lastly the authors have provided solution to the existing problem.

INTRODUCTION

Indian economy has witnessed tremendous growth in recent times, with tertiary sector surpassing primary sector as the biggest contributor. Interestingly half of the Indian work-force is still employed in the agricultural sector. However, it is pertinent to note that the Agricultural Sector in India contributes just 17% of the Gross Domestic Product. This in turn shows us the stark reality of the Agricultural Sector

DOI: 10.4018/979-8-3693-4330-2.ch016

in India. Hence, it is the need of the hour that we take steps to mitigate income inequality and provide for equitable distribution of resources (Ashwani, 2017)

However, when we delve into the reasons to understand that why Indian Agricultural Sector is lagging behind then we discover that the biggest deficiency is the lack of synchronization in Agricultural Supply Chain System. Furthermore, a number of research in this field bring to light the fact that we are way behind the Western Counterparts in regard to the application of supply chain mechanism. As a result, it is necessary that we improve the same. However, in order to do the same, we first need to understand that what does Supply Chain Management System actually means? (Routroy and Behera, 2017).

Supply Chain Management System consists of activities undertaken by both individuals and institutions who are associated with the activity of agricultural production and distribution. It also includes activities related to the sale and marketing of agricultural products. It's a long process which starts with the procurement of raw agricultural products and ends with the delivery of final and finished product to the consumer. It all starts with the acquiring of raw materials and ends with delivering goods to final consumer. In this 21st century globalized world, SCM plays a very crucial role in enhancing efficiency, minimizing production and distribution costs, and expanding market access. Unfortunately, Indian Agriculture Sector faces innumerable challenges, especially in the preliminary stages of SCM development, including digitalization and the utilization of Artificial Intelligence (Routroy and Behera, 2017).

Research Design/ Methodology

The methodology for the present study is descriptive and analytical. The researchers have critically analyzed the application of artificial intelligence in Agricultural SCM and the possibility of enhancement in the Indian Agricultural sector with the advancement in SCM.

Research Questions

1. What are the present challenges faced by the supply chain mechanisms in Indian Agricultural Sector and how can Artificial Intelligence (AI) be integrated into various stages of the agricultural supply chain in India?
2. What are the potential benefits of incorporating AI in supply chain management for Indian agriculture?
3. What are the regulatory and policy challenges associated with the adoption of AI in Indian agricultural supply chains?

4. What will be the socioeconomic impacts of integrating AI into agricultural supply chains in India?

Originality

The researchers have gone through various articles and report and has found out that no significant research has been done on the application of A.I. in Agricultural SCM in India. As a result, the researchers have undertaken a comparative study upon the legal and policy implications of such application of A.I. and the benefit it may generate for the Indian Agricultural Sector in coming times.

REVIEW OF LITERATURE

In order to identify and fully comprehend the topic in hand, existing literature has been consulted before we started writing this chapter. An attempt had been made to provide recommendations by pointing out the gaps and discontinuities found in the existing sources. Based on this underlying reading, the authors have attempted to build a basis for the current research after identifying a few concerns and gaps linked to the topic.

In an article titled, ***"Artificial Intelligence in Extended Agri-Food Supply Chain: A Short Review Based on Bibliometric Analysis"*** the authors have highlighted that a digital shift in agriculture is being driven by population explosion and climate change. Due to this, agri-food supply chains are become increasingly sophisticated, generating enormous volumes of data, and expanding the parameters of conventional food lifecycle. However, it is pertinent to note that the application of artificial intelligence (AI) to the extended agri-food supply chain is relatively a new concept. A brief assessment of the literature on the intelligent agri-food supply chain covering eighteen studies is presented in this article. Key study areas and contemporary developments in the AI-enabled phases of food production, distribution, and sustainable consumption are identified by the bibliometric analysis of this piece. Intelligent planning for demand uncertainty and customized end-user needs, storage optimization, waste reduction in the post-production phase (Distribution and Recycling), and boundary-spanning analytics are additional crucial advancements of AI that need to be added to the production process. In terms of theory, this study outlines prospects for further research in the extended agri-food supply chain and emphasizes areas where AI use in agri-food is mature. In terms of application, the review's conclusions can motivate incumbents and startups interested in expanding agri-food ecosystems to pursue pilot projects for the intelligent and sustainable agri-food digital transition. AI methods can help reduce the gap in agri-food supply chains that are sustainable.

In a piece titled, ***"Agri-Tech Revolution: Next-Generation Supply Chain in America's Agriculture,"*** the authors have argued that the 21st century world has presented before us a new phase of technical development which has caused a significant impact upon innumerable sectors, including agriculture. In the background of the AgriTech wave which has revolutionized American agricultural sector, this essay highlights the important role that these kinds of upcoming technologies play in restructuring the agricultural supply chain. The authors have further discussed about the methodology employed and the amalgamation of innovative technology in the agricultural sector. Post this the author has highlighted the advantages of using these technologies in the agricultural sector.

Apart from that the author has also raised concerns pertaining to the limitations of the same. The author has pointed out issues like skill gap, financial difficulties, digital divide etc.... Hence, it is necessary that we work upon the same for improving the effectiveness. Furthermore, the author has highlighted the promising start of AgriTech in the American Agricultural Sector and if things go in the prescribed manner, then it can be said that AgriTech would revolutionize the way agriculture is done in the United States of America. Lastly in concluding remarks the authors have given important insights on the dynamic interactions between technology and agriculture in the modern era for a wide range of stakeholders, including farmers, agribusinesses, technology providers, policymakers, and consumers.

In an article titled, ***"Revolutionizing Vegetable Value Chains: A Comprehensive Review of Digital Technologies and their Impact on Agricultural Transformation,"*** the authors have contended that with the application of digital tools, the agricultural industry has undergone a significant change, particularly in the arena of supply chain mechanism. This piece addresses the significant effects of the emerging technologies in the agricultural sector with a special emphasis on how these technologies are being applied along the supply chain. The study also comprises of an in-depth examination of a range of technological innovations. Furthermore, a thorough analysis of Ninjacart, a well-known platform for agricultural technology, is included in the paper as a case study that shows how these technologies are applied in real-world situations. It is pertinent to note that these advances have completely changed inventory management. IoT devices are used for tracking crop health and maximizing resource use, and blockchain technology guarantees supply chain transparency and dependability. Additionally, due to AI-driven solutions the food waste has greatly decreased via logistics routes that were effectively planned and supply chain intricacies that were handled. Lastly this paper also offers a thorough examination of how digital technologies have impacted the agriculture value chain for good.

In an essay titled, ***"A Review of Application of Machine Learning in Agriculture Supply Chain to improve Food and Agriculture Industry,"*** the authors have contended that Artificial Intelligence (AI) is a creative tool that lets machines simulate human intelligence and capabilities. This technology thus operates by mimicking the tasks and functions carried out by people. Interestingly Artificial Intelligence has now been used extensively in the agricultural arena. By 2050, it is predicted that there will be 9 billion people on this planet Earth and due to this the agricultural sector would face innumerable challenges like the need for increased food production. A number of factors, including resource constraint, the COVID-19 pandemic, climate change, and socioeconomic supposition, has made it even more challenging to project the amount of food and agricultural production that would be required by the population in large in 2050. In such a scenario Artificial Intelligence would play a very significant role for enhancing production and improving supply-chain mechanism. The food and agriculture industries could thus become more profitable and efficient with the use of AI and machine learning. The authors in this paper have suggested the measures which if implemented would help us in drawing maximum benefits of the technological advancements.

In an article titled, ***"Harnessing Artificial Intelligence for Sustainable Agricultural Development in Africa: Opportunities, Challenges, and Impact,"*** the author has shown how artificial intelligence (AI) could revolutionize sustainable agricultural growth in a least developed African territory. The paper highlights the impact, difficulties, and opportunities as it moves through the ever-changing field of AI applications in agriculture. Prospects like climate-resilient techniques, crop monitoring, and precision farming have also been explored, along with issues with data accessibility, technical infrastructure, and skill gaps. The author in length has discussed about the impact generated by Artificial Intelligence upon supply chain, inclusive growth and in the lives of farmers with small landholdings. Furthermore, the author has also pointed out the ethical issues pertaining to the incorporation of AI and the subsequent policy outcome. Hence, it can be said that the article gives us a comprehensive idea about the usage of AI in Agricultural Sector in Africa.

Furthermore, in a piece titled, "AI in supply chain optimization: A comparative review of USA and African Trends" the authors have made a comparative study of how AI is used and employed in the Agricultural Sectors in USA and Africa. The authors have started their essay with an argument that Artificial Intelligence in the field of Agriculture has turned out as a game-changer. It is because it affects both productivity and competitiveness in the International Market. The authors have also pointed out the limitations and difficulties around the deployment of artificial intelligence in agricultural sector. It is pertinent to note that AI is heavily used in the United States for the purpose of supply chain optimization. However, the focal point of the same is utilization of machine learning, automation technologies etc.

Apart from that the American Business Houses also make the use of AI with the objective of ameliorating the inventory management, logistics etc. This in turn has reduced lead times and improved demand prediction accuracy. Due to this the organizations have been able to reserve their expenses. Furthermore, the usage of AI algorithms for the purpose of route optimization has helped in improving the delivery productivity. The customers are also happy with the services. Hence, it can be said that the deployment of AI has actually done wonders in the United States.

However, the same is not the case in Africa. Even though we find the usage of AI for the purpose of supply chain optimization in Africa, but it is being done on a limited scale. It is because there is an acute shortage of both infrastructure and investment in Africa. Due to this there are a lot of hinderances to the same. However, the Organizations are working hard and are developing new strategies to deal with the challenges posed before them. The promise of artificial intelligence to revolutionize supply chain management is acknowledged by both American and African governments. Even though the United States of America is leading the way in the usage of AI, but interestingly Africa is also pushing forward with customized solutions that fit its unique circumstances in terms of execution. Cooperation and knowledge sharing amongst these areas may open the door to a worldwide strategy for supply chain optimization using AI. This review emphasizes how crucial it is to comprehend local differences when implementing AI technologies, to promote cooperation for shared advantages, and to further the worldwide advancement of supply chain management powered by AI.

The authors have also gone through a report prepared by National Institute of Agricultural Extension Management for completing this chapter. The report was extremely helpful to the authors because it helped us understand the Supply Chain Management in multiple sectors like Horticulture, Poultry and Dairy. Lastly the authors have also gone through the report prepared by the FICCI which highlighted the challenges posed to the Indian Agricultural Sector and how Artificial Intelligence can help us in improving the supply chain mechanism.

INDIAN AGRICULTURE AND SUPPLY CHAIN MECHANISM

Within the Indian economy, the agriculture sector holds a significant place. Approximately 58% of India's population earn their livelihood from agriculture, making it thus a high-priority sector of the national economy. Additionally, it plays an important role in the world economy and is second only to China in terms of the 11.9% gross value added (GVA) in global agriculture. Despite a 50% increase in crop production over the past ten years, the supply-and-demand analysis for food indicates that this rate of growth in food production will not be enough to feed the world's growing population, which is expected to reach about ten billion people by the year 2050. Therefore, there is a need to accelerate the productivity of the agriculture sector. Modernizing the industry is also important for ensuring that it expands steadily and gives farmers the opportunity to profit as much as possible. It is important that we turn the sector's numerous problems into possibilities. It is pertinent to note that the need for food and nutrition security has evolved since COVID-19. As a result, in order to guarantee a sustainable supply to the nation and the rest of the world, the agricultural industry must be positioned appropriately. This will undoubtedly need for intelligent technologies that would help in anticipating the unexpected, as the knowledge report's title emphasizes. Given the complexity and diversity of Indian agriculture, any reforms must prioritize the needs of the country's farmers. Digital technologies and artificial intelligence can help our farmers to overcome a number of problems. If our farmers can recognize, accept, and use digital technologies to improve agriculture, that will be a huge accomplishment (PTI, 2021).

A recent report of FICCI suggests that although agriculture continues to be a vital sector supporting the livelihood of around 58% of the nation's population, technological adoption in the sector is in its infancy and thus pose numerous challenges and obstacles along the value chain. Agriculture in India is facing a number of obstacles along the value chain, while being in a transitional stage with an emphasis on integrating technologies for improved operations. It is stated that the system as a whole must undergo modification and must adopt to a more holistic approach that is based upon native and traditional agricultural knowledge combined with innovative smart farming techniques, such as the use of artificial intelligence (AI) tools and methods. The IT Industry Body NASSCOM has recently stated in a research that legislative and financial assistance are required for India to fully utilize artificial intelligence (AI) in the agriculture sector, which still deals with numerous obstacles. According to the report, AI has the potential to be a major factor in releasing the industry from the majority of its taxing input conditions and accelerating the transition to data-driven farming. Furthermore, the estimates from the Food and Agricultural Organization (FAO), India wastes over 40% of its food production. Its expenses can reach Rs 92,000 crore, of which Rs 40,000 crore comes from the waste of fruits and vegetables alone. Their availability in the market is reduced by loss and waste, which raises prices and limits the ability of low-income consumers to obtain the same. Water, land, energy, and other natural resources that are used to create food are wasted as a result of this loss (*Farmers in India Are Using AI for Agriculture – Here's How They Could Inspire the World*, 2024).

By addressing these losses and waste, food availability will increase and natural and financial resources will be used as efficiently as possible. Technology can be very helpful in this, both in terms of supply chain optimization and effective long-term storage and processing. How might artificial intelligence (AI) enhance the supply chain, close the large knowledge gap that exists between growers and retailers, and raise the price of produce? The farmer is at the beginning of the simplest supply chain, which terminates at the local grocery store. This is how it happens. Farmer Ò Village collection agent Ò Transport agent Ò Adhatiya in Mandi Ò Wholesaler Ò Vegetable retail store with a neighborhood location. Farmers in this system face price risk, information asymmetry on demand, inefficient distribution, and delayed payments. Conversely, the retailers deal with issues like as increased prices, poor and unclean produce, excessive price fluctuations, and the daily headache of having to visit the market extremely early in the morning. As a result, the conventional "Supply Chain" is very unorganized, inefficient, and prone to food waste. The primary cause is a lack of accurate information among different parties, which allows middlemen to manipulate prices in order to offset their risk. Farmers and consumers are the ones that ultimately lose out. How therefore does AI assist? The technology generates projections based on historical data analysis, which are sent to farmers via SMS in their native tongue one week in advance. After five days, this advice is revised again in light of market rates, supply, and demand over the preceding several days. Farmers deliver their goods to collection centers in rural areas in accordance with the SMS messages they get (Kesavan et al. (2022).

Produce is put in plastic crates after being sorted by weight and quality. These are then shipped to sizable fulfillment centers, where crates are sorted according to orders received from different merchants and a second round of quality control is completed. The routes that the AI system plans for the vehicles carrying the crates are prearranged. Delivering the boxes, the truck also retrieves the empty crates and money from the shops. This eliminates knowledge asymmetry, minimizes loss or waste, and eliminates several levels of middlemen, resulting in stable demand, better prices for producers, and reasonable prices for consumers.

Retailers have fresh produce delivered to their stores at reasonable costs. Consequently, this aids in the development of dependable, economical, and quick logistics and infrastructure.

The assessment of fresh fruit and vegetable quality, another major issue facing this ecosystem, has been resolved thanks to the power of AI and computer vision. Using cameras and sensors, the AI-powered system visualizes food in a manner similar to that of consumers. These algorithms can determine the quality and then arrange fruits and vegetables based on their training on color, size, and other characteristics. AI optimizes quality by eliminating human bias from quality assessment and providing instantaneous, objective data. The sorting system can be integrated with the automated packing equipment to pack vegetables in nylon nets in quantities ranging from 500 grams to one kilogram or more (Kesavan et al. (2022).

This makes it possible to print lot and source numbers, among other information, resulting in whole packet traceability. In the event of contamination or disease, traceability makes it possible to quickly identify the source. Maintaining a complete history of production and distribution enables farmers to respond promptly to problems such as poor-quality seed or improper use of pesticides, among other things. Farm production history can also help farmers make better decisions and better direct government assistance, particularly for small farmers. Better planning of future processing and storage facilities will also benefit from this data. Cameras can be used to watch employees in quality-cum-sorting centers, and object recognition software can be used to detect whether employees are wearing gloves and masks as needed by food safety laws (Kumar, 2021).

In order for India to fully utilize AI, it will take the cooperation of the government, business community, and start-ups to provide the required infrastructure and policy support, to foster AI innovation across industries, and to mentor and fund entrepreneurs. According to Nasscom, revenue growth, innovation, and the farmer/end-customer experience are critical areas, with "Trust in AI" serving as the fundamental tenet. The poll was performed to determine where agribusinesses are in the AI journey. A dedicated AI strategy and budget are also being seen by businesses as a crucial requirement for scaling AI activities across the entire corporation after COVID-19. Many AI-led start-ups are currently working on creative AI-led solutions that address certain value chain problems. The three main areas of attention are crop disease management, precision farming, and produce sorting and grading.

To understand this problem in great detail we have to first understand the meaning of supply chain in agriculture. Supply chain simply means the procedure that are part of production of goods to the supply of goods to the final consumer. In this procedure every method shall be the part of supply chain. It will include the seeding of crops by the farmer to final purchase of the goods by the end consumer. If technology can be included in all this process, then only supply chain mechanism can be improved. Many may argue that what is the best practice of use of A.I. around the world should be implemented in India. However, we did a comprehensive case study of United States. There the central issue was limited viability of labour forces to be indulged in agriculture therefore, their innovation is restricted to labour supply enhancement. However, research shows that in India problem lies with distribution and last mile connectivity of foodgrains with the final consumer. Hence, the A.I application must be on the lines of distribution and supply of goods. This aspect has been missed by many researchers. There is very limited growth on this issue as the focus of use of A.I. rest with the production activity of foodgrains. Production activity is important. Nevertheless, distribution shall also be enhanced. It is being said that around 30% of the foodgrains are being lost in the process of supply. Hence. A.I. should be implemented here.

Agriculture and Artificial Intelligence: A Case Study of United of States of America

The story of American agriculture's evolution and adaptation is one of constant change. Originally based on the subsistence farming of the first settlers, it underwent significant transformation in the 19th and 20th centuries, mostly as a result of advances in technology and shifts in the socioeconomic landscape. Agricultural methods were transformed and productivity was greatly increased with the introduction of machinery such as the cotton gin and the mechanical reaper in the 19th century (Hurt, R. D., 2002). The 20th century saw a transition from traditional farming to industrial agriculture, symbolized by the Green Revolution that brought high-yield crop varieties, chemical pesticides, and synthetic fertilizers. The agricultural terrain was drastically changed at this time, opening the door for extensive mechanized farming.

AgriTech is currently driving yet another significant shift in American agriculture. This cutting-edge innovation includes tools like AI-powered predictive analytics for effective farm management and precision agriculture, which employs GPS and IoT for more precise planting and harvesting. Drone technology is being used for crop monitoring and spraying, and robotics and automation are becoming more and more popular for jobs like harvesting. Through the reduction of resource use and environmental effect, these technologies target sustainability in addition to efficiency. Thus, the next-generation supply chain in American agriculture represents a synergy between traditional farming and technology that has the potential to address some of the most urgent issues facing the sector. These difficulties include the lack of resources, maintaining the environment, and meeting (Becker, 2024).

The lack of labor-force in the agricultural sector is threatening the industry's ability to remain profitable and produce sufficient food. The world's food systems cannot continue to be supported by the current number of labor-force. The United States, which produces the third-largest agricultural output after China and India, is severely affected by this issue. The ageing of farmers is one of the major problems. Many people are unable of putting in the long hours of physical labor required to manage a farm. Even though many of these businesses have been family-owned for years, younger generations are choosing to work in less labor-intensive but higher-paying fields rather than taking over as leaders as they once did. These tendencies are also observed among the hired laborers, who are primarily immigrants and comprise the remainder of the agricultural workforce.

According to Emily Buckman labor is a major concern as today the average age of farmer is 60 years old. As a result, some of the farmers in the US are also looking to new robots and AI solutions to help solve the problem, especially because agricultural yields are declining and the worker pool is getting fewer and more expensive.

Furthermore, it is pertinent to note that the application of artificial intelligence in the agriculture sector is not a novel concept; for example, auto-steering guidance systems for row crops like corn have been in usage for around 20 years. However, the adoption of AI has been huge in current times; as of 2021, 87% of US agricultural enterprises were estimated to be utilizing AI in some or the other capacity. The federal government is also accelerating the agricultural sector's transition to technology by offering financial incentives to hasten AI development and implementation across the nation (Becker, 2024).

It is interesting to note that the rest of the globe may be significantly impacted if the almost two million American farmers that are pushing for AI adoption are successful. This is particularly true as the world's population rises and climate change jeopardizes the agricultural ecosystems of some of the world's greatest producers, including the United States, which exported goods worth over $100 billion (£78.9 billion) to China, Mexico, Canada, and Europe in the previous year alone. According to Buckman, the industry wants to create and implement new technologies on a large scale that is accessible and inexpensive so that farmers can meet the demands of the people.

A broader approach known as precision agriculture is the need of the hour which implies using AI to increase crop yields while using less resources. The plan is for farmers to precisely target problem regions by utilizing cutting-edge technologies, such as drones that can measure a field's water or pesticide levels. Precision agriculture helps reduce water waste, be more efficient and do more with less. It can be claimed that US agriculture would've needed 100 million more acres 30 years ago to match today's production levels. The adoption of precision agriculture techniques in conjunction with AI and other technologies is primarily responsible for the output gains.

The most common AI technologies available in the market are drones and GPS devices. However, there are a ton of other options as well, such self-driving tractors and combines, as well as high-quality sensors on those vehicles, which can assist farmers in determining which portions of their crops require more or less care. Additionally, robots are capable of performing automated sorting jobs that previously required human vision. For example, some farms are using AI technology to sift potatoes and search for particular flaws. Others are planting seeds and pulling weeds from fields with AI-enhanced cultivators, which can essentially replace dozens of human laborers.

While lowering expenses and raising yields are clearly important to US farms and agricultural businesses, the advantages are far-reaching and may have global effects. Higher field productivity translates into higher farmer profits, a stronger supply chain, and lower consumer prices. The US government has therefore invested in and implemented a number of grants for AI research institutes that are centered on agricultural, demonstrating their recognition of the potential of AI in farming. Developing easy-to-use, AI-aided decision support tools to help land stewards identify win-win choices across environment, economy, and society is the main objective of these programs. Sophisticated new techniques abound, with an emphasis on precision resource distribution, soil surveillance, pathogen contact tracking to enhance food safety, and robotics to aid in harvesting. Iowa State researchers have used AI to create apps for smartphones that spot pests, saving them hours of laborious work. The instruments under development might also assist farmers in cutting costs, which is essential to making agriculture financially sustainable given the narrow profit margins that the sector now enjoys. When these technologies are in the hands of significant numbers of farmers, experts are cautiously enthusiastic about how they will help the agricultural sector combat changing risks from the environment, the unstable labor market, and other sources. It indeed has a huge potential (Atadoga et al., 2024).

Agriculture and Artificial Intelligence: A Case Study of Africa

Mobile technologies, cloud-based solutions, and infrastructural problems all have an impact on the adoption of artificial intelligence (AI) in African nations. The widespread deployment of artificial intelligence (AI) has been hampered by infrastructure issues in Africa, such as poor power supplies and internet access. However, since mobile technologies are now widely used throughout the continent, they have created a platform for the delivery of AI-powered solutions, which has opened up chances for the adoption of AI. Because cloud-based solutions offer scalable and reasonably priced computer resources, they have also significantly contributed to the overcome of infrastructural constraints. Artificial intelligence has penetrated many industries and sectors of African society and has influenced the supply chain mechanisms in Africa. This includes manufacturing, mining and agriculture. In agriculture A.I is being used for crop yield and resource management. In mining it is being used to enhance the extraction process and productive maintenance of resources. In manufacturing A.I based supply chain is being used in quality control and other ancillary activities. All of this may paint a positive picture of A.I. supply chain in Africa. Yet significant structural problems remain pertinent on the whole continent. The limitations are various, but significant of them being limited availability of financial and technological resources on the continent. This has developed into a tremendous obstacle in growth of the A.I supply chain in Africa. Due to this, Africa job markets lack requisite A.I skills. Furthermore, cultural factors are also significant. A.I. The skill set could not be developed in Africa without bringing in a cultural shift in favor of technology in the minds of the local population (*Har-*

nessing Artificial Intelligence for Sustainable Agricultural Development in Africa: Opportunities, Challenges, and Impact, n.d).

In many regions of Africa farmers have small land holdings. The proportion of these farmers in the African value chain is pretty significant. The use of A.I. the based supply chain mechanism has tremendously influenced the lives of the farmers in resolving problems that have been long standing in the sector. The A.I. based agriculture value chain system has not only helped farmers in basic farm-based activities, rather it suggested solutions to the problem of logistic and decision-making process. The A.I. has supported small farmers by providing them access to market data and insights related to them. It further helped them in the decision-making process. If the growth in A.I. remains consistent specially for small farm holders than the agriculture sector could see a revolutionary success, rarely seen in this sector. Furthermore, the small farmers can gain a lot by use of A.I. in agriculture. The tools of artificial intelligence are not merely restricted to the Activities of active farming. These tools provide information and market insights accessible to the farmer. This helps farmers in the decision-making process. However, it can also help them in activities that are ancillary to basic agricultural activities such as finance and insurance. A.I. technologies help farmers in choosing the suitable methods of finance that will support agriculture. Cutting age technology has helped farmers in determining the suitable market condition and has potential to write an unprecedented success story of the African smallholding farmers. A story exceptionally seen in the global south (*Harnessing Artificial Intelligence for Sustainable Agricultural Development in Africa: Opportunities, Challenges, and Impact*, n.d).

The African agricultural sector rests upon the idea of inclusive growth. It simply means growth and benefit for each and every party that is being involved in the process. The A.I. tools efficiently help in fulfilling this idea of inclusive growth. The tools address disparities. Attempts to move towards sustainable growth and empower underrepresented groups. These tools by providing small landholding farmers insight and market knowledge attempts to end disparity in the structure. By advising the farmers on important aspect of decision making it essentially increases the yield and growth of farmers.

As Africa charts its way for sustainable agricultural development, AI integration becomes even more important. It is pertinent to note that in order to fully realize the AI's potential to transform agriculture and ensure food security for future generations, it is necessary to overcome existing challenges and take advantage of the possibilities, and establish ethical and regulatory frameworks. Case studies and future trends show us that how integrating AI can revolutionize African agriculture. Two such illustrations of these innovations that have the potential to empower smallholder farmers, boost productivity, and encourage sustainable practices are precision farming and blockchain-enabled supply chain transparency. However, successful adoption necessitates a focus upon ethical issues, transparent AI methods, and inclusive laws in order to ensure equitable rewards and minimize potential risks. In the future, cooperative efforts under the direction of responsible AI standards can create a more robust system (Kividhami (2021).

Nearly a quarter of Africa's GDP is derived from agriculture, which employs more than 60% of smallholder farmers across the continent. Even while agriculture already fulfills a vital role in the area, there is still much space for advancement and modernization. According to one research, grains and cereals might be produced in Africa two or three times more than they are at the moment. The production of cattle and horticultural crops can both be predicted using the same methodology. Increasing productivity would logically result in more robust farms, higher household incomes, and improved food security. AI has the potential to be a significant contributor to this higher output target.

Additionally, this technology can reduce the need for pesticides, fertilizers, and irrigation, all of which are good for the environment and human health. In 2020, there were 75 million AI devices employed by farmers worldwide; by 2050, the average farm will get 4.1 million data points daily. Artificial Intelligence (AI) has specific applications in agriculture, including soil content sensing, irrigation, crop monitoring, weeding, and crop establishment. The consequences of climate change in Africa make more sophisticated technology necessary. Africa experienced half of the droughts brought on by human-caused climate change between 2001 and 2011. Artificial intelligence has the potential to mitigate the effects of global warming on African farming communities and increase the accuracy of drought forecasts. Due to their heavy reliance on subsistence farming, small-scale rural farmers need to ensure that their crops can withstand drought and other erratic weather conditions. AI may be able to meet these requirements (Kividhami (2021).

Precision farming is the most well-known application of AI technology in agriculture. AI technology is utilized to identify plant nutrients, diseases, and pests. The system is able to recognize and target weeds, ascertain the type and quantity of herbicide required, and spray it in the appropriate zone. This extremely effective method keeps pesticides from being overused, reducing the number of pollutants that end up in food that people eat. Although precision agriculture is not widely used in Africa, efforts are being made to boost capacity. The African Conference on Precision Agriculture brought together pertinent parties in December 2020 to expedite the development of precision agriculture capability in Africa. Africa has already used numerous AI techniques. For instance, some farmers in Kenya are identifying illnesses and pests among other things by utilizing near-infrared cameras installed on drones. Nuru, an AI tool under beta testing that can be used on Android smartphones with or without internet access, is also available in Kenya. Specifically, armyworm infections in maize and mite and viral diseases in cassava are diagnosed using this method. Kenya's 28,000 cassava farmers are already making use of this instrument. African farmers could benefit much from artificial intelligence in terms of their ability to withstand the agricultural risks posed by the climate issue. Lastly, it would increase the crop production and will also pave way for sustainable management of resources.

CONCLUSION: THE WAY AHEAD

The application of Artificial intelligence in the agricultural supply chain mechanism has led to a paradigm shift of the whole sector. The A.I. technologies provide support for complex decision-making processes. It greatly increases the efficiency at different levels of the supply chain. It optimizes and streamlines the whole supply chain system. The other two important aspects of the supply chain are distribution and logistics. A.I. provides support in this as well. By the use of predictive tools, time taken to access the markets can be reduced. This reduction will result to reduction on the cost and more importantly it will limit the wastage in post-harvest distribution. Furthermore, A.I. also helps in tracking the real time demand and inventory levels. The access to the inventory level is quite significant in logistics as it leads to reduction in wastage of yields.

A.I. plays an important role improving the precision and caliber of the whole decision-making process. It uses machine learning technology to understand the past trend of the market. This helps in determining the crops to be grown. This reduces the possibility of overstocking and understocking. Furthermore, it provides financial security to the farmer. A.I. also improves the transparency in the whole process of supply chain mechanism as it allows farmers as well as other stakeholders in accessing the journey of food from field to plate. This in turn leads to price transparency, safety transparency as well as reduction in wastage to agricultural supply (Monteiro and Barata (2021)).

The A.I. becomes more significant as it promotes inclusivity through its A.I. powered financial services and market technology. The use of technology in finance is always inclusivity based. It provides access to the weakest of farmers and small landholders to the financial market. It is a fact that use of technology enhances knowledge and creates awareness in the minds of the general public. The use of A.I. in the financial market leads to similar results. It will benefit those farmers as well who have limited to no access to the resources. A.I. can also help in gender parity. It may provide support to those women agriculturists who lack behind because of lack of availability of resources. If the technologies are being used properly then it will eradicate disparity and poverty. It will lead to better access to resources and the decision-making process for women. That would result in growth of yields and higher inclusivity.

Artificial intelligence is quite beneficial as it provides access to resources even to the most vulnerable members of the community. It leads to better inclusivity in the society. For example, we can consider the role played by women or underprivileged sections of the society in agriculture if Artificial Intelligence and other Technology could be used in decision making process as well as for generating financial inclusivity. However, there are some ethical dilemmas in the whole process as well. Artificial intelligence relies upon data and with data comes the concern of data privacy. How data can be used and who has access to these data. Therefore, with the use of artificial intelligence in agriculture the concerns related to data remain pertinent. In African scenario as well, the questions remained the same. The central issue is how the data will be protected and if data is not protected properly then even though artificial intelligence is pretty helpful for the agricultural community, yet it would fail to fulfill its purpose and it would affect the lives of the people in an inappreciable manner (Monteiro and Barata (2021)).

The AI will also lead to employment displacement. That simply means that the excessive use of AI in technology will lead to people losing their jobs. It remains a significant ethical dilemma in relation to the usage of artificial intelligence in agricultural sector. Therefore, attempts should be made to increase education and knowledge. This in turn would create jobs that would be more technology based. Legislation has to be made in such a manner that would ensure that even with the arrival of technology people have access to jobs. This means that the most vulnerable sections of the society and women who have generally faced significant disparity in the society should have access to technology. If access to technology is being provided to those people, then only artificial intelligence could become successful in our society. Hence, even in agriculture the women and the marginalized farmer should be on priority basis provided with access to the technology. Only then they could use this A.I. technology in order to ensure their active participation and benefit from the use of technology.

There is lack of regulation on the usage of AI and similar appears true for the agriculture sector as well. Therefore, there always remains a concern for regulating artificial intelligence in the agriculture sector. Hence, if any regulation is being made for the agriculture sector it needs to ensure that those regulations create accountability, fairness and transparency in the system. Policy makers should make policies in such a manner that it would enhance creativity in the system as well as will lead to long term use of artificial intelligence in the agricultural sector. Rules should be made in such a manner that it would answer both ethical concerns and enhancement of agriculture by the use of technology. Only then it can be said that artificial intelligence is important for the growth of agriculture.

Today a new wave of agricultural start-ups and AgriTechs are investing in digital projects. These technologies are being used to improve crop monitoring, assess plant health, practice precision farming, and determine the best course of action. Farmers are able to make innovative and well-informed decisions thanks to the modern-day technology. Therefore, it will be essential for the future to guarantee a fast-track approach towards the adoption rate of such technology within the farming community. It is important to note that these technologies have the potential to bring about a huge transformation in food management by reducing losses along the whole value chain, from the farm to the end user (Monteiro and Barata (2021).

Although there has been progress in developing an AI-enabled ecosystem in the agriculture sector, but still there are still a lot of challenges which must be resolved in order to allow for a seamless transition. The two main obstacles to AI innovation and technology in the agriculture sector are the small pool of AI and sector-specific knowledge, gaps in the literature currently covering AI research, low data quality and restricted access, a lack of coordination, and a lack of international cooperation. There are barriers to scalability, IoT devices, data annotation, data security and privacy, and technical knowledge when it comes to the adoption and deployment of AI in agriculture.

An ICAR-CIPET report suggests that there have been losses of around INR 92,651 crore across all commodities. The money that farmers can make from preventing such food loss will likewise increase with the use of new technologies. In order to create an environment that is favorable for Indian agriculture, the Government of India is taking every possible measure. The Government of India has issued a significant declaration permitting the usage of drone in agricultural sector. The Government of India has been consistently developing and executing policies and programs to encourage digital transformation in agriculture, with a focus on bolstering ecosystem participants, such as agri-tech startups. Additionally, the government is supporting AgriTech startups by creating an institutional environment through incubators and speed bumps. Furthermore, it has taken up the motto "AI for All" and established a set of general guidelines for developing the AI ecosystem in India via the NITI Aayog.

It is important that AI technology must adopt a variety of locally appropriate methods in Indian agriculture, which not only acquires knowledge from local systems but also promotes innovation along the entire value chain. In order to turn AI into Agri intelligence, we need to employ a 3S strategy that uses the levers of scale, skill, and service. This will allow important actors to work together as change agents and accelerate the technology's widespread acceptance throughout the ecosystem. Furtehrmore interventions in the areas of regulation, policy, awareness, and capacity building can also lead to scale. The development of AgriTech innovation centers through public-private partnership (PPP) models, the revision of agricultural education curricula to include AI, and increased outreach through the network of agricultural universities (AUs) and Krishi Vigyan Kendras (KVKs) are some ways to achieve the required target.

The central issue in Indian agriculture is with regards to distribution. There is no proper distribution of products to the markets and mandis. The government has recently passed few Acts with regards to agriculture. Although these Acts were taken back due to lack of support from the side of the farmer. The were focused upon making the distribution better. Many new electronic and online resources have come up. They focus upon distribution. In this regard an interesting case study could be of Ninjakart. Like any other initiative of A.I and agriculture this too keeps farmers at the lamplight. Yet what they do is much more then that. They do not merely use tech to focus upon enhancement of agriculture production. Rather they extend further. They provide support to the farmers of A.I. to ensure that these farmers can take their proceeds to the mandis. Thereafter they use technology to ensure that they get proper warehousing. By these steps they ensure that appropriate food and quality agriculture proceed reaches to the final consumer. It helps both the farmer as well as final consumer. It further leads to the beneficial result for the whole agricultural economy.

Ultimately, implementing straightforward farmer-friendly technologies, forming long-term alliances, and boosting funding for open-source, inexpensive cloud platform solutions can all lead to improved service. These levers, which are operated by the government, academic and research institutions, and commercial entities, including start-ups, will lead to better sector-specific application, more access to the newest technology, improved acceptance, and an improved innovation environment. Since the government is now contributing, it is the responsibility of the other parties to cooperate in order for farmers in particular and the value chain as a whole to benefit from the Government's policy decisions. There are specific roles for the agriculture industry, startups, academics, farmers, scientists, and regulators on whom it depends whether such attempts will be successful or not.

REFERENCES

Ashwani, D. G. &. M. (2017). *Indian Economy, 72nd Edition*. S. Chand Publishing.

Atadoga, N. A., Obi, N. O. C., Osasona, N. F., Onwusinkwue, N. S., Daraojimba, N. I., & Dawodu, N. S. O. (2024). AI in supply chain optimization: A comparative review of USA and African Trends. *International Journal of Science and Research Archive*, 11(1), 896–903. DOI: 10.30574/ijsra.2024.11.1.0156

Becker, S. (2024, April 4). *US farms are making an urgent push into AI. It could help feed the world*. https://www.bbc.com/worklife/article/20240325-artificial-intelligence-ai-us-agriculture-farming

Farmers in India are using AI for agriculture – here's how they could inspire the world. (2024, January 16). World Economic Forum. https://www.weforum.org/agenda/2024/01/how-indias-ai-agriculture-boom-could-inspire-the-world/#:~:text=Artificial%20Intelligence,-Follow&text=The%20initiative%20transformed%20the%20chili,that%20they%20doubled%20their%20income

Harnessing artificial intelligence for sustainable agricultural development in Africa: Opportunities, challenges, and impact. (n.d.). https://arxiv.org/html/2401.06171v1#:~:text=The%20adoption%20of%20AI%20in%20agriculture%20in%20Africa%20has%20a,solutions%20to%20long%2Dstanding%20challenges

Job, I. A., Essien, E. N., & Ododo, E. P. (2022). A Review of Application of Machine Learning in Agriculture Supply Chain to improve Food and Agriculture Industry. Zenodo *(CERN European Organization for Nuclear Research)*. https://doi.org/DOI: 10.5281/zenodo.6258922

Kesavan, T. R., Mathur, H., & Harnal, J. (2022). Redefining agriculture through artificial intelligence: Predicting the unpredictable. In *PwC*.

Kividhami. (2021, June 16). An overview of AI technologies in African agriculture. *KDHI Agriculture*. https://www.kdhi-agriculture.com/single-post/an-overview-of-ai-technologies-in-african-agriculture

Kumar, P. (2021, September 2). AI-enabled supply chain: A win-win for farmers, consumers. *Business Today*. https://www.businesstoday.in/opinion/columns/story/ai-enabled-supply-chain-a-win-win-for-farmers-consumers-305739-2021-09-02

Monteiro, J., & Barata, J. (2021). Artificial Intelligence in Extended Agri-Food Supply Chain: A short review based on bibliometric analysis. *Procedia Computer Science*, 192, 3020–3029. DOI: 10.1016/j.procs.2021.09.074

Pti. (2021, March 25). India farm sector needs policy, financial support to realise full potential of AI: Nasscom. *Business Today*. https://www.businesstoday.in/industry/it/story/india-farm-sector-needs-policy-financial-support-to-realise-full-potential-of-ai-nasscom-291741-2021-03-25

Routroy, S., & Behera, A. (2017). Agriculture supply chain: A systematic review of literature and implications for future research. *Journal of Agribusiness in Developing and Emerging Economies*, 7(3), 275–302. DOI: 10.1108/JADEE-06-2016-0039

Chapter 17
Driving Agricultural Performance by Leveraging Digital Technologies in Agriculture 4.0

Sandeep Singh
https://orcid.org/0000-0002-8141-3960
Chitkara Business School, Chitkara University, Punjab, India

Anurag Tewari
Chitkara University Institute of Engineering and Technology, Chitkara University, Punjab, India

Aparna Sharma
https://orcid.org/0009-0003-3979-9905
Government College, Barwala, India

Ajit Bansal
Chitkara Business School, Chitkara University, Punjab, India

ABSTRACT

Agri-food sustainability has grown to be the globe's most important issue throughout the years. Technologies that promote environmentally friendly agricultural farming have proven effective in reducing poverty brought on by increased food demand. Agri-food system uses have been significantly altering the global landscape recently as consequence of internal and external variables. The forthcoming contribution

DOI: 10.4018/979-8-3693-4330-2.ch017

of study is that we can utilize it for enhancing economic efficiency, better control mechanism and improved decision making. This study will prove beneficial to the users and other stakeholders like technology providers, experts and society at large. Furthermore, this study provided guidelines for further research on the proposed concept. The conclusions drawn in the study will be helpful to researchers, experts, smart technology suppliers and policy makers who ultimately are responsible for supporting the evolution to extra sustainable and cultured agri-business practices.

1. INTRODUCTION

The farming industry is a decisive industry which upsets all facets of humanity because it has contributed to meeting global food needs and ensuring adequate nutrition.

Moreover, the agricultural output is very important for the economic wellbeing of a country, happiness and wellness of its citizens. Farmers face lots of challenges in producing crop like lack of stable prices of agricultural product, scary an unpredictable weather conditions, the exposure of agricultural output to insects and diseases.

A cutting-edge technology called digital agriculture (DA) is assisting in meeting the aggregate mandate for ecologically cognizant agriculture on a worldwide scale. Reduce waste, maximize farming components, and improve crop yields by incorporating several DA sub-branches—like artificial intelligence, automation and robots, detectors, Web of Things (IoT), and statistical analytics—into agricultural procedures. By utilizing the geospatial information technique crop functions can be managed in the best manner to reduce the cost of production and improve the productivity. One more benefit which we can derive by utilizing this technique is the optimum utilization of resources and limitation of wastage.

This research paper will emphasize upon the requirement of digitizing farming procedures and practices so that it may ultimately result in augmenting the overall productivity. This paper also aims to check the economic and financial feasibility of farming products.

To guarantee food security and dependability, a remarkably undeveloped agricultural sector must also modernize its technological infrastructure. Additionally, the implementation of environmentally friendly e-agricultural supply chain administration (e-AgriSCM) is greatly aided by blockchain technology. Products websites (web design) in the era of blockchain innovations in electronic advertising need to match supply chain collaborators' expectations for guaranteeing effective processes.

For fulfilling the requirements of a growing world population and address current difficulties including labor shortages, resource limited availability, global warming, and ecological issues like the condition of the soil, the environment, and the cleanliness of water, agrarian industries want to boost production. Simultaneously, the agriculture industry is perceived to benefit from digital change.

The application of contemporary digital technology enables the formation of ideal soil-agro scientific and organizational-territorial circumstances to produce agricultural commodities throughout the whole life cycle, as demonstrated by worldwide practice and the experiences of prosperous local farming families.

Digital transformations are the key points that trigger the wealth in farming throughout the process of reconstruction. These techniques empower a distinguished development in superiority and throughput in the commencement of harvests, services, and supplementary products as associated to traditional schemes of ascendency. These countries appreciate a modest advantage since their frugalities are constructed on the newest know-how.

2. DIGITAL AGRICULTURE (DA)

Agro-food system implementations have been significantly altering the global landscape recently because of internal as well as outside forces. A cutting-edge technique called digital farming (DA) is supporting in satisfying the growing requirement for maintainable agricultural products on a global scale. Decreasing waste, optimizing agri inputs, and accumulative harvest productivity can be attained by integrating several sub-branches of digital farming technique, like artificial intelligence, robotics, detectors, Internet of Things (IoT), and data statistical analysis, into agricultural procedures. Due to the accountability of both goods and processes made possible by this, production in agriculture can rise as labor-intensive procedures can be replaced with continually computerized procedures. Agro-food producers can obtain precise and up-to-date measurements on various factors impacting their profitability, including the wellbeing of plants, the condition of the soil, and meteorological conditions, by utilizing DA (Shalender & Yadav, 2019; Sridhar et al., 2023).

The foundation of DA is the conventional of five rendezvous levels that, whenever united, indorse originality: The propagation of sensors, vacillating from nanotechnology instruments to satellites that has created a spare information and abridged broadcast and conservation. Furthermore, an improvement in calculating resources has empowered the employment of networks and novel exhibiting technologies. Other developments include communication and insights trade connections, organizing information and science and technology for agricultural decision-making, technology, authority, and automation, and networking and communication. Over the past three decades, farming practices have changed dramatically due to the widespread application of technological advancements like as computerized management and machinery, internet-based statistical tools, and smartphone apps. Increasing crop yield and resource productivity has been the primary goal of these improvements.

3. AMAZING TECHNOLOGICAL ADVANCEMENTS IN DIGITAL FARMING

Thanks to technology, farmers can now more efficiently manage their crop cultivation and harvesting operations from remote locations.

The actuators, detectors, and gadgets linked to the Internet of Things (IoT) are going to be equipped to autonomously interact, operate, and make choices in real time with agricultural machinery. Due to this innovative technology, there is fewer requirement of manpower which results in maximizing profits. Smart Agriculture and further wireless farming administration systems pursue to assimilate this information from different spaces into their policy making procedure. For such reasons, farmers can use data to conduct ongoing strategic assumptions.

3.1 Robotics and Electronic Vehicles Used in Farming

The constant need to reduce loudness and emission levels is the primary force behind the creation of electrical infrastructure. As cutting-edge machinery, DA lowers expenses for inputs and promotes durability by using automation and electrical cars that are fueled by renewable energy sources. These types of tools and machinery can be used on numerous farms to produce power using renewable energies. Bots and humans can collaborate in an environment of cooperation to achieve common goals in the farming sector. Managing weeds with this co-robotic approach has proven to be successful, reducing the requirement for workers.

3.2. Automated Operations Based on Sensor Technology and IoT

In digital agriculture, an improved consideration of the interactions between soil, crop production, and surroundings can result in higher yields and profits. A field-wide network of interconnected replication, consumers, and sensor-equipped nodes makes up an automated system that utilizes the Internet of Things.

By integrating the physical environment into the data collection system using sophisticated, networked, and creative structures, the application of IoT to farming will enhance the performance of already available equipment. IoT offers countless possible use areas, ranging from livestock and crop machinery surveillance to conservatory surveillance and combining data from multiple cameras and RFID tags (Kaloxylos et al. 2012).

By connecting to the Internet of Things, connected sensors enable agribusiness to be built despite geographic distances between the researchers, cultivators, and harvests. By empowering manufacturers to make prompt and suitable decisions based on actual time and precise information, IoT technology facilitates the effective utilization of natural assets (Jeffrey & Bommu, 2024; Liu, et al., 2020; Moreno et al., 2024).

3.3. Cloud Technology and Big Data for Agriculture

The digital transformation in agriculture is largely dependent on big data technology. These days, algorithms for deep learning and computer responses can be produced from the study of the various sensors that machines are outfitted with to measure data from their surroundings. This is known as the "digital agriculture era." Big data, according to Hashem et al. (2015), is the complement of methods that need integration forms to separate unknown values from vast, diverse, and intricate data sets. Farmers can monitor all production metrics of in-the-moment activities and enhance their decision-making processes with the help of big data (IDEAGRO 2015). When combined with information gathered from outside big data sources of information, such rivals' or market-specific data, internal information sources can have an impact on leverage.

3.4. Application of Artificial Intelligence in Agriculture

Using cutting-edge machine learning (AI) technology in agriculture has become a viable way to increase yield prospects and crop management strategies. The adoption of cutting-edge artificial intelligence (AI) technologies in agriculture marks a turning point in the development of contemporary farming methods and the beginning of an age of organic farming and precise agriculture.

The confluence of AI technology with farming methods presents unparalleled prospects to optimize cultivation techniques, improve yield the results, and decrease environmental concerns in the face of security problems, environmental uncertainty, and resource restrictions (Gabriel & Gandorfer, 2023; Idoje et al. 2021).

3.5. The Evaluation of Data

The data has considerably been increased for the clouds and the Web of Things' (WoT) swift advances in processing. This information, also known as massive data (BD), consists of multimedia materials (such as videos, audio recordings, and picture files) and written material (such as organized, semi- organized, and unorganized information). It is the procedure of looking through this data to find previously unidentified relationships, trends in the market, customer needs, and other important information. The agricultural BD is typically produced by a variety of agricultural industries and phases. Data could be gathered from these sources employing a variety of techniques, including ground detectors, aerial automobiles, and ground-based vehicles equipped with specialized cameras and gauges; notifications and laws from government-run bodies; on the internet service providers from private companies; interviews to gather information from the farming community; and social networking platforms. These days, a lot of agriculture-related organizations use sophisticated machine learning and statistical methods to glean information from big, complicated data sets like satellite imaging, soil testing, and meteorological data. Regarding crop administration, such knowledge can assist farmers in making better choices.

4. BENEFITS OF USING DIGITAL TECHNOLOGIES IN AGRICULTURE

All the innovations are being developed, used, and iterated in the farming industry within various spatial settings as part of the DA (Chergui et al., 2020).

With the application of ICT, AI, and data analysis, DA's solutions based on data enable farmers to implement these strategies for their enterprises. These approaches might be different and encompass a range of farming tasks, such as generating prediction models, evaluating vulnerabilities and catastrophes, and so forth.

There are so many benefits that can persuade producers and other players to support the use of electronic devices in the agriculture industry. These advantages have an opportunity to maximize agricultural output, enhance the quality of products, and eventually increase security of food (Chouhan et al., 2021; Colaço et al., 2024; Dibbern et al., 2024). We can enumerate the advantages of DA for farming as follows:

4.1. Improved Agility

Digital technologies enhance the flexibility and responsiveness of farm operations. By leveraging real-time monitoring and predictive systems, farmers or agricultural specialists can promptly respond to any potential changes in environmental and water conditions to protect crops from stress.

4.2 Green Methodology

By maximizing the amount and quality of produce from agriculture while lowering the use of water, electricity, chemical fertilizers, and pesticides, technological tools can improve the use of resources effectiveness.

4.3 Saves Money and Time

Digital tools can save a significant amount of time and money by automating tasks like planting, harvesting, watering, and controlling the use of manure and insecticides.

4.4 Asset Leadership

Thanks to digital methods, agricultural property and machinery may be instantly monitored to thwart theft, repair parts more quickly, and perform regular upkeep.

4.5 Product Security

The use of electronic devices deters fraudulent activity linked to tampering imitations, and synthetic modifications, ensuring adequate farm output as well as a secure and nutritious supply of agro-food commodities. It gives the farmer helpful knowledge that helps them make decisions about the amount, timing, and location of applying fertilizer, seedlings, nourishment, water, as well as additional pesticides and inputs for farming. It protects the natural world and contributes to the creation of healthy goods by enabling the precise application of resources in each field of view, such as seeds, fertilizers, chemicals, and farming, as well as by allowing for the variation in input component amounts.

Farmers can now access advanced management strategies against warming temperatures, additional environmental problems, and natural disasters thanks to information by data technologies. Through the help of these technologies, farmers can keep a close eye on the health of their crops and receive timely alerts about potential pests, illnesses, or even environmental issues. In terms of a marketing perspective, farmers can gain from complicated algorithms that provide information about the economy and which items could increase their earnings.

5. AGRICULTURAL SUPPLY CHAIN

Because the agricultural sector is so undeveloped technological advancements are necessary to ensure the accuracy and security of food. Blockchain technology is being used in a digital environment to successfully implement sustained e-agricultural supply chain administration (e-Agri-SCM).

Because of the increasing concern of the degradation of resources, changing of climate, and overpopulation, companies in the agricultural industry must rethink their present supply chain structures to incorporate economic and ecological factors to make them sustainable. Over the past 40 years, the supply chain for agricultural products has seen a dramatic change (Hovelaque et al., 2009). Authorities and agricultural businesses have placed a strong emphasis on technology for farming and marketing, which have been critical to the upgrading of the supply chain for farming. Moreover, advancements in data, its dissemination, and modes of mobility have changed. At present, they are enmeshed in extensive internet usage across multiple platforms around the globe (Javalgi et al., 2014).

By creating social networks that are essential for locating, identifying, connecting with, and servicing customers, agricultural businesses, particularly when combined with contemporary digital technologies like blockchain, has completely transformed the way business is done with clients globally (Glavas et al., 2014). Businesses may see significant returns on the money they spend on technology, including increased productivity, reduced expenses, and superior service (Kumar et al., 2021). Furthermore, the online purchasing industry has been bolstered by technological growth. In just one year, China's gross online product volume increased by 64.5 percent to reach 4.8 trillion Yuan, according to a National Bureau of Statistics data from 2016. According to Akram et al., 2018, the increase was observed to be four times quicker than the overall sale of items.

6. APPLICATIONS OF DIGITAL TECHNOLOGY IN AGRICULTURAL SUPPLY CHAIN

Digital technology has transformed the agriculture supply chain in various ways, enhancing efficiency, transparency, and sustainability. By leveraging digital technology, the agriculture supply chain can become more resilient, responsive, and responsible, ultimately contributing to global food security and sustainability.

The effectiveness and excellence of agricultural output have significantly grown in the past few decades due to the increasing digitization of humanity and the incorporation of technology into agriculture. Nevertheless, there are now additional issues and obstacles because of this emancipation which must be resolved. Notwithstanding these obstacles, technology and agriculture together have a lot of potential to contribute to agricultural sustainability in future generations.

Here are some key areas where digital technology is making a significant impact:

6.1 Precision Agriculture

Using digital tools such as GPS, aerial vehicles, satellite imagery, and Internet of Things sensors, farmers can accurately track and control weather, agricultural products, and soil health, resulting in higher yields and less waste. Moisture of the soil and health of crops can be monitored by the help of sensor technologies. Advanced GPS enabled equipments can be used for optimal crop administration. Further data processing could be utilized for forecasts of weather, pests, and diseases. For improving precision and saving labor cost autonomous farming can be utilized.

6.2 Supply Chain Access

Real-time surveillance and tracking of produce transportation is made possible by digital tools, which shorten lead times and help participants make informed choices.

The use of computer technology in farming has made it possible to keep tabs on food movements, excellence, and status in real time along the entire supply chain. A wide range of technologies, including mobile applications, cloud-based systems, blockchain, GPS and RFID monitoring, data analytics, and electronic data transfer, enable partners to obtain accurate and up-to-date data about the location, movements, and state of the produce. Improved quality control, fewer losses and waste products, improved security of food, increased production and efficiency, easier regulatory compliance, greater openness and stakeholder confidence are all facilitated by this digital transparency.

Agriculture supply networks may attain full transparency, streamline processes, and guarantee that consumers receive fresh, premium food by utilizing this digital technology. In the end, this transparency and responsibility help buyers, customers logistics suppliers, and farmers make educated decisions that improve the efficiency, sustainability, and customer-focused nature of the farming supply chain.

6.3. Forecasting Analytics

Data-driven individual decisions, increased reliability of forecasts, less waste and overproduction, and improved transportation and logistical services are all made possible by the application of statistical analysis to farming supply chain administration. Predictive analytics is a tool that assists with demand forecasting, forecasting crop yields, quality problem identification, and pricing prediction by utilizing complex algorithms and artificial intelligence approaches. Supply chain participants can reduce risks, improve inventory control, and optimize processes with this preemptive strategy, which eventually boosts production, revenue, and effectiveness. Agriculture supply networks may remain far ahead of the competition by utilizing statistical analysis to adapt to changing market circumstances and guarantee that premium food is delivered to end users in the shortest possible time.

6.4 Digital Marketplaces

By utilizing technological advances, agricultural online marketplaces are revolutionizing the ways in which consumers and sellers interact, do business, and obtain essential services. Commerce platforms enable the online purchase and sale of goods and services, including produce, while electronic payment methods guarantee safe and easy operations. Logistics administration platforms improve monitoring and shipping, while quality testing and accreditation technologies evaluate and confirm the quality of produced goods. Statistics and market data offer important perspectives on demand, pricing, and trends. Online agreements and contracts guarantee privacy and security, and technologies for farmer identification and profile authenticate individuals and production metrics. Furthermore, product categorization and search functions make it easier to locate specific goods, and confidence is fostered via feedback and ratings.

6.5 Modern Logistics

Technological advances expedite shipping times, cut fuel costs, and optimize routes for transport. The delivery, storage, and transportation of product have all changed because of the use of digital technology into agricultural operations. Although warehouse administration programs simplify monitoring and simplify the administration of goods, transportation scheduling systems improve routes to save energy expenditure and improve timeliness for delivery. Real-time delivery information and updates are provided by supplier visibility channels, and IoT sensors keep an eye on environmental conditions while shipments are being transported and stored. Techniques that automate identifying and data collection, like RFID and barcodes, make traceability and monitoring more effective. Reactive management of logistics is made possible by analytics that is predictive, which estimates demand and identifies possible disruptions. Shippers and carriers are connected via digital freight platforms, which automate reservation and rate haggling.

Farm logistics are made more dependable, efficient, and economical by utilizing these electronic devices, while guaranteeing that the product is delivered to buyers in an unspoiled and excellent state.

6.6. Smart Warehousing

The integration of digital technology in smart warehousing for the agriculture supply chain has revolutionized the way goods are stored, managed, and shipped. Automated Storage and Retrieval Systems (AS/RS) optimize storage capacity and reduce labor costs, while Warehouse Management Systems (WMS) streamline inventory management, tracking, and reporting (Du et al., 2023). IoT sensors monitor temperature, humidity, and other environmental factors to ensure optimal storage conditions, and RFID and barcode scanning enable efficient tracking, tracing, and inventory management. Additionally, drone-based inventory management uses drones for aerial inventory scanning and tracking, and predictive analytics forecasts storage needs, detects potential issues, and optimizes warehouse operations. Digital twin technology creates virtual replicas of warehouses for simulation, analysis, and optimization, and robotics and automation automate tasks such as palletizing, de-palletizing, and packaging. Real-time monitoring and alerts provide instant notifications for inventory levels, expiration dates, and storage conditions, and data analytics and reporting offer insights on inventory turnover, storage capacity, and supply chain efficiency. By leveraging these digital technologies, smart warehouses become efficient, data-driven hubs that enhance the agriculture supply chain's overall productivity and responsiveness.

6.7 Quality Control

Real-time production quality tracking is made possible by digital solutions, which also lower the chance of spoiling and guarantee adherence to standards.

The use of electronic equipment in the farming supply chain improves quality control by tracking both transportation and storage circumstances and instantly identifying flaws and impurities. Additionally, it facilitates the automation of grading and sorting procedures, and the analysis of goods produced through electronic spectroscopic and hyperspectral photography.

Blockchain technology can be used to provide transparent and safe quality control and data can be shared with stakeholders. As a result, supervision of quality becomes more effective, precise and dependable, guaranteeing that consumers receive products of the highest caliber while lowering wasteful consumption and enhancing safety.

6.8 Farm-to-Table Traceability

End-to-end accountability, made possible by technological advances, lets customers follow the food's path from its source to its destination. It can be done by using the digital payment methods for secure payments and creating online avenues for selling directly to consumers. Data analytics can be utilized for enabling pricing optimization and assessment of market trends

6.9. Extension Functions

Farmers can improve their abilities and efficiency by using digital platforms to get instruction, exchange of expertise, and consulting services. For this the digital farming consulting services can be started. Mobile applications for getting information and experience in agriculture. Data analytics can be applied for generating customized suggestions

A new technological age is dawning for agriculture. Science-fiction techniques are starting to appear on farms all around the planet. The use of aerial vehicles, robotics, detectors, analytics for large amounts of data, and precise cultivation equipment has completely changed the way crops are produced and managed. The agriculture services sector and supply chains are undergoing an exciting shift thanks to digital technology. It pledges to feed the world's expanding population, improve long-term sustainability, and increase financial stability.

The modern era efficiently uses digital advertising to promote agricultural products by improving website design to draw in clients. The goal is to give agricultural enterprises an opportunity for creating technological policies that will ensure agricultural availability and reliability. In order to successfully adopt global agri-SCM, scientists are attempting to improve agricultural web design; nevertheless, the web layout is currently not technologically accessible to clientele using the blockchain technology (Alkahtani et al., 2021).

7. CONCLUSION

Digital techniques for farming are crucial for improving the effectiveness of product manufacturing and handling, as agricultural output has grown into the fundamental link influencing patterns of growth for associated sectors in the wake of internationalization. It is essential to transform the agriculture sector and steer it in the direction of innovative development depending on the internet economy. Digital technology is accompanied by the difficulties of identifying the target audience and adapting digital strategies for various agriculture business models. Problems with technological improvements include determining the target audience and matching up technological solutions for different kinds of agriculture. Intentional needs for the development of the largest machine-building firms in the globe include the computerization and automation of as many farming processes as feasible. The productivity of agricultural products is among the most endangered industries due to its dependence on ecological conditions and natural catastrophes.

Supply chain management has been greatly reinforced by the growth of e-companies in recent decades, which has also provided enterprises with an edge in marketing potential. Using DM-SCM, agribusiness companies are looking for intelligent solutions to market their goods and persuade consumers. Because blockchain technology makes transactions easier and enhances safety as well as simplicity for product accountability, automation with a focus on sustainability is crucial for the creation of a responsible agri-SCM.

Industry 4.0's enhancing adoption of cutting-edge technology has made it possible to oversee the supply chain with greater effectiveness and economically. When we discuss the "digitalization of the supply chain," we mean the shift toward a more intelligent model incorporating technological advances like blockchain, the Internet of Things, machine learning, etc. The preparation, procurement, and buying tactics can now be optimized to a greater extent thanks to these developments in technology.

In this setting, digital technologies including edge and cloud computing, blockchain technology, and the Internet of Things (IoT) are extensively utilized by the public (Balkrishna et al, 2023; Centobelli et al, 2020). Adding more smart gadgets in addition to using data effectively will boost the productivity of production. These innovations, when used in conjunction with satellite imaging and communication technology (ICT), have the potential to assist farmers by improving agricultural systems' performance and efficiency and by utilizing the best farm management methods. The tools facilitate integrated, immediate time decision-making processes and transform convoluted data into useful data (Abiri et al., 2023; Bansal and Bansal 2012; Zhang, X., & Fan, 2024). Using digitization scenarios to analyze and comprehend farm data could aid in the development of healthier, more profitable, and environmentally friendly farming methods.

REFERENCES

Abiri, R., Rizan, N., Balasundram, S. K., Shahbazi, A. B., & Abdul-Hamid, H. (2023). Application of digital technologies for ensuring agricultural productivity. *Heliyon*, 9(12), e22601. Advance online publication. DOI: 10.1016/j.heliyon.2023.e22601 PMID: 38125472

Akram, U., Hui, P., Khan, M. K., Tanveer, Y., Mehmood, K., & Ahmad, W. (2018). How website quality affects online impulse buying: Moderating effects of sales promotion and credit card use. *Asia Pacific Journal of Marketing and Logistics*, 30(1), 235–256. DOI: 10.1108/APJML-04-2017-0073

Alkahtani, M., Khalid, Q. S., Jalees, M., Omair, M., Hussain, G., & Pruncu, C. I. (2021). E-agricultural supply chain management coupled with blockchain effect and cooperative strategies. *Sustainability (Basel)*, 13(2), 816. DOI: 10.3390/su13020816

Balkrishna, A., Pathak, R., Kumar, S., Arya, V., & Singh, S. K. (2023). A comprehensive analysis of the advances in Indian Digital Agricultural architecture. *Smart Agricultural Technology*, 5, 100318. DOI: 10.1016/j.atech.2023.100318

Bansal, A., Tewari, A., Sharma, A., & Bansal, A. (2024). Implementations and Rationale for Blockchain Technique in Agriculture. In *Applications of Computer Vision and Drone Technology in Agriculture 4.0* (pp. 83–92). Springer Nature Singapore. DOI: 10.1007/978-981-99-8684-2_6

Bansal, A. K., & Bansal, A. (2012). Micro finance and poverty reduction in India. *Integral Review-A Journal of Management*, 5(1), 31–35.

Centobelli, P., Cerchione, R., Esposito, E., & Shashi, . (2020). Evaluating environmental sustainability strategies in freight transport and logistics industry. *Business Strategy and the Environment*, 29(3), 1563–1574. DOI: 10.1002/bse.2453

Chergui, N., Kechadi, M. T., & McDonnell, M. (2020, February). The impact of data analytics in digital agriculture: a review. In *2020 International Multi-Conference on: "Organization of Knowledge and Advanced Technologies" (OCTA)* (pp. 1-13). IEEE. DOI: 10.1109/OCTA49274.2020.9151851

Chouhan, S. S., Singh, U. P., Sharma, U., & Jain, S. (2021). Leaf disease segmentation and classification of Jatropha Curcas L. and Pongamia Pinnata L. biofuel plants using computer vision based approaches. *Measurement*, 171, 108796. DOI: 10.1016/j.measurement.2020.108796

Colaço, A. F., Whelan, B. M., Bramley, R. G. V., Richetti, J., Fajardo, M., McCarthy, A. C., Perry, E. M., Bender, A., Leo, S., Fitzgerald, G. J., & Lawes, R. A. (2024). Digital strategies for nitrogen management in grain production systems: Lessons from multi-method assessment using on-farm experimentation. *Precision Agriculture*, 25(2), 983–1013. DOI: 10.1007/s11119-023-10102-z

Dibbern, T., Romani, L. A. S., & Massruhá, S. M. F. S. (2024). Main drivers and barriers to the adoption of Digital Agriculture technologies. *Smart Agricultural Technology*, 8, 100459. DOI: 10.1016/j.atech.2024.100459

Du, X., Wang, X., & Hatzenbuehler, P. (2023). Digital technology in agriculture: A review of issues, applications and methodologies. *China Agricultural Economic Review*, 15(1), 95–108. DOI: 10.1108/CAER-01-2022-0009

Gabriel, A., & Gandorfer, M. (2023). Adoption of digital technologies in agriculture—An inventory in a european small-scale farming region. *Precision Agriculture*, 24(1), 68–91. DOI: 10.1007/s11119-022-09931-1

Glavas, C., Pike, S., & Mathews, S. (2014). Internet-enabled tourism entrepreneurs. International entrepreneurial values elicited through repertory test and laddering analysis. *International Journal of Tourism Research*, 16(1), 44–55. DOI: 10.1002/jtr.1896

Hovelaque, V., Duvaleix-Tréguer, S., & Cordier, J. (2009). Effects of constrained supply and price contracts on agricultural cooperatives. *European Journal of Operational Research*, 199(3), 769–780. DOI: 10.1016/j.ejor.2008.08.005

Idoje, G., Dagiuklas, T., & Iqbal, M. (2021). Survey for smart farming technologies: Challenges and issues. *Computers & Electrical Engineering*, 92, 107104. DOI: 10.1016/j.compeleceng.2021.107104

Javalgi, R. G., Hall, K. D., & Cavusgil, S. T. (2014). Corporate entrepreneurship, customer-oriented selling, absorptive capacity, and international sales performance in the international B2B setting: Conceptual framework and research propositions. *International Business Review*, 23(6), 1193–1202. DOI: 10.1016/j.ibusrev.2014.04.003

Jeffrey, L., & Bommu, R. (2024). Innovative AI Solutions for Agriculture: Enhancing CropManagement and Yield. *International Journal of Advanced Engineering Technologies and Innovations*, 1(3), 203–221.

Kumar, V., Ramachandran, D., & Kumar, B. (2021). Influence of new-age technologies on marketing: A research agenda. *Journal of Business Research*, 125, 864–877. DOI: 10.1016/j.jbusres.2020.01.007

Liu, Y., Ma, X., Shu, L., Hancke, G. P., & Abu-Mahfouz, A. M. (2020). From industry 4.0 to agriculture 4.0: Current status, enabling technologies, and research challenges. *IEEE Transactions on Industrial Informatics*, 17(6), 4322–4334. DOI: 10.1109/TII.2020.3003910

Moreno, J. C., Berenguel, M., Donaire, J. G., Rodríguez, F., Sánchez-Molina, J. A., Guzmán, J. L., & Giagnocavo, C. L. (2024). A pending task for the digitalisation of agriculture: A general framework for technologies classification in agriculture. *Agricultural Systems*, 213, 103794. DOI: 10.1016/j.agsy.2023.103794

Shalender, K., & Yadav, R. K. (2019). Strategic flexibility, manager personality, and firm performance: The case of Indian Automobile Industry. *Global Journal of Flexible Systems Managment*, 20(1), 77–90. DOI: 10.1007/s40171-018-0204-x

Sridhar, A., Balakrishnan, A., Jacob, M. M., Sillanpää, M., & Dayanandan, N. (2023). Global impact of COVID-19 on agriculture: Role of sustainable agriculture and digital farming. *Environmental Science and Pollution Research International*, 30(15), 42509–42525. DOI: 10.1007/s11356-022-19358-w PMID: 35258730

Zhang, X., & Fan, D. (2024). Can agricultural digital transformation help farmers increase income? An empirical study based on thousands of farmers in Hubei Province. *Environment, Development and Sustainability*, 26(6), 14405–14431. DOI: 10.1007/s10668-023-03200-5 PMID: 37362968

Chapter 18
Applications, Advantages, and Challenges of Artificial Intelligence in Agriculture

Aparna Sharma
https://orcid.org/0009-0003-3979-9905
Government College, Barwala, India

Anurag Tewari
Chitkara University Institute of Engineering and Technology, Chitkara University, Punjab, India

Ajit Bansal
Chitkara Business School, Chitkara University, Punjab, India

ABSTRACT

The economy is significantly influenced by agriculture. Globally, the primary issue and burgeoning topic is agriculture mechanization. The population is growing at an exponential rate, which is simultaneously driving up demand for jobs and food. For the farmers to meet these needs, the conventional techniques they employed were insufficient. New automated techniques were consequently introduced. The utilization of Artificial intelligence technique has protected agricultural productivity from various threats, such as rising populations, changing climates, employment difficulties, and problems with food safety. Architectures based on machine learning have been confirmed to be among the best regarding precision and resilience. By using systems based on AI, crop yields can be increased while using less water, fertilizer, and pesticides. Besides improving worker safety, intelligent expertise can support in

DOI: 10.4018/979-8-3693-4330-2.ch018

reducing its consequence on expected ecosystems, which will help keep food costs low and ensure more crops are grown to feed the world's expanding population.

1. INTRODUCTION

Among the earliest and biggest industries in the entire globe is agriculture. The need for food and jobs is rising due to the worldwide increase in population. Therefore, to meet the globe's food requirements—which also happen to create jobs for billions of people—new automated ways are being implemented in place of the conventional farming methods.

Due to a lack of labor, tighter laws, rising worldwide population, and dwindling farming population, producers are compelled to look for new alternatives. Almost every sector is impacted by technologies like the Internet of Things, Big Data & Analytics, Artificial Intelligence (AI), and Machine Learning (ML). Research and efforts are being made to enhance the caliber and output of food products by utilizing "smart farming," which involves connecting and adaptive farming practices (Jha et al, 2019, Smith, 2018).

By 2050, there will likely be close to 10 billion people on the planet, which would increase agricultural productivity by almost 50% compared to 2013—all while maintaining modest levels of economic growth (FAO, 2017). Currently, crops are grown on around 37.7% of the land surface. Agriculture has a momentous impression on the nation's economy and creates jobs. It is actively influencing the economies of developing countries while also heavily adding to the financial stability of industrialized nations. The farming community's disposable revenue has considerably augmented because of the extension of agriculture.

Hence, it makes sense and is suitable to concentrate more on the agriculture sector. In countries like India, the agricultural sector employs half of the labour force and contributes 18% of the GDP. Agriculture-related expansion will accelerate rural growth, which will then propel change in the countryside and, ultimately, a structural shift (Mogili and Deepak, 2018; Shah et al., 2019).

Numerous segments worldwide have seen a noteworthy shift in response to the introduction of know-how (Kakkad et al., 2019). Interestingly, regardless of being the most digitalized industry, agriculture has experienced a surge in the creation and monetization of agricultural innovations. AI has started to become a significant part of what we do every day, expanding our perspectives and giving us the potential to alter our surroundings. As a result, a system for harvest scheduling that combines crop allocation with routing of vehicles is provided. The workforce, who was once limited to a few business sectors, has since contributed to many industries thanks to growing technologies. Creating a device that works like an individual's brain is the fundamental idea behind artificial intelligence.

Intelligent systems and programs are created by investigating how the human brain functions, including how populace acquire knowledge, make choices, and come together to resolve issues. Similar to the human brain, this software is fed learning data, and such machines give us the desired response for each valid input.

Artificial Intelligence is an emerging innovation in agriculture. AI-powered machinery and equipment have raised the bar for today's agricultural sector. Crop production has augmented accompanied by continuous tracking, collection, processing, and marketing thanks to this technology (Yanh et al., 2007). The agro dependent industry has furthered importantly from the most recent automated system innovations, which use drones and farm robots. Many advanced computerized systems are made to identify numerous crucial factors, such as quality of crops, production, and weed identification, among many other methods.

It is challenging to make inferences in the changing field of agriculture that could lead to a thorough explanation. Thanks to artificial intelligence techniques, we are now able to understand the nuances of every circumstance and provide the optimal answer for each issue. Different artificial intelligence systems are gradually solving increasingly complex and challenging tasks. According to Baker et al. (1983), machine learning was mainly utilized in agriculture in 1983. Ample procedures have been extinguished to handle the present issues facing agriculture, ranging from databases to managerial support systems (Thorpe et al., 1992).

In recent decades, agricultural manufacturing processes have suffered greatly due to factors such as changing climates, rising farming costs, diminishing water for irrigation equipment, and a decrease in the number of farm workers (Jung et al., 2021). Moreover, the COVID-19 pandemic threatens to disrupt the cultivation and distribution of food chains (Outlaw et al., 2020).

Significant discoveries are always needed to stay ahead of **ongoing** climate change (Hatfield et al., 2014). The apparent problem in this situation is how to gather sufficient food to nourish the world's growing population. Investigators are always putting state-of-the-art information to use and coming up with fresh approaches to integrate it into the system of agriculture.

1.2 The Role of AI in the Agriculture Sector

Agriculture is a labor-intensive industry, and technology is playing a bigger role as the need for agricultural output rises along with the population. When it comes to parts, technology, and applications, AI greatly helps farmers. Crop availability and quality are ensured by enhanced farm and crop management practices and statistical analysis. Businesses calculate acreage and monitor the condition of crops in real time using satellite pictures and meteorological data. (Subeesh and Mehta 2021). Big data, artificial intelligence, and machine learning are tools that businesses may employ to predict prices, estimate tomato yield and output, and spot pest and disease outbreaks. They can offer farmers advice on how much demand there will be, what crop variations to plant to maximize profits, how to use pesticides, and potential price trends.

As AI significantly lessens the labor and resource deficit, it will be a powerful tool for organizations to deal with the increasing complexity of modern agriculture. Big businesses should be making investments in this field long overdue. AI technology is being used by many industries to increase efficiency and production (Awasthi, 2020; Skvortsov, 2020). AI technologies are assisting individuals across many industries in overcoming traditional challenges. A few industries that use AI applications are accounting or finance, transit, medical care, and the agricultural sector (Shadrin et al., 2019)

1.2.1 Why AI is Suitable to be Used in Agriculture Sector?

Farmers frequently struggle to get good yields under unfavorable weather or during crop illnesses. In a comparable way, the absence of digitization amid worldwide epidemics renders overseeing diverse procedures more difficult.

Concurrently, continual upsurge in the global population and development has been noticed. Growing financial resources and changing patterns of spending put producers under tremendous strain to provide the growing global food demand. This problem makes it hard to develop methods to increase output because there isn't as much organic matter available, which means that conventional farming methods must be abandoned.

Investigating the application of machine learning (AI) in agricultural globally shows enormous potential for reducing risks and improving management. Artificial Intelligence has the potential to completely change the agriculture industry through offering farmer's additional throughput with less exertion and besides this it also offers various other benefits at least cost. Yet it's important to recognize that Intelligence is an additional instrument that can enhance current agricultural techniques rather than an independent method.

We'll immediately examine the potential that artificial intelligence (AI) offers the agricultural industry and discover how this state-of-the-art technology can support overcome current challenges.

1.2.2 What Role can AI Play in the Agricultural Industry?

Numerous mechanical steps and procedures are needed in agricultural. Intelligent machines (AI) can assist by improving current gadgets and making difficult but routine tasks simpler. When collective with countless other techniques, it may gathcr and analyzc cnormous amounts of figures from a website, determine the most effective course of action, and sometimes start things. Farmers will profit from the technology's many applications, which include analyzing farm data to improve crop yield and accuracy; using AI sensors to identify weeds; and detecting illnesses in vegetation, pests, and other things.

Agriculture robots, which are meant to assist farmers, are the answer to this. This bot collects crops more quickly and in larger quantities. Intelligent machines and agriculture working together can help with several procedures:

1.2.2.1 Soil Grounding

Farmers challenge vagueness, impulsiveness, enquiring, and much more while treating the soil in the absence of technological resolutions. The method of concocting soil involved various judgements, like ploughing, adding nutrients, generating elevations and reductions etc. Since the farmers have experience of plenty of years so the producers have faith in them as far as soil preparation is concerned. Hence this task is not exactly science, but it is an art.

The crop bears the price of procrastination. Their choices are informed by precise information as opposed to guesswork thanks to AI.

AI-driven technology with smart sensors are utilized to obtain current time soil information. Further artificial intelligence algorithms analyze the biological composition of soil specimens, revealing important details about the soil's composition, moisture content, and nutrient content. Moreover, for analyzing the historical weather conditions, Machine Leaning techniques can be used.

The producers can promote utilization of this evidence for soil preparation and for improving the largely excellence of the soil. It also helps them to analyze soil quality scientifically.

1.2.2.2 Improving Automated Drenching Organisms

With the help of modern techniques like Artificial Intelligence, farming can be done unconventionally. The quantity of water needed for irrigation could be found out accurately by utilizing IOT detectors and algorithms. Hence it will help in improving the excellence of manufacture and overall productivity.

1.2.2.3 Finding Outflow or Irrigation Scheme Impairments

The main threat which the developing countries are facing is low productivity, lack of irrigation facilities and automation techniques. Being an agrarian economy, we must introduce automation in agriculture so as to resolve various challenges associated with farming sector. For increasing food demand and export opportunities the Government should plan to implement automation techniques.

1.2.2.4 Organizing Pests and Infections

Machine vision can identify diseases and pests along with monitoring crop growth and soil conditions. This works by employing machine learning to examine images for pests, rot, mold, and various other crop health hazards. This enables producers to respond quickly to remove bugs or isolate vegetation to stop the propagation of disease when combined with safety measures.

Researchers had to first collect images of these beetles to obtain the required quantity of data for the task set used for teaching the algorithm's principles.

1.2.2.5 Recognizing the Fitness of Livestock

It might seem easier to spot health problems in livestock than in agricultural products, but it's really very hard for one to do so. Thankfully, AI can help with this. In January 2019, Terry Canning, a dairy farmer from Co Armagh, Northern Ireland, and Adam Askew, a senior architect, founded CattleEye and this company has developed a system that marks it possible to track the condition of cattle remotely using recording devices, unmanned aircraft, and machine vision. It can identify aberrant cow activity and tell apart normal behaviors from birthing.

Using AI and ML innovations, CattleEye analyzes how nutritional and external variables affect animals, yielding valuable information. Farmers can improve the physical condition of their cattle and increase the production of milk by using this knowledge.

1.2.2.6 Reduction in Cost

Artificial Intelligence helps the farmers in reducing the input costs and hence minimizing the overall cost of production. Through modern techniques we can increase productivity, efficiency and achieve sustainability. Farmers can minimize the wastage and input costs by utilizing insecticides, composts, and water for irrigation facilities.

If we introduce automation techniques in agriculture, labour cost, efficiency and productivity will increase.

1.2.2.7 Improving Automatic Watering Systems

Algorithms used in AI in agriculture allow for autonomous crop management. When coupled with the IoT, sensors that monitor soil moisture and rainfall patterns, computers can determine in real-time how critical it is to provide water to crops. The goal of an independent agricultural irrigation apparatus is to support sustainable farming methods whilst preserving water.

1.2.2.8 Finding Dumps or Irrigation System Failure

Artificial Intelligence is essential for identifying water system leaks. Algorithms can test data to find patterns and abnormalities that point to implicit leaks. Machine learning (ML) models can be trained to recognize leak indicators, such as temperature or inflow variations in water

1.3 Artificial Intelligence (AI) Uses for Agriculture

1.3.1 Management of Crops

The very first step in managing the crop is seeding, which also includes harvesting, storing, and distributing the crop. It can be defined as methods that amplify the output and growth of agricultural creations. A good awareness of the various crop classes and when to plant them about the type of soil that thrives would surely increase crop productivity. Agricultural gathering is a field that appeals to automation for several reasons.

Shortly, it looks to be both technically as well as economically possible to automate agriculture harvesting machinery (Kakkad et al., 2019; Bansal et al., 2023).

1.3.2 Soil Management

An essential component of agricultural activities is soil management. A deep grasp of the various types and characteristics of soil will enhance crop output and preserve soil resources. It is the process of applying procedures, methods, and treatments to enhance soil productivity. Increasing nitrate concentrations in groundwater have been attributed to inadequate handling of nitrogen (N). To develop and test a model that strives for the best nitrogen management to: (i) reduce nitrate leaching; (ii) maximize production; and (iii) maximize profits, Li and Yost (2000) conducted research (Ahngar et al., 2022).

1.3.3 Analytics that Predict

The results of climate revolution mean that producers can no longer rely only on traditional wisdom, making forecasts more important for crop production. If projections are more accurate, farmers might be able to select the ideal days for planting or harvesting. AI techniques use reinforced learning to learn from past predictions and outcomes. The process of weather forecasting involves feeding data into an algorithm that uses methods of deep learning to learn from past data and provide forecasts (Nihalani, 2024).

1.3.4 Weather Prediction

Artificial intelligence for agricultural purposes can be utilized to anticipate weather patterns, analyze crop sustainability, and inspect farms for diseases and pests (Chukkapalli et al., 2020; Sharma and Bansal, 2023). It can also be integrated with satellite data.

1.3.5 Management of Insect Pests and Diseases

For agricultural harvests to yield as much as possible, prevention of disease is necessary. Diseases of plants and animals are a major barrier to raising productivity. These diseases that affect plants and animals are influenced by a variety of factors, including wind, temperature, rain, drought, soil type, and genetics. Particularly in agriculture on a large scale, regulating the effects of these elements and the unpredictable causal influences of certain illnesses is a significant problem. To effectively control diseases and reduce losses, farmers should employ a comprehensive approach to disease prevention and control that combines physical, chemical, and biological approaches (Javaid et al., 2023). With the assistance of the internet, this technology helps users diagnose ailments caused by mushrooms and offers beneficial suggestions.

1.3.6 Control of Weeds

Weeds are unique of the foremost things that reduce a farmer's expected earnings. For instance, if weed incursion is not managed, dried bean and maize crops may lose 52% of their yield and wheat production may be reduced by 47% due to weed competition. Though certain weeds are destructive and potentially pretense a health risk to the public, weeds nevertheless contend with crops over resources like nutrients, water, and sunshine. Although spray is regularly used to prevent weed growth, overuse of the product can contaminate the natural world and hurt public health. Therefore, to conclude the accurate quantity of sprinkle to use and to correctly sprinkle on the target spot, computerized weed-detecting technologies have been developed in laboratories (Barenkamp, 2020).

1.3.7 Aid in Producing Healthier Crops

For making a diversity of agriculture-related jobs better throughout the food chain, control pests, assess soil and growth illnesses, organize data for farmers, produce healthier crops, and assist with workloads. The agricultural sector is becoming increasingly reliant on AI technologies. Farmers find it challenging to pinpoint the ideal time to plant seeds due to changing seasons and rising pollution levels. Farmers can schedule when to plant seeds and what kind of crop to produce by combining weather forecasts with machine learning (AI) to evaluate the climate. By streamlining the crop choice procedure, AI can assist farmers in determining which crops will bring in the most money. AI can assist farmers in determining which crops will bring in the most money and also in eliminating non profitable crops (Orn et al., 2020; Lin et al., 2021).

1.3.8 Forecast Plant Diseases

Artificial Intelligence has the capability of recognizing and eradicating weeds, anticipating plant disease symptoms, and suggesting effective pest management strategies. Artificial Intelligence facilitates the forecasting of the optimal mix of agronomic goods, scheduling of cultivation, and timing of nutrient delivery. AI allows for the automation of harvesting and even the prediction of the best time to do it. Predictive analytics applications have the potential to drastically change sectors. More data can be collected by framers with the help of machine learning. Farmers can also seek the support of AI for seeking the solutions of so many problems like forecasting of prices, prediction of market demand testing of soil quality and for agricultural planting and harvesting schedules (Partel et al., 2021).

1.3.9 Predict Plant Ailments

Artificial intelligence can identify and eliminate weeds, predicting the signs of plant illnesses, and making practical insect control recommendations. Artificial intelligence makes it easier to predict the ideal combination of horticultural products, schedule irrigation, and provide nutrients at the accurate time. AI makes it conceivable to automate gathering and to forecast when it will be furthermost effective. Applications of statistical analysis have the power to fundamentally alter industries. With AI, farmers can analyze and collect more data than was conceivable in the past. Farmers can employ AI to address critical issues such as consumer need assessments, pricing estimates, and establishing and harvesting patterns (Songol et al., 2021; Santangeli et el., 2020).

1.3.10 Finding Trends of Unproductive Usage of Sources

Through data analysis and interpretation on the distribution of resources and utilization artificial intelligence (AI) systems possess the capability to detect trends in inefficient resource utilization and offer optimization alternatives. The expenditure of veterinarian services and equipment maintenance is reduced by proactive measures including monitoring animal health as well as performance of equipment. If farmers implement AI into agriculture, it will help them in enhancing yields without spending much amount on the additional resources and on the other hand it will also help farmers in lowering the overall cost related to different stages of agricultural processes (Sánchez et al., 2020; Raman et al. 2021).

1.3.11 Give appropriate directions for managing water resources.

Programs that give farmers the best guidance on bug diseases, crop rotation, managing water, timely picking, and other subjects are the result of using AI in agriculture. We can assess agricultural sustainability, control nutrition, and predict weather patterns with machine-learning algorithms and images from satellites and drones. Reliable data inputs are used in precise agriculture to optimize agricultural yields. Using artificial intelligence and smart phones, producers may obtain a personalized agricultural plan (Zhang, 2020).

1.4 Artificial Intelligence's Challenges in Agriculture

Many believe that artificial intelligence is an item that is exclusive to the digital realm and has little bearing on actual farming chores. Famers are by and large not very educated and due to this issue, they are not able to understand the benefits of introducing modern techniques in agriculture. Most of the farmers still want to use the conventional methods of farming.

1.4.1 Substantial Upfront Overheads

The utilization of modern automated techniques like artificial intelligence in the field of agriculture involves huge expenditure and so it may not suit small and marginal farmers. Nevertheless, the authorities can guide the farmers about the probable advantages of using modern techniques and can provide subsidies to small and marginal farmers so that they can afford such inventions.

1.4.2 Insufficient Hands-on Knowledge with Emerging Technologies

Globally, there are differences in the level of technological improvement in multiple facets of agriculture. While some areas could fully gain from AI, otherwise there are several obstacles in nations where modern agriculture equipment is rare. Technology businesses might have to implement a visionary strategy if they hope to conduct business in areas where agriculture is becoming more and more prevalent. Apart from providing their goods, businesses also need to provide farmers and agribusiness owners who are prepared to take on novel outcomes with education and continuous assistance (Bali et al., 2023).

1.4.3 Infrastructure Needs

Artificial intelligence in agriculture cannot be successfully implemented without a state-of-the-art and dependable technology platform. Before implementing ML solutions, consumers should consider electricity, internet access, and availability of devices. Unfortunately, a lot of rural places lack sufficient access to cutting-edge technology, which can make gathering and analyzing data quite difficult.

1.4.4 Cultural Divides

The biggest agricultural enterprises in the globe and tiny family-owned farms can take advantage of utilizing artificial intelligence to agriculture. Individuals in the latter group, on the other side, are less inclined to change their farming for a complex web of tenacious beliefs, customs, and presumptions. For instance, a lot of farmers are adamant that the most significant condition for creating excellent outcomes is the depth of their psychological bond with their cattle (Solanki et al., 2024).

Notwithstanding these difficulties, AI is being used more and more in agriculture, and producers that use these novel instruments should see gains in productivity, lower expenses, and higher yields. Farmers face lots of challenges for harvesting, irrigation and other tasks. In order to enable the farmers to face these challenges properly and execute their task effectively organizations, authorities must work together to provide suitable environment to farmers so that they can incorporate AI into farming. It means not just offering guidance, support, and affordable technology, but also establishing a friendly regulatory setting that encourages investment and innovation in this field.

2. CONCLUSION

Many different manual jobs and processes are needed in agriculture. Innovative techniques like machine learning and artificial intelligence help in improving the processes and upgrading the quality of production. By utilizing such innovations farmers can check the weather conditions and make customized plans as per their requirements.

Artificial Intelligence. also helps the farmers in increasing their productivity without increasing costs of production. Hence the increasing domestic demand for food grains can be met and surplus production can be utilized for earning valuable foreign exchange.

Farmers can also check through automated mechanisms about the area which require major attention for pesticide, fertilizers and irrigation. It aids farmers in understanding what they learn by gathering data on wind speed, temperatures, sun irradiation and precipitation. AI technology could potentially assist producers in addressing issues such as decreased yields caused by pest and plant diseases, as well as rising temperatures. AI will be used in agriculture to improve the industry's overall success.

REFERENCES

Ahngar, T. A., Bahar, F. A., Singh, L., Bhat, M. A., Singh, P., Mahdi, S. S., & Ahmad, L. (2022). Artificial intelligence in agriculture, applications, benefits and challenges: A review. *The Pharma Innovation Journal*, 11(3), 1407–1414.

Awasthi, Y. (2020). Press â€ œAâ€ for Artificial Intelligence in Agriculture: A Review. *JOIV: International Journal on Informatics Visualization*, 4(3), 112–116. DOI: 10.30630/joiv.4.3.387

Bali, A., Chouhan, S. S., Kumar, G., Kumar, R., & Singh, U. P. (2023). Adaptive Fault-Tolerant Control for Pure-Feedback Stochastic Nonlinear Systems with Sensor and Actuator Faults. *Circuits, Systems, and Signal Processing*, 42(10), 5838–5867. DOI: 10.1007/s00034-023-02366-7

Bansal, A., Bansal, R., Bansal, A., Kumar, M., & Kumar, P. (2023). Normative Concerns in Neuromarketing. In *Promoting Consumer Engagement Through Emotional Branding and Sensory Marketing* (pp. 234–243). IGI Global.

Bansal, A., Tewari, A., Sharma, A., & Bansal, A. (2024). Implementations and Rationale for Blockchain Technique in Agriculture. In Chouhan, S. S., Singh, U. P., & Jain, S. (Eds.), *Applications of Computer Vision and Drone Technology in Agriculture 4.0*. Springer., DOI: 10.1007/978-981-99-8684-2_6

Barenkamp, M. (2020, June). A new IoT gateway for artificial intelligence in agriculture. In *2020 International Conference on Electrical, Communication, and Computer Engineering (ICECCE)* (pp. 1-5). IEEE. DOI: 10.1109/ICECCE49384.2020.9179418

Chouhan, S. S., Singh, U. P., & Jain, S. (2024). Introduction to Computer Vision and Drone Technology. In Chouhan, S. S., Singh, U. P., & Jain, S. (Eds.), *Applications of Computer Vision and Drone Technology in Agriculture 4.0*. Springer., DOI: 10.1007/978-981-99-8684-2_1

Chukkapalli, S. S. L., Mittal, S., Gupta, M., Abdelsalam, M., Joshi, A., Sandhu, R., & Joshi, K. (2020). Ontologies and artificial intelligence systems for the cooperative smart farming ecosystem. *IEEE Access : Practical Innovations, Open Solutions*, 8, 164045–164064. DOI: 10.1109/ACCESS.2020.3022763

Fennimore, S. A., Slaughter, D. C., Siemens, M. C., Leon, R. G., & Saber, M. N. (2016). Technology for automation of weed control in specialty crops. *Weed Technology*, 30(4), 823–837. DOI: 10.1614/WT-D-16-00070.1

Food and Agriculture Organization (FAO). (2017). *The State of Food and Agriculture: Leveraging Food Systems for Inclusive Rural Transformation*. Food and Agriculture Organization of the United Nations.

Jha, K., Doshi, A., Patel, P., & Shah, M. (2019). A comprehensive review on automation in agriculture using artificial intelligence. *Artificial Intelligence in Agriculture*, 2, 1–12. DOI: 10.1016/j.aiia.2019.05.004

Kakkad, V., Patel, M., & Shah, M. (2019). Biometric authentication and image encryption for image security in cloud framework. *Multiscale and Multidisciplinary Modeling. Experiments and Design*, 2(4), 233–248. DOI: 10.1007/s41939-019-00049-y

Li, M., & Yost, R. S. (2000). Management-oriented modeling: Optimizing nitrogen management with artificial intelligence. *Agricultural Systems*, 65(1), 1–27. DOI: 10.1016/S0308-521X(00)00023-8

Lim, W. M., Gupta, S., Aggarwal, A., Paul, J., & Sadhna, P. (2021). How do digital natives perceive and react toward online advertising? Implications for SMEs. *Journal of Strategic Marketing*, •••, 1–35.

Mogili, U. R., & Deepak, B. B. V. L. (2018). Review on application of drone systems in precision agriculture. *Procedia Computer Science*, 133, 502–509. DOI: 10.1016/j.procs.2018.07.063

Mumtaz, N., & Nazar, M. (2022). Artificial intelligence robotics in agriculture: See & spray. *Journal of Intelligent Pervasive and Soft Computing*, 1(01), 21–24.

Nihalani, R., Chouhan, S. S., Mittal, D., Vadula, J., Thakur, S., Chakraborty, S., ... & Saxena, A. (2024). Long Short-Term Memory (LSTM) model for Indian sign language recognition. *Journal of Intelligent & Fuzzy Systems*, (Preprint), 1-19. .DOI: 10.3233/JIFS-233250

Orn, D., Duan, L., Liang, Y., Siy, H., & Subramaniam, M. (2020, October). Agro-AI education: artificial intelligence for future farmers. In *Proceedings of the 21st Annual Conference On Information Technology Education* (pp. 54-57). DOI: 10.1145/3368308.3415457

Panchal, B., Pranjal, P., Patel, R. K., Sharma, A., & Chouhan, S. S. (2024). Assessing the Quantity of a Crop Field Using Aerial Images. In Chouhan, S. S., Singh, U. P., & Jain, S. (Eds.), *Applications of Computer Vision and Drone Technology in Agriculture 4.0*. Springer., DOI: 10.1007/978-981-99-8684-2_11

Partel, V., Costa, L., & Ampatzidis, Y. (2021). Smart tree crop sprayer utilizing sensor fusion and artificial intelligence. *Computers and Electronics in Agriculture*, 191, 106556. DOI: 10.1016/j.compag.2021.106556

Raman, D. R., Saravanan, D., Parthiban, R., Palani, D. U., David, D. D. S., Usharani, S., & Jayakumar, D. (2021). A study on application of various artificial intelligence techniques on internet of things. *European Journal of Molecular and Clinical Medicine*, 7(9), 2531–2557.

Sánchez, J. M., Rodríguez, J. P., & Espitia, H. E. (2020). Review of artificial intelligence applied in decision-making processes in agricultural public policy. *Processes (Basel, Switzerland)*, 8(11), 1374. DOI: 10.3390/pr8111374

Santangeli, A., Chen, Y., Kluen, E., Chirumamilla, R., Tiainen, J., & Loehr, J. (2020). Integrating drone-borne thermal imaging with artificial intelligence to locate bird nests on agricultural land. *Scientific Reports*, 10(1), 10993. DOI: 10.1038/s41598-020-67898-3 PMID: 32665596

Saxena, A., Chouhan, S. S., Aziz, R. M., & Agarwal, V. (2024). A comprehensive evaluation of Marine predator chaotic algorithm for feature selection of COVID-19. *Evolving Systems*, 15(4), 1235–1248. Advance online publication. DOI: 10.1007/s12530-023-09557-2

Shadrin, D., Menshchikov, A., Ermilov, D., & Somov, A. (2019). Designing future precision agriculture: Detection of seeds germination using artificial intelligence on a low-power embedded system. *IEEE Sensors Journal*, 19(23), 11573–11582. DOI: 10.1109/JSEN.2019.2935812

Shah, G., Shah, A., & Shah, M. (2019). Panacea of challenges in real-world application of big data analytics in healthcare sector. *Journal of Data. Information & Management*, 1(3-4), 107–116. DOI: 10.1007/s42488-019-00010-1

Shalender, K., & Yadav, R. K. (2019). Strategic flexibility, manager personality, and firm performance: The case of Indian Automobile Industry. *Global Journal of Flexible Systems Managment*, 20(1), 77–90. DOI: 10.1007/s40171-018-0204-x

Sharma, A., & Bansal, A. (2023). Digital Marketing in the Metaverse: Beginning of a New Era in Product Promotion. In *Applications of Neuromarketing in the Metaverse* (pp. 163-175). IGI Global.

Sharma, A., Patel, R. K., Pranjal, P., Panchal, B., & Chouhan, S. S. (2024). Computer Vision-Based Smart Monitoring and Control System for Crop. In Chouhan, S. S., Singh, U. P., & Jain, S. (Eds.), *Applications of Computer Vision and Drone Technology in Agriculture 4.0*. Springer., DOI: 10.1007/978-981-99-8684-2_5

Skvortsov, E. A. (2020). Prospects of applying artificial intelligence technologies in the regional agriculture. *Ekonomika Regiona= Economy of Regions*, (2), 563.

Smith, M. J. (2018). Getting value from artificial intelligence in agriculture. *Animal Production Science*, 60(1), 46–54. DOI: 10.1071/AN18522

Solanki, S., Singh, U. P., Chouhan, S. S., & Jain, S. (2024). A systematic analysis of magnetic resonance images and deep learning methods used for diagnosis of brain tumor. *Multimedia Tools and Applications*, 83(8), 23929–23966. DOI: 10.1007/s11042-023-16430-6

Songol, M., Awuor, F., & Maake, B. (2021). Adoption of artificial intelligence in agriculture in the developing nations: A review. *Journal of Language. Technology & Entrepreneurship in Africa*, 12(2), 208–229.

Subeesh, A., & Mehta, C. R. (2021). Automation and digitization of agriculture using artificial intelligence and internet of things. *Artificial Intelligence in Agriculture*, 5, 278–291. DOI: 10.1016/j.aiia.2021.11.004

Zhang, J. (2020, October). Research on digital image processing and recognition technology of weeds in maize seedling stage based on artificial intelligence. In *Journal of Physics*. Conference. DOI: 10.1088/1742-6596/1648/4/042058

Compilation of References

Abass, A. B., Ndunguru, G., Mamiro, P., Alenkhe, B., Mlingi, N., & Bekunda, M. (2014). Post-harvest food losses in a maize-based farming system of semi-arid savannah area of Tanzania. *Journal of Stored Products Research*, 57, 49–57. DOI: 10.1016/j.jspr.2013.12.004

Abiri, R., Rizan, N., Balasundram, S. K., Shahbazi, A. B., & Abdul-Hamid, H. (2023). Application of digital technologies for ensuring agricultural productivity. *Heliyon*, 9(12), e22601. Advance online publication. DOI: 10.1016/j.heliyon.2023.e22601 PMID: 38125472

Acharya, S. S., & Agarwal, N. L. (2021). Agricultural Marketing. In *India* (7th Revised edition). CBS Publishers & Distributors.

Ada, N., Kazancoglu, Y., Sezer, M. D., Ede-Senturk, C., Ozer, I., & Ram, M. (2021). Analyzing barriers of circular food supply chains and proposing industry 4.0 solutions. *Sustainability (Basel)*, 13(12), 1–29. DOI: 10.3390/su13126812

Adebanjo, D. (2009). Understanding demand management challenges in intermediary food trading: A case study. *Supply Chain Management*, 14(3), 224–233. DOI: 10.1108/13598540910954566

Adhikari, K. B., Regmi, P. P., Gautam, D. M., Thapa, R. B., & Joshi, G. R. (2017). Value chain analysis of orthodox tea: Evidence from Ilam district of Nepal. *Journal of Agriculture and Forestry University*, 1(1), 61–68.

Adjognon, G. S., Gassama, A., Guthoff, J., & Pouliquen, V. (2019). Implementing Effective Warehouse Receipt Financing Systems: Lessons from a pilot WRS project in the Senegal River Valley. The World Bank, 139445, 1-28. retrieved from https://documents1.worldbank.org/curated/en/173921563521093845/pdf/Implementing-Effective-Warehouse-Receipt-Financing-Systems-Lessons-from-a-Pilot-WRS-Project-in-the-Senegal-River-Valley.pdf

AfDB. (2023). *Perspectives économiques au Togo*. Retrieved from https://www.afdb.org/fr/countries/west-africa/togo/togo-economic-outlook

Agribots – How Robots Made Their Way onto Farms – Pigeonsfarm.co.uk. (2019, February 3).https://pigeonsfarm.co.uk/agribots-how-robots-made-their-way-onto-farms/

Agriculture Dive. (n.d.). Israel-Hamas war: Fertilizer exports and shipping through Gaza. https://www.agriculturedive.com/news/israel-hamas-war-fertilizer-exports-shipping-gaza/696567/

Agriculture Sector - India GHG Program. India GHG Program. Retrieved from https://indiaghgp.org/agriculture-sector

Ahmad, F., Zakaria, N. H., & Osman, W. R. S. (2008, April). Transforming Information-Based Agricultural Portal to Knowledge-Based Agricultural Hub. In 2008 3rd International Conference on Information and Communication Technologies: From Theory to Applications (pp. 1-4). IEEE. DOI: 10.1109/ICTTA.2008.4529914

Ahngar, T. A., Bahar, F. A., Singh, L., Bhat, M. A., Singh, P., Mahdi, S. S., & Ahmad, L. (2022). Artificial intelligence in agriculture, applications, benefits and challenges: A review. *The Pharma Innovation Journal*, 11(3), 1407–1414.

Ajila, C., Brar, S. K., Verma, M., & Prasada Rao, U. (2012). Sustainable Solutions for Agro Processing Waste Management: An Overview. In Malik, A., & Grohmann, E. (Eds.), *Environmental Protection Strategies for Sustainable Development. Strategies for Sustainability*. Springer., DOI: 10.1007/978-94-007-1591-2_3

Akkartal, G. R., & Mızrak, F. (2024). Operational Efficiency and Competitiveness in the Global Logistics Industry: An Examination of Human Resources Management Strategies. In *Strategic Innovations for Dynamic Supply Chains* (pp. 71-102). IGI Global.

Akkucuk, U. (Ed.). (2020). *Handbook of research on sustainable supply chain management for the global economy*. IGI Global. DOI: 10.4018/978-1-7998-4601-7

Akram, U., Hui, P., Khan, M. K., Tanveer, Y., Mehmood, K., & Ahmad, W. (2018). How website quality affects online impulse buying: Moderating effects of sales promotion and credit card use. *Asia Pacific Journal of Marketing and Logistics*, 30(1), 235–256. DOI: 10.1108/APJML-04-2017-0073

Aksoy, M. A., & Beghin, J. C. (2005). *Global agricultural trade and developing countries*. The World Bank., Retrieved from https://cisp.cachefly.net/assets/articles/attachments/01518_agriculturaltrade.pdf#page=98

Alabi, M. O., & Ngwenyama, O. (2023). Food security and disruptions of the global food supply chains during COVID-19: Building smarter food supply chains for post COVID-19 era. *British Food Journal*, 125(1), 167–185. DOI: 10.1108/BFJ-03-2021-0333

Alazzai, W. K., Abood, B. S. Z., Al-Jawahry, H. M., & Obaid, M. K. (2024). Precision Farming: The Power of AI and IoT Technologies. In *E3S Web of Conferences* (Vol. 491, p. 04006). EDP Sciences.

Aldrich, D. P. (2012). *Building resilience: Social Capital in post-disaster recovery*. University of Chicago Press. DOI: 10.7208/chicago/9780226012896.001.0001

Ali, I., Nagalingam, S., & Gurd, B. (2017). Building resilience in SMEs of perishable product supply chains: Enablers, barriers and risks. *Production Planning and Control*, 28(15), 1236–1250. DOI: 10.1080/09537287.2017.1362487

Alkahtani, M., Khalid, Q. S., Jalees, M., Omair, M., Hussain, G., & Pruncu, C. I. (2021). E-agricultural supply chain management coupled with blockchain effect and cooperative strategies. *Sustainability (Basel)*, 13(2), 816. DOI: 10.3390/su13020816

Allard-Poesi, F. (2003). Coder les données. In Y. Giordano (Ed.), *Conduire un projet de recherche une perspective qualitative* (pp. 320). Edition EMS Management & société.

Amolegbe, K. B., & Adewumi, M. O. (2022). Agribusiness Firms and Rural Dairy Development. A Case of FrieslandCampina Dairy Development Programme in Nigeria. *AGRIS On-Line Papers in Economics and Informatics*, 14(1), 3–18. DOI: 10.7160/aol.2022.140101

Amorim, P., Meyr, H., Almeder, C., & Almada-Lobo, B. (2013). Managing perishability in production-distribution planning: A discussion and review. *Flexible Services and Manufacturing Journal*, 25(3), 389–413. DOI: 10.1007/s10696-011-9122-3

Amrapali, F. P. O. Kairo. (2020). Amrapali FPO Kairo. Retrieved September 10, 2024,from https://www.facebook.com/search/top/?q=amrapali%20fpo%20kairo

Anastasiadis, F., Apostolidou, I., & Michailidis, A. (2020). Mapping sustainable tomato supply chain in Greece: A framework for research. *Foods*, 9(5), 539. DOI: 10.3390/foods9050539 PMID: 32357573

Anderson, K., & Gaston, K. J. (2013). Lightweight unmanned aerial vehicles will revolutionize spatial ecology. In *Frontiers in Ecology and the Environment* (Vol. 11, Issue 3, pp. 138–146). DOI: 10.1890/120150

Anderson, D. P., & Hanselka, D. (2009). *Adding value to agricultural products.* Texas FARMER Collection.

Antonucci, F., Figorilli, S., Costa, C., Pallottino, F., Raso, L., & Menesatti, P. (2019). A review on blockchain applications in the agri-food sector. *Journal of the Science of Food and Agriculture*, 99(14), 6129–6138. DOI: 10.1002/jsfa.9912 PMID: 31273793

Anupindi, R., & Sivakumar, S. (2007). Supply chain reengineering in agri-business: A case study of ITC's e-Choupal. In *Building supply chain excellence in emerging economies* (pp. 265–307). Springer US. DOI: 10.1007/978-0-387-38429-0_12

Apriliyanti, I. D., & Alon, I. (2017). Bibliometric analysis of absorptive capacity. *International Business Review*, 26(5), 896–907. DOI: 10.1016/j.ibusrev.2017.02.007

Aramyan, L., Grainger, M., Logatcheva, K., Piras, S., Setti, M., Stewart, G., & Vittuari, M. (2021). Food waste reduction in supply chains through innovations: A review. *Measuring Business Excellence*, 25(4), 475–492. DOI: 10.1108/MBE-11-2019-0105

Aria, M., & Cuccurullo, C. (2017). bibliometrix: An R-tool for comprehensive science mapping analysis. *Journal of Informetrics*, 11(4), 959–975. DOI: 10.1016/j.joi.2017.08.007

Aribi, H. B., Ghidaoui, M. H., & Fattouch, S. (2023). Monitoring Environmental Performance of Agricultural Supply Chains Using Internet of Things. In Integrating Intelligence and Sustainability in Supply Chains (pp. 273-292). IGI Global.

Arishenbagam, P., Tamil Selvi, S., & Ganapathy Ram, B. (2024). Wireless Geofenced Inventory Management for Agricultural Warehouse. *Journal of the Institution of Engineers (India): Series B, 105*, 375–383. https://doi.org/DOI: 10.1007/s40031-023-00967-z

Ashok Kumar, A., Reddy, B.V.S., Blummel, M., Anandan, S., Reddy, Y.R., & Reddy, Ch.R., Srinivasa Rao, P., Reddy, P.S. (2010). On-farm evaluation of elite sweet sorghum genotypes for grain and stover yields and fodder quality. *Animal Feed Science and Technology*, 10S, 69–78.

Ashwani, D. G. &. M. (2017). *Indian Economy, 72nd Edition*. S. Chand Publishing.

Astill, J., Dara, R. A., Campbell, M., Farber, J. M., Fraser, E. D. G., Sharif, S., & Yada, R. Y. (2019). Transparency in food supply chains: A review of enabling technology solutions. *Trends in Food Science & Technology*, 91(July), 240–247. DOI: 10.1016/j.tifs.2019.07.024

Atadoga, N. A., Obi, N. O. C., Osasona, N. F., Onwusinkwue, N. S., Daraojimba, N. I., & Dawodu, N. S. O. (2024). AI in supply chain optimization: A comparative review of USA and African Trends. *International Journal of Science and Research Archive*, 11(1), 896–903. DOI: 10.30574/ijsra.2024.11.1.0156

Audilakshmi, S., Mall, A. K., Swarnalatha, M., & Seetharama, N. (2010). Inheritance of sugar concentration in stalk (brix), sucrose content, stalk and juice yield in sorghum. *Biomass and Bioenergy*, 34(6), 813–820. DOI: 10.1016/j.biombioe.2010.01.025

Aung, M. M., & Chang, Y. S. (2014). Traceability in a food supply chain: Safety and quality perspectives. *Food Control*, 39, 172–184. DOI: 10.1016/j.foodcont.2013.11.007

Awasthi, Y. (2020). Press â€ œAâ€ for Artificial Intelligence in Agriculture: A Review. *JOIV: International Journal on Informatics Visualization*, 4(3), 112–116. DOI: 10.30630/joiv.4.3.387

Ayyaz, S., Bonney, L. B., & Adhikari, R. (2023). Consumers' insight and behaviour towards mango quality and safety attributes: Implications for integrated value chain development. *British Food Journal*, 125(12), 4518–4538. DOI: 10.1108/BFJ-01-2023-0022

Ba, M. N. (2016). Strategic agricultural commodity value chains in Africa for increased food: the regional approach for food security.

Bai, C., & Sarkis, J. (2020). A supply chain transparency and sustainability technology assessment model for blockchain technology. *Journal of Cleaner Production*, 258, 120614. DOI: 10.1016/j.jclepro.2020.120614

Balaji, M., Arshinder, K., 2016. Modeling the causes of food wastage in Indian perishable food supply chain. Resour. Conserv. Recycl. 114, 153–167. https://doi.org/ 25, 2272–2317. https://doi.org/DOI: 10.1016/j.resconrec.2016.07.016Benchmarking

Bali, A., Chouhan, S. S., Kumar, G., Kumar, R., & Singh, U. P. (2023). Adaptive Fault-Tolerant Control for Pure-Feedback Stochastic Nonlinear Systems with Sensor and Actuator Faults. *Circuits, Systems, and Signal Processing*, 42(10), 5838–5867. DOI: 10.1007/s00034-023-02366-7

Balié, J., Del Prete, D., Magrini, E., Montalbano, P., & Nenci, S. (2019). Does trade policy impact food and agriculture global value chain participation of sub-Saharan African countries? *American Journal of Agricultural Economics*, 101(3), 773–789. DOI: 10.1093/ajae/aay091

Balkrishna, A., Pathak, R., Kumar, S., Arya, V., & Singh, S. K. (2023). A comprehensive analysis of the advances in Indian Digital Agricultural architecture. *Smart Agricultural Technology*, 5, 100318. DOI: 10.1016/j.atech.2023.100318

Bammann, H. (2019). Participatory value chain analysis for improved farmer incomes, employment opportunities and food security.

Bangladesh Bureau of Statistics (BBS) 2019. Report on Agriculture and Rural Statistics, Statistics Division, Ministry of Planning, Govt. of Bangladesh.

Bansal, A. K., & Bansal, A. (2012). Micro finance and poverty reduction in India. *Integral Review-A Journal of Management*, 5(1), 31–35.

Bansal, A., Bansal, R., Bansal, A., Kumar, M., & Kumar, P. (2023). Normative Concerns in Neuromarketing. In *Promoting Consumer Engagement Through Emotional Branding and Sensory Marketing* (pp. 234–243). IGI Global.

Bansal, A., Tewari, A., Sharma, A., & Bansal, A. (2024). Implementations and Rationale for Blockchain Technique in Agriculture. In *Applications of Computer Vision and Drone Technology in Agriculture 4.0* (pp. 83–92). Springer Nature Singapore. DOI: 10.1007/978-981-99-8684-2_6

Baralla, G., Ibba, S., Marchesi, M., Tonelli, R., & Missineo, S. (2018). A blockchain-based system to ensure transparency and reliability in food supply chain. In Proceedings of the Euro-Par 2018: Parallel Processing (pp. 379-391). https://doi.org/DOI: 10.1007/978-3-319-99106-1_27

Barbosa, M. W. (2021). Uncovering research streams on agri-food supply chain management: A bibliometric study. *Global Food Security*, 28, 100517. DOI: 10.1016/j.gfs.2021.100517

Barenkamp, M. (2020, June). A new IoT gateway for artificial intelligence in agriculture. In *2020 International Conference on Electrical, Communication, and Computer Engineering (ICECCE)* (pp. 1-5). IEEE. DOI: 10.1109/ICECCE49384.2020.9179418

Barnes, J., & Higginson, L. (2020). *Practical implications for the South African food processing industry. Digital Industrial Policy Brief 4*. Industrial Development Think Tank.

Basavaraj, G., Rao, P. P., Basu, K., Reddy, C. R., Kumar, A. A., Rao, P. S., & Reddy, B. V. S. (2012). Assessing viability of bio-ethanol production from sweet sorghum (No. 30, p. 20). Working paper series.

Bastein, A. G. T. M., Roelofs, E., Rietveld, E., & Hoogendoorn, A. (2013). *Opportunities for a circular economy in the Netherlands*. TNO.

Bateman, M. (2010). *Why Doesn't Microfinance Work? The Destructive Rise of Local Neoliberalism*. Bloomsbury Publishing. DOI: 10.5040/9781350223974

Baumüller, H. (2015). Assessing the role of mobile phones in offering price information and market linkages: The case of m-farm in Kenya. *The Electronic Journal on Information Systems in Developing Countries*, 6(68), 1–16. DOI: 10.1002/j.1681-4835.2015.tb00492.x

Bayramova, A., Edwards, D. J., & Roberts, C. (2021). The role of blockchain technology in augmenting supply chain resilience to cybercrime. *Buildings*, 11(7), 283. Advance online publication. DOI: 10.3390/buildings11070283

Becker, S. (2024, April 4). *US farms are making an urgent push into AI. It could help feed the world*. https://www.bbc.com/worklife/article/20240325-artificial-intelligence-ai-us-agriculture-farming

Beck, T., Demirgüç-Kunt, A., & Levine, R. (2007). Finance, inequality, and the poor. *Journal of Economic Growth*, 12(1), 27–49. DOI: 10.1007/s10887-007-9010-6

Beckwith, S. (2019, June 7). We're Back in Bhutan. *In Pursuit of Tea*. Retrieved September 22, 2024, from https://inpursuitoftea.com/blogs/the-ipot-journal/thunder-dragon-green-tea-bhutan

Ben Ayed, R., & Hanana, M. (2021). Artificial Intelligence to Improve the Food and Agriculture Sector. *Journal of Food Quality, 2021*(Ml). DOI: 10.1155/2021/5584754

Ben-Daya, M., Hassini, E., Bahroun, Z., & Banimfreg, B. H. (2021). The role of internet of things in food supply chain quality management: A review. *The Quality Management Journal*, 28(1), 17–40. https://doi.org/https://doi.org/10.1080/10686967.2020.1838978. DOI: 10.1080/10686967.2020.1838978

Béné, C. (2020). Resilience of Local Food Systems and Links to Food Security – A Review of Some Important Concepts in The Context of COVID-19 and Other Shocks. *Food Security*, 12(4), 805–822. https://link.springer.com/article/10.1007/s12571-020-01076-1. DOI: 10.1007/s12571-020-01076-1 PMID: 32837646

Berube, V. (2012). *How Co-operatives Grow*. McKinsey and Company. Retrieved from https://www.mckinsey.com/~/media/mckinsey/dotcom/client_service/Strategy/McKinsey%20on%20Cooperatives/PDFs/McK_on_Cooperatives-How_cooperatives_grow.ashx

Bibi, F., Guillaume, C., Gontard, N., & Sorli, B. (2017). A review: RFID technology having sensing aptitudes for food industry and their contribution to tracking and monitoring of food products. *Trends in Food Science & Technology*, 62, 91–103. DOI: 10.1016/j.tifs.2017.01.013

Bisheko, M. J., & Rejikumar, G. (2023, March). A study on farmers' perceptions about the scope of the Kisan Suvidha App in improving agricultural sustainability. In *2023 Conference on Information Communications Technology and Society (ICTAS)* (pp. 1-5). IEEE. DOI: 10.1109/ICTAS56421.2023.10082741

Black, A., Edwards, L., Gorven, R., & Mapulanga, W. (2020). Agro-processing, value chains, and regional integration in Southern Africa. WIDER Working Paper 2020/36. Helsinki: UNU-WIDER. https://doi.org/DOI: 10.35188/UNU-WIDER/2020/793-4

Blackburn, J., & Scudder, G. (2009). Supply chain strategies for perishable products: The case of fresh produce. *Production and Operations Management*, 18(2), 129–137. DOI: 10.1111/j.1937-5956.2009.01016.x

Bonin, C. L., Heaton, E. A., Cogdill, T. J., & Moore, K. J. (2016). Management of sweet sorghum for biomass production. *Sugar Tech*, 18(2), 150–159. DOI: 10.1007/s12355-015-0377-y

Borzaga, C., & Roger, S. (2004). *Trends and challenges for co-operatives and social enterprises in developed countries.* Trento Italy. Retrieved from https://emes.net/content/uploads/publications/trends_and_challenges.pdf

Bosona, T., & Gebresenbet, G. (2013). Food traceability as an integral part of logistics management in food and agricultural supply chain. *Food Control*, 33(1), 32–48. DOI: 10.1016/j.foodcont.2013.02.004

Boukachour, J., & Benaini, A. (Eds.). (2023). *Transport and Logistics Planning and Optimization.* IGI Global. DOI: 10.4018/978-1-6684-8474-6

Bourguignon, F. (2004). Trade exposure and income volatility in cash-crop exporting developing countries. *European Review of Agriculture Economics*, 31(3), 369–387. DOI: 10.1093/erae/31.3.369

Bowersox, D. J., Closs, D. J., Cooper, M. B., & Bowersox, J. C. (2020). *Supply chain logistics management.* Mcgraw-hill.

Burrell, D. N. (2022). Telehealth as an innovative supply chain and logistics Management approach. [IJHSTM]. *International Journal of Health Systems and Translational Medicine*, 2(1), 1–9. DOI: 10.4018/IJHSTM.306971

By Staff Reporter; New Era Live; Namibia: Two Beef Value Chains and the Shifting of the 'Red Line'; https://neweralive.na/namibia-two-beef-value-chains-and-the-shifting-of-the-red-line/; as retrieved on 19th August 2024

Cacho, J. L., Marques, L., & Nascimento, Á. (2020). Customer-oriented global supply chains: port logistics in the era of globalization and digitization. In *Anthropological Approaches to Understanding Consumption Patterns and Consumer Behavior* (pp. 82–103). IGI Global. DOI: 10.4018/978-1-7998-3115-0.ch005

Caro, M. P., Ali, M. S., Vecchio, M., & Giaffreda, R. (2018). Blockchain-based traceability in agri-food supply chain management: A practical implementation. In *IoT Vertical and Topical Summit on Agriculture* (pp. 1–4). IEEE., DOI: 10.1109/IOT-TUSCANY.2018.8373021

Carrus, P. P., & Pinna, R. (2011). Information technology and supply chain management coordination: The role of third party logistics providers. [IJESMA]. *International Journal of E-Services and Mobile Applications*, 3(4), 21–36. DOI: 10.4018/jesma.2011100102

Casino, F., Kanakaris, V., Dasaklis, T. K., Moschuris, S., & Stachtiari, E. (2020). Blockchain-based food supply chain traceability: A case study in the dairy sector. *International Journal of Production Research*, 58(3), 707–721. DOI: 10.1080/00207543.2019.1668116

Centobelli, P., Cerchione, R., Esposito, E., & Shashi, . (2020). Evaluating environmental sustainability strategies in freight transport and logistics industry. *Business Strategy and the Environment*, 29(3), 1563–1574. DOI: 10.1002/bse.2453

Centre starts releasing onions from buffer stock to rein in prices (2023). The Economic Times. Retrieved from https://economictimes.indiatimes.com/news/economy/foreign-trade/centre-starts-releasing-onions-from-buffer-stock-to-rein-in-prices/articleshow/102641853.cms?from=mdr

Ceyhun, G. Ç. (Ed.). (2019). *Handbook of research on the applications of international transportation and logistics for world trade*. IGI Global.

Chamlee-Wright, E., & Storr, V. H. (2011). Social Capital as collective narratives and post-disaster community recovery. *The Sociological Review*, 59(2), 266–282. DOI: 10.1111/j.1467-954X.2011.02008.x

Chandrasekaran, N., & Raghuram, G. (2014). Agribusiness supply chain management. CRC Press.

Charlebois, S., Latif, N., Ilahi, I., Sarker, B., Music, J., & Vezeau, J. (2024). Digital traceability in Agri-Food Supply Chains: A Comparative analysis of OECD member countries. *Foods*, 13(7), 1075. DOI: 10.3390/foods13071075 PMID: 38611378

Chauhan, A., Brouwer, B., & Westra, E. (2022). Robotics for a Quality-Driven Post-harvest Supply Chain. *Current Robotics Reports*, 3(2), 39–48. DOI: 10.1007/s43154-022-00075-8

Chauhan, A., Debnath, R. M., & Singh, S. P. (2018). Modelling the drivers for sustainable Agri- food waste management. Benchmarking. *International Journal (Toronto, Ont.)*, 25, 981–993.

Chauhan, A., Kaur, H., Yadav, S., & Jakhar, S. K. (2020). A hybrid model for investigating and selecting a sustainable supply chain for agri-produce in India. *Annals of Operations Research*, 290(1-2), 621–642. DOI: 10.1007/s10479-019-03190-6

Chengappa, P. G. (2004). Emerging trends in agro-processing in India. *Indian Journal of Agricultural Economics*, 59(1).

Chengappa, P. G. (2018). Development of agriculture value chains as a strategy for enhancing farmers' income. *Agricultural Economics Research Review*, 31(1), 1–12. DOI: 10.5958/0974-0279.2018.00001.0

Chenhall, R. H., Hall, M., & Smith, D. (2010). Social capital and management control systems: A study of a non-government organization. *Accounting, Organizations and Society*, 35(8), 737–756. DOI: 10.1016/j.aos.2010.09.006

Chen, R. Y. (2017). An intelligent value stream-based approach to collaboration of food traceability cyber physical system by fog computing. *Food Control*, 71, 124–136. DOI: 10.1016/j.foodcont.2016.06.042

Chen, S., Brahma, S., Mackay, J., Cao, C., & Aliakbarian, B. (2020). The role of smart packaging system in food supply chain. *Journal of Food Science*, 85(3), 517–525. DOI: 10.1111/1750-3841.15046 PMID: 32056210

Chergui, N., Kechadi, M. T., & McDonnell, M. (2020, February). The impact of data analytics in digital agriculture: a review. In *2020 International Multi-Conference on: "Organization of Knowledge and Advanced Technologies"(OCTA)* (pp. 1-13). IEEE. DOI: 10.1109/OCTA49274.2020.9151851

Chinaka, M. (2016). Blockchain technology: Applications in improving financial inclusion in developing economies. Case study for small scale agriculture in Africa. *Journal of African Business*, 17(1), 149–159. DOI: 10.1080/15228916.2016.1138709

Chisoro-Dube, S., Das Nair, R., Nkhonjera, M., & Tempia, N. 2018. Structural transformation in agriculture and agro-processing value chains. Centre for Competition, Regulation and Economic Development. Working Paper 8/2018

Chkanikova, O. (2016). Sustainable Purchasing in Food Retailing: Interorganizational Relationship Management to Green Product Supply. *Business Strategy and the Environment*, 25(7), 478–494. DOI: 10.1002/bse.1877

Chouhan, S. S., Singh, U. P., Sharma, U., & Jain, S. (2021). Leaf disease segmentation and classification of Jatropha Curcas L. and Pongamia Pinnata L. biofuel plants using computer vision based approaches. *Measurement*, 171, 108796. DOI: 10.1016/j.measurement.2020.108796

Chowdhury, S. (2020, August 7). *Heifer International*. Retrieved from Bangladesh Cooperatives Care for Communities Struggling During COVID-19 Lockdown: https://www.heifer.org/blog/newsworthy/bangladesh-cooperatives-care-for-communities-struggling-during-covid-19-lockdown.html

Chukkapalli, S. S. L., Mittal, S., Gupta, M., Abdelsalam, M., Joshi, A., Sandhu, R., & Joshi, K. (2020). Ontologies and artificial intelligence systems for the cooperative smart farming ecosystem. *IEEE Access : Practical Innovations, Open Solutions*, 8, 164045–164064. DOI: 10.1109/ACCESS.2020.3022763

Ciccullo, F., Cagliano, R., Bartezzaghi, G., & Perego, A. (2021). Implementing the circular economy paradigm in the agri-food supply chain: The role of food waste prevention technologies. *Resources, Conservation and Recycling*, 164, 105114. DOI: 10.1016/j.resconrec.2020.105114

Ciulli, F., Kolk, A., & Boe-Lillegraven, S. (2020). Circularity brokers: Digital platform organizations and waste recovery in food supply chains. *Journal of Business Ethics*, 167(2), 299–331. DOI: 10.1007/s10551-019-04160-5

Climate finance for agriculture. Retrieved from https://ccafs.cgiar.org/sites/default/files/202010/Module%203%20Presenter%20Guide.pdf

Coffé, H., & Geys, B. (2007). Toward an empirical characterization of bridging and bonding social capital. *Nonprofit and Voluntary Sector Quarterly*, 36(1), 121–139. DOI: 10.1177/0899764006293181

Co, H. C., & Barro, F. (2009). Stakeholder theory and dynamics in supply chain collaboration. *International Journal of Operations & Production Management*, 29(6), 591–611. DOI: 10.1108/01443570910957573

Colaço, A. F., Whelan, B. M., Bramley, R. G. V., Richetti, J., Fajardo, M., McCarthy, A. C., Perry, E. M., Bender, A., Leo, S., Fitzgerald, G. J., & Lawes, R. A. (2024). Digital strategies for nitrogen management in grain production systems: Lessons from multi-method assessment using on-farm experimentation. *Precision Agriculture*, 25(2), 983–1013. DOI: 10.1007/s11119-023-10102-z

Coral, C., & Mithöfer, D. (2023). The backbone of agrifood value chain resilience: Innovation in the Ecuadorian banana value chain from a historical perspective. *World Development Perspectives*, 29, 100476. DOI: 10.1016/j.wdp.2022.100476

Costa, C., & Delgado, C. (2019). The Cashew Value Chain in Mozambique.

Coyle, J. J., Novack, R. A., Gibson, B. J., & Langley, C. J. (2021). *Supply chain management: a logistics perspective*. Cengage Learning.

CRISIL. the Onion Paradox (2022). Retrieved from https://www.crisil.com>documents>2022/11.

Damasceno, B. M. C., Schaffert, E. R., & Dweikat, I. (2014). Mining genetic diversity of sorghum as a bioenergy feedstock, plants and bioenergy. *Adv. Plant Biol.*, 4, 81–106.

Daqiq, M. H. (2021). Analysis of Pomegranate Value Chain in Kandahar Province of Afghanistan: Issues and Prospects, https://repository.lahoreschool.edu.pk/xmlui/bitstream/handle/123456789/17448/05%20Hasham.pdf?sequence=1&isAllowed=y

Das Nair, R., & Nontenja, N. (2023). Digital Technology Adoption in Agro-processing Value Chains.

Dash, R. K., Nguyen, T. N., Cengiz, K., & Sharma, A. (2021). Fine-tuned support vector regression model for stock predictions. *Neural Computing & Applications*, 33(3), 959–972. DOI: 10.1007/s00521-020-05323-3

Dayıoğlu, M. A., & Turker, U. (2021). Digital transformation for sustainable future-agriculture 4.0: A review. *Journal of Agricultural Sciences*, 27(4), 373–399.

Day, J. M., Melnyk, S. A., Larson, P. D., Davis, E. W., & Whybark, D. C. (2012). Humanitarian and disaster relief supply chains: A matter of life and death. *The Journal of Supply Chain Management*, 48(2), 21–36. DOI: 10.1111/j.1745-493X.2012.03267.x

De Koster, R., Le-Duc, T., & Roodbergen, K. J. (2007). Design and control of warehouse order picking: A literature review. *European Journal of Operational Research*, 182(2), 481–501. DOI: 10.1016/j.ejor.2006.07.009

de Sousa, M. A., Addison, T., & Ekman, B. (2001). From humanitarian assistance to poverty reduction in Angola.

De, A., & Singh, S. P. (2021). Analysis of fuzzy applications in the agri-supply chain: A literature review. *Journal of Cleaner Production*, 283, 124577. DOI: 10.1016/j.jclepro.2020.124577

Deepika, M. (2024). *Nurturing India's Agriculture: A Closer Look at Onion Price Dynamics*. BigHaat.

Deng, X., Yang, X., Zhang, Y., Li, Y., & Lu, Z. (2019). Risk propagation mechanisms and risk management strategies for a sustainable perishable products supply chain. *Computers & Industrial Engineering*, 135, 1–13. DOI: 10.1016/j.cie.2019.01.014

Department of Agriculture, Government of Andhra Pradesh. (2017). *Agriculture action plan 2017-2018*. https://apagrisnet.gov.in/2017/Agri%20Action%20Plan%20 2017-18%20(English)/5%20Fertilizers.pdf

Department of Fertilizers, Government of India. (2022-2023). *Annual report 2022-2023*. https://www.fert.nic.in/sites/default/files/2020-082024-01/Annual%20Report %20fertilizer%20new%20new17%20%20%2029-07-2023.pdf

Department of Fertilizers, Government of India. (2023). *Annual report 2022-2023*. https://www.fert.nic.in/sites/default/files/2020-082024-01/Annual%20Report%20 fertilizer%20new%20new17%20%20%2029-07-2023.pdf

Department of Fertilizers, Government of India. (n.d.). *Official website*. https:// www.fert.nic.in/

Department of Fertilizers, Government of India. (n.d.). *Phosphatic and potassic (P&K) policy*. https://www.fert.nic.in/phosphatic-and-potassic-pk-policy

Department of Food and Public Distribution, Government of India. (n.d.). *National food security act*. https://dfpd.gov.in/WriteReadData/Other/nfsa_1.pdf

Derens-Bertheau, E., Osswald, V., Laguerre, O., & Alvarez, G. (2015). Cold chain of chilled food in France. *International Journal of Refrigeration*, 52, 161–167. DOI: 10.1016/j.ijrefrig.2014.06.012

Dethier, J.-J., & Effenberger, A. (2012, March 12). Agriculture and development: A brief review of the literature. *Economic Systems*, 36(2), 175–205. DOI: 10.1016/j.ecosys.2011.09.003

Dhaliwal, A., & Arora, S. (2021). Rapid response logistics: Strategies, models, and technologies for effective supply chain management. [IJSSMET]. *International Journal of Service Science, Management, Engineering, and Technology*, 12(6), 73–88. DOI: 10.4018/IJSSMET.2021110105

Diagne, A., & Zeller, M. (2001). Access to credit and its impact on welfare in Malawi (Research Report No. 116, 168 p.). International Food Policy Research Institute.

Díaz-Valderrama, J. R., Njoroge, A. W., Macedo-Valdivia, D., Orihuela-Ordóñez, N., Smith, B. W., Casa-Coila, V., Ramírez-Calderón, N., Zanabria-Gálvez, J., Woloshuk, C., & Baributsa, D. (2020). Postharvest practices, challenges and opportunities for grain producers in Arequipa, Peru. *PLoS One*, 15(11), e0240857. DOI: 10.1371/journal.pone.0240857 PMID: 33147234

Dibbern, T., Romani, L. A. S., & Massruhá, S. M. F. S. (2024). Main drivers and barriers to the adoption of Digital Agriculture technologies. *Smart Agricultural Technology*, 8, 100459. DOI: 10.1016/j.atech.2024.100459

Diezma, B., Barreiro, P., Baltazar, P., & Correa, E. C. (2021). A general procedure for predicting the remaining shelf life of nectarines and peaches for virtualization of the value chain. *Postharvest Biology and Technology*, 181, 1–7. DOI: 10.1016/j.postharvbio.2021.111677

District Census Handbook (2011). Lohardaga, Series 21, Part XII B. Directorate of Census Operations, Jharkhand. *Pp 18-19. Retrieved 1 Aug* 2024.

Dixit, P., & Ravichandran, R. (2022). Bridging the skill gap in the Indian food processing sector: A review. *FoodSci Indian Journal of Research in Food Science and Nutrition*, 9(2), 85. DOI: 10.15613/fijrfn/2022/v9i2/217902

Do, D. T., Nguyen, T. T. T., Nguyen, T. N., Li, X., & Voznak, M. (2020). Uplink and downlink NOMA transmission using full-duplex UAV. *IEEE Access : Practical Innovations, Open Solutions*, 8, 164347–164364. DOI: 10.1109/ACCESS.2020.3023163

Domínguez, I. P., Fellmann, T., Chatzopoulos, T., Pieralli, S., Jensen, H., Barreiro-Hurle, J., & Micale, F. (2018). EU commodity market development: Medium-term agricultural outlook. In *Proceedings of the October 2017 workshop* (p. 133).

Drago, E., Campardelli, R., Pettinato, M., & Perego, P. (2020). Innovations in smart packaging concepts for food: An extensive review. *Foods*, 9(11), 1628. Advance online publication. DOI: 10.3390/foods9111628 PMID: 33171881

Dube, S., Das Nair, R., Nkhonjera, M., & Tempia, N. (2018). Structural transformation in agriculture and agro-processing value chains.

Duong, L. N. K., Al-Fadhli, M., Jagtap, S., Bader, F., Martindale, W., Swainson, M., & Paoli, A. (2020). A review of robotics and autonomous systems in the food industry: From the supply chains perspective. *Trends in Food Science & Technology*, 106(May), 355–364. DOI: 10.1016/j.tifs.2020.10.028

Dutta, P., Choi, T. M., Somani, S., & Butala, R. (2020). Blockchain technology in supply chain operations: Applications, challenges and research opportunities. *Transportation research part e: Logistics and transportation review, 142*, 102067.

Du, X., Wang, X., & Hatzenbuehler, P. (2023). Digital technology in agriculture: A review of issues, applications and methodologies. *China Agricultural Economic Review*, 15(1), 95–108. DOI: 10.1108/CAER-01-2022-0009

Ekemezie, I. O., & Digitemie, W. N.Ifeanyi Onyedika EkemezieWags Numoipiri Digitemie. (2024). Carbon Capture and Utilization (CCU): A review of emerging applications and challenges. *Engineering Science & Technology Journal*, 5(3), 949–961. DOI: 10.51594/estj.v5i3.949

Ekumankama, O., Ezeoha, A., & Uche, C. (2019). The role of multinational corporations in local dairy value chain development: Case of Friesland Campina WAMCO (FCW) in Nigeria. *The International Food and Agribusiness Management Review*, 23(1), 55–69. DOI: 10.22434/IFAMR2018.0108

El Bilali, H. (2020). Transition heuristic frameworks in research on agro-food sustainability transitions. *Environment, Development and Sustainability*, 22(3), 1693–1728. DOI: 10.1007/s10668-018-0290-0

El Din, M. S., & Reason, M. (2024). Fair Trade and Ethical Sourcing in Agriculture: Paving the Way for Sustainable Supply Chains. In Sustainable Practices for Agriculture and Marketing Convergence (pp. 80-104). IGI Global.

Elgar, F. J., Davis, C. G., Wohl, M. J., Trites, S. J., Zelenski, J. M., & Martin, M. S. (2011). Social capital, health and life satisfaction in 50 countries. *Health & Place*, 17(5), 1044–1053. DOI: 10.1016/j.healthplace.2011.06.010 PMID: 21784694

El-Ramady, H. R., Domokos-Szabolcsy, É., Abdalla, N. A., Taha, H. S., & Fári, M. (2015). Postharvest Management of Fruits and Vegetables Storage. In: Lichtfouse, E. (eds) *Sustainable Agriculture Reviews*. Sustainable Agriculture Reviews, vol 15. Springer, Cham. https://doi.org/DOI: 10.1007/978-3-319-09132-7_2

Esmaeilian, B., Wang, B., Lewis, K., Duarte, F., Ratti, C., & Behdad, S. (2018). The future of waste management in smart and sustainable cities: A review and concept paper. *Waste Management (New York, N.Y.)*, 81, 177–195. DOI: 10.1016/j.wasman.2018.09.047 PMID: 30527034

Essential Commodities Amendment. Bill 2020 (5th June 2020). Retrieved from: https://loksabhadocs.nic.in/Refinput/New_Reference_Notes/English/10092020_163243_102120474.pdf

EU Green deal. (2020). Farm to Fork Strategy. In *European Comission*. https://ec.europa.eu/food/sites/food/files/safety/docs/f2f_action-plan_2020_strategy-info_en.pdf Accessed: 28. 06. 2024

Eum, H. (2017). *Cooperatives and Employment: Second Global Report.* Retrieved from https://www.cicopa.coop/wp-content/uploads/2018/01/Cooperatives-and-Employment-Second-Global-Report-2017.pdf

European Central Bank. (2022). *Economic bulletin article: The Russia-Ukraine war and its impact on fertilizers.* https://www.ecb.europa.eu/press/economic-bulletin/articles/2022/html/ecb.ebart202204_01~7b32d31b29.en.html

European Comission CAP. (2021). *A Greener and Fairer CAP (Common Agricultural Policy).* https://agriculture.ec.europa.eu/system/files/2022-02/factsheet-newcap-environment-fairness_en_0.pdf Accessed: 28. 06. 2024

European Commission. (2017). *EU Agricultural Outlook for the Agricultural Markets and Income 2017-2030.* European Commission.

European Parliamentary Research Service. (2019). *Cooperatives: Characteristics, activities, status, challenges.*https://www.europarl.europa.eu/RegData/etudes/BRIE/2019/635541/EPRS_BRI(2019)635541_EN.pdf

FAO. (2017). *Report on The Future of Food and Agriculture: Trends and Challenges.* Food and Agriculture Organization, Rome. Retrieved from https://www.fao.org/3/a-i6583e.pdf

FAO. (2018). *TRANSFORMING FOOD AND AGRICULTURE TO ACHIEVE THE SDGs 20 interconnected actions to guide decision-makers.* https://www.fao.org/3/I9900EN/i9900en.pdf Accessed: 28. 06. 2024

FAO. (2020). Food and Agriculture Organization of the United Nations. *The State of Food and Agriculture*, 2020, •••. https://www.fao.org/interactive/state-of-food-agriculture/2020/en/

FAOSTAT. (2022). *Commodities Production by country.* https://www.fao.org/faostat/en/#rankings/commodities_by_country Accessed: 28. 06. 2024

Farmers in India are using AI for agriculture – here's how they could inspire the world. (2024, January 16). World Economic Forum. https://www.weforum.org/agenda/2024/01/how-indias-ai-agriculture-boom-could-inspire-the-world/#:~:text=Artificial%20Intelligence,-Follow&text=The%20initiative%20transformed%20the%20chili,that%20they%20doubled%20their%20income

Fennimore, S. A., Slaughter, D. C., Siemens, M. C., Leon, R. G., & Saber, M. N. (2016). Technology for automation of weed control in specialty crops. *Weed Technology*, 30(4), 823–837. DOI: 10.1614/WT-D-16-00070.1

Fertilizer Association of India. (2023). *Fertilizer Statistics* (68th ed.).

Fertilizer Association of India. (2023). *Fertilizer statistics* (68th ed.). Fertilizer Association of India.

Fiksel, J., Sanjay, P., & Raman, K. (2021). Steps toward a resilient circular economy in India. *Clean Technologies and Environmental Policy*, 23(1), 203–218. DOI: 10.1007/s10098-020-01982-0 PMID: 33199980

Filippi, M. (2012). *Support for Farmers' Cooperatives- country report France.* University of Bordeaux INRA, France. Wageningen: Wageningen UR. Retrieved from https://edepot.wur.nl/244795

Filippi, A. C. G., Guarnieri, P., Carvalho, J. M., Reis, S. A., & da Cunha, C. A. (2020). New configurations in Brazilian agribusiness: Rural warehouse condominiums. *Journal of Agribusiness in Developing and Emerging Economies*, 10(1), 41–63. DOI: 10.1108/JADEE-12-2018-0178

Food and Agriculture Organization (FAO). (2017). *The State of Food and Agriculture: Leveraging Food Systems for Inclusive Rural Transformation.* Food and Agriculture Organization of the United Nations.

Food and Agriculture Organization (FAO). (2023). Climate-Smart Agriculture. Retrieved from https://www.fao.org/climate-smart-agriculture

Food and Agriculture Organization of the United Nations. 2019. Food Loss and Food Waste [WWW Document]. URL https://www.fao.org/food-loss-and-food-waste/en/ (accessed 6.12.19).

Food and Agriculture Organization of United Nations. (2021). https://www.fao.org/faostat/en/#data/FBS, as retrieved on 20th September 2024.

Food and Agriculture Organization. (n.d.). *Title unknown.* https://openknowledge.fao.org/server/api/core/bitstreams/9ddf4aab-bc18-4d48-aeb6-ab14efaed1fd/content

Fortunato, A. & ENCISO VALDIVIA, S. H. E. Y. L. A. (2023). Food for Growth: A Diagnostics of Namibia's Agriculture Sector. CID Research Fellows and Graduate Student Working Paper Series.

Foschi, E. (2020). System innovation and life cycle thinking in packaging value chain: the circularity of plastics.

Franklin-Johnson, E., Figge, F., & Canning, L. (2016). Resource duration as a managerial indicator for Circular Economy performance. *Journal of Cleaner Production*, 133, 589–598. DOI: 10.1016/j.jclepro.2016.05.023

Freund, A. (2023). Effects of digitalisation on food safety. *Studies in Agricultural Economics (Budapest)*, 125(3), 154–168. DOI: 10.7896/j.2498

FrieslandCampina. (2024). *Nourishing By Nature*. Retrieved from https://www.frieslandcampina.com/ Accessed on 1st November,2024

Gabriel, A., & Gandorfer, M. (2023). Adoption of digital technologies in agriculture—An inventory in a european small-scale farming region. *Precision Agriculture*, 24(1), 68–91. DOI: 10.1007/s11119-022-09931-1

Gad, A. G., Mosa, D. T., Abualigah, L., & Abohany, A. A. (2022). Emerging Trends in Blockchain Technology and Applications: A Review and Outlook. *Journal of King Saud University. Computer and Information Sciences*, 34(9), 6719–6742. DOI: 10.1016/j.jksuci.2022.03.007

Gadge, S. S., & Lawande, K. E. (2012). Crop Damage Due to Climatic Change: A Major Constraint in Onion Farming. *Indian Research Journal of Extension Education*, 2(Special Issue), 38–41.

Galli, B. J. (2018). How continuous improvement can support logistics: A reflection of best practices. [IJoSE]. *International Journal of Strategic Engineering*, 1(1), 1–23. DOI: 10.4018/IJoSE.2018010101

Galvez, J. F., Mejuto, J. C., & Simal-Gandara, J. (2018). Future challenges on the use of blockchain for food traceability analysis. *Trends in Analytical Chemistry*, 107, 222–232. DOI: 10.1016/j.trac.2018.08.011

Gandino, F., Montrucchio, B., Rebaudengo, M., & Sanchez, E. R. (2009). On improving automation by integrating RFID in the traceability management of the agri-food sector. *IEEE Transactions on Industrial Electronics*, 56(7), 2357–2365. DOI: 10.1109/TIE.2009.2019569

Ganeshkumar, C., Jena, S. K., Sivakumar, A., & Nambirajan, T. (2023). Artificial intelligence in agricultural value chain: Review and future directions. *Journal of Agribusiness in Developing and Emerging Economies*, 13(3), 379–398. DOI: 10.1108/JADEE-07-2020-0140

Gardas, B. B., Raut, R. D., & Narkhede, B. (2018). Evaluating critical causal factors for post- harvest losses (PHL) in the fruit and vegetables supply chain in India using the DEMATEL approach. *Journal of Cleaner Production*, 199, 47–61. DOI: 10.1016/j.jclepro.2018.07.153

Garg, M., Singhal, S., Sood, K., Rupeika-Apoga, R., & Grima, S. (2023). Price discovery mechanism and volatility spillover between National Agriculture Market and National Commodity and Derivatives Exchange: The study of the Indian agricultural commodity market. *Journal of Risk and Financial Management*, 16(2), 62. DOI: 10.3390/jrfm16020062

Gebre, G. G., Rik, E., & Kijne, A. (2020). Analysis of banana value chain in Ethiopia: Approaches to sustainable value chain development. *Cogent Food & Agriculture*, 6(1), 1742516. DOI: 10.1080/23311932.2020.1742516

Getahun, W., Tesfaye, A., Mamo, T., & Ferede, S. (2018). Apple value chain analysis in the Central Highlands of Ethiopia. *International Journal of Agriculture Innovation and Research*, 7(1), 2319–1473.

Geyer, H. S. (2019). Global Value Chain Participation and Trade Barriers in Sub-Saharan Africa. In Scholvin, S., Black, A., Diez, J. R., & Turok, I. (Eds.), *Value Chains in Sub-Saharan Africa: Challenges 21 of Integration Into the Global Economy* (pp. 13–27). Springer., DOI: 10.1007/978-3-030-06206-4_2

Gharde, Y., Singh, P. K., Dubey, R. P., & Gupta, P. K. (2018). Assessment of yield and economic losses in agriculture due to weeds in India. *Crop Protection (Guildford, Surrey)*, 107, 12–18. DOI: 10.1016/j.cropro.2018.01.007

Ghosh, A. (2015). 16| EMBEDDEDNESS AND THE DYNAMICS OF GROWTH: THE CASE OF THE AMUL COOPERATIVE, INDIA. *Social and Solidarity Economy: Beyond the Fringe*, 284.

Giampietri, E., Verneau, F., Del Giudice, T., Carfora, V., & Finco, A. (2018). A Theory of Planned behaviour perspective for investigating the role of trust in consumer purchasing decision related to short food supply chains. *Food Quality and Preference*, 64, 160–166. DOI: 10.1016/j.foodqual.2017.09.012

Glavas, C., Pike, S., & Mathews, S. (2014). Internet-enabled tourism entrepreneurs. International entrepreneurial values elicited through repertory test and laddering analysis. *International Journal of Tourism Research*, 16(1), 44–55. DOI: 10.1002/jtr.1896

Gligor, D., Tan, A., & Nguyen, T. N. T. (2018). The obstacles to cold chain implementation in developing countries: Insights from Vietnam. *International Journal of Logistics Management*, 29(3), 942–958. DOI: 10.1108/IJLM-02-2017-0026

Goedhals-Gerber, L. L., & Khumalo, G. (2020). Identifying temperature breaks in the export cold chain of navel oranges: A Western Cape case. *Food Control*, 110, 107013. DOI: 10.1016/j.foodcont.2019.107013

Gokarn, S., & Kuthambalayan, T. S. (2017). Analysis of challenges inhibiting the reduction of waste in food supply chain. *Journal of Cleaner Production*, 168, 595–604. DOI: 10.1016/j.jclepro.2017.09.028

Gordon, A., DeVlieger, D., Vasan, A., & Bedard, B. (2020). Technical considerations for the implementation of food safety and quality systems in developing countries. *Food Safety and Quality Systems in Developing Countries*, 1-40. https://doi.org/ DOI: 10.1016/B978-0-12-814272-1.00001-2

Gorjian, S., Fakhraei, O., Gorjian, A., Sharafkhani, A., & Aziznejad, A. (2022). Sustainable food and agriculture: Employment of renewable energy technologies. *Current Robotics Reports*, 3(3), 153–163. DOI: 10.1007/s43154-022-00080-x PMID: 35600253

Gorse, L., Chambas, G., & Raynaud, E. (2008). *Microfinance et agriculture en développement: Études de cas.* FinDev Gateway. Retrieved from https://www.findevgateway.org/sites/default/files/publications/files/mfg-fr-etudes-de-cas-microfinance-et-agriculture-en-developpement-09-2008.pdf

Goyal, S. B., Pradeep, N., Shukla, P. K., Ghonge, M. M., & Ravi, R. V. (Eds.). (2022). *Utilizing blockchain technologies in manufacturing and logistics management.* IGI Global. DOI: 10.4018/978-1-7998-8697-6

Goyal, S., Esposito, M., & Kapoor, A. (2018). Circular economy business models in developing economies: Lessons from India on reduce, recycle, and reuse paradigms. *Thunderbird International Business Review*, 60(5), 729–740. DOI: 10.1002/tie.21883

Goyal, S., Garg, D., & Luthra, S. (2021). Analyzing critical success factors to adopt sustainable consumption and production linked with circular economy. *Environment, Development and Sustainability*, •••, 1–30.

Greenhouse Gas Emission from Indian Agriculture. krishi.icar.gov.in. Retrieved from https://krishi.icar.gov.in/jspui/bitstream/123456789/32431/1/GreenhouseGasEmissionfromIndianAgricultureHPathaketal%20(1).pdf

Greentech Americas boosts horticultural productivity. (2023). Www.greentech.nl. Retrieved June 20, 2024, from https://www.greentech.nl/press-releases/americas/greentech-americas-boosts-horticultural-productivity

Grey, B. T. (2023). *Decoding FrieslandCampina Strategy: Dairy Industry Competitive Analysis*. GreyB. https://www.greyb.com/blog/frieslandcampina-strategy/

Grootaert, C., Narayan, D., Jones, V. N., & Woolcock, M. (2004). Measuring social Capital: An integrated questionnaire. The World Bank., Retrieved February 2019, from. DOI: 10.1596/0-8213-5661-5

GS1. 2013. White Paper: The need for global standards and solutions to combat counterfeiting. Online. Available: https://bit.ly/3ZdTojX

GS1. n. d. -a. How traceability standards work. Online. Available: https://www.gs1.org/howtraceability-standards-work. (Accessed: 11 March 2023)

Gupta, V. (2017). Building the hyperconnected future on blockchains. In World Government Summit (pp. 1-14). Retrieved from http://internetofagreements.com/files/WorldGovernmentSummitDubai2017.pdf

Gupta, N., Bhattacharjee, M., & Saha, A. R. (2022). SUSTAINABILITY AND DIETARY CHANGE: AN ANALYSIS OF INDIAN FOOD CONSUMPTION PATTERNS. *INDIAN JOURNAL OF APPLIED ECONOMICS AND BUSINESS*, 5(1), 85–105. DOI: 10.47509/IJAEB.2023.v05i01.05

Guru, P. N., Mridula, D., Dukare, A. S., Ghodki, B. M., Paschapur, A. U., Samal, I., Nikhil Raj, M., Padala, V. K., Rajashekhar, M., & Subbanna, A. R. (2022). A comprehensive review on advances in storage pest management: Current scenario and future prospects. *Frontiers in Sustainable Food Systems*, 6, 993341. DOI: 10.3389/fsufs.2022.993341

Gu, V. C., Cao, R. Q., Black, K., & Zeng, H. (2017). Managing collaborative relationships in third party logistics: An empirical study. [IJISSCM]. *International Journal of Information Systems and Supply Chain Management*, 10(2), 42–65. DOI: 10.4018/IJISSCM.2017040103

Haggblade, S. (2011). Modernizing African agribusiness: Reflections for the future. *Journal of Agribusiness in Developing and Emerging Economies*, 1(1), 10–30. DOI: 10.1108/20440831111131532

Halloran, A., Clement, J., Kornum, N., Bucatariu, C., & Magid, J. (2014). Addressing food waste reduction in Denmark. *Food Policy*, 49, 294–301. DOI: 10.1016/j.foodpol.2014.09.005

Hamel, D., Rozman, V., & Liška, A. (2020). Storage of Cereals in Warehouses with or without Pesticides. *Insects*, 11(12), 846. DOI: 10.3390/insects11120846 PMID: 33260599

Hao, J., Shi, H., Shi, V., & Yang, C. (2020). Adoption of automatic warehousing systems in logistics firms: A technology–organization–environment framework. *Sustainability (Basel)*, 12(12), 5185. DOI: 10.3390/su12125185

Harnessing artificial intelligence for sustainable agricultural development in Africa: Opportunities, challenges, and impact. (n.d.). https://arxiv.org/html/2401.06171v1#:~:text=The%20adoption%20of%20AI%20in%20agriculture%20in%20Africa%20has%20a,solutions%20to%20long%2Dstanding%20challenges

Harshitha, M. S., Shashidhar, R., & Roopa, M. (2021). Blockchain-based agricultural supply chain: A review. *Global Transitions Proceedings*, 2(2), 220–226. DOI: 10.1016/j.gltp.2021.08.041

Hash Technologies Article. (2021), The Agriculture Industry Is Being Transformed By The Internet Of Things, accessed on 23 March 2024, https://hashstudioz.com/blog/how-internet-of-things-is-transforming-the-agriculture-sector/https://primathon.in/blog/ai-agriculture-revolutionizing-the-future-of-sustainable-farming/

Hashem, I. A. T., Yaqoob, I., Anuar, N. B., Mokhtar, S., Gani, A., & Ullah Khan, S. (2015). The rise of "big data" on cloud computing: Review and open research issues. In *Information Systems* (Vol. 47, pp. 98–115). Elsevier Ltd., DOI: 10.1016/j.is.2014.07.006

Hassoun, A., Prieto, M. A., Carpena, M., Bouzembrak, Y., Marvin, H. J. P., Pallarés, N., Barba, F. J., Punia Bangar, S., Chaudhary, V., Ibrahim, S., & Bono, G. (2022). Exploring the role of green and Industry 4.0 technologies in achieving sustainable development goals in food sectors. *Food Research International*, 162(August), 112068. Advance online publication. DOI: 10.1016/j.foodres.2022.112068 PMID: 36461323

Hayami, Y., & Ruttan, V. W. (1998). *Agriculture et Développement, une approche international*. QUAE Editions.

Hazen, B. T., Overstreet, R. E., & Cegielski, C. G. (2012). Supply chain innovation diffusion: Going beyond adoption. *International Journal of Logistics Management*, 23(1), 119–134. DOI: 10.1108/09574091211226957

Helmuth Stehn; Rangeland Management, Farmers' Support Project (FSP); Joint Presidency Committee (NAU and the NNFU) Private Bag 13255, Windhoek, Namibia; https://cms.my.na/assets/documents/p1bk4fuusf1eud1qcifql1e5s1rer2.pdf; as retrieved on 19th August 2024

Hidayat, S. E., & Musari, K. (2022). ASEAN towards a global halal logistics through the digitally enabled community. [IJABIM]. *International Journal of Asian Business and Information Management*, 13(2), 1–15. DOI: 10.4018/IJABIM.20220701.oa1

Hillman, A. J., & Dalziel, T. (2003). Boards of directors and firm performance: Integrating agency and resource dependence perspectives. *Academy of Management Review*, 28(3), 383–396. DOI: 10.5465/amr.2003.10196729

Hitt, M.A., L. Bierman, K. Uhlenbruck, and K. Shimizu. "The Importance of Resources in the Internationalization of Professional Service Firms: The Good, the Bad, and the Ugly," Academy of Management Journal, (49:1), 2006, pp. 1137- 1157

Hoffmann-Thoma, G., Hinkel, K., Nicolay, P., & Willenbrink, J. (1996). Sucrose accumulation in sweet sorghum stem internodes in relation to growth. *Physiologia Plantarum*, 97(2), 277–284. DOI: 10.1034/j.1399-3054.1996.970210.x

Höllinger, F., Rutten, L., & Kiriakov, K. (2009). The use of warehouse receipt finance in agriculture in transition countries. Presented at the World Grain Forum 2009, FAO. https://www.fao.org/4/i3339e/i3339e.pdf

Hollinger, F. (2004). *Financing agriculture term investments*. Agricultural Finance.

Honja, T. (2014). Review of mango value chain in Ethiopia. *Journal of Biology, Agriculture and Healthcare*, 4(25), 230–239.

Horticultural Statistics at a Glance, GOI. (28th May 2024). Retrieved from https://agriwelfare.gov.in/Documents/CWWGDATA/Agricultural_Statistics_at_a_Glance_2022_0.pdf

Horton, D., Devaux, A., Bernet, T., Mayanja, S., Ordinola, M., & Thiele, G. (2023). Inclusive innovation in agricultural value chains: Lessons from use of a systems approach in diverse settings. *Innovation and Development*, 13(3), 517–539. DOI: 10.1080/2157930X.2022.2070587

Hossain, T. (2024). *Exporter Guide Annual*. USDA Foreign Agriculture Service., Retrieved from https://apps.fas.usda.gov/newgainapi/api/Report/DownloadReportByFileName?fileName=Exporter%20Guide%20Annual_Dhaka_Bangladesh_BG2023-0025.pdf

Hovelaque, V., Duvaleix-Tréguer, S., & Cordier, J. (2009). Effects of constrained supply and price contracts on agricultural cooperatives. *European Journal of Operational Research*, 199(3), 769–780. DOI: 10.1016/j.ejor.2008.08.005

https://growthlab.hks.harvard.edu/sites/projects.iq.harvard.edu/files/growthlab/files/2023-10-cid-fellows-wp-154-namibia-agriculture.pdf, as retrieved on 24th July 2024

Hua, J., Wang, X., Kang, M., Wang, H., & Wang, F. Y. (2019). Blockchain-based provenance for agricultural products: A distributed platform with duplicated and shared bookkeeping. *IEEE Access : Practical Innovations, Open Solutions*, 7, 32439–32448. DOI: 10.1109/ACCESS.2019.2903156

Humphrey, J., & Schmitz, H. (2001). Governance in global value chains. *IDS Bulletin*, 32(3), 19–29. DOI: 10.1111/j.1759-5436.2001.mp32003003.x

Hunsigi, G., Yekkeli, N. R., & Kongawad, B. Y. (2010). Sweet stalk sorghum: An alternative sugar crop for ethanol production. *Sugar Tech*, 12(1), 79–80. DOI: 10.1007/s12355-010-0017-5

Hunter, E. L., & Anderson, I. C. (1997). Sweet sorghum. In Janick, J. (Ed.), *Horticultural Reviews* (Vol. 21, pp. 73–104). John Wiley and Sons. DOI: 10.1002/9780470650660.ch3

Hussain, A. (2016). *A STUDY ON INDIAN FOOD PROCESSING INDUSTRIES (FPI)*. EPRA Journals. https://eprajournals.com/IJES/article/8721/

Huss, M., Brander, M., Kassie, M., Ehlert, U., & Bernauer, T. (2021). Improved storage mitigates vulnerability to food-supply shocks in smallholder agriculture during the COVID-19 pandemic. *Global Food Security*, 28, 100468. DOI: 10.1016/j.gfs.2020.100468 PMID: 36568028

Idoje, G., Dagiuklas, T., & Iqbal, M. (2021). Survey for smart farming technologies: Challenges and issues. *Computers & Electrical Engineering*, 92, 107104. DOI: 10.1016/j.compeleceng.2021.107104

IFC. (n.d.). Global Warehouse Finance Program (GWFP). Retrieved from https://www.ifc.org/en/what-we-do/sector-expertise/financial-institutions/global-trade/global-warehouse-finance-program

IIMB Manag. Rev. 30, 270–282. https://doi.org/DOI: 10.1016/j.iimb.2018.04.001

India Environment Portal. (2016). Solid waste management rules. https://www.indiaenvironmentportal.org.in/content/427824/solid-waste-management-rules-2016/

India Export of Agro Food Products. (2021). APEDA Agri Exchange. (2nd June 2024). Retrieved from https://agriexchange.apeda.gov.in/IndExp/PortNew.aspx

India Forum. (n.d.). Looming crisis: Fertilizers. https://www.theindiaforum.in/article/looming-crisis-fertilisers

Indian agriculture Archives | AgriBazaar. (2019). AgriBazaar. https://blog.agribazaar.com/tag/indian-agriculture/

Informatics, A. (2020), Understanding Agro Informatics and the Way it is Transforming Agriculture, Posted On September 18 2020 / Agro Informatics, accessed 23rd March 2024, https://www.kisangates.com/agro-informatics.html

Institute for Public Policy Research to the Namibia National Farmers Union. Final report: Value Chain Analysis of Livestock and Livestock Products in the Northern Communal Areas of Namibia. Windhoek, Namibia, May 31 2019

Intelligence, H. E. Beef Industry in Namibia June 2023; https://hei.com.na/wp-content/uploads/2023/08/Updated-beef-report.pdf, as retrieved on 24th July 2024

International Food Policy Research Institute (IFPRI). (2023). Climate Change and Food Security. Retrieved from https://www.ifpri.org/publication/climate-change -and-food-security

International Food Policy Research Institute (IFPRI). (2023). Global fertilizer trade 2021-2023: What happened after war-related price spikes? https://www.ifpri.org/blog/ global-fertilizer-trade-2021-2023-what-happened-after-war-related-price-spikes

Islam, J. (n.d.). The Problems and Prospects of Co-operative Society: A case study on its evolution and future possibilities in context of Bangladesh.

Iyer, L. S. (2021). AI enabled applications towards intelligent transportation. *Transportation Engineering*, 5, 100083. DOI: 10.1016/j.treng.2021.100083

Izmailov, A. Y. (2019). Intelligent technologies and robotic means in agricultural production. *Herald of the Russian Academy of Sciences*, 89(2), 209–210. DOI: 10.1134/S1019331619020072

Jackson, D. R., Arthur, M. F., Davis, M., Kresovich, S., Lawhon, E. S., Lipinsky, W. T., & Rudolph, A. 1980. Research Report on Development of Sweet Sorghum as an Energy Crop, vol. 1. Agricultural Task. US DOE, US Department of Commerce, Battele, Columbus.

Jaffee, S., Siegel, P., & Andrews, C. (2010). Rapid agricultural supply chain risk assessment: A conceptual framework. *Agriculture and rural development discussion paper, 47*(1), 1-64.

Jámbor, Z., & Nagy, J. (2022). Resilience of Food Supply Chains – a Dynamic Capability Approach. *Ekonomska Misao i Praksa*, 31(2), 473–486. DOI: 10.17818/ EMIP/2022/2.6

Javalgi, R. G., Hall, K. D., & Cavusgil, S. T. (2014). Corporate entrepreneurship, customer-oriented selling, absorptive capacity, and international sales performance in the international B2B setting: Conceptual framework and research propositions. *International Business Review*, 23(6), 1193–1202. DOI: 10.1016/j.ibusrev.2014.04.003

Jeanneaux, P. (2023). Toward a more profitable value chain for New York state onions. *Journal of Agriculture, Food Systems, and Community Development*, 13(1), 1–7. DOI: 10.5304/jafscd.2023.131.022

Jeffrey, L., & Bommu, R. (2024). Innovative AI Solutions for Agriculture: Enhancing CropManagement and Yield. *International Journal of Advanced Engineering Technologies and Innovations*, 1(3), 203–221.

Jena., ... (2023). financing climate-smart agriculture for sustainable agri-food systems. Retrieved from https://www.nitk.ac.in/document/attachments/5611/T20_PB_TF6_6162_Jena.pdf

Jena., ... (2023). *How to mobilise climate-smart agriculture finance-World Economic Forum-* Retrieve from https://www.weforum.org/agenda/2023/05/climate-smart-agriculture-finance/

Jha, S. N., Vishwakarma, R. K., Ahmad, T., Rai, A., & Dixit, A. K. (2015). Report on assessment of quantitative harvest and post-harvest losses of major crops/commodities in India. ICAR-All India Coordinated Research Project on Post-Harvest Technology. ICAR-CIPHET, PO: PAU, Ludhiana, India.

Jha, K., Doshi, A., Patel, P., & Shah, M. (2019). A comprehensive review on automation in agriculture using artificial intelligence. *Artificial Intelligence in Agriculture*, 2, 1–12. DOI: 10.1016/j.aiia.2019.05.004

Jhawar, A., & Garg, S. K. (2016). System dynamics modelling to study the effects of investment in information technology on logistics performance: A case study from India. [IJSDA]. *International Journal of System Dynamics Applications*, 5(2), 19–40. DOI: 10.4018/IJSDA.2016040102

Jiayi, L., Yang, W., & Jun, C. (2019). Blockchain-based agricultural product quality supervision system. *IEEE Access : Practical Innovations, Open Solutions*, 7, 88969–88980. DOI: 10.1109/ACCESS.2019.2925718

Job, I. A., Essien, E. N., & Ododo, E. P. (2022). A Review of Application of Machine Learning in Agriculture Supply Chain to improve Food and Agriculture Industry. Zenodo *(CERN European Organization for Nuclear Research)*. https://doi.org/DOI: 10.5281/zenodo.6258922

Joensuu, T., Edelman, H., & Saari, A. (2020). Circular economy practices in the built environment. *Journal of Cleaner Production*, 276, 124215. DOI: 10.1016/j.jclepro.2020.124215

Johnson, D. G. (1975). World Agriculture, Commodity Policy, and Price Variability. *American Journal of Agricultural Economics*, 57(5), 823–828. DOI: 10.2307/1239087

Jonkman, J., Barbosa-Póvoa, A. P., & Bloemhof, J. M. (2019). Integrating harvesting decisions in the design of agro-food supply chains. *European Journal of Operational Research*, 276(1), 247–258. DOI: 10.1016/j.ejor.2018.12.024

Joshi, P., & Visvanathan, C. (2019). Sustainable management practices of food waste in Asia: Technological and policy drivers. *Journal of Environmental Management*, 247, 538–550. DOI: 10.1016/j.jenvman.2019.06.079 PMID: 31260920

Joshi, S. (Ed.). (2015). *Designing and Implementing Global Supply Chain Management*. IGI Global.

Joshi, S., Singh, R. K., & Sharma, M. (2023). Sustainable agri-food supply chain practices: Few empirical evidences from a developing economy. *Global Business Review*, 24(3), 451–474. DOI: 10.1177/0972150920907014

Kakkad, V., Patel, M., & Shah, M. (2019). Biometric authentication and image encryption for image security in cloud framework. *Multiscale and Multidisciplinary Modeling. Experiments and Design*, 2(4), 233–248. DOI: 10.1007/s41939-019-00049-y

Kalamkar, S. S., Ahir, K., & Bhaiya, S. R. (2019). Status of implementation of electronic National Agriculture Market (eNAM) in selected APMCs of Gujarat. PROGRESS, 112.

Kamble, S. S., Gunasekaran, A., & Sharma, R. (2019). Modeling the blockchain enabled traceability in agriculture supply chain. *International Journal of Information Management*, 52, 101967. DOI: 10.1016/j.ijinfomgt.2019.05.023

Kamilaris, A., Fonts, A., & Prenafeta-Boldú, F. X. (2019). The rise of blockchain technology in agriculture and food supply chains. *Trends in Food Science & Technology*, 91, 640–652. DOI: 10.1016/j.tifs.2019.07.034

Kang, Z., Zhao, Y., Chen, L., Guo, Y., Mu, Q., & Wang, S. (2022). Advances in machine learning and hyperspectral imaging in the food supply chain. *Food Engineering Reviews*, 14(4), 596–616. DOI: 10.1007/s12393-022-09322-2

Kannadhasan, S., Nagarajan, R., Srividhya, G., & Wang, X. (2022). Recent Trends in Logistics Management: Past, Present, and Future. In *Utilizing Blockchain Technologies in Manufacturing and Logistics Management* (pp. 234-249). IGI Global.

Kasemsap, K. (2017). The roles of logistics service providers and supply chain integration in global supply chain. In *Global intermediation and logistics service providers* (pp. 265–285). IGI Global. DOI: 10.4018/978-1-5225-2133-4.ch014

Katiyar, V., & Ghosh, T. (2021). *Nanotechnology in edible food packaging*. Springer., DOI: 10.1007/978-981-33-6169-0

Katunze, M., Kuteesa, A., Mijumbi, T., & Mahebe, D. (2017). Uganda Warehousing Receipt System: Improving Market Performance and Productivity. *African Development Review*, 29(S2), 135–146. DOI: 10.1111/1467-8268.12268

Kayapınar, Ö., & Lorcu, F. (2020). The role of technology level and logistics performance on the relationship between logistics service quality and firm performance. In *Handbook of Research on Sustainable Supply Chain Management for the Global Economy* (pp. 107–135). IGI Global. DOI: 10.4018/978-1-7998-4601-7.ch006

Kayikci, Y. (2019). E-Commerce in logistics and supply chain management. In *Advanced Methodologies and Technologies in Business Operations and Management* (pp. 1015–1026). IGI Global. DOI: 10.4018/978-1-5225-7362-3.ch076

Kayikci, Y., Kazancoglu, Y., Gozacan-Chase, N., & Lafci, C. (2022). Analyzing the drivers of smart sustainable circular supply chain for sustainable development goals through stakeholder theory. *Business Strategy and the Environment*, 31(7), 3335–3353. DOI: 10.1002/bse.3087

Kendro, S. (2012). Développement du secteur financier et financement de l'activité agricole dans un contexte de crise alimentaire: Quelle place pour la micro-finance? *Développement Durable & Territoires*, 3(3), 1–19.

Kenis, P., & Raab, J. (2020). Back to the future: Using organization design theory for effective organizational networks. *Perspectives on Public Management and Governance*, 3(2), 109–123. DOI: 10.1093/ppmgov/gvaa005

Kesavan, T. R., Mathur, H., & Harnal, J. (2022). Redefining agriculture through artificial intelligence: Predicting the unpredictable. In *PwC*.

Khan, S. H., Chowdhury, M., Das, P., Mandal, N., Ahuja, V., & Baset, M. (2014). *Milk Vita –Realities and reaching out to dreams of millions* [Conference Paper]. Dairy Asia: Towards Sustainability Held in Bangkok During 21 to 24 May, 2014, Bangkok, Thailand. https://www.researchgate.net/publication/306378001_DDairyasia-pmpaperdraft-khan blri_Page1 _Milk_Vita_-realities_and_reaching_out_to_dreams_of_millions_1

Khang, A. (Ed.). (2023). *Handbook of Research on AI-Equipped IoT Applications in High-Tech Agriculture*. IGI Global. DOI: 10.4018/978-1-6684-9231-4

Khatri-Chhetri, A., Sapkota, T. B., Sander, B. O., Arango, J., Nelson, K. M., & Wilkes, A. (2021). Financing climate change mitigation in agriculture: Assessment of investment cases. *Environmental Research Letters*, 16(12), 124044. DOI: 10.1088/1748-9326/ac3605

Khemka, M. (2022). Operation Flood: The Successful Rural Development Vision. *Indian Business@ 75—Different Aspects*, 65.

Kherallah, M., & Kirsten, J. F. (2002). The New Institutional Economics: Applications for Agricultural Policy Research in Developing Countries: "New institutional economists are the blue-collar guys with a hearty appetite for reality." Oliver Williamson, 2000a. *Agrekon*, 41(2), 110–133.

Kim, M., & Day, D. (2011). Composition of sugar cane, energy cane, and sweet sorghum suitable for ethanol production at Louisiana sugar mills. *Journal of Industrial Microbiology & Biotechnology*, 38(7), 803–807. DOI: 10.1007/s10295-010-0812-8 PMID: 20803247

Kim, M., Hilton, B., Burks, Z., & Reyes, J. (2018). Integrating blockchain, smart contract-tokens, and IoT to design a food traceability solution. In *Proceedings of the 9th International Conference on Information Systems and Technology* (pp. 1-10). Claremont Graduate University, CA, USA: IEEE. DOI: 10.1109/IEMCON.2018.8615007

Kitole, F. A., Mkuna, E., & Sesabo, J. K. (2024). Digitalization and agricultural transformation in developing countries: Empirical evidence from Tanzania agriculture sector. *Smart Agricultural Technology*, 7, 100379. DOI: 10.1016/j.atech.2023.100379

Kividhami. (2021, June 16). An overview of AI technologies in African agriculture. *KDHI Agriculture*. https://www.kdhi-agriculture.com/single-post/an-overview-of-ai-technologies-in-african-agriculture

Klugman, J. (1999). Social and economic policies to prevent complex humanitarian emergencies: lessons from experience.

Kovacevic, V., Rajnovic, L., & Vukovic, P. (2023). Warehouse receipt finance in the agricultural sectors: Lessons learned in Serbia. *Ekonomika (Nis)*, 69(1), 59–68. DOI: 10.5937/ekonomika2301059K

Kpenekuu, F., Antwi-Agyei, P., Nimoh, F., Amoakowaah Ose, M., & Guodaar, L. (2024). Understanding drought risk management in vulnerable communities in dryland farming systems: Evidence from northwest Ghana. *International Journal of Disaster Risk Reduction*, 100, 104186. DOI: 10.1016/j.ijdrr.2023.104186

Krishna, A. (2002). Enhancing Political Participation in Democracies: What is the Role of Social Capital? *Comparative Political Studies*, 35(4), 437–460. DOI: 10.1177/0010414002035004003

Kulkarni, P. M., Pote, S., Gokhale, P., Lakshminarayana, K., & Tigadi, B. (2024). Integrating IoT for Enhancing Supply Chain Agility. In Practical Approaches to Agile Project Management (pp. 183-206). IGI Global. DOI: 10.4018/979-8-3693-3318-1.ch011

Kulwijila, M. (2021). Socio-economic determinants of postharvest losses in the grape value chain in Dodoma Municipality and Chamwino District, Tanzania. *African Journal of Economic Review*, 9(2), 288–305.

Kulwijila, M., Makindara, J., & Laswai, H. (2018). Grape value chain mapping in Dodoma region, Tanzania. *Journal of Economics and Sustainable Development*, 9(2), 171–182.

Kumar, K. B. S., & Perepu, I. (2023). Twiga-A B2B Agritech Startup's Innovative Solution to Improve Food Security in Kenya. IUP Journal of Supply Chain Management, 20(4).

Kumar, P. (2021, September 2). AI-enabled supply chain: A win-win for farmers, consumers. *Business Today*. https://www.businesstoday.in/opinion/columns/story/ai-enabled-supply-chain-a-win-win-for-farmers-consumers-305739-2021-09-02

Kumar, C., Ram, C. L., Jha, S. N., & Vishwakarma, R. K. (2021). Warehouse Storage Management of Wheat and Their Role in Food Security. *Frontiers in Sustainable Food Systems*, 5, 675626. DOI: 10.3389/fsufs.2021.675626

Kumar, D., & Kalita, P. (2017). Reducing Postharvest Losses during Storage of Grain Crops to Strengthen Food Security in Developing Countries. *Foods*, 6(1), 8. DOI: 10.3390/foods6010008 PMID: 28231087

Kumaresan, S. J., Vikneswaran, M., Surai, P., & Vignesh, S. (2022). Agri Warehouse Management System. In *2022 International Interdisciplinary Humanitarian Conference for Sustainability (IIHC)* (pp. 705-709). IEEE. DOI: 10.1109/IIHC55949.2022.10059944

Kumar, G., Engle, C., & Tucker, C. (2018). Factors driving aquaculture technology adoption. *Journal of the World Aquaculture Society*, 49(3), 447–476. DOI: 10.1111/jwas.12514

Kumari, S., Venkatesh, V. G., Deakins, E., Mani, V., & Kamble, S. (2023). Agriculture value chain sustainability during COVID-19: An emerging economy perspective. *International Journal of Logistics Management*, 34(2), 280–303. DOI: 10.1108/IJLM-04-2021-0247

Kumar, M., Srai, J., Pattinson, L., & Gregory, M. (2013). Mapping of the UK food supply chains: Capturing trends and structural changes. *Journal of Advances in Management Research*, 10(2), 299–326. DOI: 10.1108/JAMR-05-2013-0034

Kumar, V., Ramachandran, D., & Kumar, B. (2021). Influence of new-age technologies on marketing: A research agenda. *Journal of Business Research*, 125, 864–877. DOI: 10.1016/j.jbusres.2020.01.007

Lacroix, R., & Varangis, P. (1996). Using warehouse receipts in developing and transition economies. *Finance & Development*, 33(3), 36–39.

Lan, Y. C., & Unhelkar, B. (Eds.). (2006). *Global integrated supply chain systems*. IGI Global. DOI: 10.4018/978-1-59140-611-2

Lapenu, C. (2008). Evolutions récentes dans l'offre et les stratégies de financement du secteur. Presented at the *conference on How to Regulate and Supervise Microfinance*, Kampala, Uganda, 22-26 November.

Laré, A., Lessoua, A., Diaw, D., Boutabba, A., & Gomado, K. (2021). Microfinance et pauvreté: Le cas des quartiers périphériques de Lomé au Togo. *Revue d'Economie du Developpement*, 29(3), 79–116. DOI: 10.3917/edd.353.0079

Latif, R. M. A., & Farhan, M. (2020). Retail level blockchain transformation for product supply chain using truffle development platform. *Springer Business Media*, LLC. Advance online publication. DOI: 10.1007/978-3-030-37629-6_11

Latino, M. E., Menegoli, M., Lazoi, M., & Corallo, A. (2022). Voluntary traceability in food supply chain: A framework leading its implementation in Agriculture 4.0. *Technological Forecasting and Social Change*, 178, 121564. DOI: 10.1016/j.techfore.2022.121564

Lazebnyk, L., & Voitenko, B. (2021). Digital technologies in agricultural enterprise management. *Financial and Credit Activity Problems of Theory and Practice*, 6(41), 203–210. DOI: 10.18371/fcaptp.v6i41.251440

Lehtinen, U. (2017). Sustainable Supply Chain Management in Agri-food Chains. In Bhat, R. (Ed.), *Sustainability Challenges in the Agrofood Sector*., DOI: 10.1002/9781119072737.ch7

Lelart, M. (2006). *De la finance informelle à la microfinance*. Agence Universitaire de la Francophonie.

Leng, K., Bi, Y., Jing, L., Fu, H. C., & Van Nieuwenhuyse, I. (2018). Research on agricultural supply chain system with double chain architecture based on blockchain technology. *Future Generation Computer Systems*, 86, 641–649. DOI: 10.1016/j.future.2018.04.061

Lezoche, M., Hernandez, J. E., Díaz, M. D. M. E. A., Panetto, H., & Kacprzyk, J. (2020). Agri-food 4.0: A survey of the supply chains and technologies for the future agriculture. *Computers in Industry*, 117, 103187. DOI: 10.1016/j.compind.2020.103187

Li Sun, S., & Liang, H. (2021). Globalization and affordability of microfinance. *Journal of Business Venturing*, 36(1), 106065. DOI: 10.1016/j.jbusvent.2020.106065

Li, J., Liu, Y., & Chen, Z. (2022, July). Role of Industry Associations in the GI Value Chain: The Case of Xinhui Orange Peel. In *Worldwide Perspectives on Geographical Indications*.

Liaghat, S., & Balasundram, S. K. (2010). A Review: The Role of Remote Sensing in Precision Agriculture. *American Journal of Agricultural and Biological Sciences*, 5(1), 50–55. DOI: 10.3844/ajabssp.2010.50.55

Liao, Y., & Xu, K. (2019). Traceability system of agricultural product based on blockchain and application in tea quality safety management. *Journal of Physics: Conference Series*, 1345, 052008. DOI: 10.1088/1742-6596/1345/5/052008

Li, M., & Yost, R. S. (2000). Management-oriented modeling: Optimizing nitrogen management with artificial intelligence. *Agricultural Systems*, 65(1), 1–27. DOI: 10.1016/S0308-521X(00)00023-8

Lim, W. M., Gupta, S., Aggarwal, A., Paul, J., & Sadhna, P. (2021). How do digital natives perceive and react toward online advertising? Implications for SMEs. *Journal of Strategic Marketing*, •••, 1–35.

Lin, J. Y. (2018). Agriculture is key for economic transformation, food security, and nutrition. Retrieved on, 21(11), 2018.

Lin, J., Shen, Z., & Zhang, A. (2020). Blockchain and IoT based food traceability for smart agriculture. *Journal of Food Quality*, 2020, 8810307. Advance online publication. DOI: 10.1155/2020/8810307

Lin, J., Zhang, A., Shen, Z., & Chai, Y. (2018). Blockchain and IoT based food traceability for smart agriculture. In *ACM International Conference Proceedings Series*. DOI: 10.1145/3265689.3265692

Lin, Y. P., Petway, J. R., Anthony, J., Mukhtar, H., Liao, S. W., Chou, C. F., & Ho, Y. F. (2017). Blockchain: The evolutionary next step for ICT e-agriculture. *Environments (Basel, Switzerland)*, 4(50), 1–14. DOI: 10.3390/environments4030050

Littlefield, E., Murdoch, J., & Hashemi, S. (2003). *Is microfinance an effective strategy to reach the Millennium Development Goals?* CGAP Focus Note.

Littlefield, E., & Rosenberg, R. (2004). Microfinance and the poor: Breaking down walls between microfinance and formal finance. *Finance & Development*, •••, 38–40.

Liu, Y., Ma, X., Shu, L., Hancke, G. P., & Abu-Mahfouz, A. M. (2020). From industry 4.0 to agriculture 4.0: Current status, enabling technologies, and research challenges. *IEEE Transactions on Industrial Informatics*, 17(6), 4322–4334. DOI: 10.1109/TII.2020.3003910

Louwaars, N. P. (2023). Potato in the Netherlands–a remarkable story. In *Impact of hybrid potato* (pp. 45–58). Academic. DOI: 10.3920/978-90-8686-946-6_4

Lucena, P. (2019). An approach for grain quality measurement using blockchain and smart contracts. *Journal of Physics: Conference Series*, 1342, 052012. DOI: 10.1088/1742-6596/1342/5/052012

Lueschen, W. E., Putnam, D. H., Kanne, B. K., & Hoversted, T. R. (1991). Agronomicpractices for production of ethanol from sweet sorghum. *Journal of Production Agriculture*, 4(4), 619–625. DOI: 10.2134/jpa1991.0619

Lusiantoro, L., Yates, N., Mena, C., & Varga, L. (2018). A refined framework of information sharing in perishable product supply chains. *International Journal of Physical Distribution & Logistics Management*, 48(3), 254–283. Advance online publication. DOI: 10.1108/IJPDLM-08-2017-0250

Lydia, J., Sherwin Vimalraj, L., Monisha, R., & Murugan, R. (2022). Automated food grain monitoring system for warehouse using IOT. *Measurement. Sensors*, 24, 100472. DOI: 10.1016/j.measen.2022.100472

Maditati, D. R., Munim, Z. H., Schramm, H. J., & Kummer, S. (2018). A review of green supply chain management: From bibliometric analysis to a conceptual framework and future research directions. *Resources, Conservation and Recycling*, 139(August), 150–162. DOI: 10.1016/j.resconrec.2018.08.004

Magalhães, A. E. V., Rossi, A. H. G., Zattar, I. C., Marques, M. A. M., & Seleme, R. (2019). Food traceability technologies and foodborne outbreak occurrences. *British Food Journal*, 121(12), 3362–3379. DOI: 10.1108/BFJ-02-2019-0143

Maheshwar, C., & Chanakwa, T. (2006). Postharvest Losses Due to Gaps in Cold Chain In India-A Solution. *Acta Horticulturae*, (712), 777–784. DOI: 10.17660/ActaHortic.2006.712.100

Maime, M. J. (2015). *Evaluation of the fertility of communal bulls in Moretele district Northwest Province in South Africa*. University of Pretoria.

Maiwashe-Tagwi, A., Mawela, A. S., & Chauke, P. K. (Eds.). (2023). *Global Agricultural and Food Marketing in a Global Context: Advancing Policy, Management, and Innovation: Advancing Policy, Management, and Innovation*. IGI Global.

Making Climate Finance Work in Agriculture - World Bank. Retrieved from https://www.worldbank.org/en/topic/agriculture/publication/making-climate-finance-work-in-agriculture

Makule, E., Dimoso, N., & Tassou, S. A. (2022). Precooling and Cold Storage Methods for Fruits and Vegetables in Sub-Saharan Africa—A Review. *Horticulturae*, 8(9), 776. DOI: 10.3390/horticulturae8090776

Malik, R. M. M. K. V. K. (2020). *Indian food processing: an untapped growth opportunity*. Invest India. https://www.investindia.gov.in/siru/indian-food-processing-sector-untapped-growth-opportunity

Manandhar, A., Milindi, P., & Shah, A. (2018). An Overview of the Post-Harvest Grain Storage Practices of Smallholder Farmers in Developing Countries. *Agriculture*, 8(4), 57. DOI: 10.3390/agriculture8040057

Manjunath, B. L., & Narayana, C. K. (2021). Boosting agro-processing industry in India. *Indian Farming*, 71(9).

Mao, D., Wang, F., Hao, Z., & Li, H. (2018). Credit evaluation system based on blockchain for multiple stakeholders in the food supply chain. *International Journal of Environmental Research and Public Health*, 15(8), 1627. DOI: 10.3390/ijerph15081627 PMID: 30071695

Marchiori, D., & Franco, M. (2020). Knowledge transfer in the context of inter-organizational networks: Foundations and intellectual structures. *Journal of Innovation and Knowledge*, 5(2), 130–139. DOI: 10.1016/j.jik.2019.02.001

Marín, M. A. F. V. (2020). The New Concept of Logistics Platforms 4.0: Creating Competitiveness Within the Paradigm of Global Sustainable Logistics. In *Handbook of Research on Sustainable Supply Chain Management for the Global Economy* (pp. 36-62). IGI Global.

Market Intelligence and Analytics. (2023). *The Onion Paradox: Output falls, but so do prices*. CRISIL An S & P Global Company.

McKenzie, D., & Woodruff, C. (2008). Experimental evidence on returns to capital and access to finance in Mexico. *The World Bank Economic Review*, 22(3), 457–482. DOI: 10.1093/wber/lhn017

McMichael, C., & Manderson, L. (2004). Somali women and well-being: Social networks and social Capital among immigrant women in Australia. *Human Organization*, 63(1), 88–99. DOI: 10.17730/humo.63.1.nwlpjdj4d4l97561

Meemken, E., Barrett, C. B., Michelson, H. C., Qaim, M., Reardon, T., & Sellare, J. (2021). Sustainability standards in global agrifood supply chains. *Nature Food*, 2(10), 758–765. DOI: 10.1038/s43016-021-00360-3 PMID: 37117971

Mehring, L. (2023). *France's Agricultural Sector Produced 1.1% of Global Agricultural Emissions in 2021.* Climate Scorecard. Retrieved from https://www.climatescorecard.org/2023/03/frances-agricultural-sector-produced-1-1-of-global-agricultural-emissions-in-2021/

Meneghetti, A., & Monti, L. (2014). Greening the food supply chain: An optimisation model for sustainable design of refrigerated automated warehouses. *International Journal of Production Research*, 53(21), 6567–6587. DOI: 10.1080/00207543.2014.985449

Mentzer, J. T., DeWitt, W., Keebler, J. S., Min, S., Nix, N. W., Smith, C. D., & Zacharia, Z. G. (2001). Defining supply chain management. *Journal of Business Logistics*, 22(2), 1–25. DOI: 10.1002/j.2158-1592.2001.tb00001.x

Mersland, R., & Strom, R. O. (2010). Microfinance mission drift. *World Development*, 38(1), 28–36. DOI: 10.1016/j.worlddev.2009.05.006

Milkvita (2024). History of Milk Vita. Retrieved October 24, 2024, from https://milkvita.org.bd/site/page/3862d7a1-be70-4b80-a575-61ac5a151443/History-of-Milkvita

MilkVita. (2018). www.facebook.com. Retrieved November 8, 2024, Retrieved September 18, 2024 from https://www.facebook.com /MilkVitaBangladesh /photos

Ministry of Agriculture & Farmers Welfare. (2024, August 02). *Farmer Producer Organisations (FPOs).* [Press release]. https://pib.gov.in/PressReleaseIframePage.aspx?PRID=2040845

Ministry of Consumer Affairs, Food and Public Distribution. (2024). Grand Challenge for development of "Technologies for Primary Processing, Storage and Valorization of Onions". Retrieved from https://doca.gov.in>Grandchallenge-Document.

Ministry of Consumer Affairs. Food & Public Distribution (2024). Government has directed NCCF and NAFED to initiate procurement of 5 lakh tonnes of onion for the buffer requirement directly from the farmers. Retrieved from https://pib.gov.in/PressReleasePage.aspx?PRID=2016405#:~:text=In%20the%20current%20year%2C%20the,started%20arriving%20in%20the%20market

Ministry of Food Processing Industries. (2024, March 14). *ODOP List of 726 districts under 35 states/UTs.*https://www.mofpi.gov.in/sites/default/files/revised_list_of_odop_for_35_states_13.03.2024_1.pdf

Mintz, A. (2019). *Integrating Sustainable Practices: Compost Tea as a Nutrient Supplement for Aquaponic Plant Production* (Doctoral dissertation, Evergreen State College).

Mirabelli, G., & Solina, V. (2020). Blockchain and agricultural supply chains traceability: Research trends and future challenges. *Procedia Manufacturing*, 42, 414–421. DOI: 10.1016/j.promfg.2020.02.054

Miranda, J., Ponce, P., Molina, A., & Wright, P. (2019). Sensing, smart and sustainable technologies for Agri-Food 4.0. *Computers in Industry*, 108, 21–36. DOI: 10.1016/j.compind.2019.02.002

Miranda, M. J., Mulangu, F. M., & Kemeze, F. H. (2019). Warehouse receipt financing for smallholders in developing countries: Challenges and limitations. *Agricultural Economics*, 50(5), 629–641. DOI: 10.1111/agec.12514

Mobolade, A. J., Bunindro, N., Sahoo, D., & Rajashekar, Y. (2019). Traditional methods of food grains preservation and storage in Nigeria and India. *Annals of Agricultural Science*, 64(2), 196–205. DOI: 10.1016/j.aoas.2019.12.003

Mogili, U. R., & Deepak, B. B. V. L. (2018). Review on application of drone systems in precision agriculture. *Procedia Computer Science*, 133, 502–509. DOI: 10.1016/j.procs.2018.07.063

Mohanraj, I., Ashokumar, K., & Naren, J. (2016). Field monitoring and automation using IOT in agriculture domain. *Procedia Computer Science*, 93, 931–939. DOI: 10.1016/j.procs.2016.07.275

Mohieldin, M., Kenewendo, B., & Wambui, R. (2023). Breaking Financing Barriers for Just Climate Transition in Africa.

Mollik, B. A. (2022, March). Cooperative based rural development in Bangladesh-the good and the bad. SSRN *Electronic Journal*. DOI: 10.2139/ssrn.4044966

Monasa Advisory & Associates; Study to investigate the impact of the cattle marketing scheme in Namibia, Windhoek, Namibia, February 2019

Monastyrnaya, E., Joerin, J., Dawoe, E., & Six, J. (2016). Assessing the resilience of the cocoa value chain in Ghana. *Case study report*, 1-137.

Mondal, S., Wijewardena, K., Karuppuswami, S., Kriti, N., Kumar, D., & Chahal, P. (2021). Blockchain inspired RFID-based information architecture for food supply chain. *IEEE Internet of Things Journal*, 8(12), 9442–9452. DOI: 10.1109/JIOT.2021.3050959

Montalbano, P., & Nenci, S. (2020). The effects of global value chain (GVC) participation on the economic growth of the agricultural and food sectors.

Monteiro, J., & Barata, J. (2021). Artificial Intelligence in Extended Agri-Food Supply Chain: A short review based on bibliometric analysis. *Procedia Computer Science*, 192, 3020–3029. DOI: 10.1016/j.procs.2021.09.074

Monthly Report of Onion. (2017). Horticultural Statistics Division, Department of Agriculture, Cooperation and Farmers Welfare, Ministry of Agriculture & Farmers Welfare, Government of India.

Morduch, J. (2000). The microfinance schism. *World Development*, 28(4), 617–629. DOI: 10.1016/S0305-750X(99)00151-5

Moreno, J. C., Berenguel, M., Donaire, J. G., Rodríguez, F., Sánchez-Molina, J. A., Guzmán, J. L., & Giagnocavo, C. L. (2024). A pending task for the digitalisation of agriculture: A general framework for technologies classification in agriculture. *Agricultural Systems*, 213, 103794. DOI: 10.1016/j.agsy.2023.103794

Morgan, R. M., & Hunt, S. D. (1994). The commitment-trust theory of relationship marketing. *Journal of Marketing*, 58(3), 20–38. DOI: 10.1177/002224299405800302

Movahed, A. B., Movahed, A. B., Aliahmadi, B., & Nozari, H. (2024). Green and Sustainable Supply Chain in Agriculture 6.0. In Advanced Businesses in Industry 6.0 (pp. 32-45). IGI Global.

Mukherjee, A. A., Singh, R. K., Mishra, R., & Bag, S. (2022). Application of blockchain technology for sustainability development in agricultural supply chain: Justification framework. *Operations Management Research : Advancing Practice Through Research*, 15(1), 46–61. DOI: 10.1007/s12063-021-00180-5

Mumtaz, N., & Nazar, M. (2022). Artificial intelligence robotics in agriculture: See & spray. *Journal of Intelligent Pervasive and Soft Computing*, 1(01), 21–24.

Murray, S. C., Rooney, W. L., Hamblin, M. T., Mitchell, S. E., & Kresovich, S. (2009). Sweet sorghum genetic diversity and association mapping for brix and height. *The Plant Genome*, 2(1), 48–62. DOI: 10.3835/plantgenome2008.10.0011

Musa, S. F. P. D., & Basir, K. H. (2021). Smart farming: Towards a sustainable agri-food system. *British Food Journal*, 123(9), 3085–3099. DOI: 10.1108/BFJ-03-2021-0325

Muscatello, J. R., Parente, D. H., & Swinarski, M. (2016). The impact of ERP alignment on logistics costs: A work system theoretical approach. [IJEIS]. *International Journal of Enterprise Information Systems*, 12(3), 1–17. DOI: 10.4018/IJEIS.2016070101

Mustapha, U. F., Alhassan, A. W., Jiang, D. N., & Li, G. L. (2021). Sustainable aquaculture development: A review on the roles of cloud computing, internet of things and artificial intelligence (CIA). *Reviews in Aquaculture*, 13(4), 2076–2091. DOI: 10.1111/raq.12559

Mvumi, B., Matsikira, L. T., & Mutambara, J. (2016). The banana postharvest value chain analysis in Zimbabwe. *British Food Journal*, 118(2), 272–285. DOI: 10.1108/BFJ-08-2014-0293

Naeem, M. A., Nguyen, T. N., Ali, R., Cengiz, K., Meng, Y., & Khurshaid, T. (2021). Hybrid cache management in IoT-based named data networking. *IEEE Internet of Things Journal*, 8(5), 3378–3389. DOI: 10.1109/JIOT.2020.3032451

Naik, G., Raghuram, G., Rajan, J., Bansal, M., Gopikuttan, G. S., Tawri, P., & Singh, R. (2022). Agricultural Warehousing in India: Trends, Constraints, and Policies. IIM Bangalore. Retrieved from https://ncdex.com/downloads/Research/casestudies-conceptpaper/Agricultural%20Warehousing%20in%20India,%202022-%20A%20report%20by%20IIMB.pdf

Naik, G., Suresh, D.N., 2018. Challenges of creating sustainable agri-retail supply chains.

Nakagawa, Y., & Shaw, R. (2004). Social Capital: A missing link to disaster recovery. *International Journal of Mass Emergencies and Disasters*, 22(1), 5–34. DOI: 10.1177/028072700402200101

Nakamoto, S. (2008). Bitcoin: A peer-to-peer electronic cash system. Retrieved from https://bitcoin.org/bitcoin.pdf

Nakano, Y., & Magezi, E. F. (2020). The impact of microcredit on agricultural technology adoption and productivity: Evidence from randomized control trial in Tanzania. *World Development*, 133, 104997. DOI: 10.1016/j.worlddev.2020.104997

Namiban Agronomic Board. (2021). Market Intelligence Report Onion. 3: pp 1-21. Retrieved from https://www.nab.com.na/wp-content/uploads/2022/04/Market-Intelligence-Report-Onion.pdf

Nanaji Deshmukh Krishi sanjeevani Prakalp (PoCRA) and Indian Institute of Technology, Bombay (2022), 300 MT Climate Agnostic Onion Storage Structure, (Project of Government of Maharashtra in Partnership with World Bank). Retrieved from https://mahapocra.gov.in≥DPR.

Nasirahmadi, A., & Hensel, O. (2022). Toward the Next Generation of Digitalization in Agriculture Based on Digital Twin Paradigm. *Sensors (Basel)*, 22(2), 498. https://doi.org/https://doi.org/10.3390/s22020498. DOI: 10.3390/s22020498 PMID: 35062459

National Agriculture in Class Room NACR. (2024), Relevancy and Engagement, Text Book Tracing the Agriculture Supply Chain. Accessed May 2-24, https://agclassroom.org/matrix/lesson/print/838/

National Association for Farmer Producer Organizations. (2022). *State of Sector Report on Farmer Producer Organisations in India: A report on FPOs in India and emerging landscape.*https://www.nafpo.in/wp-content/uploads/2022/07/SOFPO_Report_(06-07-22)_Hi-res_WCM.pdf

National Co-operative Database. (n.d.). Retrieved from Ministry of Co-operation, GOI: https://cooperatives.gov.in/en

National Mission for Sustainable Agriculture (NMSA). (2023). Climate Change and Agriculture. Retrieved from https://www.nmsa.gov.in/climate-change-and-agriculture

Nations, F. a. (2021). *Trade of Agricultural Commodities (2005-21).* FAOSTAT Analytical Brief. Retrieved from https://openknowledge.fao.org/server/api/core/bitstreams/75eaf30a-b1b0-4e7b-8e21-d8970ff5833b/content

Nattassha, R., Handayati, Y., Simatupang, T. M., & Siallagan, M. (2020). Understanding circular economy implementation in the agri-food supply chain: The case of an Indonesian organic fertiliser producer. *Agriculture & Food Security*, 9(1), 1–16. DOI: 10.1186/s40066-020-00264-8

Nayak, A. (2022a). Case Studies of FPOs in India 2019-2021. In NABARD (No. 38–48). Department of Economic Analysis & Research, NABARD. Retrieved September 15, 2024, from https://www.nabard.org/auth/writereaddata/tender/0306220035case-studies-of-fpos-in-india-2019-21.pdf

Nayak, A. (2022b). Case Studies of FPOs in India 2019-2021. In NABARD (No. 113-129). Department of Economic Analysis & Research, NABARD. Retrieved September 15, 2024, from https://www.nabard.org/auth/writereaddata/tender/0306220035case-studies-of-fpos-in-india-2019-21.pdf

Nayal, K., Raut, R., Priyadarshinee, P., Narkhede, B. E., Kazancoglu, Y., & Narwane, V. (2022). Exploring the role of artificial intelligence in managing agricultural supply chain risk to counter the impacts of the COVID-19 pandemic. *International Journal of Logistics Management*, 33(3), 744–772. DOI: 10.1108/IJLM-12-2020-0493

NCCD, 2015. All India Cold-chain Infrastructure Capacity (Assessment of Status & Gap).

Nedacadmin. (2022). BMPCUL, Bangladesh - NEDAC - Network for the Development of Agricultural Cooperatives in Asia and the Pacific. *NEDAC - Network for the Development of Agricultural Cooperatives in Asia and the Pacific - Network for the Development of Agricultural Cooperatives in Asia and The Pacific*. https://nedac.info/2022/02/24/bmpcul-bangladesh/

Negi, S., & Anand, N. (2015). Issues and Challenges in the Supply Chain of Fruits & Vegetables Sector in India: A Review. *Int. J. Manag. Value Supply Chain.*, 6(2), 47–62. DOI: 10.5121/ijmvsc.2015.6205

Negi, S., & Anand, N. 2016. Factors Leading to Losses and Wastage in the Supply Chain of Fruits and Vegetables Sector in India. Energy, Infrastruct. Transp. Challenges W. Forw. I, 80–105. https://doi.org/DOI: 10.13140/RG.2.1.2395.5607

Nematollahi, M., & Tajbakhsh, A. (2020). Past, present, and prospective themes of sustainable agricultural supply chains: A content analysis. *Journal of Cleaner Production*, 271, 122201. DOI: 10.1016/j.jclepro.2020.122201

Nemes, S. A., Szabo, K., & Vodnar, D. C. (2020). Applicability of agro-industrial by-products in intelligent food packaging. *Coatings*, 10(6), 1–26. DOI: 10.3390/coatings10060550

Neven, D., & Reardon, T. (2004). The rise of Kenyan supermarkets and the evolution of their horticulture product procurement systems. *Development Policy Review*, 22(6), 669–699. DOI: 10.1111/j.1467-7679.2004.00271.x

Nguyen, G. T., & Kim, K. (2018). A survey about consensus algorithms used in blockchain. *Journal of Information Processing Systems*, 14(1), 101–128. DOI: 10.3745/JIPS.03.0095

Nihalani, R., Chouhan, S. S., Mittal, D., Vadula, J., Thakur, S., Chakraborty, S., ... & Saxena, A. (2024). Long Short-Term Memory (LSTM) model for Indian sign language recognition. *Journal of Intelligent & Fuzzy Systems*, (Preprint), 1-19..DOI: 10.3233/JIFS-233250

Nozari, H. (2024). Supply Chain 6.0 and Moving Towards Hyper-Intelligent Processes. In Information Logistics for Organizational Empowerment and Effective Supply Chain Management (pp. 1-13). IGI Global.

Nozari, H., & Szmelter-Jarosz, A. (2024). An Analytical Framework for Smart Supply Chains 5.0. In *Building Smart and Sustainable Businesses With Transformative Technologies* (pp. 1–15). IGI Global.

OEDC Report - Agricultural Policies in India, 2018. Accessed March 2024 https://www.oecd.org/regional/agricultural-policies-in-india-9789264302334-en.htm

Olabi, A. G., Shehata, N., Sayed, E. T., Rodriguez, C., Anyanwu, R. C., Russell, C., & Abdelkareem, M. A. (2023). Role of microalgae in achieving sustainable development goals and circular economy. *The Science of the Total Environment*, 854, 158689. DOI: 10.1016/j.scitotenv.2022.158689 PMID: 36108848

Olibile, O. Okapuka Feedlot mentorship student, Beef Production Systems, MEATCO Namibia, https://www.meatco.com.na/news/714/Beef-Production-Systems/, as retrieved on 19th August 2024

Olomu, M. O., Ekperiware, M. C., & Akinlo, T. (2020). Agricultural sector value chain and government policy in Nigeria: Issues, challenges and prospects. *African Journal of Economic and Management Studies*, 11(3), 525–538. DOI: 10.1108/AJEMS-03-2019-0103

Olorunfemi, B. J., & Kayode, S. E. (2021). Post-Harvest Loss and Grain Storage Technology- A Review. *Turkish Journal of Agriculture -. Food Science and Technology (Campinas)*, 9(1), 75–83. DOI: 10.24925/turjaf.v9i1.75-83.3714

Opara, U. L., & Mazaud, F. (2001). Food traceability from field to plate. *Outlook on Agriculture*, 30(4), 239–247. DOI: 10.5367/000000001101293724

Orn, D., Duan, L., Liang, Y., Siy, H., & Subramaniam, M. (2020, October). Agro-AI education: artificial intelligence for future farmers. In *Proceedings of the 21st Annual Conference On Information Technology Education* (pp. 54-57). DOI: 10.1145/3368308.3415457

Óskarsdóttir, K., & Oddsson, G. V. (2019). Towards a decision support framework for technologies used in cold supply chain traceability. *Journal of Food Engineering*, 240, 153–159. DOI: 10.1016/j.jfoodeng.2018.07.013

Osvald, A., & Stirn, L. Z. (2008). A vehicle routing algorithm for the distribution of fresh vegetables and similar perishable food. *Journal of Food Engineering*, 85(2), 285–295. DOI: 10.1016/j.jfoodeng.2007.07.008

Oyedijo, A., Kusi-Sarpong, S., Mubarik, M. S., Khan, S. A., & Utulu, K. (2024). Multi-tier sustainable supply chain management: A case study of a global food retailer. *Supply Chain Management*, 29(1), 68–97. DOI: 10.1108/SCM-05-2022-0205

Paalosmaa, T., & Shafie-khah, M. (2021). Feasibility of innovative smart mobility solutions: A case study for vaasa. *World Electric Vehicle Journal*, 12(4), 188. DOI: 10.3390/wevj12040188

Pacheco, C. and Paul, B. (2023), Applying complexity theory perspective to knowledge management in the innovation context, VINE Journal of Information and Knowledge Management Systems, Vol. ahead-of-print No. ahead-of-print.

Pal, K. (2023). A Review of Big Data Analytics for the Internet of Things Applications in Supply Chain Management. Applied AI and Multimedia Technologies for Smart Manufacturing and CPS Applications, 221-245.

Pal, K., & Karakostas, B. (2015). The use of cloud computing in shipping logistics. In *Cloud Technology: Concepts, Methodologies, Tools, and Applications* (pp. 1080-1100). IGI Global. DOI: 10.4018/978-1-4666-6539-2.ch050

Pal, K. (2023). Drivers of sustainable supply chain management using internet of things-based blockchain technology. In *Government Impact on Sustainable and Responsible Supply Chain Management* (pp. 171–201). IGI Global. DOI: 10.4018/978-1-6684-9062-4.ch010

Pandey, P. R., & Bhandari, H. (2018). *Rice Technological Innovation and Value Chain Development in South Asia: Current Status and Future Directions: SAARC Agriculture Centre* (p. 164). Bangladesh.

Pandey, B. K., Kanike, U. K., George, A. S., & Pandey, D. (Eds.). (2024). *AI and machine learning impacts in intelligent supply chain*. IGI Global. DOI: 10.4018/979-8-3693-1347-3

Pandian, S. D. (2023). SEEDS Annual Report 2023-24. In *SEEDS Annual Report*. https://seedsvo.org/wp-content/uploads/2024/08/SEEDS-Annual-Report-2023-24.pdf

Pant, S. C., & Kansra, P. (2023). Indian Agriculture Sector and Blockchain Technology: A Systematic Review Envisaging Business Opportunities. *Empirical Economics Letters, 22* (Special Issue 1), 21-35. https://doi.org/DOI: 10.5281/zenodo.8266156

Papadopoulos, T., Gunasekaran, A., Dubey, R., & Balta, M. (2018). Big data and RFID in supply chain and logistics management: A review of the literature and applications for data driven research. *Operations and Service Management: Concepts, Methodologies, Tools, and Applications*, 662-674.

Papa, S. F. (2017, June). Use of blockchain technology in agribusiness: Transparency and monitoring in agricultural trade. In *2017 International Conference on Management Science and Management Innovation (MSMI 2017)* (pp. 38-40). Atlantis Press. DOI: 10.2991/msmi-17.2017.9

Parameshachari, B. D., & Panduranga, H. T. (2021). Secure transfer of images using pixel-level and bit-level permutation based on knight tour path scan pattern and Henon map. In *Cognitive Informatics and Soft Computing* (pp. 271–283). Springer., DOI: 10.1007/978-981-33-6041-0_24

Partel, V., Costa, L., & Ampatzidis, Y. (2021). Smart tree crop sprayer utilizing sensor fusion and artificial intelligence. *Computers and Electronics in Agriculture*, 191, 106556. DOI: 10.1016/j.compag.2021.106556

Patidar, S. (2024, March 28). AI Agriculture: Revolutionizing the Future of Sustainable Farming.

Patil, K., Ojha, D., Struckell, E. M., & Patel, P. C. (2023). Behavioral drivers of blockchain assimilation in supply chains–A social network theory perspective. *Technological Forecasting and Social Change*, 192, 1–16.

Paul, S., & Ace, M. Comparisons of the beef value chain in Namibia, Botswana and Australia; Namibia Economist, https://economist.com.na/63512/agriculture/towards-value-chain-equity-in-the-beef-sector/, as retrieved on 24th July 2024

Paul, S., & Ace, M. Namibia: Towards Value Chain Equity in the beef Sector; https://allafrica.com/stories/202108090805.html, as retrieved on 24th July 2024

Paul, R. K., & Garai, S. (2021). Performance comparison of wavelets-based machine learning technique for forecasting agricultural commodity prices. *Soft Computing*, 25(20), 12857–12873. DOI: 10.1007/s00500-021-06087-4

Paul, R. K., Yeasin, M., Kumar, P., Paul, A. K., & Roy, H. S. (2023). Deep learning technique for forecasting the price of cauliflower. *Current Science*, 124(9), 1065. DOI: 10.18520/cs/v124/i9/1065-1073

Phillips, P. W. B., Relf-Eckstein, J.-A., Jobe, G., & Wixted, B. (2019). Configuring the new digital landscape in Western Canadian agriculture. *NJAS Wageningen Journal of Life Sciences*, 90(1), 100295. DOI: 10.1016/j.njas.2019.04.001

Piekkari, R., & Westney, D. E. (2017), Language as a Meeting Ground for Research on the MNC and Organization Theory, Multinational Corporations and Organization Theory: Post Millennium Perspectives (Research in the Sociology of Organizations, Vol. 49), Emerald Publishing Limited, Bingley, pp. 193-232.

Pienaar, L., & Barends-Jones, V. (2021). The economic contribution of South Africa's pomegranate industry. *Agriprobe*, 18(4), 57–64.

Piñeiro, V., Arias, J., Dürr, J., Elverdin, P., Ibáñez, A. M., Kinengyere, A., Opazo, C. M., Owoo, N., Page, J. R., Prager, S. D., & Torero, M. (2020). A scoping review on incentives for adoption of sustainable agricultural practices and their outcomes. *Nature Sustainability*, 3(10), 809–820. DOI: 10.1038/s41893-020-00617-y

Pinstrup-Andersen, P., Rahmanian, M., Allahoury, A., Guillou, M., Hendriks, S., Hewitt, J., Iwanaga, M., Kalafatic, C., Kliksberg, B., Maluf, R., Murphy, S., Oniang'o, R., Pimbert, M., Sepúlveda, M., Tang, H., 2014. Food losses and waste in the context of sustainable food systems.

Plant Genetic Resources and Crop Improvement, Onion (2023). National Horticultural Research & Development Foundation. Retrieved from: www.nhrdf.ora/Onion.php

Porter, M. E. (1979). *Decision Support Tools: Porter's Value Chain*. Cambridge University, Institute for Manufacturing.

Posthouwer, W. Posthouwer - Van Leeuwen, Faems, D., & WHU Otto Beisheim School of Management. (2020). Managing the fuzzy Front-End of joint business development: the case of FrieslandCampina [Article]. *WHU Otto Beisheim School of Management*. https://www.researchgate.net/publication/342212616

Poudel, K. L. (2017). Contract Farming and its Prospective in Nepal. https://www.fao.org/fileadmin/user_upload/contract_farming/presentations/Contract_farming_in_Nepal.pdf

Prakash, S., Soni, G., Rathore, A. P. S., & Singh, S. (2017). Risk analysis and mitigation for perishable food supply chain: A case of dairy industry. *Benchmarking*, 24(1), 2–23. DOI: 10.1108/BIJ-07-2015-0070

Press Information Bureau. (2020). [Press release]. https://pib.gov.in/PressReleasePage.aspx?PRID=1601273

Press Information Bureau. (2024). [Press release]. https://pib.gov.in/PressReleasePage.aspx?PRID=1968931

Press Information Bureau. (n.d.). [Press release]. https://pib.gov.in/newsite/PrintRelease.aspx?relid=191212

Press Information Bureau. (n.d.). [Press release]. https://pib.gov.in/PressReleasePage.aspx?PRID=1886054

Priyadarshini, P., & Abhilash, P. C. (2020). Circular economy practices within energy and waste management sectors of India: A meta-analysis. *Bioresource Technology*, 304, 123018. DOI: 10.1016/j.biortech.2020.123018 PMID: 32087547

Priyadarshini, P., & Abhilash, P. C. (2020). Policy recommendations for enabling transition towards sustainable agriculture in India. *Land Use Policy*, 96, 104718. DOI: 10.1016/j.landusepol.2020.104718

PRS Legislative Research. (n.d.). *System of fertilizer subsidy*. https://prsindia.org/policy/report-summaries/system-of-fertilizer-subsidy

Pti. (2021, March 25). India farm sector needs policy, financial support to realise full potential of AI: Nasscom. *Business Today*. https://www.businesstoday.in/industry/it/story/india-farm-sector-needs-policy-financial-support-to-realise-full-potential-of-ai-nasscom-291741-2021-03-25

Qader, G., Shahid, Z. A., Junaid, M., Shaikh, I. M., & Qureshi, M. A. (2023). The role of diffusion of innovation theory towards the adoption of halal meat supply chain. *Journal of Islamic Marketing*, 14(5), 1211–1228. DOI: 10.1108/JIMA-01-2021-0032

Qu, C. C., Sun, X. Y., Sun, W. X., Cao, L. X., Wang, X. Q., & He, Z. Z. (2021). Flexible Wearables for Plants. *Small*, 17(50), 2104482. Advance online publication. DOI: 10.1002/smll.202104482 PMID: 34796649

Quintero, J. A., Montoya, M. I., Sanchez, O. J., Giraldo, O. H., & Cardona, C. A. (2008). Fuel ethanol production from sugarcane and corn: Comparative analysis for a Colombian case. *Energy*, 33(3), 385–399. DOI: 10.1016/j.energy.2007.10.001

Rahman, L. F., Alam, L., Marufuzzaman, M., & Sumaila, U. R. (2021). Traceability of sustainability and safety in fishery supply chain management systems using radio frequency identification technology. *Foods*, 10(10), 2265. Advance online publication. DOI: 10.3390/foods10102265 PMID: 34681313

Rahmaty, M. (2024). IoE-Based Supply Chain Management. In Information Logistics for Organizational Empowerment and Effective Supply Chain Management (pp. 14-26). IGI Global.

Rai, B. K. (2023). IoT based humidity and temperature control system for smart warehouse. *Gazi University Journal of Science*, 36(1), 173–188. DOI: 10.35378/gujs.993959

Rajan Selin, A., & Shanmugasundram, S. (2024). *Traditional approaches of agriculture soil remediation: boon or bane for global food security. In Bioremediation of Emerging Contaminants from Soils*. Elsevier.

Rajan, R. G., & Lamba, R. (2023). *Breaking the Mould: Reimagining India's Economic Future*. Penguin Random House India Private Limited.

Raj, M., Gupta, S., Chamola, V., Elhence, A., Garg, T., Atiquzzaman, M., & Niyato, D. (2021). A survey on the role of Internet of Things for adopting and promoting Agriculture 4.0. *Journal of Network and Computer Applications*, 187(April), 103107. DOI: 10.1016/j.jnca.2021.103107

Raman, D. R., Saravanan, D., Parthiban, R., Palani, D. U., David, D. D. S., Usharani, S., & Jayakumar, D. (2021). A study on application of various artificial intelligence techniques on internet of things. *European Journal of Molecular and Clinical Medicine*, 7(9), 2531–2557.

Ramasamy, T. (2020). *NABARD promoted "SEEDS Farmers Producer Company Ltd" is a standing testimony and an agent of change that takes food from farm to family*. Retrieved September 18, 2024, from https://www.linkedin.com/pulse/nabard-promoted-seeds-farmers-producer-company-ltd-agent-ramasamy/

Rana, N. P., Chatterjee, S., Dwivedi, Y. K., & Akter, S. (2022). Understanding dark side of artificial intelligence (AI) integrated business analytics: Assessing firm's operational inefficiency and competitiveness. *European Journal of Information Systems*, 31(3), 364–387. DOI: 10.1080/0960085X.2021.1955628

Rao, S. S., Patil, J. V., Rao, B. D., Ratnavathi, C. V., Umakanth, A. V., Mishra, J. S., & Chapke, R. R. (2014). 14. Current status of sweet sorghum biofuel industrial experiences for pilot–production of bioethnol. Proc. of "Model Training Course on "Improved sorghum cultivation and value-addition perspectives, 22, 29.

Raut, R. D., Gardas, B. B., Kharat, M., & Narkhede, B. (2018). Modeling the drivers of post- harvest losses – MCDM approach. *Computers and Electronics in Agriculture*, 154, 426–433. DOI: 10.1016/j.compag.2018.09.035

Rayhan, M. J., Rahman, S. M., Mamun, A. A., Saif, A. N. M., Islam, K. A., Alom, M. M., & Hafiz, N. (2024). FinTech solutions for sustainable agricultural value chains: A perspective from smallholder farmers. *Business Strategy & Development*, 7(2), e358. DOI: 10.1002/bsd2.358

RDCD, Bangladesh. (n.d.). Retrieved from Network for the Development of Agricultural Co-operatives in Asia and the Pacific: https://nedac.info/2022/02/23/rdcd-bangladesh/

Rehberger, E., West, P. C., Spillane, C., & McKeown, P. C. (2023). What climate and environmental benefits of regenerative agriculture practices? an evidence review. *Environmental Research Communications*, 5(5), 052001. DOI: 10.1088/2515-7620/acd6dc

Reinecke, J., & Donaghey, J. (2017). Collective action and social partnership in global supply chains: Lessons from the Bangladesh accord. *Annual Review of Social Partnerships*, 12, 32–37.

Rejeb, A., Keogh, J. G., & Treiblmaier, H. (2020). Leveraging blockchain technology in supply chains: A case study of European agri-food supply chains. *Journal of Business Research*, 123, 493–502. DOI: 10.1016/j.jbusres.2020.10.002

Retail onion price up 57 pc; Centre steps up buffer onion sale to provide relief to consumers (2023).The Indian Express. Retrieved from https://www.newindianexpress.com/business/2023/Oct/27/retail-onion-price-up-57-pc-centre-steps-up-buffer-onion-sale-to-provide-relief-to-consumers-2627557.html#:~:text=NEW%20DELHI%3A%20As%20the%20all,to%20provide%20relief%20to%20consumers

Rhyne, E. (1998). The yin and yang of microfinance: Reaching the poor and sustainability. *Microbanking Bulletin, 2*, July.

Riccaboni, A., Neri, E., Trovarelli, F., & Pulselli, R. M. (2021). Sustainability-oriented research and innovation in 'farm to fork'value chains. *Current Opinion in Food Science*, 42, 102–112. DOI: 10.1016/j.cofs.2021.04.006

Richards, G. (2017). *Warehouse management: A complete guide to improving efficiency and minimizing costs in the modern warehouse*. Kogan Page Publishers.

Rijpkema, W. A., Rossi, R., & van der Vorst, J. G. A. J. (2014). Effective sourcing strategies for perishable product supply chains. *International Journal of Physical Distribution & Logistics Management*, 44, 494–510. DOI: 10.1108/IJPDLM-01-2013-0013

Rise in prices of Domestic Products (2021). Ministry of Consumer Affairs and Public Distribution. Retrieved from www.pib.press.release.

Ritthaisong, Y. M., Johri, L., & Speece, M. (2014). Sources of sustainable competitive advantage: The case of rice-milling firms in Thailand. *British Food Journal*, 116(2), 272–291. DOI: 10.1108/BFJ-01-2012-0003

Robinson, M. (2001). *The microfinance revolution: Sustainable banking for the poor*. The World Bank. DOI: 10.1596/0-8213-4524-9

Ronaghi, M. H. (2021). A blockchain maturity model in agricultural supply chain. *Information Processing in Agriculture*, 8(3), 398–408. DOI: 10.1016/j.inpa.2020.10.004

Routroy, S., & Behera, A. (2017). Agriculture supply chain: A systematic review of literature and implications for future research. *Journal of Agribusiness in Developing and Emerging Economies*, 7(3), 275–302. DOI: 10.1108/JADEE-06-2016-0039

S&P Global Ratings. (2022). *The Russia-Ukraine war is reshaping the fertilizer industry*. https://www.spglobal.com/ratings/en/research/articles/220912-the-russia-ukraine-war-is-reshaping-the-fertilizer-industry-12484018

Sagar, N. A., & Pareek, S. (2020). Safe Storage and Preservation Techniques in Commercialized Agriculture. *Natural Remedies for Pest, Disease and Weed Control*, 221-234. https://doi.org/DOI: 10.1016/B978-0-12-819304-4.00019-1

Saha, P., Akash, F. A., Shovon, S. M., Monir, M. U., Ahmed, M. T., Khan, M. F. H., Sarkar, S. M., Islam, M. K., Hasan, M. M., Vo, D.-V. N., Aziz, A. A., Hossain, M. J., & Akter, R. (2024). Grey, blue, and green hydrogen: A comprehensive review of production methods and prospects for zero-emission energy. *International Journal of Green Energy*, 21(6), 1383–1397. DOI: 10.1080/15435075.2023.2244583

Sajjad, A., Eweje, G., Tappin, D., 2019. Managerial perspectives on drivers for and barriers to sustainable supply chain management implementation: Evidence from New Zealand. Bus. Strateg. Environ. 1–13. https://doi.org/DOI: 10.1002/bse.2389

Sakina, M. (2019). The food processing industry in India: challenges and prospects. *Gap Interdisciplinaries, 2*(3).

Salah, K., Nizamuddin, N., Jayaraman, R., & Omar, M. (2019). Blockchain-based soybean traceability in agricultural supply chain. *IEEE Access : Practical Innovations, Open Solutions*, 7, 73295–73305. DOI: 10.1109/ACCESS.2019.2918000

Salam, M., Viantika, N. M., Amiruddin, A., Pinontoan, F. M., & Rahmatullah, R. A. (2021, March). Value chain analysis of Toraja coffee. [). IOP Publishing.]. *IOP Conference Series. Earth and Environmental Science*, 681(1), 012115. DOI: 10.1088/1755-1315/681/1/012115

Samanta, K. K., Pandit, P., Samanta, P., & Basak, S. (2019). Water consumption in textile processing and sustainable approaches for its conservation. In *Water in textiles and fashion* (pp. 41–59). Woodhead Publishing. DOI: 10.1016/B978-0-08-102633-5.00003-8

Sánchez, J. M., Rodríguez, J. P., & Espitia, H. E. (2020). Review of artificial intelligence applied in decision-making processes in agricultural public policy. *Processes (Basel, Switzerland)*, 8(11), 1374. DOI: 10.3390/pr8111374

Sanjana Reddy, P., Reddy, B. V. S., & Srinivasa Rao, P. (2011). Genetic analysis of traits contributing to stalk sugar yield in sorghum. *Cereal Research Communications*, 39(3), 453–464. DOI: 10.1556/CRC.39.2011.3.15

Santangeli, A., Chen, Y., Kluen, E., Chirumamilla, R., Tiainen, J., & Loehr, J. (2020). Integrating drone-borne thermal imaging with artificial intelligence to locate bird nests on agricultural land. *Scientific Reports*, 10(1), 10993. DOI: 10.1038/s41598-020-67898-3 PMID: 32665596

Sasakawa Africa Association. (2024), retrieved from https://saa-safe.org/newsletter/saa-new-technology-helps-farmers-improve-onion-shelf-life/

Satheeshkumar, D. R., Krishnan, D. S. G., & Rawath, S. S., & S J, A. K. (2022). A Study on Distribution Channel Strategy: Retailers' Perspective. *Rangasamy, S., Krishnan, SG, Rawath, S., Rakshith, M., & Ajith Kumar, S. (2022, September 27). A STUDY ON DISTRIBUTION CHANNEL STRATEGY: RETAILERS'PERSPECTIVE.International Journal for Innovative Engineering and Management Research*, 11(9), 240–252.

Saunders, M., Lewis, P., & Thornhill, A. (2009). *Research methods for business students*. Pearson education.

Saurabh, S., & Dey, K. (2021). Blockchain technology adoption, architecture, and sustainable agri-food supply chains. Journal of Cleaner Production, p. 284, 124731.

Saurabh, S., & Dey, K. (2021). Blockchain technology adoption, architecture, and sustainable agri-food supply chains. *Journal of Cleaner Production*, 284, 124731. DOI: 10.1016/j.jclepro.2020.124731

Saurabh, S., & Dey, K. (2021). Blockchain technology adoption, architecture, and sustainable agri-food supply chains: A case study of smallholder farmers in India. *International Journal of Information Management*, 52, 101981. DOI: 10.1016/j.ijinfomgt.2020.101981

Saxena, R., Pant, D. K., Pant, S. C., Joshi, L., Paul, R. K., & Singh, R. (2023a). Artificial intelligence-led innovations for agricultural transformation: A scoping study. *Agricultural Economics Research Review, 36* (Conference Number), 1-18.

Saxena, A., Chouhan, S. S., Aziz, R. M., & Agarwal, V. (2024). A comprehensive evaluation of Marine predator chaotic algorithm for feature selection of COVID-19. *Evolving Systems*, 15(4), 1235–1248. Advance online publication. DOI: 10.1007/s12530-023-09557-2

Saxena, R., & Kumar, R. (2020). *Commodity Profile on Onion*. ICAR-National Institute of Agricultural Economics and Policy Research.

Saxena, R., Pant, D. K., Pant, S. C., & Singh, R. (2023b). Mapping the Global Research Landscape: Bibliometric Analysis of Agri-food Systems and Nutritional Security. *Indian Journal of Agricultural Economics*, 78(3), 328–342. https://isaeindia.org/wp-content/uploads/2024/02/03-RAKA-SAXENA.pdf. DOI: 10.63040/25827510.2023.03.003

Schauberger, B., Kato, H., Kato, T., Watanabe, D., & Ciais, P. (2022). French crop yield, area and production data for ten staple crops from 1900 to 2018 at county resolution. *Scientific Data*, 9(1), 38.

Schimke, D., & Buechel, D. (2021). Socio-economic factors influencing blockchain adoption in developing countries: A comprehensive review. *Journal of Technology Management & Innovation*, 16(2), 45–57. DOI: 10.4067/S0718-27242021000200045

Schultz, H. W. (1981). United States Warehouse Act. In *Food Law Handbook*. Springer., DOI: 10.1007/978-94-011-7373-5_17

Seeds F-Mart. (2023). Seeds NGO F-Mart [Photograph]. Instagram. Retrieved September 18, 2024, from https://www.instagram.com/seeds.ngo?igsh=MTV2OXUxYmFmc2d0bA

Serna-Saldívar, S. O., Chuck-Hernández, C., Pérez-Carrillo, E., & Heredia-Olea, E. (2012). Sorghum as a multifunctional crop for the production of fuel ethanol: Current status and future trends. *Bioethanol*, 2012, 51–74.

Serpantié, G. (2009). L'agriculture de conservation à la croisée des chemins en Afrique et à Madagascar. *Vertigo: La revue en sciences de l'environnement, 9*(3), 21.

Sgarbossa, F., & Russo, I. (2017). A proactive model in sustainable food supply chain: Insight from a case study. *International Journal of Production Economics*, 183, 596–606. DOI: 10.1016/j.ijpe.2016.07.022

Sgroi, F., & Marino, G. (2022). Environmental and digital innovation in food: The role of digital food hubs in the creation of sustainable local agri-food systems. *The Science of the Total Environment*, 810, 152257. DOI: 10.1016/j.scitotenv.2021.152257 PMID: 34902402

Shabani, A., Saen, R. F., & Torabipour, S. M. R. (2012). A new benchmarking approach in Cold Chain. *Applied Mathematical Modelling*, 36(1), 212–224. DOI: 10.1016/j.apm.2011.05.051

Shadrin, D., Menshchikov, A., Ermilov, D., & Somov, A. (2019). Designing future precision agriculture: Detection of seeds germination using artificial intelligence on a low-power embedded system. *IEEE Sensors Journal*, 19(23), 11573–11582. DOI: 10.1109/JSEN.2019.2935812

Shafi, U., Mumtaz, R., García-Nieto, J., Hassan, S. A., Zaidi, S. A. R., & Iqbal, N. (2019). Precision agriculture techniques and practices: From considerations to applications. *Sensors (Basel)*, 19(17), 3796. DOI: 10.3390/s19173796 PMID: 31480709

Shah, G., Shah, A., & Shah, M. (2019). Panacea of challenges in real-world application of big data analytics in healthcare sector. *Journal of Data. Information & Management*, 1(3-4), 107–116. DOI: 10.1007/s42488-019-00010-1

Shalender, K., & Yadav, R. K. (2019). Strategic flexibility, manager personality, and firm performance: The case of Indian Automobile Industry. *Global Journal of Flexible Systems Managment*, 20(1), 77–90. DOI: 10.1007/s40171-018-0204-x

Shamsuddoha, A. K., Edwards, G., & School of Business, La Trobe University. (2000). Dairy Industry in Bangladesh: Problems and Prospects. In *AARES 2000 Conference* [Conference-proceeding].

Shankar Tumuluru, J. (2023). *Introductory Chapter: Food Processing, Preservation, and Packaging – A Brief Overview*. IntechOpen., DOI: 10.5772/intechopen.110229

Sharma, A., & Bansal, A. (2023). Digital Marketing in the Metaverse: Beginning of a New Era in Product Promotion. In *Applications of Neuromarketing in the Metaverse* (pp. 163-175). IGI Global.

Sharma, D., & Chauhan, A. (2022). Kharif Onion Production in India-Present Status and Future Potential: A Review. *Agricultural Reviews*. Retrieved from https://arccjournals.com/journal/agricultural-reviews/R-2455

Sharma, D., Banyal, S. K. and Chauhan, A. (2020). Kharif Onion Cultivation – a boon for the farmers of Northern India. *Indian Horticulture*, ICAR. PP 20-23.

Sharma, N., & Dutta, N. (2023). Omnichannel retailing: exploring future research avenues in retail marketing and distribution management. *International Journal of Retail & Distribution*

Sharma, A. K., Chanderwal, N., & Khan, R. (Eds.). (2023). *Convergence of cloud computing, AI, and agricultural science*. IGI Global. DOI: 10.4018/979-8-3693-0200-2

Sharma, H. K., & Kumar, N. (2022). Agro processing: scope and importance. In *Agro-Processing and Food Engineering: Operational and Application Aspects* (pp. 1–22). Springer Singapore. DOI: 10.1007/978-981-16-7289-7_1

Sharma, Y. K., Mangla, S. K., Patil, P. P., & Liu, S. (2019). When challenges impede the process: For circular economy-driven sustainability practices in food supply chain. *Management Decision*, 57(4), 995–1017. DOI: 10.1108/MD-09-2018-1056

Shepherd, A. (2007). *Approaches to linking producers to markets*. https://vtechworks.lib.vt.edu/handle/10919/68275

Shepherd, A. W., & Farina, E. M. (2013). Value chain finance: Taking stock of recent learning and looking ahead. *World Development*, 44, 181–193.

Shiferaw, B., & Teklewold, H. (2007). Structure and functioning of chickpea markets in Ethiopia: Evidence based on analyses of value chains linking smallholders and markets. *IPMS Working Paper*.

Shukla, M., & Jharkharia, S. (2013). Agri-fresh produce supply chain management: A state-of- the-art literature review. *International Journal of Operations & Production Management*, 33(2), 114–158. DOI: 10.1108/01443571311295608

Siddh, M. M., Soni, G., Jain, R., & Sharma, M. K. (2018). Structural model of perishable food supply chain quality (PFSCQ) to improve sustainable organizational performance. *Benchmarking*, 25(7), 2272–2317.

Singh, A. S., Bharti, N., & Kumar, P. (2023). Is Micro-warehousing a Solution to Agricultural Commodity Storage? *Asian Journal of Management Cases*. https://doi.org/DOI: 10.1177/09728201231195441

Singh, R., & Shabani, A. (2016). the Identification of Key Success Factors in Sustainable Cold Chain Management : Insights From the. *J. Oper. Supply Chain Manag.*, 9(2), 1–16. DOI: 10.12660/joscmv9n2p1-16

Singh, S., Singh, P., Kumar, A., Baheliya, A. K., & Patel, K. K. (2024). Promoting environmental Sustainability through Vertical Farming: A review. *Journal of Advances in Biology & Biotechnology*, 27(5), 210–219. DOI: 10.9734/jabb/2024/v27i5780

Sinha, D. (Ed.). (2019). Global Supply Chains and Multimodal Logistics: Emerging Research and Opportunities: Emerging Research and Opportunities.

Sinha, A., Jha, S., & Amritesh, A. (2019). Green Agrevolution Pvt Ltd: Delivering 360°"seed-to-market" solution. *Emerald Emerging Markets Case Studies*, 9(1), 1–23. DOI: 10.1108/EEMCS-06-2018-0148

Size, F. A. M. (2020). Share & Industry Analysis. By Type (Class F and Class C), by Application (Cement & concretes, Fills & embankments, Waste Stabilization, Mining, Oil field Service and Road Stabilization and Others), and Regional Forecast, 2027.

Sjah, T., & Zainuri, Z. (2020). Agricultural supply chain and food security. In *Zero Hunger* (pp. 79–88). Springer International Publishing. DOI: 10.1007/978-3-319-95675-6_82

Skvortsov, E. A. (2020). Prospects of applying artificial intelligence technologies in the regional agriculture. *Ekonomika Regiona= Economy of Regions*, (2), 563.

Smidt, H. J., & Jokonya, O. (2022). Factors affecting digital technology adoption by small-scale farmers in agriculture value chains (AVCs) in South Africa. *Information Technology for Development*, 28(3), 558–584. DOI: 10.1080/02681102.2021.1975256

Smith, G. (2009). *Interaction of Public and Private Standards in the Food Chain. OECD Food, Agriculture and Fisheries Papers, No. 15.* OECD Publishing., DOI: 10.1787/18156797

Smith, G. A., Bagby, M. O., Lewellan, R. T., Doney, D. L., Moore, P. H., Hills, F. J., Campbell, H. J., Hogaboam, G. J., Coe, G. E., & Freeman, K. (1987). Evaluation of sweet sorghum for fermentable sugar production potential. *Crop Science*, 27(4), 788–793. DOI: 10.2135/cropsci1987.0011183X002700040037x

Smith, G. A., & Buxton, D. R. (1993). Temperate zone sweet sorghum ethanol production potential. *Bioresource Technology*, 43(1), 71–75. DOI: 10.1016/0960-8524(93)90086-Q

Smith, M. J. (2018). Getting value from artificial intelligence in agriculture. *Animal Production Science*, 60(1), 46–54. DOI: 10.1071/AN18522

Socio-Economic Statistical Information about India. (15th May 2024). Retrieved from www.indiastat.com

Solanki, S., Singh, U. P., Chouhan, S. S., & Jain, S. (2024). A systematic analysis of magnetic resonance images and deep learning methods used for diagnosis of brain tumor. *Multimedia Tools and Applications*, 83(8), 23929–23966. DOI: 10.1007/s11042-023-16430-6

Solér, C., Bergstrom, K., & Shanahan, H. (2010). Green Supply Chains and the Missing Link Between Environmental Information and Practice. *Business Strategy and the Environment*, 25(1), 14–25. DOI: 10.1002/bse.655

Solodovnik, A. I., Savkin, V. I., & Amelina, A. V. (2021, September). The role of the internet of things as direction for the development of agriculture 4.0 for rural areas. [). IOP Publishing.]. *IOP Conference Series. Earth and Environmental Science*, 839(3), 032040. DOI: 10.1088/1755-1315/839/3/032040

Solomou, A. D., Germani, R., & Georgakopoulou, M. (2020). Ecological value, cultivation, utilization and commercialization of Aloe vera in Greece. *JAPS: Journal of Animal & Plant Sciences*, 30(4), 1013–1023.

Solutions, A. (2024). Application technology and digital agriculture, article on" The best agribusiness solutions for your property!" Accessed on 23 March Agrocision. com.br. https://www.agrocision.com.br/

Somashekhar, I. C., Raju, J. K., & Patil, H. (2014). Agriculture supply chain management: A scenario in India. *Research Journal of Social Science and Management*, 4(07), 89–99.

Songol, M., Awuor, F., & Maake, B. (2021). Adoption of artificial intelligence in agriculture in the developing nations: A review. *Journal of Language. Technology & Entrepreneurship in Africa*, 12(2), 208–229.

Sridhar, A., Balakrishnan, A., Jacob, M. M., Sillanpää, M., & Dayanandan, N. (2023). Global impact of COVID-19 on agriculture: Role of sustainable agriculture and digital farming. *Environmental Science and Pollution Research International*, 30(15), 42509–42525. DOI: 10.1007/s11356-022-19358-w PMID: 35258730

Srivastava, R., Meena, K., Tiwari, A., Singh, N., & Behra, T. K. (2022). Yield and Economics of Kharif Onion (Allium cepa L.) under Front Line Demonstration in Eastern Plain Zone of Uttar Pradesh, India. *International Journal of Plant and Soil Science*, 34(23), 1034–1040. DOI: 10.9734/ijpss/2022/v34i232513

Srivastava, S. K., Chaudhuri, A., & Srivastava, R. K. (2015). Propagation of risks and their impact on performance in fresh food retail. *International Journal of Logistics Management*, 26(3), 568–602. DOI: 10.1108/IJLM-02-2014-0032

Srivastava, S., & Mishra, H. N. (2021). Ecofriendly nonchemical/nonthermal methods for disinfestation and control of pest/fungal infestation during storage of major important cereal grains: A review. *Food Frontiers*, 2(1), 93–105. DOI: 10.1002/fft2.69

Subeesh, A., & Mehta, C. R. (2021). Automation and digitization of agriculture using artificial intelligence and internet of things. *Artificial Intelligence in Agriculture*, 5, 278–291. DOI: 10.1016/j.aiia.2021.11.004

Subervie, J., & Galtier, F. (2017). L'information sur les prix agricoles par la téléphonie mobile: Le cas du Ghana. Retrieved from https://hal.archives-ouvertes .fr/hal-01594670/

Swain, P. K., Jamir, C., Dinerstien, M., & Narula, S. (2022). Impact of development of National Agriculture Market (e-NAM) on farmers. *Indian Journal of Agricultural Sciences*, 92(12), 1484–1489. DOI: 10.56093/ijas.v92i12.127701

Syed, T., Goodman, P., Mataruka, Z., Evans, D. E., & Pachena, H. T. Roadmap for Sustainable Livestock Value Chains in Southern Africa (English). Washington, D.C.: World Bank Group. https://documents.worldbank.org/curated/en/099430006082232802/P174621055c7850b80808d09c38c18d5ed9, as retrieved on 24th July 2024

Taghipour, A. (2018). A contemporary approach to plan independent logistics actors. In *Contemporary Approaches and Strategies for Applied Logistics* (pp. 337–364). IGI Global. DOI: 10.4018/978-1-5225-5273-4.ch014

Tallontire, A., Opondo, M., Nelson, V., & Martin, A. (2011). Beyond the vertical? Using value chains and governance as a framework to analyse private standards initiatives in agri-food chains. *Agriculture and Human Values*, 28(3), 427–441. DOI: 10.1007/s10460-009-9237-2

Tanveer, M. (2022). Supply chain and logistics operations management under the era of advanced technology. In *Integrating Blockchain technology into the circular economy* (pp. 126–136). IGI Global. DOI: 10.4018/978-1-7998-7642-7.ch008

Tashi, S., Yangchen, U., Dahal, Y., & Gurung, D. B. (Eds.). (2022). *Case Studies of Successful Farmers, Agri-enterprises and Farmers' Groups and Cooperatives in Bhutan*

Taye, B. M. (2020). Coffee: The Driver of Economic Development and Socio-Cultural Integration in Ethiopia.

Technology, F. M. (2021, November 24). What is Global Food Supply Chain - FMT Magazine. Food, Marketing Technology. https://fmtmagazine.in/what-is-global-food-supply-chain/

Thakur, A. K., Singh, R., & Ashwani, K. (2014). "The science behind the system of rice intensification (SRI)." Research Bulletin No. 69, Directorate of Water Management (ICAR), Bhubaneswar, Odisha.

Thakur, A. K., Mohanty, R. K., Paul, D. U., & Kumar, A. (2014). Impact of water management on yield and water productivity with system of rice intensification (SRI) and conventional transplanting system in rice. *Paddy and Water Environment*, 12(4), 413–424. DOI: 10.1007/s10333-013-0397-8

The Economic Times News. (2023). How the onion rolls: The politics and economy of India's kitchen staple. Retrieved from https://economictimes.indiatimes.com/news/economy/agriculture/how-the-onion-rolls-the-politics-and-economy-of-indias-kitchen-staple/articleshow/104498132.cms?from=mdr

The Hindu Business Line. (2024). Fertilizer subsidy crosses 70% of budget allocation. https://www.thehindubusinessline.com

Thiell, M., & Hernandez, S. (2010). Logistics services in the 21st century: supply chain integration and service architecture. In *Service science and logistics informatics: Innovative perspectives* (pp. 359–378). IGI Global. DOI: 10.4018/978-1-61520-603-2.ch019

Tian, F. (2016). An agri-food supply chain traceability system for China based on RFID & blockchain technology. In *Proceedings of the 13th International Conference on Service Systems and Service Management* (pp. 1-6). DOI: 10.1109/ICSSSM.2016.7538424

Tian, F. (2017). A supply chain traceability system for food safety based on HACCP, blockchain & Internet of Things. In *Proceedings of the International Conference on Service Systems and Service Management* (pp. 1-6). DOI: 10.1109/ICSSSM.2017.7996119

Tibus, C. A., & Brennan, L. L. (2010). RFID and labor management systems selection in the logistics industry. [JCIT]. *Journal of Cases on Information Technology*, 12(1), 31–49. DOI: 10.4018/jcit.2010010103

Till, E. R. (2022, July 02). The Role of Agriculture in Economic Development. *Agriculture for Economic Development in Africa*, 9-17. Retrieved from https://link.springer.com/chapter/10.1007/978-3-031-07901-6_2

Togo First. (2024). Panorama de l'agriculture au Togo: Aujourd'hui et demain.... Retrieved from https://www.togofirst.com/fr/panorama-agriculture/1301-4703-panorama-de-l-agriculture-au-togo-aujourd-hui-et-demain

Tolsma Grisnich-Onion Storage. (2024). retrieved from https:/www.tolsmagrisnich.com/pages/onion-storge

Trademap. (n.d.). Retrieved from https://r.search.yahoo.com/_ylt=Awrx_NfKKdtmpwQAv7O7HAx.;_ylu=Y29sbwNzZzMEcG9zAzEEdnRpZAM Ec2VjA3Ny/RV=2/RE=1726848715/RO=10/RU=https%3a%2f%2fwww.trademap.org%2f/RK=2/RS=mU.k8jfCNVje4vV4iX7w6PEH32M-

Tran, L. A. D., & Pham, V. K. (2021). Accessing the Logistics Service Competitiveness: A Study on Four Emerging Nations in Southeast Asia. [IRMJ]. *Information Resources Management Journal*, 34(2), 29–42. DOI: 10.4018/IRMJ.2021040102

Tribis, Y., El Bouchti, A., & Bouayad, H. (2018). Supply chain management based on blockchain: A systematic mapping study. MATEC Web of Conferences, 200, 00020. DOI: 10.1051/matecconf/201820000020

Tripoli, M., & Schmidhuber, J. (2018, August). Emerging opportunities for the application of blockchain in the agri-food industry. Food and Agriculture Organization of the United Nations (FAO). Retrieved from https://www.fao.org/3/CA2920EN/ca2920en.pdf

Troise, C., Tani, M., Dinsmore, J.Jr, & Schiuma, G. (2021). Understanding the implications of equity crowdfunding on sustainability-oriented innovation and changes in agri-food systems: Insights into an open innovation approach. *Technological Forecasting and Social Change*, 171, 120959. www.iifiir.org. DOI: 10.1016/j.techfore.2021.120959

Tse, D., Zhang, B., Yang, Y., Cheng, C., & Mu, H. (2017). Blockchain application in food supply information security. In *Proceedings of the IEEE International Conference on Industrial Engineering and Engineering Management (IEEM)* (pp. 1357-1361). DOI: 10.1109/IEEM.2017.8290114

Tsolakis, N., Aivazidou, E., & Srai, J. S. (2019). Sensor applications in agrifood systems: Current trends and opportunities for water stewardship. *Climate (Basel)*, 7(3), 44. Advance online publication. DOI: 10.3390/cli7030044

U. S. Department of Agriculture. (n.d.). Retrieved from Foreign Agriculture Service: https://fas.usda.gov/regions/bangladesh

UNCTAD. (n.d.), United Nations Conference on Trade and Development. Published Reports, https://unctad.org/, as retrieved on 10th May 2024.

UNFCCC. (2023). Retrieved from: https://unfccc.int/sites/default/files/resource/Session%201b_Introduction%20to%20carbon%20pricing_MENA.pdf

Ungerman, O., & Dědková, J. (2020). Model of the circular economy and its application in business practice. *Environment, Development and Sustainability*, 22(4), 3407–3432. DOI: 10.1007/s10668-019-00351-2

United Nations. (2022). [Press release]. https://press.un.org/en/2022/sgsm21350.doc.htm

United States Department of Agriculture. (2021). *Agricultural Cooperatives Statistics Summary, 2020.*https://content.govdelivery.com/accounts/USDARD/bulletins/300bab5

USAID-KAVES. (2015). French Bean value chain Analysis: An online publication prepared by Fintrac Inc. for review by the United States Agency for International Development, https://erepository.uonbi.ac.ke/bitstream/handle/11295/164474/Kabugua%20L_Assessment%20of%20Household%20Expenditure%20Patterns%20and%20Their%20Effects%20on%20Food%20Security-%20the%20Case%20of%20French%20Beans%20Producing%20Households%20in%20Mbooni%20Sub-county,%20Kenya.pdf?sequence=1, as retrieved on 10th May 2024.

USDA. (n.d.). USWA Licensed Warehouse. Agricultural Marketing Service 2022. Available online: https://publicdashboards.dl.usda.gov/t/ MRP_PUB/views/WC-MDDashboard/WCMDDashboard?:isGuestRedirectFromVizportal=y&:embed=y (accessed on 15 August 2024).

USDA. Foreign Agricultural Service, U.S. Department of Agriculture, https://fas.usda.gov/data/production/commodity/0459200

Vachon, S., & Klassen, R. D. (2006). Green project partnership in the supply chain: The case of the package printing industry. *Journal of Cleaner Production*, 14(6/7), 661–671. DOI: 10.1016/j.jclepro.2005.07.014

Van Der Vorst, J. G. A. J., & Beulens, A. J. M. (2002). Identifying sources of uncertainty to generate supply chain redesign strategies. *International Journal of Physical Distribution & Logistics Management*, 32(6), 409–430. DOI: 10.1108/09600030210437951

Villalba, R., Venus, T. E., & Sauer, J. (2023). The ecosystem approach to agricultural value chain finance: A framework for rural credit. *World Development*, 164, 106177. DOI: 10.1016/j.worlddev.2022.106177

Vinutha, K. S., Rayaprolu, L., Yadagiri, K., Umakanth, A. V., & Srinivasa Rao, P. (2014). Sweet sorghum research and development in India: Status and Prospects. *Sugar Tech*, 16(2), 133–143. DOI: 10.1007/s12355-014-0302-9

Vivek, T. R., & Mohammed, F. (2024). Why India Faces a Constant Onion Crisis. The Plate. Retrieved from https://theplate.in/why-india-faces-a-constant-onion-crisis/

Wachtendorf, T., & Kendra, J. M. "Considering Convergence, Coordination and Social Capital in Disasters," 2004 Presentation to the *Canadian Risk and Hazards Networks 1st Annual Symposium*. Winnipeg, Manitoba, November 19.

WAI. (2022). e-Handbook on Warehousing Standards. Retrieved from https://warehousingindia.org/wp-content/uploads/2022/09/Handbook-on-Warehousing-Standards.pdf

Wampfler, B., & Lapenu, C. (2002). *Le financement de l'agriculture familiale dans un contexte de libéralisation. Quelle contribution de la microfinance? Actes du Séminaire de Dakar, 21-24 janvier*. CIRAD-CERISE.

Wang, G., Zhang, Z., Li, S., & Shin, C. (2023). Research on the influencing factors of sustainable supply chain development of agri-food products based on cross-border live-streaming e-commerce in China. *Foods*, 12(17), 3323.

Wang, S., Zhang, Y., & Zhang, Y. (2018). A blockchain-based framework for data sharing with fine-grained access control in decentralized storage systems. *IEEE Access : Practical Innovations, Open Solutions*, 6, 41640–41649. DOI: 10.1109/ACCESS.2018.2851611

Wang, Y., Naim, M., & Potter, A. (2010). Electronic logistics marketplaces. In *Encyclopedia of e-business development and management in the global economy* (pp. 218–226). IGI Global. DOI: 10.4018/978-1-61520-611-7.ch022

Wautelet, Y. (2012). Representing, modeling and engineering a collaborative supply chain management platform. [IJISSCM]. *International Journal of Information Systems and Supply Chain Management*, 5(3), 1–23. DOI: 10.4018/jisscm.2012070101

WDRA. (2015). Report on warehousing in India Study commissioned by the Warehousing Development and Regulatory Authority. Retrieved from https://wdra.gov.in/documents/32110/38476/Report12.pdf/fcfa2f56-a673-a5f6-a870-f73292eab04a

Weber, D. M., & Kauffman, R. J. (2011). What drives global ICT adoption? Analysis and research directions. *Electronic Commerce Research and Applications*, 10(6), 683–701. DOI: 10.1016/j.elerap.2011.01.001

Westerkamp, M., Victor, F., & Küpper, A. (2019). Tracing manufacturing processes using blockchain-based token compositions. *Digital Communications and Networks*, 5(4), 193–202. DOI: 10.1016/j.dcan.2019.01.007

Whitfield, M. B., Chinn, M. S., & Veal, M. W. (2012). Processing of materials derived from sweet sorghum for biobased products. *Industrial Crops and Products*, 37(1), 362–375. DOI: 10.1016/j.indcrop.2011.12.011

Wilson, H., & Daniel, E. (2007). The multi-channel challenge: A dynamic capability approach. *Industrial Marketing Management*, 36(1), 10–20. DOI: 10.1016/j.indmarman.2006.06.015

Wolfert, S., Ge, L., Verdouw, C., & Bogaardt, M. J. (2017). Big Data in Smart Farming – A review. *Agricultural Systems, 153*, 69–80. rrrrrhttps://DOI: 10.1016/j.agsy.2017.01.023

Wood, L. C., & Duong, L. N. (Eds.). (2021). *Logistics and Supply Chain Management in the Globalized Business Era*. IGI Global. DOI: 10.4018/978-1-7998-8709-6

Woolcock, M. (1998). Social Capital and Economic Development: Toward a Theoretical Synthesis and Policy Framework. *Theory and Society*, 27(2), 151–208. DOI: 10.1023/A:1006884930135

World Bank. (2023). Climate-Smart Agriculture. Retrieved from https://www.worldbank.org/en/topic/climate-smart-agriculture

World Bank. (2024). *Togo - Vue d'ensemble*. Retrieved from https://www.banquemondiale.org/fr/country/togo

World Bank. 2003. Social Capital for Development. Available at: http://www.worldbank.org/ poverty/scapital/. Accessed in February 2019.

World Integrated Trade Solution. (n.d.). Retrieved from HH Market Concentration Index: https://wits.worldbank.org/CountryProfile/en/country/by-country/startyear/LTST/endyear/LTST/indicator/HH-MKT-CNCNTRTN-NDX

Wu, X., Staggenborg, S., Propheter, J. L., Rooney, W. L., Yu, J., & Wang, D. (2010). Features of sweet sorghum juice and their performance in ethanol fermentation. *Industrial Crops and Products*, 31(1), 164–170. DOI: 10.1016/j.indcrop.2009.10.006

Xie, C., Sun, Y., & Luo, H. (2017). Secured data storage scheme based on blockchain for agricultural products tracking. In *Proceedings of the 3rd International Conference on Big Data Computing and Communications* (pp. 45-50). Chengdu, China: IEEE. DOI: 10.1109/BIGCOM.2017.28

Xue, L., Cao, Z., Scherhaufer, S., Östergren, K., Cheng, S., & Liu, G. (2021). Mapping the EU tomato supply chain from farm to fork for greenhouse gas emission mitigation strategies. *Journal of Industrial Ecology*, 25(2), 377–389. DOI: 10.1111/jiec.13080

Xu, J., Guo, S., Xie, D., & Yan, Y. (2019). Blockchain: A new safeguard for agri-foods. *Artificial Intelligence in Agriculture*, 4, 153–161. DOI: 10.1016/j.aiia.2020.08.002

Yadav, A. R., & Singh, R. K. (2019). Use of blockchain to solve select issues of Indian farmers. *AIP Conference Proceedings*, 2148, 040008. DOI: 10.1063/1.5123972

Yadav, V. S., Singh, A. R., Raut, R. D., & Govindarajan, U. H. (2020). Blockchain technology adoption barriers in the Indian agricultural supply chain: An integrated approach. *Resources, Conservation and Recycling*, 161, 104877. DOI: 10.1016/j.resconrec.2020.104877

Yang, J., Li, Z., & Wang, T. (2020). Price discovery in Chinese agricultural futures markets: A comprehensive look. *Journal of Futures Markets*, 41(4), 536–555. DOI: 10.1002/fut.22179

Yang, L., Sathishkumar, V. E., & Manickam, A. (2023). Information retrieval and optimization in distribution and logistics management using deep reinforcement learning. *International Journal of Information Systems and Supply Chain Management*, 16(1), 1–19. DOI: 10.4018/IJISSCM.316166

Yang, P. (2022). Information Management in the Logistics and Distribution Sector Using Metaheuristic Techniques. [IJISSCM]. *International Journal of Information Systems and Supply Chain Management*, 15(4), 1–21. DOI: 10.4018/IJISSCM.305850

Yara. (2023). One year on: The impact of the war in Ukraine. [Press release]. https://www.yara.com/news-and-media/news/archive/2023/one-year-on-the-impact-of-the-war-in-ukraine/

Yashada, (2019), International Symposium on Edible Alliums: Challenges and Opportunities, Feb 9-12. Retrieved from https://krishi.icar.gov.in>jspui>bitstream.

Yohe, H. S. (1924). The Banker and the United States Warehouse Act. Retrieved from https://archive.org/details/CAT31297878/page/4/mode/2up

Yousefi, H., Su, H.-M., Imani, S. M., Alkhaldi, K., Filipe, C. D. M., & Didar, T. F. (2019). Intelligent Food Packaging: A Review of Smart Sensing Technologies for Monitoring Food Quality. *ACS Sensors*, 4(4), 808–821. https://doi.org/https://doi.org/10.1021/acssensors.9b00440. DOI: 10.1021/acssensors.9b00440 PMID: 30864438

Yuen, J. S., Choy, K. L., Lam, H. Y., & Tsang, Y. P. (2018). An intelligent-internet of things (IoT) outbound logistics knowledge management system for handling temperature sensitive products. [IJKSS]. *International Journal of Knowledge and Systems Science*, 9(1), 23–40. DOI: 10.4018/IJKSS.2018010102

Zailani, S., & Keogh, J. G. (2020). Blockchain technology in the food industry: A review of potentials, challenges, and future research directions. *Logistics (Basel)*, 4(4), 27. DOI: 10.3390/logistics4040027

Zaridis, A., Vlachos, I., & Bourlakis, M. (2021). SMEs strategy and scale constraints impact on agri-food supply chain collaboration and firm performance. *Production Planning and Control*, 32(14), 1165–1178.

Zhai, T., Wang, D., Zhang, Q., Saeidi, P., & Raj Mishra, A. (2023). Assessment of the agriculture supply chain risks for investments of agricultural small and mediumsized enterprises (SMEs) using the decision support model. Economic research-. *Ekonomska Istrazivanja*, 36(2). Advance online publication. DOI: 10.1080/1331677X.2022.2126991

Zhang, J. (2020, October). Research on digital image processing and recognition technology of weeds in maize seedling stage based on artificial intelligence. In *Journal of Physics*. Conference. DOI: 10.1088/1742-6596/1648/4/042058

Zhang, X., & Fan, D. (2024). Can agricultural digital transformation help farmers increase income? An empirical study based on thousands of farmers in Hubei Province. *Environment, Development and Sustainability*, 26(6), 14405–14431. DOI: 10.1007/s10668-023-03200-5 PMID: 37362968

Zhang, Y. (2005). A review of food traceability in food supply chain. *Journal of Food Engineering*, 70(3), 381–392. DOI: 10.1016/j.jfoodeng.2004.08.007

Zhang, Z. (2022). Internet of Things-Enabled Logistic Warehouse Scheduling Management With Human Machine Assistance. [IJISSCM]. *International Journal of Information Systems and Supply Chain Management*, 15(4), 1–17. DOI: 10.4018/IJISSCM.305852

Zhao, G., Vazquez-Noguerol, M., Liu, S., & Prado-Prado, J. C. (2024). Agri-food supply chain resilience strategies for preparing, responding, recovering, and adapting in relation to unexpected crisis: A cross-country comparative analysis from the COVID-19 pandemic. *Journal of Business Logistics*, 45(1), e12361. DOI: 10.1111/jbl.12361

Zhong, B., Yang, F., & Chen, Y. L. (2015). Information empowers vegetable supply chain: A study of information needs and sharing strategies among farmers and vendors. *Computers and Electronics in Agriculture*, 117, 81–90. DOI: 10.1016/j.compag.2015.07.009

Zoundi, J., Hitimana, L., & Hussein, K. (2005). Économie familiale et innovation agricole en Afrique de l'ouest: Vers de nouveaux partenariats. *SAH / D, 550*, 106.

About the Contributors

Sneha Kumari is an Assistant professor at Symbiosis School of Economics. She is an Associate Researcher at Circular Economy and Territories for the European Chair of Excellence. She has received the Scientific High-Level Visiting Fellowship from The French Institute in India - Embassy of France in India. Prior to that, she worked as an Assistant professor at Vaikunth Mehta National Institute of Cooperative Management (VAMNICOM), Ministry of Cooperation, Government of India, Pune. She has been a Visiting Faculty at the National Institute of Food Technology and Entrepreneurship Management (NIFTEM), Ministry of Food Processing Industries, Symbiosis Institute of Operations Management, and Symbiosis School for Online and Digital Learning (SSODL). She completed her Ph.D. as J.R.F from Symbiosis International Deemed University. She is interested in the agriculture supply chain, agriculture value chain, and Sustainability. She is also associated with different institutes for various educational and research-related project assignments. She has published several research papers in the area of big data, agriculture, and sustainability in ABDC and Scopus journals and has attended several national and international conferences. She has published three international books with Emerald publishers and 2 national books.

Hemachandra Padhan completed his PhD and Post-Doc Equivalent Fellowship (PDEF) from Department of Humanities and Social Sciences, IIT Madras. MA in Applied Economics and PG Diploma in Statistics and Research Methodology from Pondicherry Central University, Pondicherry. He was awarded as Post-Graduate Merit Fellowship for University Rank Holder (2014-2016); National Fellowship (2018) and CBSE-UGC NET. Before joining IIM Sambalpur, he has been associated with Indian Institute of Management (IIM) Mumbai and Symbiosis School of Economics (SSE) Pune as an Assistant Professor in Economics. His research interest is in the areas of Energy Economics, Finance & Strategy, Environmental Sustainability, Macroeconomics and Applied Econometrics. His work has appeared in top-tier journals including Energy Economics, Energy Policy, Resources Policy, Renewable Energy, Frontier in Energy Research, Environmental Science and Pollution Research, Energy and Environment, Environment, Development & Sustainability, International Review of Economics and Journal of Public Affairs. He received many rewards from national and international level. He is also working with projects from the ICSSR.

K. K. Tripathy, an officer of Indian Economic Service (IES) of 1999 batch, is presently Economic Adviser in the Ministry of Health and Family Welfare, Government of India. During the past 24 years of public service, he has served in the Ministries of Cooperation, Education, Agriculture and Farmers Welfare, Rural Development, Food Processing Industries, Planning Commission besides the United Nations Development Programme. Prior to joining IES, he served the Government of Odisha. He has obtained his PhD on micro-finance management from the Department of Management Studies of Indian Institute of Technology Delhi. He has published many peer-reviewed research articles, books and policy notes in the field of Agriculture, Rural Development and Cooperation.

V.G. Venkatesh is a corporate practitioner/ an academic specializing in Supply Chain Management and Logistics domain with versatile experience and background. He is a Certified Supply Chain Professional (CSCP) from APICS-USA and having International industry experience from Honduras, Sri Lanka, USA, Hong Kong and Bangladesh in international supply chain practices. He has over 17 years of experience from Industry and academics. His background is supported by Masters in Manufacturing Management from BITS-Pilani and MFTech from National Institute of Fashion Technology (NIFT), India and a qualified trainer in the commonwealth countries with PGCert Higher Education from NTU (UK) and He is also the visiting faculty for Auckland University of Technology – MBA and Professional master's programs for operations pathway courses.Professional Associations: Institute of Supply Management (ISM-USA) Chartered Institute of Logistics and Transport (CILT-UK) – Life Time member American Production and Inventory control society (APICS- USA) International Supply Chain Education Alliance (ISCEA)

***

Binod Anand is a seasoned Cooperator and serves as the Secretary General of the Confederation of NGOs of Rural India (CNRI), and Founder of World Cooperation Economic Forum and have addressed General assembly of the United nations during High level political forum and Summit to the Future on SDG issues. He is particularly working to achieve the SDG target and incubating the civil society to prepare themselves when the world is transforming itself to ESG from SDG. With these efforts Government of India have appointed in the Prime Minister Minimum Support prices and Agri Reform Committee to take his broader vision of making Agricultural value chain more farmer centric and transforming Indian economic reforms accordingly. He supports the idea that farmers' cooperatives can play a central role in rural development, providing a platform for collective bargaining, access to better resources, and market connections. Anand's approach aligns with principles of equitable growth, reducing economic inequality through collective ownership and shared prosperity. Binod Anand has engaged with the United Nations (UN) on matters of sustainability, particularly in agriculture. He has worked on global policies with a focus on sustainable agricultural practices, aiming to achieve goals like net-zero emissions by 2047 or 2070. Anand emphasizes the integration of local agricultural practices with global trends, recognizing the importance of community involvement in addressing environmental challenges. His advocacy includes promoting sustainable innovations such as biofuels and gender-specific agricultural tools. Additionally, Anand has participated in multilateral forums where he has pushed for cooperative economic frameworks to reduce inequality. His involvement also includes initiatives like creating sustainable frameworks for farming communities, which align with broader UN Sustainable Development Goals (SDGs), such as Climate Action. Anand 's effort has also made him appointed as Expert Committee to review Kunming – Montreal Global Biodiversity framework, Multilateral protocol to create Community led, Other effective Conservation measures in India under Ministry of Environment, forest and Climate Change. These efforts are part of his vision to link local and global agricultural transformations. Binod works to integrate sustainability, advocating for reforms that help farmers adapt to global market dynamics while supporting local agricultural practices. By promoting cooperatives, Anand seeks to create a "solidarity economy," where community-driven economic activities contribute to both local and global development.

Ajit Bansal is a Professor and Assistant Dean at Chitkara Business School, Chitkara University, Punjab, India, responsible for mentoring and administering courses since 2018. He was awarded a doctorate by HPU Shimla in 2011, and he has master's degrees in management and commerce. He has over 25 years of experience working with various reputed industries and educational institutions like Graphic Era University, Dehradun, Shoolini University Solan, MAU, Solan, etc., imparting management education in India. His research interests are in the fields of Corporate Finance, Taxation, Cultural Tourism, and Microfinance. He has published various research papers in reputed journals and participated in conferences/seminars on critical issues relating to management.

Dharmik Borisagar is a Research Scholar at department of Agricultural Economics, Institute of Agricultural Sciences, Banaras Hindu University.

Mukul Kumar Choudhary completed a 5-year B.A. LL.B. degree from Chanakya National Law University, Patna (2016-2021). This was followed by L.L.M. (2021-2022), which he pursued and successfully completed from the National Law School of India University (NLSIU), Bangalore. Furthermore, he cleared the UGC NET Exam in February 2022 for which examination was held in November 2021. Currently he is an Assistant Professor of Law in Galgotias Law School, Gautam Budh Nagar. Apart from that he is also pursuing PhD Degree from Faculty of Law, University of Delhi.

Roland Condor is an Associate Professor of Entrepreneurship. He joined EM Normandie in 2001. He has a PhD in management science from the Université de Caen-Normandie, awarded in 2002. His thesis is on project design in small enterprises. His research interests are collective entrepreneurship, namely in agricultural fields. He is the Chairholder of the "Entrepreneurial Models in Agriculture" Chair.

Ashok Dalwai, as a member of the Indian Administrative Service of the 1984 batch (Odisha Cadre) has served in different capacities at both field and policy formulation stages in the States of Odisha and Karnataka, as also the Government of India (GoI). At the senior level, he has served as Secretary and Principal Secretary in the Departments of Industries, Steel, Mines, and Textiles, besides holding various tenures as Chairman and Managing Director of Public Sector Undertakings, Research, and Academic Institutions. He served Government of India from 2010 to 2023 in different capacities. Since 2015, he was with the Union Ministry of Agriculture and Farmers Welfare initially as Addnl. Secretary and subsequently as CEO of National Rainfed Area Authority for seven long years, on elevation as Secretary. He was the Chairman of the Committee on Doubling Farmers Income and submitted fourteen volume reports to the GoI in September 2018. His report recommended secondary agriculture as one of the ways to enhance farm incomes. From 2018, he was tasked to supervise and guide the implementation of DFI strategy as the Chairman of the Empowered Body. He continues to be an active promoter of agriculture and farmers' welfare across the country in his capacity as the Chairman of Commodities Derivative Advisory Committee (CDAC) by the Securities Exchange Board of India, Co-Chairman, Indian Chamber of Food and Agriculture, member of Jammu and Kashmir Holistic Agriculture Development Program (HADP), amongst others. He holds a Masters degree in Agricultural Economics, and a Ph.D. in Economics. To his credit are several articles, research papers, and book chapters.

His passion for academia is no less than administration. He has won several awards including conferment of Honorary Doctorates of Science, Doctorate of Literature and Professorship by several Universities, besides Honorary Fellowship by Agricultural Economics Research Association, India

Madhavi Damle has completed her PhD in Innovation and Knowledge Management studies from NMIMS, Mumbai, India with Masters in business management and B.Sc.Tech. Her experience for over 16 years, and in education for 14 years in academics in the role of teaching post graduates Her interest areas primarily align with technological areas as Digital convergence & e-Business and Broad Areas are Information Technology, Strategic Management & Financial management.

Pravin Dange is an Associate Professor in Marketing. He did his B.Tech. in Agriculture Engineering. His area of interest are Higher Education, and Corporate Social Responsibility.

Soumyadeep Das is currently working as Senior Research Fellow under Australian Centre for International Agricultural Research (ACIAR)-funded multi-country project "Rupantar" and pursuing PhD in the Department of Agricultural Economics at Uttar Banga Krishi Viswavidyalaya, West Bengal, India. He is a recipient of the prestigious Australia Awards Fellowship funded by Australian Department of Foreign Affairs and Trade, and has research experience of working with IFPRI as Principal Invstigator and Collaborator. Holding an MS degree in Agricultural Economics from Dr. Rajendra Prasad Central Agricultural University, he has vast work experience in the fields of farm diversification, financial institutions, and food systems along with several research publications and awards to his credit.

Anna Freund, PhD candidate is an assistant lecturer at the Department of Supply Chain Management of the Corvinus University of Budapest since September 2020. Her research activity covers the areas of the digitalisation of supply chain management at the field of agri-food industry and the sustainability of public catering.

Rahul B. Hiremath holds a Master's degree from BITS Pilani and a PhD from the Indian Institute of Science, Bangalore, followed by a post-doctoral research fellowship at the London School of Economics and Political Science (LSE). He has actively collaborated with several renowned international organisations, including DFID UK, SDC, Global Environmental Fund (GEF), and World Bank, on numerous research projects. He has several publications in peer-reviewed journals.

Gurudevi Hiremath is Assistant Professor at Punyashlok Ahilyadevi Holkar Solapur University, Solapur.

Ishwari Jadhav is an Executive Trainee at Rallis India Limited, a leading agrochemical company dedicated to serving the farming community. She holds a postgraduate degree in Agri-Business Management, which has enabled her to work closely with farmers, addressing crop, disease, pest, and nutrient-related challenges at the grassroots level. With a year of hands-on training in crops like maize, cotton, soybean, onion, and pomegranate, she has honed her communication and marketing skills. Her research interests focus on sustainability and cooperatives, reflecting her commitment to creating a positive impact in agriculture and empowering farming communities.

Zsófia Jámbor did PhD program from Corvinus Doctoral School. She has taught major subjects on Business Economics (BSC), Operations Management (BSC), Logistics Models (MSC).

Labanya Prakash Jena is a doctoral scholar at Xavier School of Management, Jamshedpur, and Senior Manager in Climate Policy Initiative

Vishita Khanna is working as Assistant Professor in the Dept. of HRD and Personnel Management at International Agribusiness Management Institute, Anand Agricultural University, Anand, Gujarat since 2017 where she is involved in teaching various subjects like Quality Management, Food Processing Technology, Human Resource Management, Business Laws, Quality Management, Strategic Management and Organizational Behavior. She has an overall experience of 13 years 6 years in Food Processing Industry and 7 years in Academics. She has served as Lead Internal Auditor and Food Safety Team Leader for 5 years in Food Industry. Her work included traceability exercise, documentation for the audits, performing internal audits and its compliance for the customers in India and Abroad. She has published 13 Research Papers and 7 book chapters in reputed books and Journals. On the education front, She is MBA- International Agribusiness Management along with B.Tech in Food Processing Technology and is currently pursuing her Ph.D. (ABM) as an In-service candidate from Anand Agricultural University

Raj Krishna completed a 5-year B.A. LL.B. degree from Chanakya National Law University, Patna (2015-2020). This was followed by L.L.M. (2020-2021), with specialization in Constitutional and Administrative Law, which he pursued and successfully completed from the National Law Institute University (NLIU), Bhopal. Furthermore, he has cleared the UGC NET Exam in February 2022 for which examination was held in November 2021. During academics his interests lay in research on current and general issues concerning Law and jurisprudence. This got reflected in the numerous research papers which got published in various law journals viz GLC Mumbai ("Is it time for India to legalize Voluntary Active Euthanasia?"), AMU Law Society Review ("Civilization has been brutal: Decriminalization of Section 377 and Way Forward") etc.

Karnatakapu Manoj Kumar holds a B.Sc (Hons) in Horticulture from Dr. YSR Horticultural University, Andhra Pradesh. He is currently pursuing a PGDM in Agri-Business Management at Vaikunth Mehta National Institute of Cooperative Management, Pune, Maharashtra. He completed Summer Internship Programme on Value Chain Mapping of Bio-Ethanol Industry in DigiGrain Solutions Private Limited.

Sonam Kumari is an enthusiastic Data Analyst with expertise in statistical analysis, data mining, and programming in Python and R. She is advanced in managing database systems and implementing machine learning algorithms, She has a keen analytical mindset. Her goal is to leverage these skills to extract valuable insights, drive strategic decision-making, and enhance organisational efficiency.

Amandine Laré is an Associate Professor of Social and Solidarity Economy at EM Normandie Business School. She holds a PhD in economics from Université de Rouen-Normandie, earned in 2014, focusing on the impact of improved access to water and sanitation on living conditions of households in developing countries from field studies in Mozambique, Togo, Mali, and Burkina Faso. Her work focuses on issues such as access to energy, water, and sanitation, agriculture, fisheries and innovations, innovative solidarity financing, as well as the implementation of sustainable development and corporate social responsibility. She is also Head of the Economics, Territories & Sustainable Development Department.

Judit Nagy is Associate Professor at Corvinus University of Budapest, Hungary. She participates and leads research projects as well as teaches in the Department of Supply Chain Management for 15 years. Her area of expertise is digitalisation, supply chain management and the food industry. She leads the Food Supply Chain Research Group at Corvinus.

Tilak Naik holds a postgraduate degree in Agri-Business Management and has a deep-rooted connection with agriculture, the farming community, and cooperatives. His academic research primarily explores themes of sustainability and cooperative development. With a keen interest in agrochemicals, he has authored research papers that emphasize sustainable practices in the sector. Currently, he serves as a Key

Account Manager for B2B Sales in agrochemicals at Titan Biotech Group, where he leverages his expertise to drive impactful solutions in the agricultural domain.

Shubhaom Panda, native of Cuttack district, Odisha, is an agri-business professional with more than five-years of academic, industrial and research expertise; especially in agribusiness-economics, agricultural input-marketing; and agricultural journalism. He has more than fifteen publications in different renowned journals, magazines and books of national and international level. He has bagged more than thirty-five certificates at different national and international level conferences, trainings, workshops, FDPs, etc. He is also associated with different academic societies at national and international level as life member, editor and reviewer. He earned Ph.D. (Agri-Business) from the Institute of Agri Business Management, SKRAU, Bikaner. He holds MBA (Agri-Business Management), MBA (Rural Management), PG Diploma (Agricultural Extension Management), Master of Journalism and Mass Communication from diversified institutions of repute. He has cleared the NET in Agricultural Business Management conducted by the ASRB, New Delhi. At present, he is working as Assistant Professor (Agri-Business) in the Faculty of Agriculture, Sri Sri University, Odisha.

Priyanka Panday has more than 10 years of experience and is currently associated as a Senior Assistant Professor with DBS Global University, Dehradun. She has more than 20 publications in national and international journals of repute. She has also published more than 16 cases with Case Center UK. Her expertise includes human resource management, sustainability studies, leadership, and gender studies.

Shaktiranjan Panigrahy, an accomplished agribusiness professional and Assistant Professor at the International Agribusiness Management Institute, Anand Agricultural University, holds a doctoral degree in Management. With over a decade of teaching experience and more than three years in various industries, including livestock, fisheries, and potato sectors, Dr. Panigrahy has made significant contributions to academia and research. His scholarly output includes numerous research papers published in national and international forums, as well as books, book chapters, popular articles, and conference proceedings. His expertise lies in agribusiness economics and policy, operations management, cooperative management, entrepreneurship, and marketing. Currently, he is actively engaged in agritourism research, focusing on infrastructure development and identifying gaps in the sector.

Satish Chandra Pant is an Associate Professor at DBS Global University, Dehradun, (India). He is an alumnus of G.B. Pant University of Agriculture and

Technology, Pantnagar, Uttarakhand (India), holding a doctorate in Management with more than decades of academic experience. Dr Pant has published over 20 research articles with reputed publishers such as Elsevier, Taylor & Francis, Emerald, Wiley, Springer, IGI etc. and presented several research papers in seminars/ conferences/ symposium organized by institutions of global repute. His area of interest includes, agribusincss, agricultural marketing, consumer behaviour, brand management, developmental issues, and cutting-edge technology for societal benefits.

Pankaj Pathak is an accomplished researcher in the field of Information Technology and Computer Science. He holds a Post Graduate degree in Information Technology, as well as a Ph.D. in Computer Science. Dr. Pathak has published numerous papers in both national and international journals, showcasing his expertise in the field. His research interests include topics such as Internet of Things, AI, and data mining. His work has been widely cited by other researchers in the field, and his papers have contributed to the advancement of knowledge in the area.

Samaya Pillai Iyengar is a full-time faculty in Symbiosis Institute of Digital and Telecom Management, Symbiosis International University, Pune (India). She holds MCA, MCM and MPhil degree under the Faculty of Management. She is Perusing her PhD with (Savitribai Phule Pune University) SPPU, Pune University under Computer-Management. She has 18 years of teaching experience with specialisation in databases and Healthcare. Her research interests focus around computer sciences and big data domains. She has taught subjects in programming -Java, Python, C. Databases –Oracle, MySQL, MongoDB, Operating Systems, MIS, Basic Statistics, and Networking etc. She has 20 publications in Scopus. 10 Papers in Conference proceedings

Suryamani Roul is an accomplished development management professional with over four decades of experience in building livelihoods strategy, leading agriculture and agri-value chain initiatives, entrepreneurship and enterprise promotion, youth and women's economic empowerment, and social development programs. Over these years, he has provided organizational leadership on strategy building, program management, quality assurance and resource mobilization. He has expertise in building and managing partnership with multi stakeholders like corporates, government, international agencies and community organisations with high level of networking proficiencies. He has worked in senior leadership positions with organizations of international and national repute namely TechnoServe, CARE International, ACCESS, Plan international project and Institute of Entrepreneurship. Currently he is associated with Global AgriSystem as Director and also holds Board positions in two organizations. He holds an MBA from Calcutta University, an

MA in Applied Economics from Utkal University and has received education and training from Asian Institute of Technology; Coady International Institute, Canada; University of East Anglia, UK; Yale School of Management, USA; CARE-Academy, Atlanta; FAO sponsored program on Livelihoods and Enterprise Development as an expert and Growing Inclusive Market organised by UNDP in Canada. He regularly speaks at conferences and other forums in both India and overseas. He has authored and presented numerous thematic papers focusing on livelihoods and agri-enterprise promotion, climate smart agriculture, micro finance, women's economic empowerment and program convergence. He is an ASPIRE Fellow.

Anup Samal works as an analyst in the Climate Policy Initiative - India

Aparna Sharma is currently working as Assistant Professor at Govt College, Barwala, Panchkula. Earlier taught at Govt College for Women, Sec-14, Panchkula. Before joining Haryana Education Services in 2017, she had taught MBA students at Department of Business Administration, NIT kurukshetra for 3 years and at University School of Management, Kurukshetra University Kurukshetra for 4 years. She did her Ph.D from Kurukshetra University Kurukshetra. She did her post graduate in the field of Commerce (M.Com) and Management (MBA). She has qualified UGC NET in General management, Personnel Management and Commerce. She has presented papers in more than 25 National or International Seminars. To her credit, 1 book and various papers have been published in journals of national repute.

Sangeeta Shroff is Professor at Vaikunth Mehta National Institute of Cooperative Management and formerly Professor & Head of Agro-Economic Research Centre, Gokhale Institute of Politics and Economics. She has conducted Research Projects for Ministry of Agriculture & Farmers Welfare, Government of India, World Bank, Industry, etc. She was nominated as a member of Working Group of erstwhile Planning Commission, Government of India, in the context of Twelfth Five Year Plan. Besides teaching experience, she has a large number of research papers to her credit, been a TEDx speaker and guided PhD students.

Parminder Singh is a distinguished agricultural economist and extension specialist with over 16 years of experience in the field of extension, research, and teaching. He is currently serving as a Senior District Extension Specialist (Agricultural Economics) at Krishi Vigyan Kendra (KVK), Sonipat, Haryana (INDIA), where he has made significant contributions to agricultural development in the region. Dr. Singh's expertise spans agricultural marketing, agribusiness, farm management, Organic/natural agri products marketing, and impact assessment. He has published over 35 research articles, 5 book chapters, 40 popular articles, and several leaflets in his domain of expertise. He has been invited to deliver radio broadcasting addressing farmers and other stakeholder to incur benefits of agricultural practices. Dr Singh has also served Haryana State Agricultural Marketing Board as an Agricultural Marketing Officer from 2007-2010.

Vaishnavi Singh is a research scholar at department of Agricultural Economics, Institute of Agricultural Sciences, Banaras Hindu University.

Sandeep Singh is Associate Professor at Chitkara Business School, Chitkara University, Punjab, India since 2019. He was awarded a doctorate by Thapar University, Patiala in 2017, and he has master's degrees in management from Punjabi University, Patiala. He has done B. Tech in Mechanical Engineering from PTU, Jalandhar. He has over 16 years of experience working with various reputed educational institutions like Thapar University, Patiala. His research interests are in the fields of Consumer Behaviour, Marketing and Advertising He has published various research papers in reputed journals and participated in conferences/seminars on critical issues relating to management.

O.P. Singh is a Professor at department of Agricultural Economics, Institute of Agricultural Sciences, Banaras Hindu University.

Ritambhara Singh is an Associate Professor (Agribusiness Management), at the School of Agribusiness and Rural Management in Dr. Rajendra Prasad Central Agricultural University, Pusa, Samastipur, Bihar, India. She holds a Master's and a Doctorate in Agricultural Economics from Govind Ballabh Pant University of Agriculture and Technology, Pantnagar. India. Besides this, she has completed a certificate course in Finance from the University of Michigan, USA, Microeconomics from the University of Illinois, USA, and Understanding value and Supply Chains from the University of Adelaide, Australia. She was one of the participants in Sathguru Cornell's Food and Agribusiness Executive Management Program, 2022 as offered by Cornell University, USA. She has around nineteen years of professional experience in various capacities with organizations of repute viz. ICAR-Indian Agricultural Research Institute, New Delhi, MCX India Ltd, Mumbai, Advanced Biofuels USA, Maryland, USA, FAS-USDA, US Embassy, New Delhi, Compound Livestock Feed Manufacturers Association, Mumbai, Anand Agricultural University, India. She is actively engaged in the teaching of U.G, PG, and Ph.D. courses in the programs of Agribusiness Management, Rural Management, and Agricultural Economics, as also acting as academic guide to students from these programmes. Besides teaching, she has handled various National and International Projects, funded by the UNCTAD, the World Bank, State Governments, ICSSR, and others. Her areas of interest are Economic Theory, International Trade, Agribusiness

Economics and Policies, Emerging marketing institutions, especially FPOs, and Secondary Agriculture. She is a member of several professional bodies and is on the editorial board of magazines. She has more than 190 publications to her credit in various forms in reputed journals, magazines, and news dailies.

Anurag Tewari is currently working as Professor and Dean, Agriculture Program at Chitkara University, Punjab since 2019. He was awarded Doctorate by National Institute of Himalayan Environment. He is having total experience of more than 30 years. He has worked with Noida International University, Indian Agricultural Research Institute, Radical Foods Limited, Tilda Ltd etc. He is leading consultant of North India. He has presented many papers in conferences of repute and published research papers.

Shubham Kumar Verma is a Research Scholar at department of Agricultural Economics, Institute of Agricultural Sciences, Banaras Hindu University.

Rohini Vilhekar has completed Ph.D. in Agricultural Economics from MPKV, Rahuri. She possesses 3 years of research experience at prestigious national institutes such as ICAR-NAARM, Hyderabad, and VAMNICOM, Pune. She has worked on various topics including impact assessment, Total Factor Productivity (TFP) analysis, and the development of cooperatives, Self-Help Groups (SHGs), and Farmer Producer Organizations (FPOs).

Hema Yadav is a distinguished academician, administrator, and professional serving as a Director of Vaikunth Mehta National Institute of Co-operative Management (VAMNICOM) located in Pune, India, from 2021. With more than two decades of rich academic background and extensive experience in agricultural marketing, administration, and management, Dr. Yadav brings a wealth of knowledge and expertise to her role. Dr Yadav has served the National Institute of Agricultural Marketing, Jaipur, India as Director till 2021. Dr Yadav has presented numerous conference papers and published research articles in the Journal of Reputes from the House of Elsevier, Springer, Wiley, etc.

Index

A

B

C

D

E

F

T

V